Elements of Biochemistry

Elements of Biochemistry

Larry G. Scheve
CALIFORNIA STATE UNIVERSITY, HAYWARD

PRENTICE HALL, Englewood Cliffs, New Jersey 07632

Library of Congress Cataloging in Publication Data

Scheve, Larry G., 1950–
Elements of biochemistry.

Bibliography: p.
Includes index.
1. Biological chemistry. I. Title.
QP514.2.S33 1984 574.19′2 83-3704
ISBN 0-205-07909-1

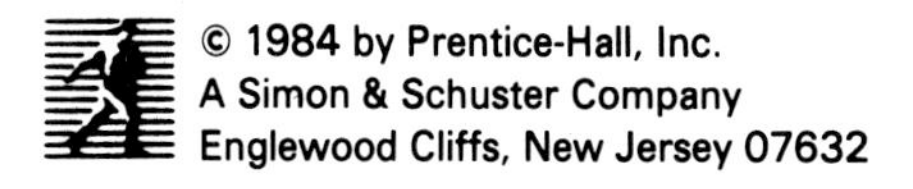

A Simon & Schuster Company
Englewood Cliffs, New Jersey 07632

Printed in the United States of America

10 9 8

ISBN 0-205-07909-1

Prentice-Hall International (UK) Limited, *London*
Prentice-Hall of Australia Pty. Limited, *Sydney*
Prentice-Hall Canada Inc., *Toronto*
Prentice-Hall Hispanoamericana, S.A., *Mexico*
Prentice-Hall of India Private Limited, *New Delhi*
Prentice-Hall of Japan, Inc., *Tokyo*
Simon & Schuster Asia Pte. Ltd., *Singapore*
Editora Prentice-Hall do Brasil, Ltda., *Rio de Janeiro*

Dedicated to my wife, Gail, and to my parents

Contents

Preface

Elements of Biochemistry was written in response to the demand for a comprehensive but concise biochemistry textbook. It is primarily designed for students majoring in nursing, medical technology, and the other allied health sciences, but can also be used in survey courses for slightly more advanced preprofessional students. It presents the core of biochemistry in clear and understandable terms to students who often have a limited background in chemistry, and it carefully interweaves examples from the clinical sciences and discussions of important disease states to motivate students and reinforce key biochemical concepts. Historical discussions and descriptions of experimental techniques have been intentionally eliminated in favor of the inclusion of more relevant examples from clinical medicine.

The book is composed of twenty chapters and presents a complete, up to date description of biochemistry. The organization follows a traditional approach. Chapter 1–7 provide a description of cells, buffers, amino acids, proteins, enzymes, and bioenergetics. Chapters 8–15 describe the basics of cellular metabolism (of carbohydrates, lipids, and amino acids). Chapters 16–18 describe the storage and transmission of genetic information. Finally, Chapter 19 covers hormones, and Chapter 20 gives a thorough but concise description of nutrition. A summary of photosynthesis and a mathematics review are found in the appendixes.

The organization is also versatile since specific chapters, or sections from a particular chapter, may be omitted without destroying the continuity of the book. To enhance student comprehension, the text is organized so that the chemistry of a particular class of biological molecule is first presented in one chapter then followed by discussions of the metabolism of that class of molecule (catabolism, then anabolism) in the next chapter.

The text contains complete descriptions of the absorption, digestion, and catabolism of carbohydrates, lipids, and amino acids. Discussions of important metabolic pathways are relatively detailed and are aided by the inclusion of many clear and unambiguous figures and charts. The compartmentalization and regulation of these pathways are also described. Minor pathways are summarized, but simple word-picture diagrams of such pathways are avoided. Rather the chemical structures of the starting materials, important intermediates, and end products are shown without clutter and confusion.

The writing style, it is hoped, will prove clear, concise and at the appropriate level. Important technical terms are defined within the sentence in which they first appear. Key words and phrases are boldfaced in order to highlight their significance, and some repetition of specific discussions and figures is intended to reiterate particular points.

Each chapter concludes with a summary, a list of suggested readings, and a comprehensive set of problems. The questions span a wide range of difficulty and include the standard but important review questions that help students define key terms and concepts. Intermediate level questions and more challenging "bonus questions," marked with an asterisk, are also provided. The Instructor's Manual provides complete answers to all end of chapter problems.

ACKNOWLEDGMENTS

Many people have made important contributions to the finished text. In particular I wish to thank Professors Delano V. Young (Boston University), Ronald Bentley (University of Pittsburgh), Donald B. Parrish (Kansas State University), Jon Robertus (University of Texas), Charles Stewart (San Diego State University), Albert Kind (University of Connecticut), Penny Gilmer (Florida State University), and Joseph Harris (Arizona State University) for their many valuable suggestions during their reading of all or a portion of the manuscript. I also wish to acknowledge the assistance provided by the staff of Allyn and Bacon, including Mary Beth Finch and James Smith, who helped make this a useful and enjoyable experience for me.

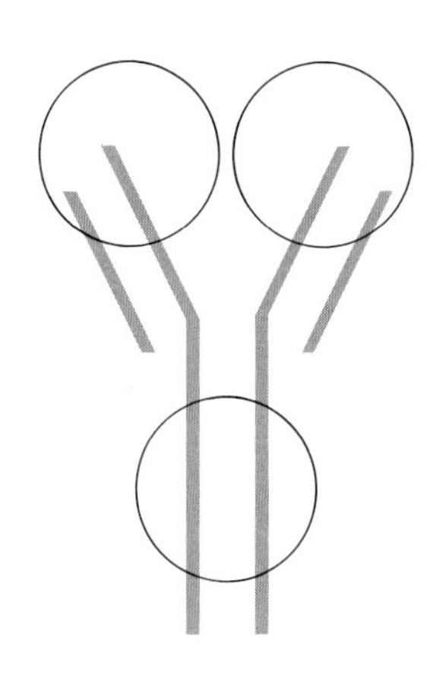

Chapter 1
The Structure of Cells

1.1 INTRODUCTION

biochemistry

Biochemistry is the study of the chemical processes associated with living organisms. This exciting science arose largely within the last 100 years, and we have gained most of our knowledge of it only within the last 30 to 50 years. Thus, biochemistry, in contrast to other areas of scientific endeavor, is a relatively new field. Our understanding of living processes is constantly changing and becoming more complete as thousands of scientists continue to unravel the complex puzzle that we call life.

Most students reading this book will enter one of the allied health professions. The study of the human body and of various diseases will occupy a significant portion of their time. Since most diseases occur as a result of a malfunction at the molecular or cellular level, *a thorough understanding of biochemistry is of paramount importance.*

In this chapter, we shall first discuss the kinds of elements found in living systems and why these elements are important for life. Second, we will discuss the criteria that distinguish living matter from inanimate matter. Finally, we will discuss the different types of cells, as well as their structure and function.

1.2 THE CHEMICAL ELEMENTS OF LIFE

One of the amazing points to consider about "living" systems is the fact that these systems are constructed from nonliving—*inanimate*—matter. Apparently, it is the complex organization of inanimate matter within cells that produces what scientists call "the living state." Interestingly, only a very few of the 90 naturally occurring elements are commonly found in living cells. Among these are carbon, hydrogen, oxygen, and nitrogen, which make up approximately 99% of the mass of the cell. From these elements are constructed amino acids, proteins, carbohydrates, lipids, and other types of

biomolecules

biomolecules (molecules found in living cells and required for various biological functions). In addition to these, phosphorus and sulfur are present in smaller amounts and are also used to construct biomolecules.

A number of ions are found in cells. Among these are Na^+, K^+, Cl^-, Mg^{2+}, Mn^{2+}, Ca^{2+}, Fe^{2+}, Zn^{2+}, and Cu^{2+}. The monovalent ions (ions with a 1+ or 1− electrical charge) are required for transmitting nerve impulses and for maintaining blood electrolyte balance. The divalent ions (ions with a 2+ electrical charge) and trivalent ions (ions with a 3+ charge)

enzymes

are often associated with **enzymes** (biological catalysts that speed up chemical reactions) and are required for enzyme activity. A number of other elements are found in cells, but in very low concentrations. These elements are called *trace* elements because only a trace amount is required for normal cellular function. Table 1–1 summarizes the different elements commonly found in most cells.

We will now discuss some of the important chemical properties of carbon, oxygen, nitrogen, and hydrogen.

Six electrons are associated with the carbon nucleus. These are arranged in the following electronic structure: $C(1s^2, 2s^2, 2p^2)$. The notation means that elemental carbon has 2 electrons in each of the $1s$, $2s$, and $2p$ energy levels, or orbitals. Carbon does not lose electrons easily from the $2s$ and $2p$

TABLE 1–1 Common Elements Found in Cells

Elements Used in Constructing Biomolecules	Monovalent, Divalent, and Trivalent Ions	Other Trace Elements
Carbon (C)	Sodium (Na^+)	Cobalt (Co)
Hydrogen (H)	Potassium (K^+)	Boron (B)
Oxygen (O)	Chlorine (Cl^-)	Aluminum (Al)
Nitrogen (N)	Magnesium (Mg^{2+})	Vanadium (V)
Phosphorus (P)	Manganese (Mn^{2+})	Molybdenum (Mo)
Sulfur (S)	Calcium (Ca^{2+})	Iodine (I)
	Iron (Fe^{2+}, Fe^{3+})	Silicon (Si)
	Copper (Cu^{2+})	Nickel (Ni)
	Zinc (Zn^{2+})	Chromium (Cr)
		Fluorine (F)
		Selenium (Se)

Source: Adapted from Lehninger, A. L., *Biochemistry*, 2nd ed. (New York: Worth, 1975), p. 17.

orbitals (that is, it does not ionize easily). However, carbon can *share* electrons with other elements, forming **covalent bonds** by electron-pair sharing, as shown in Equation 1–1.

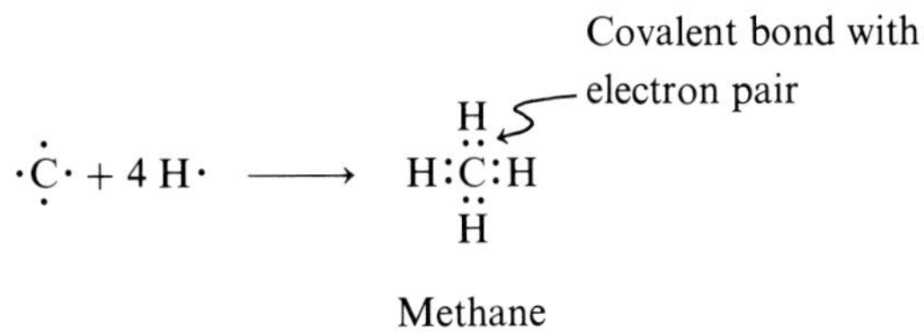

Covalent bond with electron pair

$$\cdot\dot{\underset{\cdot}{C}}\cdot + 4\,H\cdot \longrightarrow H:\overset{\textstyle H}{\underset{\textstyle H}{\ddot{\underset{..}{C}}}}:H \qquad (1\text{–}1)$$

Methane

$—\text{C}—\text{C}—$

Carbon-carbon single bond

$\text{C}=\text{C}$

Carbon-carbon double bond

$—\text{C}\equiv\text{C}—$

Carbon-carbon triple bond

$\text{C}=\text{O}$

Carbon-oxygen double bond

$\text{C}=\text{N}—$

Carbon-nitrogen double bond

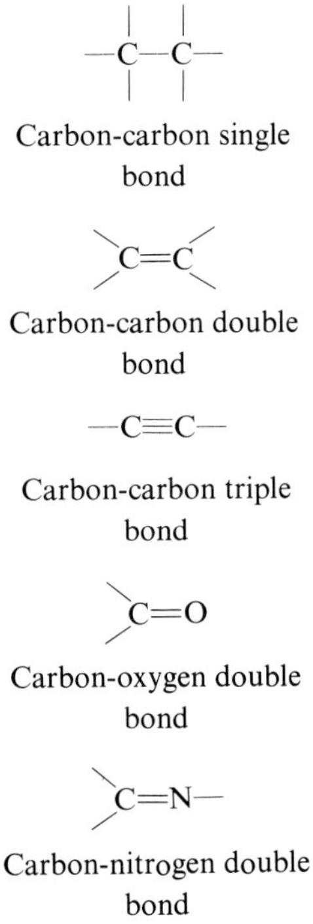

FIGURE 1–1 Multiple bonding arrangements.

Thus, carbon can easily form strong and stable covalent bonds with itself and with hydrogen, oxygen, or nitrogen. It should be noted that carbon can bond a maximum of *four* other atoms to itself. In addition, carbon can participate in single, double, and triple bonds, as shown in Figure 1–1. Because of these unique features of carbon, a diversity of covalent bonding arrangements is possible, leading to literally millions of different substances. Figure 1–2 summarizes a few of the bonding arrangements in which carbon can participate. It should be obvious that carbon is a versatile element, well suited for serving as the basis for life.

Oxygen, nitrogen, and hydrogen can also form strong and stable covalent bonds by electron-pair sharing. Oxygen and nitrogen usually form single or double bonds; hydrogen can form only single bonds (remember, hydrogen has only 1 electron available for bond formation).

Oxygen is a very **electronegative** element (electronegativity can be defined as the tendency of an atom to attract electrons to itself). As we shall see in the next chapter, this feature is the basis for the unusual properties of water and for the special type of bond called the *hydrogen bond.* Oxygen is also important to life because it serves as the final *electron acceptor* during energy production within the cells of our bodies (see Chapter 12).

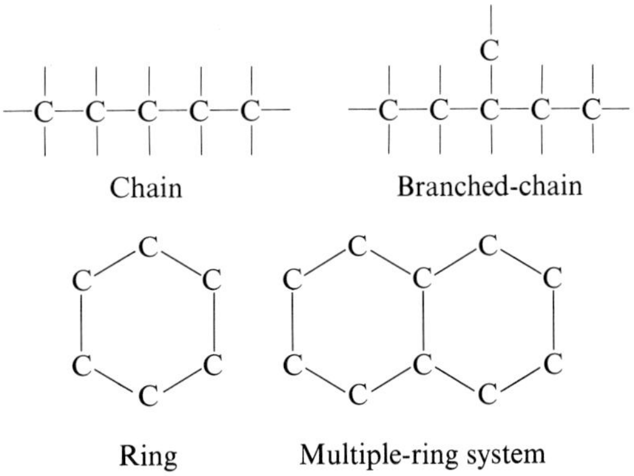

$—\text{C}—\text{C}—\text{C}—\text{C}—\text{C}—$

Chain

Branched-chain

Ring

Multiple-ring system

FIGURE 1–2 Some of the possible chemical bonding arrangements for carbon.

peptide bond

Nitrogen is an important element for living systems because it bonds with carbon, forming what is known as the **peptide bond**.

$$\mathrm{O{=}C{-}NH}$$

Peptide bond

This bond is the prime structural bond for all proteins (see Chapter 3, Amino Acids and Proteins). Nitrogen is also used in constructing *nucleic acids*, the molecules that store and transfer genetic information.

Finally, hydrogen participates in hydrogen bonding and is associated with electron transport processes within cells (see Chapter 12).

1.3 WHAT CONSTITUTES THE "LIVING" STATE?

As we have already mentioned, living cells are ultimately constructed from nonliving matter. What, then, distinguishes living matter from nonliving matter? Scientists and philosophers living before the mid-19th century believed that living matter differed significantly from inanimate matter and that it obeyed a different set of physical laws. These scientists argued that a *vital force* within living systems was responsible for the generation and maintenance of life. Today, we realize that the vital force theory cannot adequately explain living systems and that the atoms and molecules within cells obey the same physical laws as do the atoms found within inanimate matter. Ultimately, it appears that living systems are "living" because of their complexity, diversity, and structural organization at the molecular level. A few generalizations about living matter may help.

First, living matter, as we have already mentioned, is *extremely complex and highly organized.* This is true on a number of different levels. The structure and organization of the organs of the body, of the cells that make up the organs, of the subcellular structures within the cells, and of the many and varied biomolecules that make up the subcellular structures, all indicate that living matter is highly organized.

Second, living systems are constructed from structures that all seem to have a *specific purpose or function.* The ultimate purpose of each component is to aid in the growth, development, replication, and biologic task of the whole organism.

Third, living systems can *extract both matter and energy from the environment* and use these for their normal growth, development, and maintenance. As we shall see in later chapters, living systems are extremely efficient in extracting matter and energy. Yet, living systems are not perfect; they produce waste matter and waste energy (heat) and, thus, pollute their environment. Still, the waste materials can often be used by other organisms, and most life systems recycle various constituents.

Fourth, living matter can undergo *self-replication* (that is, make an exact copy of itself). Specifically, living matter contains molecules that can store genetic information **(deoxyribonucleic acid, DNA)** and transfer that information from one generation to another. This information-storage system is extremely complex and versatile. It is essentially error-free and, when mistakes are made, usually self-correcting. Just think of it: Everything that you are—the color of your eyes, the color of your hair and skin, your sex—are all coded in the DNA contained within the nucleus of every cell in your body. The DNA within the nucleus may represent the *maximum amount of information that can be stored in the minimum amount of space.*

deoxyribonucleic acid (DNA)

> In essence, the living cell is a self-assembling, self-regulating, self-replicating, . . . open system of organic molecules operating on the principle of maximum economy of parts and processes; it fosters many consecutive linked organic reactions which are necessary for the extraction and transfer of energy and for the synthesis of its own components by means of enzyme catalysts that it produces itself.[1]

1.4 THE TYPES OF CELLS

Although there are many different kinds of cells, all cells can be classified according to two rather simple classification schemes. The first scheme classifies cells according to their size and complexity. The second scheme classifies cells according to how they extract matter and energy from the environment (or, simply, how they "feed themselves"). Using the first classification scheme, we find two basic types of cells: (1) **prokaryotes** ("before a nucleus") and (2) **eukaryotes** ("good nucleus"). The differences between the two cell types are summarized in Table 1–2. Using the second classification scheme, we find two basic types of cells: (1) **autotrophs** ("self-feeding") and (2) **heterotrophs** ("feeding-on-others"). The differences between these two types of cells are summarized in Table 1–3. One should note that the

prokaryotes
eukaryotes

autotrophs
heterotrophs

TABLE 1–2 The Differences Between Prokaryotic and Eukaryotic Cells

Criteria	Prokaryote	Eukaryote
Size	Small (1–2 μm)[a]	Large (20–30 μm diameter)
Complexity	Simple internal structure	Complex internal structure (organelles present)
Nucleus	None (the DNA is found in the cytoplasm)	Present (the DNA is contained in the nucleus)
Examples	Bacteria, blue-green algae	Cells of higher animals, plants, fungi, and most algae

[a] 1 μm = 1 micrometer = 1×10^{-4} cm (alternatively, 1 micron (μ).

TABLE 1–3 The Differences Between Autotrophic and Heterotrophic Cell Types

Criteria	Autotrophic (self-feeding)	Heterotrophic (feeding-on-others)
Carbon source	CO_2 (air)	Complex organic molecules
Energy source	Sunlight	Complex organic molecules
Examples	Photosynthetic (plant) cells	Animal cells, bacteria

two classification schemes are based on entirely different criteria. Thus, a prokaryotic cell (a bacterium) can be heterotrophic with respect to feeding style and a eukaryotic cell (a plant cell) can be autotrophic.

1.5 CELL STRUCTURE AND FUNCTION

Having discussed the various types of cells, we need to explore the internal structure of both prokaryotic and eukaryotic cells. Our understanding of the internal structure of cells would not have advanced very far if the *electron microscope* had not been invented in the 1940s. With this instrument, scientists have been able to take photographs of cells at magnifications

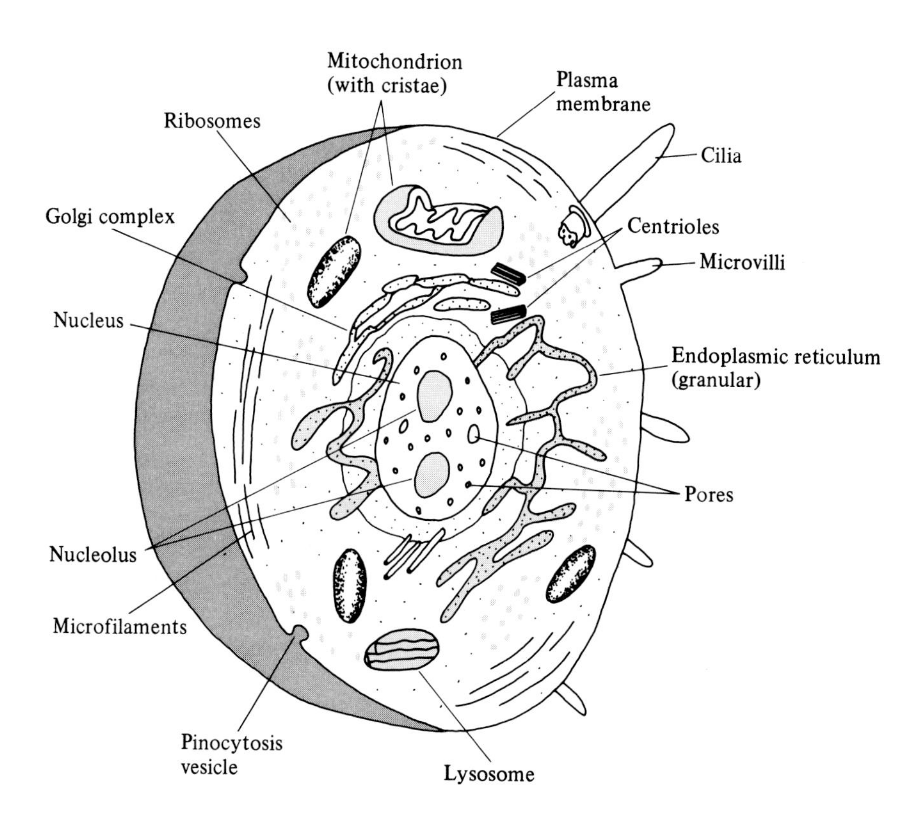

FIGURE 1–3
A 3-dimensional representation of the cell. (Adapted from Gore, R., "The New Biology: I, The Awesome Worlds Within a Cell," *National Geographic, 150*, 3 (September 1976): 362–364. Paintings by Davis Meltzer.) (Adapted from Montgomery, R., et al., *Biochemistry: A Case-Oriented Approach*, 3rd ed. (St. Louis: C. V. Mosby, 1980), p. 100.)

often greater than 100 000 ×. These photographs have revealed extremely complex internal substructures within the cell. It should be noted that the electron microscope uses a beam of electrons (instead of a light beam) to magnify the object under study. Before a cell can be analyzed under the electron microscope, it must be sliced into very thin sections. Thus, the photographs shown in Figures 1–4 to 1–6 are not photographs of the entire cell, but rather of very thin slices from the cell. By analyzing sequential slices of a cell, scientists can ultimately construct a 3-dimensional model of the cell such as that shown in Figure 1–3.

A new technique called scanning electron microscopy bounces an electron beam off the outer surface of cells to reveal the outer contours of the cell surface. A 3-dimensional image is obtained, but no internal structure is revealed. This method has been extensively used in determining how the surfaces of cells change during the various phases of cellular growth. Abnormalities in shape can also be observed using this method.

PROKARYOTIC CELLS

Prokaryotic cells are small, single-cell organisms such as bacteria and blue-green algae (see Figure 1–4). They consist of a rigid cell wall surrounding the cell cytoplasm. These cells do not contain a nucleus or complex subcellular organelles (such as mitochondria and Golgi complexes). Note the rather simple internal structure and compare it to the corresponding structure in a eukaryotic cell (see Figure 1–5, p. 9).

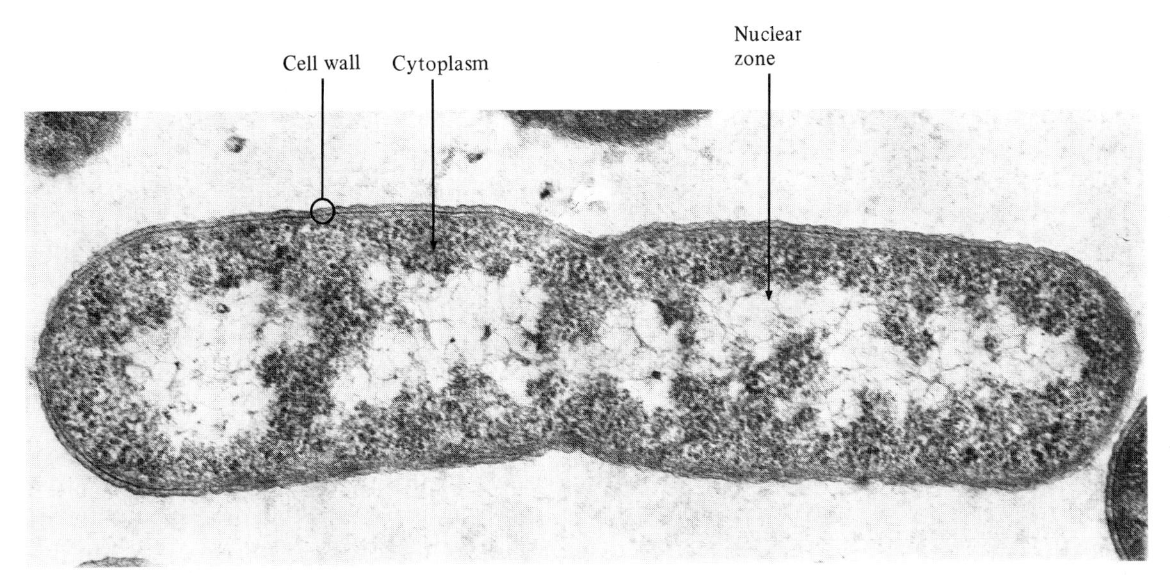

FIGURE 1–4 Electron micrograph of a typical prokaryotic cell. (Dividing bacterium, magnification 84 000 ×. Photograph courtesy of Ms. Jackie Ervin, Dept. of Biological Sciences, California State University, Hayward.)

Cell Wall. The cell wall is a rigid structure that surrounds the cytoplasm of this type of cell. It is composed of both carbohydrate and protein and resists osmotic changes in the outer environment. Associated with the inside portion of the cell wall is the *cell membrane*. This membrane is composed of lipid and protein and transports various constituents into and out of the cell.

Cytoplasm. The cytoplasm (cytosol) is the water-soluble portion of the cell. It is a very viscous solution of proteins and other cellular metabolites. Many metabolic reactions occur in the cytoplasm, including glycolysis and fatty acid synthesis.

Nuclear Zone. The nuclear zone is a structureless region in which tightly coiled DNA (deoxyribonucleic acid) exists. The DNA is double stranded (forming a helix), and the ends are linked, making it circular. The DNA contains all of the genetic information required for the normal functioning of this type of cell.

EUKARYOTIC CELLS

Eukaryotic cells are significantly more complex than are prokaryotic cells. They are larger than the prokaryotes (often 20–30 times larger) and possess a number of different and rather complex subcellular organelles. Note in Figure 1–5 the complex internal structure of this type of cell.

Cell Membrane. The cell membrane (plasma membrane) is composed of 45–50 percent protein and 45–55 percent lipid. The lipid is organized as a double layer (bilayer) in which various proteins are embedded. The cell membrane keeps the contents of the cell inside and controls the flow of nutrients into the cell and of waste products out of the cell. The membrane also contains complex carbohydrates and proteins that promote cellular adhesion and also give the cell specific immunologic properties.

Cytoplasm. The cytoplasm is a solution of proteins, enzymes, nucleic acids, carbohydrates, and other biochemical metabolites. A variety of metabolic reactions occur in the cytoplasm, including glycolysis and fatty acid synthesis.

Nucleus. The nucleus is a spherical structure (4–6 μm in diameter) surrounded by a porous nuclear membrane. The nucleus contains DNA organized (along with histone proteins) into tightly coiled chromosomes. The nucleus stores and transfers genetic information contained in the DNA.

Mitochondria. The mitochondria are rod-shaped structures (approximately 2 μm long) that are sometimes called the "powerhouses" of the

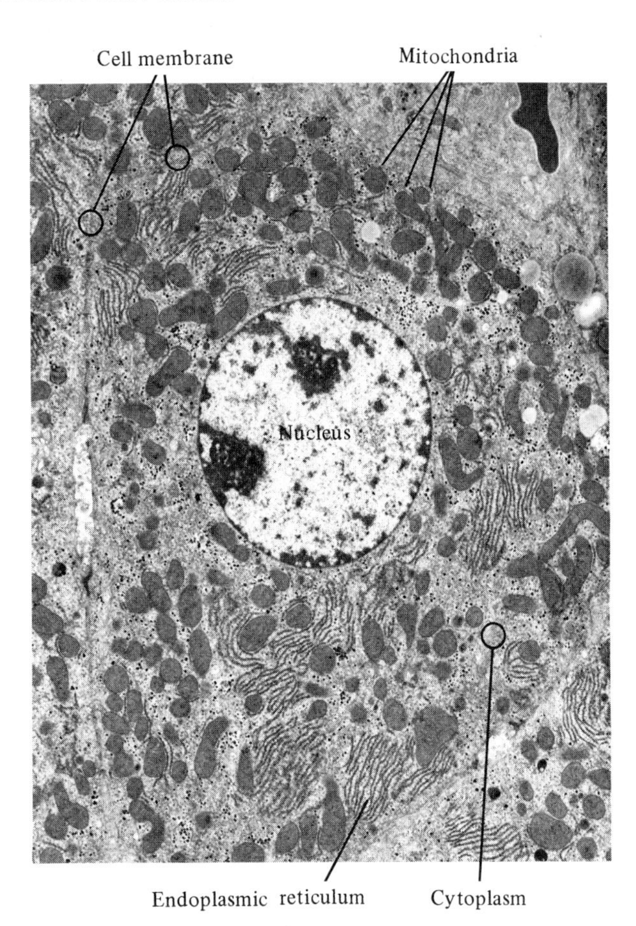

FIGURE 1–5
Electron micrograph of a typical eukaryotic cell. (Rat liver cell, magnification 13 700 ×. Photograph courtesy of Mrs. Nancy Crise-Benson, Dept. of Biological Sciences, California State University, Hayward.)

cell. Inside them, energy is extracted from various compounds and transferred to the rest of the cell as ATP. The mitochondria contain the enzymes used for the Krebs cycle, for fatty acid degradation, and for the electron transport system. They also contain a certain amount of their own DNA.

Endoplasmic Reticulum. The endoplasmic reticulum (ER) is a complex, membranous system located throughout the cytoplasm that synthesizes various proteins and cellular metabolites. Ribosomes are often associated with the endoplasmic reticulum.

Ribosomes. The ribosome is a small structure required in protein synthesis. It is constructed from proteins and ribonucleic acid (RNA).

Golgi Complex. The Golgi complex is an organization of flattened disks constructed out of an extension of the endoplasmic reticulum. It "packages" various substances into small, membrane-enclosed vesicles for transport out of the cell.

Lysosomes. Lysosomes are small, membranous vesicles containing various hydrolytic enzymes.

Microtubules. Microtubules are long fibers constructed from the protein *tubulin*. They provide the cell with an internal "skeleton" and may be responsible for cellular movement.

EUKARYOTIC PLANT CELLS

Eukaryotic plant cells are essentially very similar to eukaryotic animal cells. However, significant differences do exist. Namely, plant cells possess a rigid cell wall and contain numerous chloroplasts; plant cells also contain one or more large vacuoles. Note the chloroplasts and the large vacuole in Figure 1–6 as well as the other structures normally found in a eukaryotic cell.

Cell Wall. The cell wall is a thick, rigid structure composed of multiple layers of cellulose fibers. The cell wall gives the plant cell tremendous physical strength.

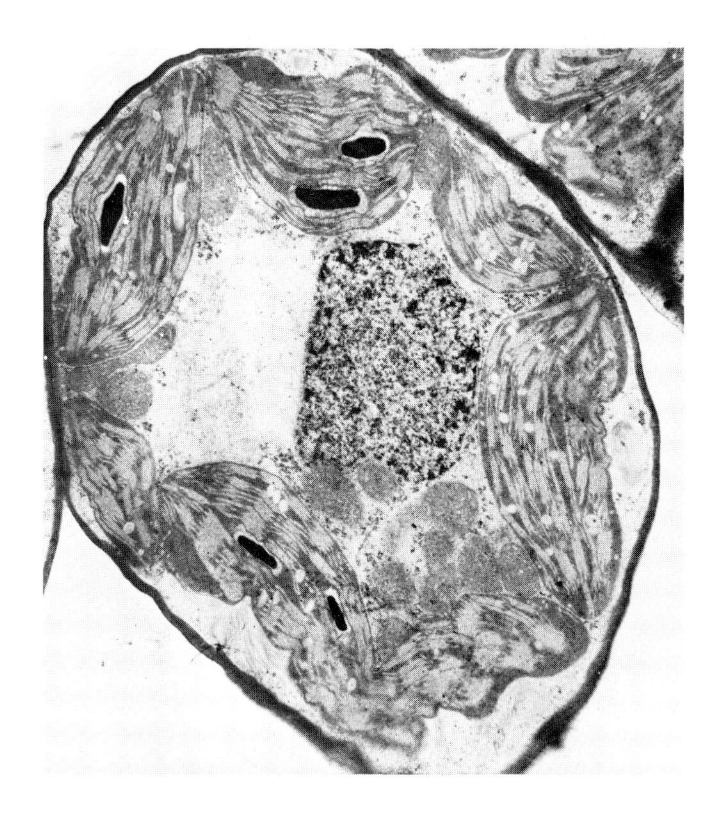

FIGURE 1–6
Electron micrograph of a typical eukaryotic plant cell. (California poppy, magnification 19 000 ×. Photograph courtesy of Mrs. Nancy Crise-Benson, Dept. of Biological Sciences, California State University, Hayward.)

Chloroplast. The chloroplast is a complex football-shaped structure consisting of a series of interconnected disks, called *thylakoids*. The thylakoids contain chlorophyll and the enzymes that are responsible for photosynthesis.

Vacuole. The vacuole is surrounded by a membrane and contains various cellular products and wastes. The vacuole expands as the cell matures; in the mature plant cell, it can occupy a very large percentage of the internal volume.

SUMMARY

Biochemistry is the study of the chemical processes associated with living systems.

Living systems are constructed from inanimate matter, largely carbon, hydrogen, oxygen, nitrogen, phosphorus, and sulfur. Inorganic ions and trace elements are also found in cells; these are required for their normal biological functioning.

The "living state" of matter can be defined as inanimate matter that is

1. complex and highly organized;
2. composed of structures that have specific functions;
3. capable of extracting matter and energy from the environment; and
4. capable of reproducing itself via an efficient and essentially error-free system.

Cells can be classified as either prokaryotic or eukaryotic and as either autotrophic or heterotrophic.

The electron microscope has contributed greatly to our understanding of cell structure. The internal structure of both prokaryotic and eukaryotic cell types was discussed. The structure and function of various subcellular organelles was presented. These included the cell wall, the cell membrane, the nucleus, the mitochondrion, the endoplasmic reticulum, the ribosome, the Golgi complex, the lysosome, the chloroplast, and the plant cell vacuole.

REVIEW QUESTIONS

1. We have presented a number of important concepts and terms in this chapter. Define (or briefly explain) the following terms:
   a. Biochemistry
   b. Biomolecule
   c. Trace element
   d. Living system
   e. Covalent bond
   f. Electronegativity
   g. Prokaryote
   h. Eukaryote
   i. Autotroph
   j. Mitochondrion
   k. Endoplasmic reticulum
2. What three factors distinguish carbon as an element that is unique and well-suited to serve as the basis for life?
3. What are the main differences between prokaryotic and eukaryotic cells?

*4. Carbon cannot ionize easily (that is, lose its electrons). Suggest a possible reason why this is so.

*5. State why a cell is referred to as an "open system."

*6. Animal (eukaryotic) cells do not possess a rigid cell wall, as do plant cells and bacteria. Suggest reasons why animal cells do not need a rigid cell wall.

SUGGESTED READING

The Chemical Elements of Life

Frieden, E. "The Chemical Elements of Life." *Scientific American* 227 (1972): 52–64.

Cell Structure and Function

Fawcett, D. W. *The Cell: An Atlas of Fine Structure.* Philadelphia: W. B. Saunders, 1966.

The Living Cell: Readings from Scientific American. San Francisco: W. H. Freeman, 1965.

NOTE

1. Lehninger, A. L. *Biochemistry* 2nd ed. (New York: Worth, 1975), p. 13.

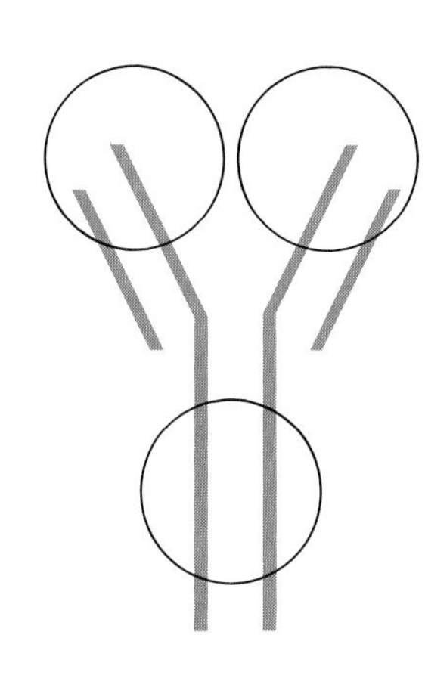

Chapter 2

The Chemistry of Water, Acids, Bases, and Buffers

2.1 INTRODUCTION

Water is essential to life; without water, life as we know it could not exist. Between 70 and 90 percent of the total weight of most cells is water. Therefore, before we can discuss the chemistry of biological systems, we must first briefly discuss the chemistry of water and explain why water is so unique and so necessary for life. In addition, we will discuss acid/base chemistry and explain how buffers work. We will present an important and relevant example by discussing the physiological buffer systems of the body.

2.2 THE CHEMISTRY OF WATER

The structure of the water molecule (H_2O) is shown in the following diagram:

$$\begin{array}{ccc} & 2\delta- & \longleftarrow \text{ Partial negative charge} \\ & \text{O} & \\ \text{H} \diagup & & \diagdown \text{H} \\ \delta+ & & \delta+ \quad \longleftarrow \text{ Partial positive charge} \end{array}$$

Since oxygen is a very electronegative element, it tends to *attract* the electrons that form the covalent bonds with the hydrogen atoms. Thus, the electrons spend most of their time closer to the oxygen atom than to the hydrogen atoms. This partially exposes the positive nucleus of each hydrogen atom. Thus, a partial positive charge ($\delta+$) resides in the vicinity of each of the hydrogen atoms. In addition, the oxygen assumes a double partial negative charge ($2\delta-$). The situation can be summarized in the following diagram:

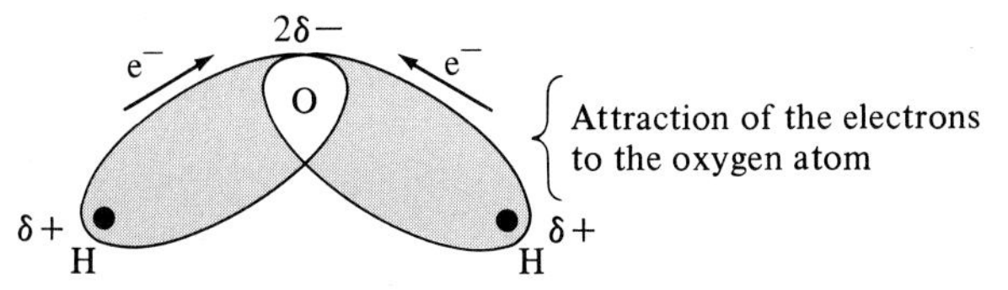

Because the water molecule has separate regions of partial electrical charge, it is referred to as a *dipole*. Since water is dipolar, it can attract other water molecules and form hydrogen bonds as shown in the following diagram:

$\delta+$ H—O $2\delta-$ (with second $\delta+$ H) $\cdots\cdots$ H ($\delta+$)—O $2\delta-$—H $\delta+$

Hydrogen bond

hydrogen bond

Notice that the $2\delta-$ oxygen attracts the $\delta+$ hydrogen from a second water molecule to form the hydrogen bond. The **hydrogen bond** is not a covalent bond in which electrons are shared between atoms; rather, it is a *dipole-dipole attraction* between separate water molecules, much like an electrostatic bond (which is based on electrical attraction between ions of opposite charge). Though the hydrogen bond is much weaker than the typical covalent bond, it provides water with some very unusual—and important—properties, as we shall see in a moment.

Because the hydrogen bond can readily form (and also break up), many water molecules can be associated with each other in a rather loose aggregation or "minicluster" as shown in Figure 2–1. These clusters can form, break up, and reform many times per second; they are thus referred to as "flickering clusters." Liquid water can be thought of as having some degree of "short-range order," but little or no "long-range order." Short-range order can be defined as a transient association of a few water molecules. Ice, on the other hand, has a high degree of long-range order in which most of the water molecules are involved in a hydrogen bonding arrangement. Because of the type of bond that water molecules form in an ice crystal, ice is less dense than liquid water and will float.

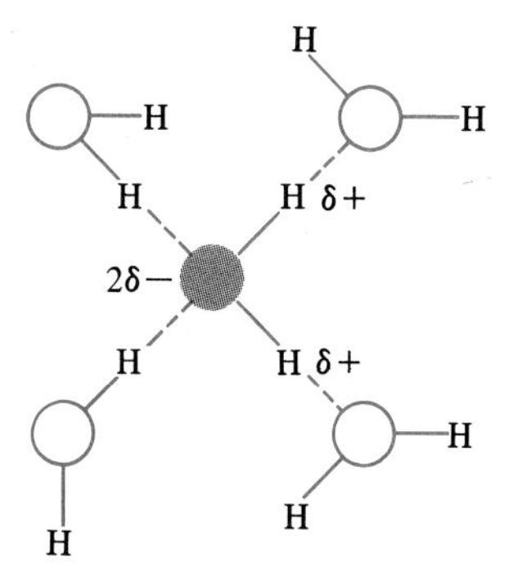

FIGURE 2–1
A "minicluster" of four water molecules.

As a result of the hydrogen bonding found in liquid water, a high degree of *internal cohesion* between water molecules is observed (water molecules tend to "stick together"). Because of this stickiness, the physical properties of water are very unusual for this size of molecule. Water boils at a higher temperature (100°C) than other substances of similar molecular weight. This is because more heat energy is required to break the numerous hydrogen bonds in water and to liberate the water as separate molecules (steam). Water also has an abnormally high freezing point (0 °C) and abnormal values for viscosity and surface tension. Again, the fact that water forms hydrogen bonds helps explain these properties. These properties are summarized in Table 2–1.

TABLE 2–1 The Physical Properties of Water and of Other Substances of Similar Molecular Weight

Property	Water (18)	Ammonia, NH_3 (17)	Methane, CH_4 (16)
Boiling point (°C)	100	−33	−161
Freezing point (°C)	0	−78	−183
Viscosity (cp)[a]	1.01	0.25	—

Note: The number in parentheses gives the molecular weight of the substance.
[a] (cp) = centipoise.

Nevertheless, why should we be so involved in these details about water? A moment of reflection will hopefully dispel this attitude. If water had a boiling point similar to that of liquid ammonia (NH_3), all the water on the earth would exist as steam. There would be no oceans, rivers, or lakes, and probably no life. Because ice floats, the ice on lakes and rivers does not sink to the bottom; thus, fish and other aquatic life can survive during the winter. Because liquid water has a large surface-tension value, water is drawn through the root system of trees and through the trunk to the leaves. This is accomplished by **capillary action** (the molecular attraction of one water molecule pulling another water molecule). Amazingly, all this is due to hydrogen bonding and the "simple" structure of water. Still, the story is not complete; water has other very important properties—namely, its ability to dissolve many different substances.

capillary action

Because water is dipolar, the partially positive or partially negative atoms of the molecule can attract atoms of unlike charge. Thus, water can *solvate* polar (**hydrophilic**, i.e., "water-loving") substances by establishing one or more hydrogen bonds with the substance, forming a "solvation shell" around that substance as shown in the following diagram:

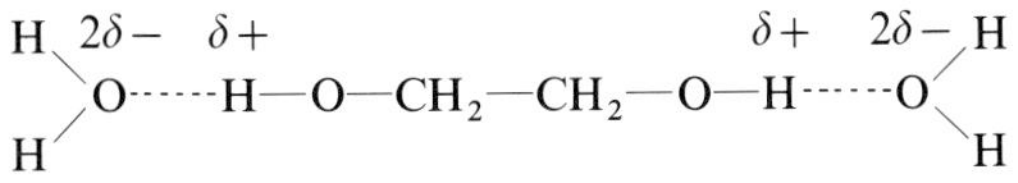

hydrophilic

$$H_2\overset{2\delta-}{O}\cdots\cdots\overset{\delta+}{H}—O—CH_2—CH_2—O—\overset{\delta+}{H}\cdots\cdots\overset{2\delta-}{O}H_2$$

Water can also form a solvation shell around ions, such as Na^+:

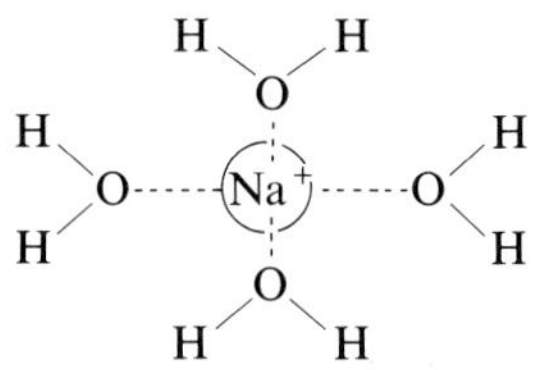

H H O H H O···(Na+)···O H H O H H

Thus, water can easily solvate a number of polar substances, such as sugars, alcohols, aldehydes, and ketones. It can solvate many positively charged and negatively charged ions as well. Since most proteins, carbohydrates, amino acids, and nucleic acids are made from polar molecules, these rather complex biomolecules are also easily dissolved by water. It is because water dissolves so many different substances that it is often referred to as the "universal solvent."

2.3 ACIDS AND BASES

acid
base

Using the Brönsted-Lowry definitions, an **acid** can be defined as a *proton donor*, while a **base** is a *proton acceptor*. With these definitions in mind, let us look at the normal dissociation (fragmentation) of water.

Water dissociates into a proton (H^+) and a hydroxyl (OH^-) ion. This occurs when water forms a hydrogen bond with another water molecule according to Equation 2–1:

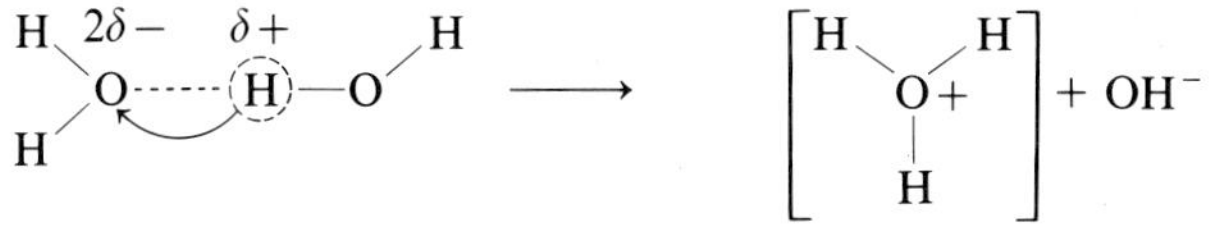

$$H_2O^{2\delta-}\cdots H^{\delta+}\text{—}OH \longrightarrow \left[H_3O^+\right] + OH^- \qquad (2\text{–}1)$$

(Hydronium ion, H_3O^+)

hydronium ion

Notice that the proton (H^+) is not really "free" in solution, but rather, is formally associated with a water molecule. This association is called the **hydronium ion** (H_3O^+). This process (and the reverse) occurs continuously and forms limited amounts of H_3O^+ and OH^- ions (approximately 10^{-7} moles/liter (M) of each in pure water). One should note that water behaves both as an acid and as a base during this process.

THE pH SCALE

Before we can discuss acids and bases in any more detail, we must first define the concept of pH. Simply stated, pH is a measure of the concentration of protons (or rather, H_3O^+ ions) in a solution. The pH is actually expressed as the negative logarithm of the H_3O^+ concentration $[H_3O^+]$.

$$pH = -\log_{10}[H_3O^+] \qquad (2\text{–}2)$$

The pH scale provides a shorthand notation for expressing the H_3O^+

concentration, consisting of a scale with numbers between 1 and 14 as follows:

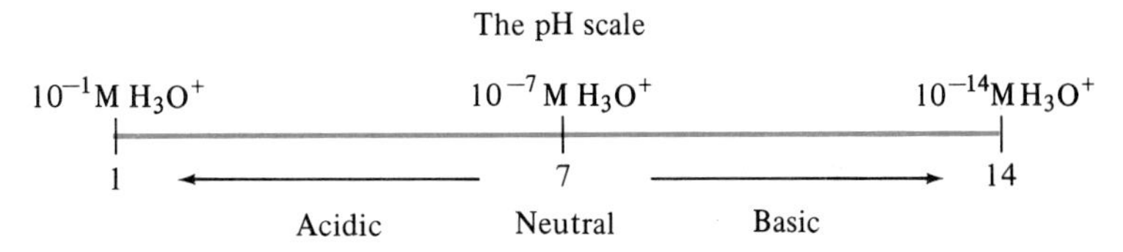

A pH of 7 is defined as neutral. This actually means that a particular solution has a H_3O^+ concentration of 10^{-7} M:

$$\begin{aligned} pH &= -\log[H_3O^+] \\ &= -\log(10^{-7}) = -(-7) = +7 \end{aligned}$$

Solutions with a H_3O^+ concentration greater than 10^{-7} M are acidic, whereas solutions with H_3O^+ concentrations less than 10^{-7} M are basic. One should note that the pH scale is a log scale. This means that one pH number may represent a H_3O^+ concentration that is 10 times as large as the next pH number. For example, a pH value of 2 = 10^{-2} M H_3O^+, whereas, a pH value of 3 = 10^{-3} M H_3O^+ (clearly, 10^{-2} M is 10 times as large as 10^{-3} M.) Table 2–2 shows the pH and H_3O^+ concentration of a few representative fluids.

strong acid

Acids (and bases) are classified according to their tendency to either donate protons (acids) or accept them (bases). For example, a **strong acid** has little affinity for its proton(s) and, therefore, will easily ionize and lose its proton(s). Hydrochloric acid (HCl) is a typical strong acid. It ionizes according to the following equation:

$$\underset{\text{(A strong acid)}}{HCl} \xrightarrow[\text{ionization}]{100\%} \underset{\text{(Actually as } H_3O^+)}{H^+ + Cl^-}$$

strong base

A **strong base** has a great affinity for protons and, therefore, readily accepts protons from the solution.

TABLE 2–2 The pH Values and H_3O^+ Concentrations of a Few Representative Solutions

Solution	pH	H_3O^+ Concentration (M)
Gastric juice	1–2	10^{-1} to 10^{-2} M
Cola drink	3	10^{-3} M
Urine	5–8	10^{-5} to 10^{-8} M
Saliva	6.4	4×10^{-7} M
Pure water	7.0	10^{-7} M
Blood	7.4	4×10^{-8} M

weak acid

A **weak acid** (HA) has a greater affinity for its own proton(s) and, therefore, does not donate its proton(s) as easily as a strong acid. Thus, a weak acid participates in limited ionization (that is, only a small percentage of the total weak-acid molecules will be ionized at any one time).

$$\underset{\text{(Weak acid)}}{\text{HA}} \xrightleftharpoons{\text{Limited ionization}} \underset{\text{(Actually as } H_3O^+)}{H^+} + A^- \qquad (2\text{–}3)$$

Most of the acid is still in the undissociated form (HA). Acetic acid is a typical weak acid.

$$\underset{\text{(HA)}}{H_3C{-}\overset{\overset{\displaystyle O}{\|}}{C}{-}O{-}\text{(H)}} \rightleftharpoons \underset{(A^-)}{H_3C{-}\overset{\overset{\displaystyle O}{\|}}{C}{-}O^-} + \underset{+(H^+)}{\text{(}H^+\text{)}}$$

In fact, if one places acetic acid in a beaker of water, most (approximately 99 percent) of the acetic acid will still be in the undissociated (fully protonated) form.

weak base

Likewise, a **weak base** is defined as a substance that has little affinity for protons and, therefore, does not easily accept protons from the solution.

One should note that a double arrow, or *equilibrium sign* ($\rightleftarrows$), has been used when discussing weak acids and bases. It means that a weak acid, for example, can undergo a certain degree of ionization, but that once a few molecules have dissociated into separate H^+ and A^- ions, the H^+ and A^- ions can reassociate and reform HA. In effect, the A^- ion has become a strong base (conjugate base) and has a relatively high affinity for proton(s). Thus, two separate reactions are occurring: (1) a dissociation reaction and (2) an association reaction. A system has come to **equilibrium** when the rate of dissociation equals the rate of association.

equilibrium

dissociation constant (K_a)

When discussing the chemistry of weak acids, the **dissociation constant** (**K_a**) must be defined:

$$\text{HA} \xrightleftharpoons{K_a} H^+ + A^- \qquad (2\text{–}4)$$

where:

$$K_a = \text{dissociation constant} = \frac{[\text{Products}]}{[\text{Reactants}]}$$

or

$$K_a = \frac{[H^+][A^-]}{[\text{HA}]} \qquad (2\text{–}5)$$

Dissociation constants are usually numbers much less than 1 (10^{-2} to 10^{-8}) and indicate the *relative strength* of the weak acid. For example, a dissociation constant of 10^{-6} means that very little of the weak acid has dissociated into H^+ and A^-.

$$K_a = 10^{-6} = \frac{[\text{Small concentrations}]}{[\text{Large concentrations}]} = \frac{(\text{Small number})}{(\text{Large number})}$$

$$= \text{Small number } (10^{-6})$$

pK_a

Since dissociation constants are relatively clumsy to handle, a shorthand notation has been developed in which **pK_a** is defined as the negative logarithm (base 10) of K_a according to Equation 2–6.

$$pK_a = -\log K_a \tag{2–6}$$

Thus, if a weak acid has a K_a of 10^{-2}, then

$$pK_a = -\log(10^{-2})$$
$$= -(-2) = +2$$

2.4 THE ACID–BASE PROPERTIES OF DRUGS

At this point, let us digress and explain how pH affects the electrical charge of drug molecules, and how this, in turn, affects the biological properties of the drug.

hydrophobic

When a drug is administered (orally, or by intradermal, intramuscular, or subcutaneous injection), it must be absorbed and transported to the target tissue before the drug can elicit the desired physiologic response. Barriers to the normal absorption and transport of the drug will obviously seriously limit its effectiveness, since little or no drug could get to the tissue where it is needed. It should be noted that each cell is surrounded by a lipid and, therefore, **hydrophobic** ("water-hating") membrane that represents a permeability barrier to many polar, *hydrophilic* ("water-loving") substances. Ionized substances cannot easily diffuse across the hydrophobic membranes of cells. Since many drugs possess functional groups that undergo ionization (they behave as weak acids or bases), the absorption of these types of drugs depends on the acidic or basic properties of the drug and on the environment in which it is present. For example, the stomach is highly acidic (pH = 1–2), while the intestinal lumen (the interior portion of the intestine) has a pH of 6.6. Blood plasma has a pH of 7.4 and, therefore, is slightly alkaline. Thus, a specific drug, if administered orally, will encounter environments with different pH values, and this may significantly influence the rate at which it is absorbed by the body.

Let us consider how aspirin (acetylsalicylic acid) is absorbed by the body. The structure of aspirin is shown in the following diagram:

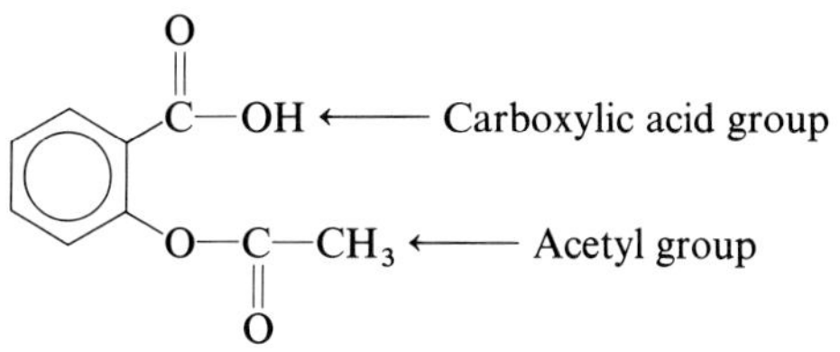

O
‖
C—OH ←—— Carboxylic acid group

O—C—CH_3 ←—— Acetyl group
‖
O

The carboxylic acid group (—COOH) has a pK_a of 3.5. This means that aspirin is a weak acid and can undergo *limited deprotonation* (that is, it does not easily lose its proton). When aspirin enters the stomach (pH = 1–2), the pH, and therefore the high H_3O^+ concentration, keeps essentially 100 percent of the aspirin molecules in the undissociated (HA) and neutral state, as shown in Figure 2–2. Since the aspirin molecule is electrically neutral (and also relatively hydrophobic), it will be readily absorbed by the gastric mucosal cells and transported into the bloodstream. Since the lower intestine and colon have a more neutral pH, some ionization of aspirin molecules would occur in those areas and limit absorption of the molecule there. Likewise, administering an excessive amount of alkali would increase the pH of the gastric contents, cause some ionization of the aspirin, and thus slow the rate of absorption by the stomach.

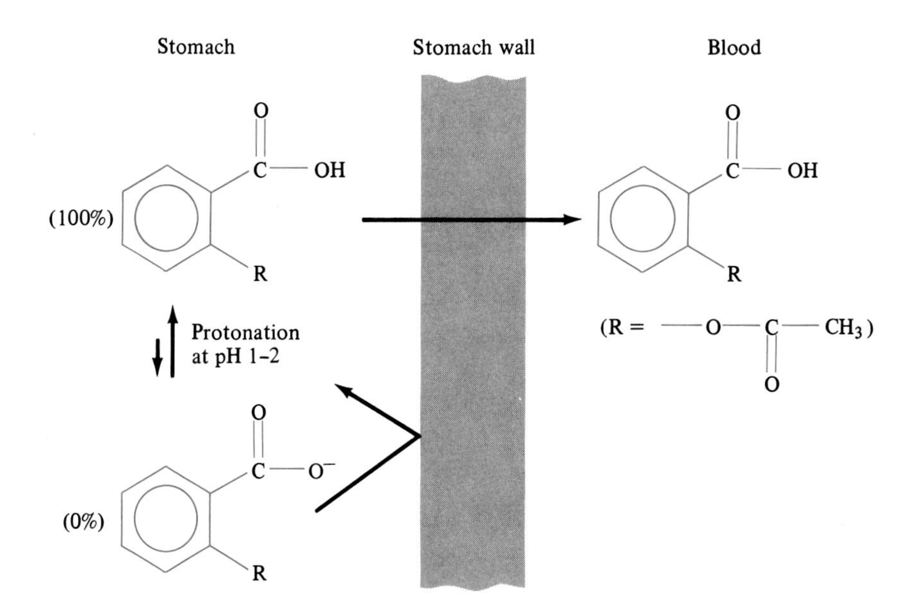

FIGURE 2–2
The absorption of aspirin.

2.5 BUFFERS

buffer

A **buffer** is a chemical system that tends to resist changes in pH when excess acid or base is added. Buffers are usually made from weak acids. To fully understand how a buffer works, consider the following example and explanation:

1. Suppose we have a weak acid (HA).

2. We know that this substance dissociates only very slightly.

$$HA \rightleftarrows H^+ \text{ and } A^-$$

3. Therefore, most of the weak acid is in the form HA and only limited amounts of H^+ and A^- are present.
4. Now, add some excess strong base (OH^-).
5. Thus,

$$HA \longrightarrow H^+ + A^-$$
$$H^+ + OH^- \text{ (excess)} \longrightarrow H_2O$$

6. As excess OH^- is added, the OH^- combines with the few H^+ ions (neutralization reaction).
7. This disturbs the equilibrium condition and causes more HA to dissociate into H^+ and A^-.
8. The newly generated H^+ reacts with more of the OH^- to form H_2O. This process continues until all of the OH^- has been neutralized.
9. Thus, the concentration of the HA decreases (as more of it dissociates into H^+ and A^-) and the concentration of A^- increases. If enough HA were originally present, then it would neutralize the excess OH^- and little or no change in the pH would be observed. A diagram of this process is found in Figure 2–3.

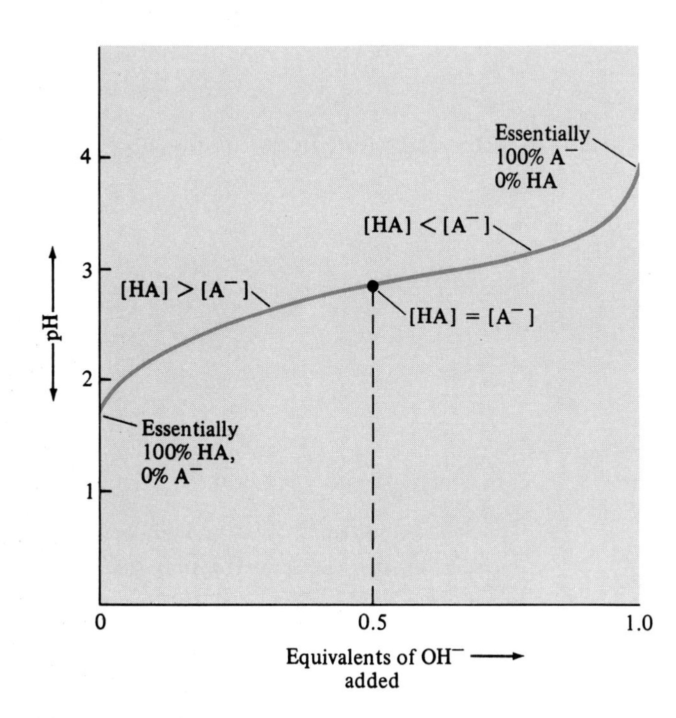

FIGURE 2–3
Titration curve for a weak acid. Notice that the pH change is minimal when [HA] = [A^-]. One equivalent of OH^- = contains 1 mole of replaceable hydroxyl.

Henderson-Hasselbalch equation

The concepts discussed in Section 2.3 and in this section can be summarized using the **Henderson-Hasselbalch equation**.

1. Consider a weak acid (HA). (See Equations 2–4 and 2–5.)

$$HA + H_2O \underset{}{\overset{K_a}{\rightleftharpoons}} H_3O^+ + A^-$$

where

$$K_a = \frac{[H_3O^+][A^-]}{[HA]}$$

and

$$[H_2O] = 1$$ (for simplicity, we will assume that $[H_2O] = 1$)

2. Rearrange the above equation.

$$K_a[HA] = [H_3O^+][A^-]$$

$$\frac{K_a[HA]}{[A^-]} = [H_3O^+]$$

3. Reverse sides.

$$[H_3O^+] = K_a \frac{[HA]}{[A^-]}$$

4. Convert to log function.

$$\log[H_3O^+] = \log(K_a) + \log\frac{[HA]}{[A^-]}$$

5. Multiply both sides by -1.

$$-\log[H_3O^+] = -\log(K_a) - \log\frac{[HA]}{[A^-]}$$

6. Substitute from Equations 2–2 and 2–6.

$$pH = -\log[H_3O^+]$$

$$pK_a = -\log(K_a)$$

7. Therefore,

$$pH = pK_a - \log\frac{[HA]}{[A^-]}$$

$$\boxed{pH = pK_a + \log\frac{[A^-]}{[HA]}} \qquad (2\text{–}7)$$

Equation 2–7 (the Henderson-Hasselbalch equation) describes the relationship between the pH, the pK_a, and the concentrations of HA and A^- for a weak-acid solution. If one knows two out of the three factors, one can solve for the unknown. As you will see in the next section, this is an extremely useful equation and one that has very practical implications when discussing blood pH and blood gas chemistry.

2.6 BLOOD pH AND THE BUFFERING CAPACITY OF THE BLOOD

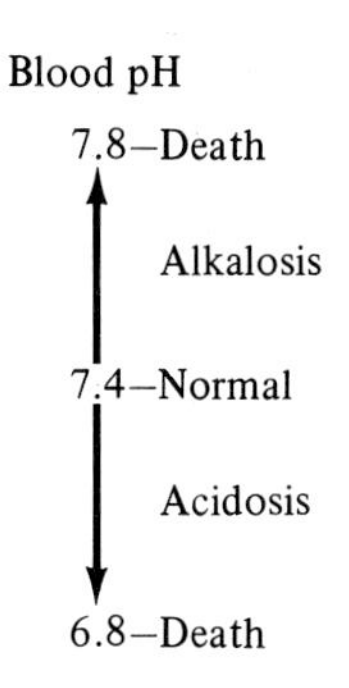

The pH of the blood is held at about pH 7.4. Very slight deviations in pH ($\pm$0.2 to 0.4 pH units) can be disastrous, as shown in the diagram on the left. If a patient exhibits blood pH values greater than 7.4, then the condition is called **alkalosis**. If the patient exhibits blood pH values less than 7.4, the condition is called **acidosis**. Our bodies have remarkable buffering systems that hold the pH of the blood at 7.4. There are four main buffer systems in the blood: (1) the carbonic acid/bicarbonate system, (2) the phosphate system, (3) the plasma protein system, and (4) the hemoglobin system. We will only consider the carbonic acid/bicarbonate (H_2CO_3/HCO_3^-) buffer system.

Carbonic acid (H_2CO_3) is generated primarily in the red blood cell by means of the enzyme carbonic anhydrase.

$$\underset{\text{(Dissolved in the red blood cell)}}{CO_2(d)} + H_2O \xrightarrow{\text{Carbonic anhydrase}} \underset{\text{(Carbonic acid)}}{H_2CO_3}$$

CO_2 dissolved in the cytoplasm of the red blood cell and in the blood plasma is noted as $CO_2(d)$. This dissolved CO_2 is actually in equilibrium with gaseous CO_2 (in the lungs) according to the following equation:

$$CO_2(\text{Gas, lungs}) \rightleftharpoons CO_2(d)$$

The partial pressure of CO_2 (in the gas phase) is constant and directly affects the concentration of dissolved CO_2 in the blood. Thus, the concentration of dissolved CO_2 within the blood must be held at a relatively constant level.

Carbonic anhydrase is probably the most active enzyme known. One enzyme molecule can combine hundreds of thousands of CO_2 and H_2O molecules per second. Once H_2CO_3 is formed, it behaves as a weak acid and donates a proton to yield the bicarbonate ion (HCO_3^-) according to the following reaction:

$$\underset{\text{(Carbonic acid)}}{H_2CO_3} \overset{pK_a = 6.1}{\rightleftharpoons} H^+ + \underset{\text{(Bicarbonate ion)}}{HCO_3^-}$$

The carbonic acid/bicarbonate system buffers blood in the following manner:

1. The buffer system must buffer excess acid (H^+): This is done by reaction of the excess acid (H^+) with the HCO_3^- to reform H_2CO_3, which, in turn, dissociates into CO_2 and H_2O. The CO_2 is liberated by the lungs.

$$HCO_3^- + \underset{\text{(Excess)}}{H^+} \longrightarrow H_2CO_3 \longrightarrow \underset{\text{(Expired)}}{CO_2(g)(\uparrow)} + H_2O$$

2. The buffer system must buffer excess base (OH^-): This is done by reaction of the excess base with H_2CO_3 according to the following reaction:

$$H_2CO_3 + \underset{\text{(Excess)}}{OH^-} \longrightarrow H_2O + HCO_3^-$$

We might ask, what concentrations of H_2CO_3 and HCO_3^- will produce a buffered system with a pH held constant at 7.4? Since the pK_a for the H_2CO_3 dissociation reaction is 6.1, we can use the Henderson-Hasselbalch equation (Equation 2–7) to solve for the concentrations of HCO_3^- (A^-) and H_2CO_3 (HA).

$$\text{pH} = \text{p}K_a + \log\frac{[A^-]}{[HA]}$$

$$7.4 = 6.1 + \log\frac{[HCO_3^-]}{[H_2CO_3]}$$

Subtract 6.1 from both sides.

$$1.3 = \log\frac{[HCO_3^-]}{[H_2CO_3]}$$

Take the antilog of both sides.

$$20 = \frac{[HCO_3^-]}{[H_2CO_3]}$$

$$\frac{[HCO_3^-]}{[H_2CO_3]} = \frac{20}{1}$$

This means that the bicarbonate concentration is 20 times that of the carbonic acid. It is because of this ratio that the system is able to buffer the blood so efficiently. If we actually analyzed a blood sample for the concentrations of HCO_3^- and H_2CO_3, we would find that the concentrations of the two components are:

$$\frac{[HCO_3^-]}{[H_2CO_3]} = \frac{27 \text{ meq/liter}}{1.35 \text{ meq/liter}} = \frac{20}{1}$$

(*Note*: meq/liter = milliequivalents/liter. One equivalent of a solution contains one mole of replaceable hydrogen or hydroxyl.)

Since the pK_a is always at a constant value of 6.1, then the concentration of either HCO_3^- or H_2CO_3 (or the HCO_3^-/H_2CO_3 ratio) must ultimately influence the final pH of the blood. This is the case in certain disease states, and the blood pH is changed from its normal value of 7.4 to some other value (as in acidosis or alkalosis).

Two examples using the Henderson-Hasselbalch equation (Equation 2–7) should help you understand this concept.

EXAMPLE 1. Patient X was admitted into General Hospital near death. She had a very elevated $[HCO_3^-]$ of 50 meq/liter; her $[H_2CO_3]$ was normal at 1.35 meq/liter. What was her blood pH?

$$pH = pK_a + \log \frac{[HCO_3^-]}{[H_2CO_3]}$$

$$pH = 6.1 + \log \frac{(50 \text{ meq/liter})}{(1.35 \text{ meq/liter})}$$

$$pH = 6.1 + \log(37)$$

$$pH = 6.1 + \log(3.7 \times 10^1)$$

$$pH = 6.1 + (\log 3.7 + \log 10^1)$$

$$pH = 6.1 + (0.57 + 1)$$

↑
(From log table)

$$pH = 6.1 + (1.57)$$

$$\boxed{pH = 7.67}$$

EXAMPLE 2. Patient Y was admitted into General Hospital near death. He had a very elevated $[H_2CO_3]$ of 5 meq/liter; his $[HCO_3^-]$ was normal at 27 meq/liter. What was his blood pH?

$$pH = pK_a + \log \frac{[HCO_3^-]}{[H_2CO_3]}$$

$$pH = 6.1 + \log \frac{(27 \text{ meq/liter})}{(5 \text{ meq/liter})}$$

$$pH = 6.1 + \log(5.4)$$

$$pH = 6.1 + (0.73)$$

↑
(From log table)

$$\boxed{pH = 6.83}$$

After examining the previous two examples, we can see how the logarithm of the HCO_3^-/H_2CO_3 ratio significantly alters the final pH value. The situation can be summarized using the "teeter-totter" example in Figure 2–4.

Various disease states do change the concentration of either the HCO_3^- component or the H_2CO_3 component and, thus, alter the ratio When a particular disease or physiological state alters the HCO_3^- concentration this condition is referred to as a **metabolic effect**, since it is abnormal cellular metabolism that alters the HCO_3^- concentration. When changes in the

metabolic effect

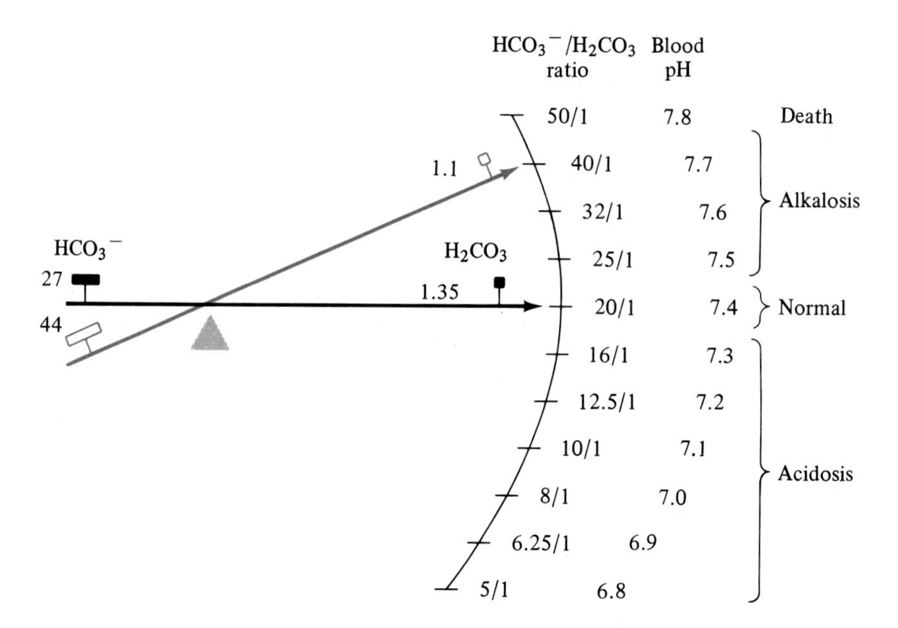

FIGURE 2–4 "Teeter-totter" relationship between HCO_3^-/H_2CO_3 ratio and blood pH. (Adapted from Tietz, N., ed., *Fundamentals of Clinical Chemistry*, 2nd ed. (Philadelphia: W. B. Saunders, 1976), p. 953.)

respiratory effect

H_2CO_3 concentration occur, this condition is referred to as a **respiratory effect**, since it is the lungs (respiration) that control the concentration of dissolved CO_2 in the blood and, thus, influence the H_2CO_3 concentration. The conditions mentioned are summarized in Table 2–3.

TABLE 2–3 A Summary of Disease States that Change the HCO_3^-/H_2CO_3 Ratio

	A. Metabolic Effects (HCO_3^-)
$\frac{(\downarrow)[HCO_3^-]}{(N)[H_2CO_3]} < \frac{20}{1}$	*Metabolic acidosis*: (excess H^+ uses up available HCO_3^- and produces decreased $[HCO_3^-]$ level) 1. Diabetic acidosis 2. Diarrhea (loss of large amounts of HCO_3^-)
$\frac{(\uparrow)[HCO_3^-]}{(N)[H_2CO_3]} > \frac{20}{1}$	*Metabolic alkalosis*: (loss of H^+ disturbs the $H_2CO_3 \rightarrow HCO_3^-$ condition) 1. Vomiting (loss of gastric H^+) 2. Excess alkali administration
	B. Respiratory Effects (H_2CO_3)
$\frac{(N)[HCO_3^-]}{(\uparrow)[H_2CO_3]} < \frac{20}{1}$	*Respiratory acidosis* : (high dissolved CO_2 leads to $(\uparrow)[H_2CO_3]$; caused by inefficient removal of CO_2 by lungs) 1. Emphysema 2. Pneumonia
$\frac{(N)[HCO_3^-]}{(\downarrow)[H_2CO_3]} > \frac{20}{1}$	*Respiratory alkalosis*: (increased removal of CO_2) 1. Hyperventilation

Symbols: (↑)—Increased; (↓)—Decreased; (N)—Normal.

In certain cases, the body is able to restore an abnormal blood pH to 7.4. This is called **compensation**. If the body is unable to restore an abnormal pH to 7.4, then the physiological state is called *uncompensation*.

compensation

Let us end our discussion of blood pH with a few words describing how blood pH and "blood gas" analyses are performed. Most hospital laboratories use an instrument called a blood-gas analyzer. When a doctor is interested in the patient's blood pH and blood gases (CO_2 and sometimes O_2), a blood sample is drawn, placed on ice in a stoppered tube (why?), and sent to the laboratory for analysis on the blood-gas analyzer. This instrument determines the pH of the blood and also its H_2CO_3 concentration. This is done by measuring the amount of dissolved CO_2 gas in the blood—dCO_2, or pCO_2. By using the *Siggaard-Andersen nomogram* (a graphical representation of the Henderson-Hasselbalch equation), the HCO_3^- concentration is then found. A very important factor called the *base excess* can also be calculated. These factors aid the physician in diagnosing the specific disease state that afflicts the patient and also in assessing the course of recovery.

Finally, one of the primary buffer systems *within the cell* should be mentioned. Within the cell, the phosphate buffer system ($HPO_4^{2-}/H_2PO_4^-$) operates and maintains a relatively constaint intracellular pH.

$$\underset{(HA)}{H_2PO_4^-} \xrightleftharpoons{pK_a = 7.2} H^+ + \underset{(A^-)}{HPO_4^{2-}}$$

The concentration of $H_2PO_4^-$ and HPO_4^{2-} are approximately equal; thus, the buffering capacity of this system is very great. Intracellular proteins also act as buffers.

SUMMARY

Water is essential to life. Between 70 and 90 percent of the total weight of most cells is water.

Water (H_2O) contains an electronegative oxygen atom that can attract electrons from the hydrogens. Partial positive and partial negative charges are thus produced in different parts of the molecule, forming a dipole. Because water is dipolar, it can participate in hydrogen bonding with other water molecules. These hydrogen bonds help determine the unique physical properties of water, including its ability to solvate salts and polar substances.

Acids are proton donors; bases are proton acceptors. Protons are associated with a water molecule and form the hydronium ion (H_3O^+). pH is the negative logarithm (base 10) of $[H_3O^+]$. Strong acids have little affinity for their protons and, therefore, easily dissociate. Weak acids have greater affinity for their protons than strong acids and, therefore, participate only in limited ionization. The extent to which a weak acid dissociates is expressed as the dissociation constant (K_a) = [Products]/[Reactants]. The pK_a is the negative logarithm (base 10) of K_a.

Drugs possess functional groups that can behave as either weak acids or bases. Electrically charged drug molecules cannot diffuse across cell membranes; neutral drug molecules can.

A buffer is a chemical system that tends to resist changes in pH when excess acid or base is added to the system. Buffers are usually made from weak acids, and are effective because of the limited dissociation of the weak acid. The Henderson-Hasselbalch equation (Equation 2–7) relates the pH of a solution to the pK_a of the chemical species and to the ratio $[A^-]/[HA]$

The pH of blood is about 7.4. Alkalosis is the physiological state in which the blood pH rises above 7.4; acidosis is the corresponding state for pH values less than 7.4. The bicarbonate/carbonic acid system keeps the pH of the blood constant and acts as a buffer. $[HCO_3^-]$ is 20 times $[H_2CO_3]$. Changes in the $[HCO_3^-]/[H_2CO_3]$ ratio cause drastic changes in the final pH of the blood. Alterations in $[HCO_3^-]$ are caused by metabolic factors; alterations in $[H_2CO_3]$ are caused by respiratory factors. The body can compensate for abnormal blood pH or remain uncompensated. The phosphate ($HPO_4^{2-}/H_2PO_4^-$) system is the primary intracellular buffer system.

REVIEW QUESTIONS

1. We have presented a number of important concepts and terms in this chapter. Define (or briefly explain) the following terms:
   a. Dipole
   b. Hydrogen bond
   c. Brönsted-Lowry acid
   d. pH
   e. Weak acid
   f. Dissociation constant (K_a)
   g. Buffer
   h. Henderson-Hasselbalch equation
   i. Bicarbonate/carbonic acid buffer system
   j. Metabolic effect
   k. Compensation
2. Discuss reasons why water is so important for life.
3. Find the logarithm of the following numbers (see the Appendix Section for a Math Review):
   a. 100 b. 2 c. 5.56 d. 40 000
4. Calculate the pH of the following solutions:
   a. 10^{-3} M HCl
   b. Pure water
   * c. 10^{-8} M HCl (this one is tricky)
   * d. 3×10^{-4} M HCl
5. Calculate the pK_a values for the following weak acids:
   a. Acetic acid ($K_a = 1.8 \times 10^{-5}$):
   b. Ammonium ion (NH_4^+) ($K_a = 5.7 \times 10^{-10}$):
6. *Using the relationship $[H_3O^+] = 10^{-pH}$, calculate $[H_3O^+]$ for:
   a. Milk (pH 6.5)
   b. Cola drink (pH 2.3)
   c. Blood (pH 6.8)
7. In terms of $[H_3O^+]$, what is the difference between a pH 3 solution and a pH 8 solution?
8. *If water dissociates into 10^{-7} M H^+ and 10^{-7} M OH^- ions, calculate the dissociation constant (K_a) for water. Assume $[H_2O] = 55.5$ M.
9. *What happens to the Henderson-Hasselbalch relationship (Equation 2–7) when $[HA] = [A^-]$? What does this tell you about the effectiveness of a buffer system?
10. Calculate the pH of a blood sample drawn from a patient with a $[HCO_3^-]$ of 35 meq/liter and a $[H_2CO_3]$ of 1.35 meq/liter (pK_a = 6.1).

SUGGESTED READINGS

The Chemistry of Water, Acids, Bases, and Buffers

Davenport, H. W. *The ABC of Acid–Base Chemistry*, 5th ed. Chicago: University of Chicago Press, 1969.

Routh, J. I. *Mathematical Preparation for the Health Sciences*, 2nd ed. Philadelphia: W. B. Saunders, 1976.

Segel, I. H. *Biochemical Calculations*, 2nd ed. New York: John Wiley, 1976.

Blood pH and Blood Gases

Tietz, N., ed. *Fundamentals of Clinical Chemistry*, 2nd ed., pp. 849–53, 887–901, and 952–74. Philadelphia: W. B. Saunders, 1976.

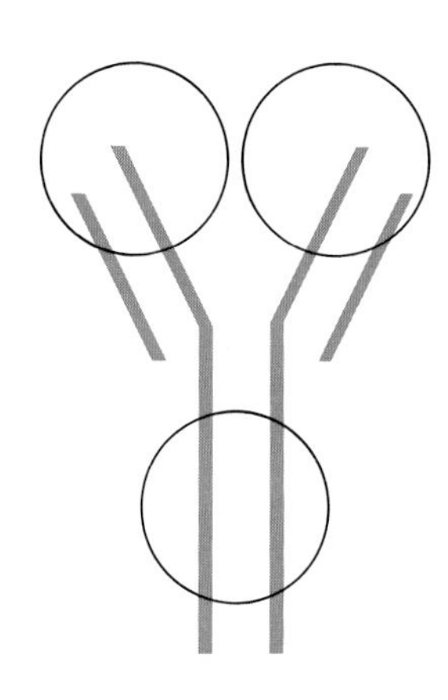

Chapter 3

Amino Acids and Proteins

3.1 INTRODUCTION

In this chapter, we shall discuss the chemistry of the amino acids. These substances are extremely important because they are the basic building blocks of all proteins. As we shall see, amino acids can link together to form long chains. These chains can enter into a diversity of folding arrangements, thus forming different three-dimensional shapes. Recent evidence suggests that the biological function of a protein molecule is determined by its specific shape. We shall discuss protein chemistry in more detail in the latter half of this chapter, where the size, shape, and functional classes of proteins will be presented. The chemistry of the peptide bond will be discussed, along with the four different "levels" of protein structure.

3.2 THE CLASSIFICATION OF THE AMINO ACIDS

amino acids

Amino acids have the following generalized structure:

```
        NH2  ←——  α-Amino group
        |
    R—CH—COOH  ←——  Carboxyl group
        ↖
          α-Carbon atom
```

The *amino group* (—NH_2^0 or —NH_3^+) resides on the *α-carbon* atom (the carbon atom next to the carbon of the carboxyl group); hence, most amino acids are called α-amino acids. Amino acids are acids because the carboxyl group can donate a proton. The R group can vary in chemical structure. There are literally hundreds of different R groups. Fortunately, only 20 amino acids with 20 different R groups are naturally found in proteins. It is with these 20 amino acids that we will deal in this section.

Since each amino acid has the same α-amino group, carboxyl group, and α-carbon atom, we can simplify our discussion by considering the differences in the 20 R groups. These 20 R groups can be classified into four main categories: (1) hydrophobic ("water-hating"); (2) uncharged hydrophilic ("water-loving"); (3) negatively charged (at pH 7) hydrophilic; and (4) positively charged (at pH 7) hydrophilic. This classification scheme is based upon the chemical structure and physical properties of the R group. The 20 amino acids with their 20 different R groups are summarized in Table 3–1. As you examine Table 3–1, try to distinguish one R group from another by chemical structure and explain why a particular R group is placed in a specific category. Also, note the names of the amino acids and their abbreviations.

TABLE 3–1 The Classification of the Amino Acids

1. The Hydrophobic Amino Acids[a]	
$H_3C-CH(NH_3^+)-COO^-$ Alanine, Ala	$C_6H_5-CH_2-CH(NH_3^+)-COO^-$ Phenylalanine, Phe
$(H_3C)_2CH-CH(NH_3^+)-COO^-$ Valine, Val	(indole ring, N–H)$-CH_2-CH(NH_3^+)-COO^-$ Tryptophan, Trp
$(H_3C)_2CH-CH_2-CH(NH_3^+)-COO^-$ Leucine, Leu	(pyrrolidine ring, N H_2^+)$-COO^-$ Proline, Pro (an *imino* acid)
$H_3C-CH_2-CH(H_3C)-CH(NH_3^+)-COO^-$ Isoleucine, Ile	$H_3C-\ddot{S}-CH_2-CH_2-CH(NH_3^+)-COO^-$ Methionine, Met

[a] The hydrophobic amino acids are arranged in order of increasing size of the R group.

TABLE 3–1 *Continued*

2. Hydrophilic (polar) Amino Acids

$H{-}CH(NH_3^+){-}COO^-$	Glycine, Gly
$NH_2{-}C(=O){-}CH_2{-}CH(NH_3^+){-}COO^-$	Asparagine, Asn
$HO{-}CH_2{-}CH(NH_3^+){-}COO^-$	Serine, Ser
$NH_2{-}C(=O){-}CH_2{-}CH_2{-}CH(NH_3^+){-}COO^-$	Glutamine, Gln
$H_3C(HO){>}CH{-}CH(NH_3^+){-}COO^-$	Threonine, Thr
$HO{-}C_6H_4{-}CH_2{-}CH(NH_3^+){-}COO^-$	Tyrosine, Tyr
$HS{-}CH_2{-}CH(NH_3^+){-}COO^-$	Cysteine, Cys

3. Negatively Charged Hydrophilic Amino Acids (at pH 7)

$^-OOC{-}CH_2{-}CH(NH_3^+){-}COO^-$	Aspartic acid, Asp
$^-OOC{-}CH_2{-}CH_2{-}CH(NH_3^+){-}COO^-$	Glutamic acid, Glu

4. Positively Charged Hydrophilic Amino Acids (at pH 7)

$^+H_3N{-}(CH_2)_4{-}CH(NH_3^+){-}COO^-$ (ε at the side-chain ^+H_3N)	Lysine, Lys
$(^+HN{=}CH{-}NH{-}CH{=}C){-}CH_2{-}CH(NH_3^+){-}COO^-$ (imidazolium ring)	Histidine, His (at pH < 6)
$(NH_2)(NH_2)^+{=}C{-}NH{-}(CH_2)_3{-}CH(NH_3^+){-}COO^-$	Arginine, Arg

The *hydrophobic* amino acids contain R groups that are composed of either hydrocarbon (aliphatic) chains or hydrocarbon (aromatic, benzene-like) rings. Notice that the size and complexity of the R group increases as one proceeds down this particular list of the hydrophobic amino acids.

Alanine is the least hydrophobic of these eight amino acids because of the small methyl group (—CH_3). Valine, leucine, and isoleucine have larger (and, therefore, more hydrophobic) R groups. In addition, these three amino acids have branched chains and are thus called *branched-chain* amino acids. Methionine is an unusual amino acid because it has a sulfur atom in the hydrocarbon chain of the R group. This modification maintains the hydrophobic character of this particular amino acid, but confers to it a special versatility. Specifically, the two pairs of electrons can bind metal ions to the amino acid. Metal ions bound to amino acids within proteins often contribute to the biological function or activity of the protein. Phenylalanine and tryptophan are more complex amino acids. Each possesses a rather large hydrophobic ring system. Phenylalanine contains an aromatic (benzene-like) *phenyl ring*; tryptophan contains a two-ring *indole* system. Proline is unique in that it does not contain a free α-amino group. Notice that the nitrogen atom of proline is actually part of the five-membered ring system. Because of this feature, proline is called an *α-imino acid*. As we shall see in Section 3.6, proline is improtant in producing "kinks" or bends in the protein chain; it thus contributes significantly to the ultimate three-dimensional shape of the protein. These eight amino acids (including proline) contribute significantly to the bulk and structural backbone of the protein molecule. Because proteins normally exist in a water environment, these hydrophobic amino acids are generally found on the inside of the protein molecule, away from water.

In contrast, the *hydrophilic* ("water-loving") amino acids generally reside on the outside of the protein molecule, in contact with water. Most hydrophilic amino acids contain an electronegative oxygen atom and, thus, can hydrogen bond with water molecules. Glycine, although classified as a hydrophilic amino acid, is rather unusual: its R group is only a single hydrogen atom. It is rather difficult, therefore, to classify glycine as a hydrophilic amino acid because its R group is not truly hydrophilic. Serine, threonine, and tyrosine all contain a polar *hydroxyl group* (—OH) that can participate in hydrogen bonding. Asparagine and glutamine both contain polar *amide* groups ($\overset{NH_2}{\underset{O}{\gt}}C-$). Cysteine possesses a sulfhydryl group (—SH). This —SH group can also participate in hydrogen bonding. In addition, two sulfhydryl groups can become oxidized (lose electrons) and link together, forming a **disulfide bond**.

disulfide bond

$$\text{⌇}-CH_2-SH \quad HS-CH_2-\text{⌇} \xrightarrow{2H} \text{⌇}-CH_2-S-S-CH_2-\text{⌇}$$

This bond is very important in holding together protein chains; thus, the disulfide bond helps determine the three-dimensional structure of the protein.

The *negatively charged* (at pH 7) hydrophilic amino acids (aspartic acid and glutamic acid) have R groups that contain a terminal carboxyl group. These carboxyl groups are acids and, therefore, easily lose their protons to the solution. At pH 6–7, the carboxyl groups of these amino acids are completely ionized and, therefore, negatively charged.

The *positively charged* (at pH 7) hydrophilic amino acids include lysine, arginine, and histidine. The R groups of these amino acids accept protons from the water environment (they are bases) and, thus, become positively charged. The terminal amino group of lysine ($—NH_3^+$) is called the *epsilon amino group* because it is on the epsilon (ε) carbon atom. The terminal region of arginine is called the *guanidinium group*; the ring system of histidine is called the *imidazolium group*.

In summary, one should remember that the chemical structure and physical properties of the R group determine the category in which a particular amino acid will be classified.

3.3 THE CHEMICAL AND PHYSICAL PROPERTIES OF THE AMINO ACIDS

THE SOLUBILITY PROPERTIES OF THE AMINO ACIDS

As you might expect, the properties of the R group influence the solubility of the amino acid in water. Although the hydrophobic amino acids "hate" water, a certain amount of a hydrophobic amino acid will still dissolve in it. This is because the α-amino and carboxyl groups of all amino acids are significantly polar and overshadow the hydrophobic properties of the R group (at least to a certain degree). Table 3–2 summarizes the solubilities of a few of the hydrophobic amino acids. Notice that the solubility *decreases* as the size (and, therefore, the hydrophobic character) of the R group increases. The hydrophilic, positively charged, and negatively charged amino acids are generally more soluble than the hydrophobic amino acids.

TABLE 3–2 Solubilities of Some Hydrophobic Amino Acids

Amino Acid	Solubility (g/liter H_2O)
Alanine	167
Valine	58
Leucine	22
Tryptophan	14

Note: These amino acids are arranged according to the increasing size and hydrophobicity of the R group.

THE ACID-BASE PROPERTIES OF THE AMINO ACIDS

All of the 20 naturally occurring amino acids possess an α-amino group (except proline) and a carboxyl group. These groups (and certain ionizable R groups) behave as either acids or bases. Specifically, the α-amino group accepts a proton from the solution, whereas the carboxyl group donates its proton to the solution. These ionization reactions are summarized in the following scheme:

$$\underset{\alpha}{R-\overset{\overset{\displaystyle NH_2}{|}}{CH}-COOH} \rightleftharpoons \underset{\alpha}{R-\overset{\overset{\displaystyle NH_3^+}{|}}{CH}-COO^-}$$

Zwitterion form (0 charge)

Since the amino acid has both a positively charged group (the protonated α-amino group) and a negatively charged group (the deprotonated carboxyl group), the net electrical charge on the amino acid is zero. This molecular species is called a **zwitterion**. In fact, most amino acids naturally exist as zwitterions when dissolved in water.

zwitterion

The ionizable groups of the amino acids can undergo *stepwise ionization reactions*. For example, consider the ionization of glycine:

$$H-\overset{\overset{\displaystyle NH_3^+}{|}}{CH}-COOH \xrightarrow{H^+} H-\overset{\overset{\displaystyle NH_3^+}{|}}{CH}-COO^- \xrightarrow{H^+} H-\overset{\overset{\displaystyle NH_2}{|}}{CH}-COO^-$$

(charge = 1+) Zwitterion (charge = 0) (charge = 1−)

Starting with the fully protonated form of glycine (electrical charge = 1+), first a proton dissociates from the carboxyl group, producing the zwitterion (electrical charge = 0). A second proton then dissociates from the α-amino group, yielding the fully deprotonated form of the amino acid (electrical charge = 1−). The reverse of these reactions can also occur. Because the carboxyl group readily donates its proton to the solution, it has a relatively small pK_a (2.3). The α-amino group, on the other hand, has a much larger pK_a (9.7); this means that the α-amino group is a very weak acid.

The stepwise ionization of the carboxyl group and the α-amino group can also be diagrammed, as shown in Figure 3–1. Starting with the fully protonated form of the amino acid, we slowly add a strong base (OH^-) and, thus, *titrate* it (dissociate the ionizable protons by a neutralization reaction). As we add base, the carboxyl proton easily dissociates. At the midpoint of the *first plateau region* (the flat portion of the curve), 50 percent of the amino acid is in fully protonated form and 50 percent is in the zwitterion form. As we continue to add base, nearly 100 percent of the amino acid is converted to the zwitterion form. Since the α-amino group has a much larger pK_a (9.7), it will not easily lose its proton until larger amounts of base are added. By adding more base, we convert the zwitterion species into the fully deprotonated form (that is, we induce the dissociation

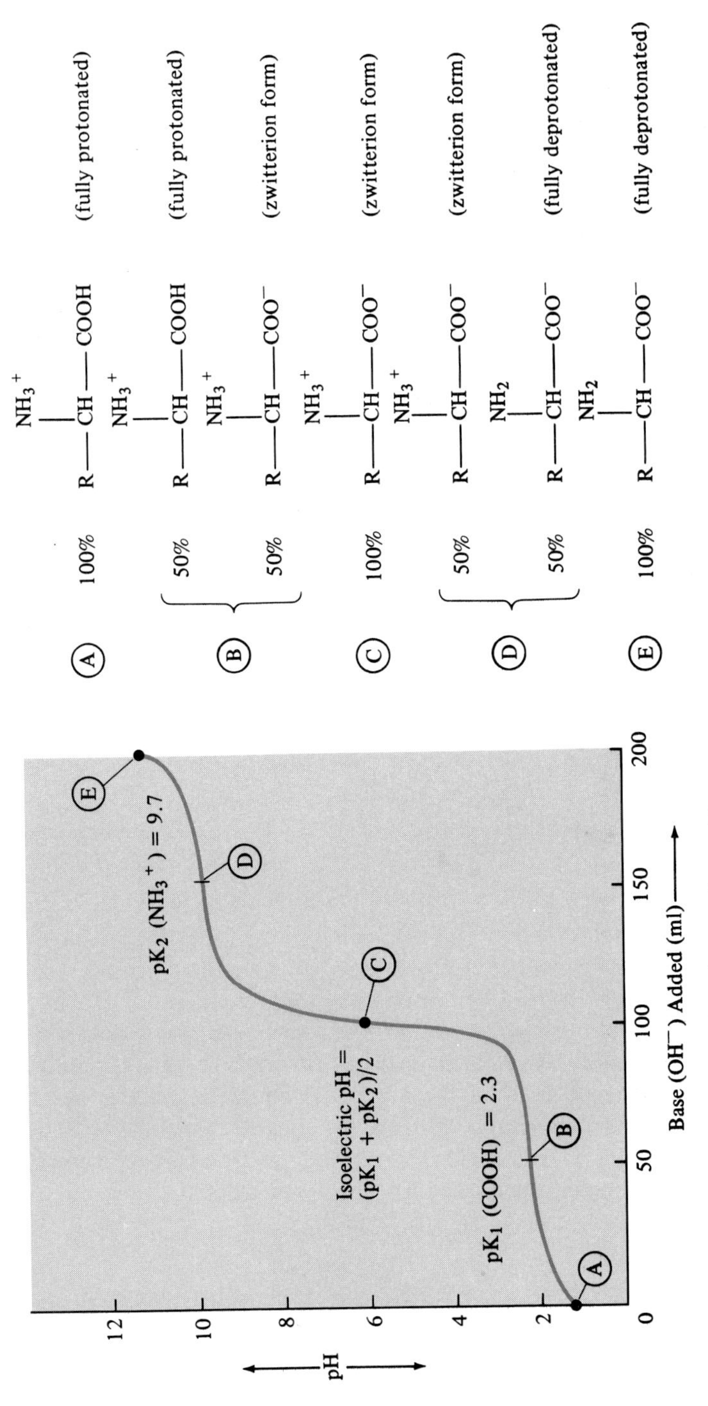

FIGURE 3–1 The stepwise titration of an amino acid.

of the α-amino group proton). At the midpoint of the *second plateau region*, 50 percent of the amino acid is in the zwitterion form and 50 percent of the amino acid is in the fully deprotonated form. By adding more base, we then generate 100 percent of the fully deprotonated form of the amino acid.

Note that this titration curve has two distinct plateau regions in which large amounts of added base change the pH of the solution only slightly. These regions are excellent buffer regions (remember, a buffer resists changes in pH when excess acid and base are added). The pK_a for a particular ionizable group is defined as the midpoint of the plateau region. The midpoint of the first plateau gives us the pK_a for the carboxyl group (2.3); the midpoint of the second plateau gives us the pK_a for the α-amino group (9.7). The region between the two plateaus is not a very good buffer region, since small amounts of excess base produce large changes in the final pH of the amino-acid solution. The arithmetic mean of the two pK_a values is the

isoelectric point

isoelectric point (pH_I):

$$pH_I = (pK_1 + pK_2)/2 \tag{3-1}$$

(*Note*: Equation 3–1 does not apply to an amino acid with three ionizable groups.) The *isoelectric* point is defined as the pH at which an amino acid is nearly 100 percent in the zwitterion form.

Amino acids with an ionizable R group exhibit more complex ionization reactions. For example, consider the complete ionization of aspartic acid:

$$\underset{\text{(Charge = 1+)}}{HOOC{-}CH_2{-}\overset{NH_3^+}{\overset{|}{C}}H{-}COOH} \xrightarrow{H^+} \underset{\substack{\text{Zwitterion form}\\ \text{(charge = 0)}}}{HOOC{-}CH_2{-}\overset{NH_3^+}{\overset{|}{C}}H{-}COO^-} \xrightarrow{H^+}$$

$$\underset{\text{(Charge = 1−)}}{{}^-OOC{-}CH_2{-}\overset{NH_3^+}{\overset{|}{C}}H{-}COO^-} \xrightarrow{H^+} \underset{\text{(Charge = 2−)}}{{}^-OOC{-}CH_2{-}\overset{NH_2^0}{\overset{|}{C}}H{-}COO^-}$$

Notice that aspartic acid has two carboxyl groups and one α-amino group. Thus, a total of three protons could dissociate, one from each of these ionizable groups. A theoretical titration curve for aspartic acid would exhibit three plateau regions with three midpoints, each corresponding to the pK_a value for one of the three ionizable groups.

THE OPTICAL ACTIVITY OF THE AMINO ACIDS

You may already have noticed that four *different* groups are bound to the α-carbon atom of an amino acid.

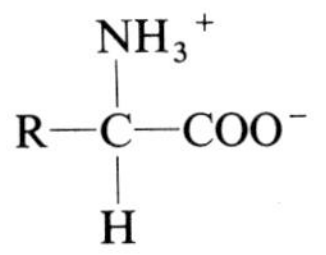

$$R-\underset{\displaystyle H}{\overset{\displaystyle NH_3^+}{C}}-COO^-$$

asymmetric enantiomers

The α-carbon atom is, therefore, what is called an **asymmetric** atom. Molecules with asymmetric atoms can exist as **enantiomers**—that is, as D- and L-isomers that are mirror images of each other and have nonsuperimposable chemical structures. The D and L forms of alanine are shown in the following diagram:

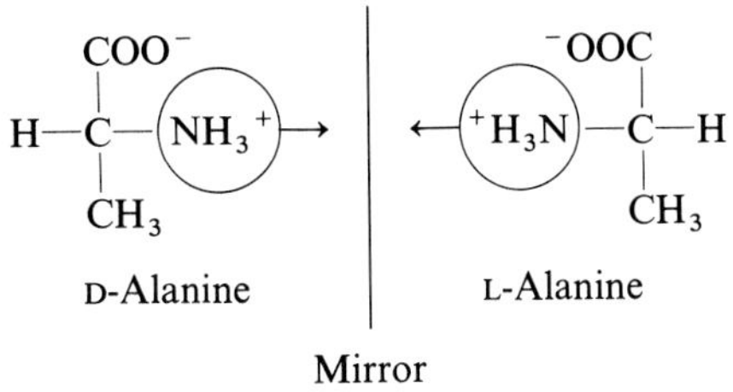

$$H-\underset{\displaystyle CH_3}{\overset{\displaystyle COO^-}{C}}-NH_3^+ \qquad | \qquad {}^+H_3N-\underset{\displaystyle CH_3}{\overset{\displaystyle {}^-OOC}{C}}-H$$

D-Alanine | L-Alanine

Mirror

optical activity

Solutions of either D or L forms of amino acids possess **optical activity**—that is, they rotate the plane of polarized light (light waves oriented in only one direction). *The proteins in your body are constructed from L-amino acids only.* Some lower life forms (bacteria) use both D and L forms in constructing various biomolecules (bacterial cell walls and antibiotics).

3.4 PEPTIDE STRUCTURE

Amino acids can link together and form chains. Specifically, the carboxyl group of one amino acid can link with the α-amino group of a second amino acid and form a **peptide bond** according to the following condensation reaction:

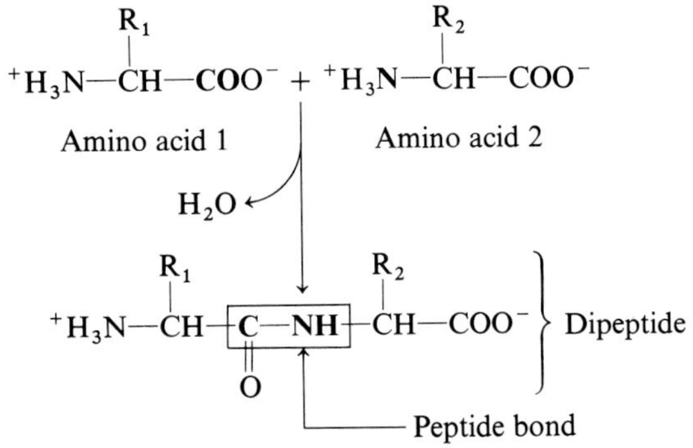

peptide bond

$${}^+H_3N-\overset{\displaystyle R_1}{CH}-COO^- + {}^+H_3N-\overset{\displaystyle R_2}{CH}-COO^-$$

Amino acid 1 | Amino acid 2

$$\downarrow \; (-H_2O)$$

$${}^+H_3N-\overset{\displaystyle R_1}{CH}-\boxed{\underset{\displaystyle \| \atop O}{C}-NH}-\overset{\displaystyle R_2}{CH}-COO^- \Big\} \text{ Dipeptide}$$

Peptide bond

The peptide bond is extremely important, since it is the main structural bond found in all proteins. Notice that the newly formed *dipeptide* still has a free α-amino group (the *N-terminal*, usually placed to the left in diagrams) and a free carboxyl group (the *C-terminal*, usually placed to the right). These two terminals can participate in other condensation reactions to yield even longer chains. Extremely long chains of amino acids can be constructed consisting of many hundreds of amino acids linked together like beads on a string. Substances consisting of multiply linked amino acids are called **polypeptides** (poly means "many" and peptide refers to the peptide bonds). Peptide chains with only a few amino acids (usually less than 10 amino acids) are called **oligopeptides** (oligo means "a few"); peptide chains with 50 or more amino acids are called **proteins.**

polypeptides

oligopeptides

proteins

The structure of a tripeptide (alanylaspartylphenylalanine, N—ala—asp—phe—C) is shown in the following diagram:

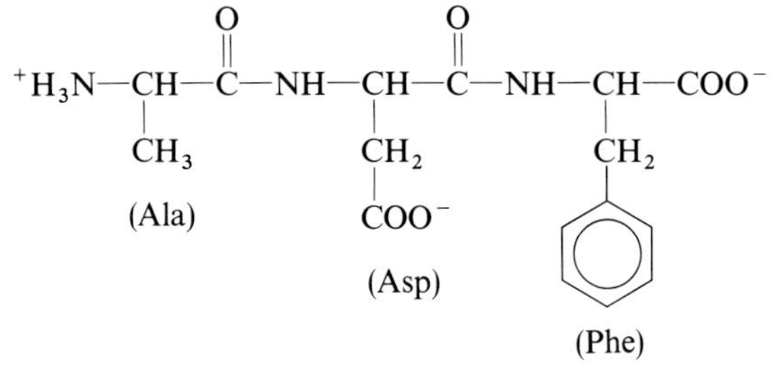

$$ {}^{+}H_3N-\underset{\underset{\text{(Ala)}}{CH_3}}{\underset{|}{CH}}-\overset{O}{\overset{\|}{C}}-NH-\underset{\underset{\underset{\text{(Asp)}}{COO^-}}{\underset{|}{CH_2}}}{\underset{|}{CH}}-\overset{O}{\overset{\|}{C}}-NH-\underset{\underset{\underset{\text{(Phe)}}{C_6H_5}}{\underset{|}{CH_2}}}{\underset{|}{CH}}-COO^- $$

The R groups, together with the free α-amino group (N-terminal) and the free carboxyl group (C-terminal) contribute to the ultimate chemical and physical properties of the peptide, including its electrical charge. This particular molecule would have an electrical charge of 1− at pH 7 (both carboxyl groups would be ionized at this pH), while the N-terminal would be protonated and have a 1+ charge.

Many smaller peptides (oligopeptides) have rather important biological functions. Researchers have recently discovered dozens of new peptides of small molecular weight, each with a specific biological function. A few of these peptides are listed in Table 3–3.

It should be stressed that the type of amino acid and the sequence of the amino acid within the polypeptide chain is critical, since the structure and chemical properties of amino acid side-chains ultimately determine the size, shape, and electrical charge of the entire peptide. These factors, in turn, influence the biological properties of the molecule.

enkephalins

Recently, a new class of small polypeptides have been discovered in brain tissue, the **enkephalins** (or endorphins). Researchers have been excited by this discovery, because the existence of such compounds in the brain may explain some aspects of brain chemistry and function. The brain both produces the enkephalins and contains specific receptors for them. Interestingly, the enkephalins are similar in chemical properties to morphine and to other opiates that kill or deaden pain. The brain contains specific receptors for the opiates. Apparently, the enkephalins bind to these receptors and block the pain response much as morphine does. In essence, the brain can produce its own pain-killer. This research has far-reaching implications. To quote from an article from *Scientific American* (1979):

TABLE 3–3 Some Important Peptides and Their Biological Function(s)

Name	Length	Amino Acid Sequence	Biological Function(s)
Thyrotropin Releasing Factor	3 amino acids	[a]Pyro—Glu—His—Pro—amide	1. Produced by the hypothalamus. 2. Stimulates the release of thyroid stimulating hormone (TSH) from the pituitary gland.
Met-enkephalin	5 amino acids	N—Tyr—Gly—Gly—Phe—Met—C	1. Produced by the brain. 2. May control pain levels.
Angiotensin	7 amino acids	N—Asp—Arg—Val—Tyr—Ile—His—Pro—C	1. Produced in the blood. 2. Increases blood pressure and conserves body fluids by decreasing fluid flow through the kidneys.
Bradykinin	9 amino acids	N—Arg—Pro—Pro—Gly—Phe—Ser—Pro—Phe—Arg—C	1. Produced in the blood. 2. May control blood pressure by dilating blood vessels; also a smooth muscle constrictor.
Oxytocin	9 amino acids	Cys—Tyr—Ile—Gly—Asn—Cys—Pro—Leu—Gly—NH_2[b] (S—S bridge between the two Cys)	1. Produced by the pituitary gland. 2. Stimulates uterine contractions and the ejection of milk.

[a] The glutamic acid is cyclic in this peptide.
[b] The C-terminal of Gly is the amide derivative $\left[-\overset{\overset{\displaystyle O}{\|}}{C}-NH_2 \right]$

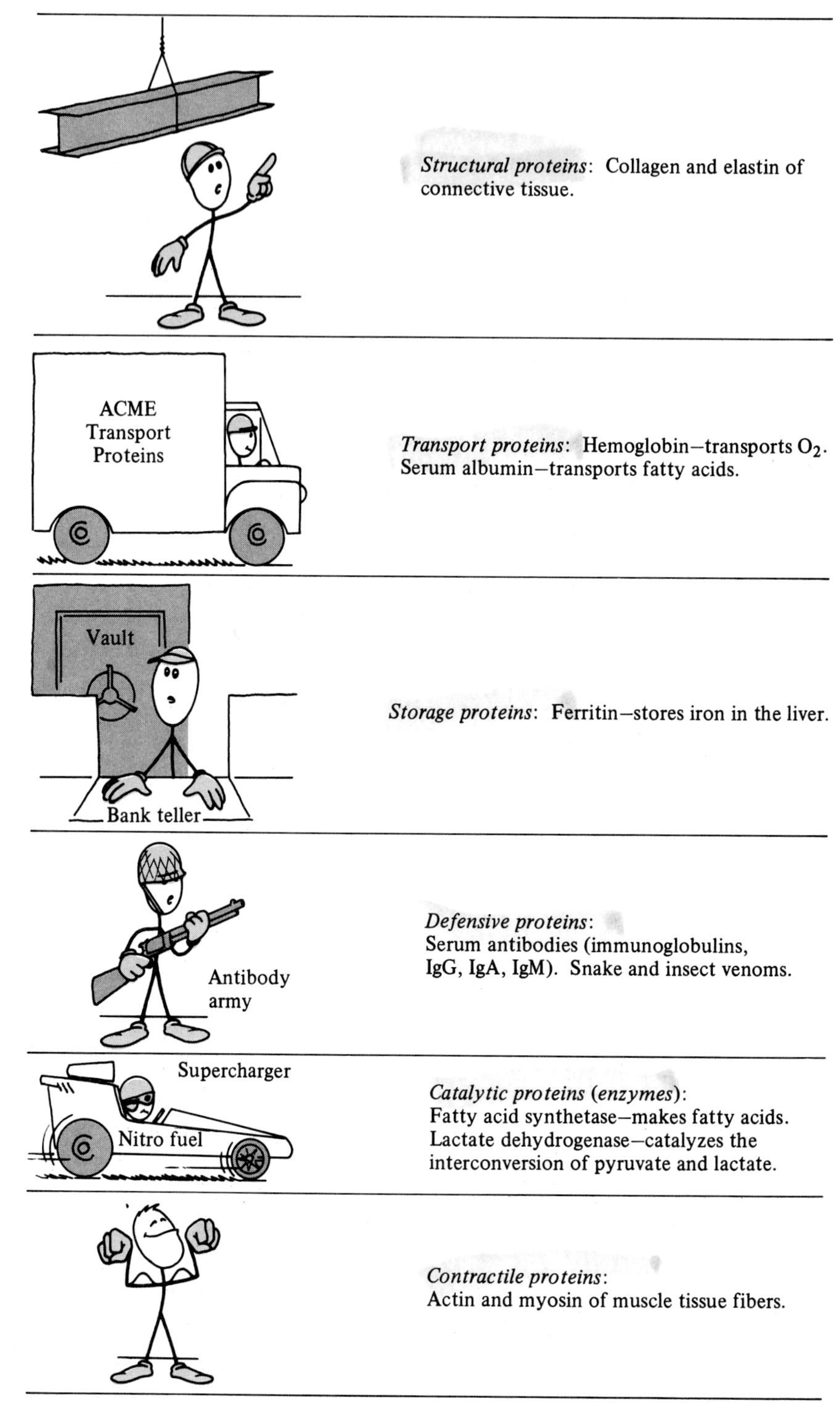

FIGURE 3–2
The classes of proteins.

> Recent experiments have suggested that several procedures employed to treat chronic pain—acupuncture, direct electrical stimulation of the brain, and even hypnosis—may act by eliciting the release of enkephalins or endorphins in the brain and spinal cord.[1]

Enkephalins may not only regulate the pain response, but may also control sleep, sexual function, and other complex physiological phenomena. Clearly, the discovery of these unusual substances has opened the door to new and exciting research regarding the chemistry and function of the brain.

3.5 THE CLASSES OF PROTEINS

Before we actually discuss the structure of protein molecules, let us first define what proteins are and how they are classified, then present some of their important biological functions.

protein

The term **protein** was first used by Gerardus J. Mulder in 1838. The word was derived from the Greek *proteios*, which means "of first importance." Mulder correctly assumed that proteins were very important and that life would be impossible without them. Proteins are long chains of amino acids (from about 50 amino acids to many thousands of amino acids in length) held together by peptide bonds. The polypeptide chain of a protein can fold in a way determined by its amino acid sequence and thus assume a specific three-dimensional shape. The shape and physical properties of the polypeptide chain contribute to the specific biological properties of the molecule. Some proteins are *structural molecules* for cells and organs. These proteins make up the connective tissue, such as tendons, elastic fibers, and structural supports for cells. Other proteins *transport*

	Molecular weight (daltons)	Diameter (Å)
Water molecule	18	4–5
Lysozyme (a protein)	13 900	30–40
Hemoglobin (a protein)	64 000	60
Pyruvate dehydrogenase enzyme complex	4.6×10^6	300

FIGURE 3–3 The approximate sizes of different protein molecules. (1 Å = 1 angstrom unit = 1×10^{-8} cm.) On this figure, a typical cell would be the size of a football field.

substances within cells or from one cell to another. Still other proteins *store* substances, while other proteins (antibodies) defend the body against foreign substances that invade it. A very important class of proteins are

enzymes

enzymes. These proteins are *biological catalysts* that speed up chemical reactions. All of the metabolic reactions that occur within the cells of the body are catalyzed by enzymes. Without enzymes, cells could not function and life, as we know it, would be impossible. Still other proteins are involved in *contraction* processes, which change the shape of a cell or move a cell (or the entire organism) from one place to another. Figure 3–2 summarizes the main functional classes of proteins.

Proteins can also be classified according to their size, shape, and chemical composition.

SIZE

Proteins vary greatly in size. Generally, they have molecular weights from about 6000 daltons (1 dalton $\approx$ the weight of 1 hydrogen atom) to several million daltons. Hemoglobin, an average protein, has a molecular weight of 64 000 daltons; thyroglobulin (a thyroxine hormone-storage protein) has a molecular weight of 660 000 daltons. Figure 3–3 graphically demonstrates the sizes of a few proteins and compares these sizes to that of a water molecule (which has a molecular weight of 18 daltons).

SHAPE

globular
fibrous

Proteins generally fall into one of two categories: (1) **globular** proteins and (2) **fibrous** proteins. Globular proteins are rather compact and, as the name implies, approximate a sphere. Some globular proteins are more elongated in shape; others are shaped like an oblate spheroid (a squashed globe). Fibrous proteins are usually very long, thin fibers.

CHEMICAL COMPOSITION

Obviously, all proteins are composed of amino acids. One can treat a protein with a strong acid and destroy the peptide bonds that hold the amino acids together. This treatment yields a solution of free amino acids. When treated with acid, some proteins not only yield amino acids, but other non–amino acid substances. These non–amino acid components (prosthetic groups) can be inorganic ions, complex organic molecules, lipids, sugars, or other substances. Some constituents derived from vitamins, known as

cofactors

cofactors, are often required for the biological activity of the protein. When a protein contains not only the polypeptide chain(s) but one of these non–amino acid cofactors, it is called a *conjugated protein*. Table 3–4 represents a few conjugated proteins. It should be noted that the polypeptide chain of a conjugated protein, called the *apoprotein*, *usually does not possess biological activity*.

TABLE 3–4 Types of Conjugated Proteins

		Prosthetic Group (cofactor)		Conjugated Protein
Apoprotein	+	Metal ion	=	*Metallo*protein
		Sugar	=	*Glyco*protein
		Lipid	=	*Lipo*protein
		Nucleic acid	=	*Nucleo*protein
		Heme	=	*Hemo*protein
		Flavin	=	*Flavo*protein

Note: *Apoprotein* refers to the polypeptide chain(s) only.

OTHER GENERALIZATIONS ABOUT PROTEIN

A few other generalizations about proteins are in order.

1. Not all proteins contain all 20 amino acids. Some proteins (such as collagen) are constructed from only a few of them.
2. Some amino acids occur less frequently in proteins than other amino acids do. For example, a protein might contain 2–3 tryptophan or tyrosine units and 20–30 glycine or alanine units.
3. The hydrophobic amino acids reside on the inside of the protein, away from the water environment.
4. The hydrophilic amino acids reside on the outside of the protein, in contact with the water environment.
5. Most globular proteins are rather compact and highly folded.
6. Globular proteins do not contain regular repeating sequences of specific amino acids (for example, . . . —ABC—ABC—ABC— . . .). Some fibrous proteins, however, do contain such regions (for example, some types of collagen have a (Gly—X—Pro) repeating sequence).

Finally, one should note that an extremely large number of different proteins is possible. This is because the 20 amino acids can be used in constructing many types of polypeptide chains, each with a different amino acid sequence. The situation is not unlike our alphabet and language. The English alphabet contains only 26 letters, yet we can construct thousands upon thousands of words and organize these words using some half dozen punctuation marks into an infinite variety of sentences and paragraphs. The amino acids can be thought of as the letters of the "alphabet" of living organisms and the proteins as the words. The variety of different cells and organisms would be the sentences and paragraphs in our language analogy.

One might ask, how many different combinations of protein sequences are possible with the 20 amino acids? In a short statement, *a lot*! Let us simplify this question by first asking how many combinations are possible using glycine (Gly), leucine (Leu), and phenylalanine (Phe) in a tripeptide in which each amino acid is used only once. We can construct six possible combinations:

1. N—Gly—Leu—Phe—C
2. N—Leu—Gly—Phe—C
3. N—Leu—Phe—Gly—C
4. N—Phe—Leu—Gly—C
5. N—Phe—Gly—Leu—C
6. N—Gly—Phe—Leu—C

As you can see, a number of combinations are possible for this rather simple tripeptide. Since most proteins contain between 100 and 300 amino acids, an extremely large number of combinations are indeed possible. For example, how many proteins 100 amino acids in length are possible? All 20 amino acids will be used in constructing the different protein sequences. We can estimate that:

The number of amino acids in the protein

$$(20)^{100} \approx (10)^{130}$$ combinations of different proteins are possible!

The 20 different amino acids

That number can be thought of as a 1 with 130 zeros after it! Clearly, the number of protein molecules is practically limitless. Still, only a small percentage of the billions of possible proteins are actually found in living systems.

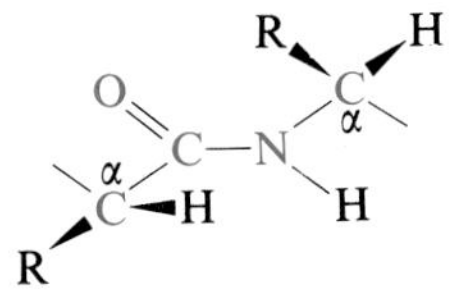

R H O C α C—N α C—H H R

FIGURE 3–4
The peptide bond. The atoms (except those in boldface) are all coplanar—that is, they all lie in the plane of the paper. The R groups and hydrogen atoms associated with the α-carbon atoms extend above and below the plane of the paper.

3.6 THE LEVELS OF PROTEIN STRUCTURE

Let us examine the three-dimensional features of a typical protein and explain what bonding forces hold protein molecules together.

THE PEPTIDE BOND

As was mentioned previously, the peptide bond results from the condensation of an α-amino group of one amino acid with the carboxyl group of a second amino acid. A detailed representation of the peptide bond is given in Figure 3–4. The peptide bond has several unique features. First of all, the atoms (except those noted in the figure with the wedge-shaped chemical bonds) all lie within the plane of the paper. That is, they are *coplanar*. Second, the carbon-nitrogen bond (C—N bond) of the peptide bond is actually *shorter* than a normal C—N bond. This is because the C—N bond has some double bond character. In essence, the C—N bond is partially a single bond and partially a double bond. This curious fact can be explained in terms of *resonance*. Resonance structures are shown in the following diagram:

$$\left[\begin{matrix} O \\ \| \\ \diagup C - \ddot{N} \diagdown \\ \quad\quad H \end{matrix} \longleftrightarrow \begin{matrix} O^{-} \\ | \\ \diagup C = N^{+} \diagdown \\ \quad\quad H \end{matrix} \right]$$

Specifically, the π-electrons associated with the C—N bond and the >C=O are shared by the O, C, and N atoms of the peptide bond and are, therefore, delocalized. Thus, the C—N bond has a partial double bond character. Because of this, the C—N single bond cannot rotate and these atoms are rigidly held in space.

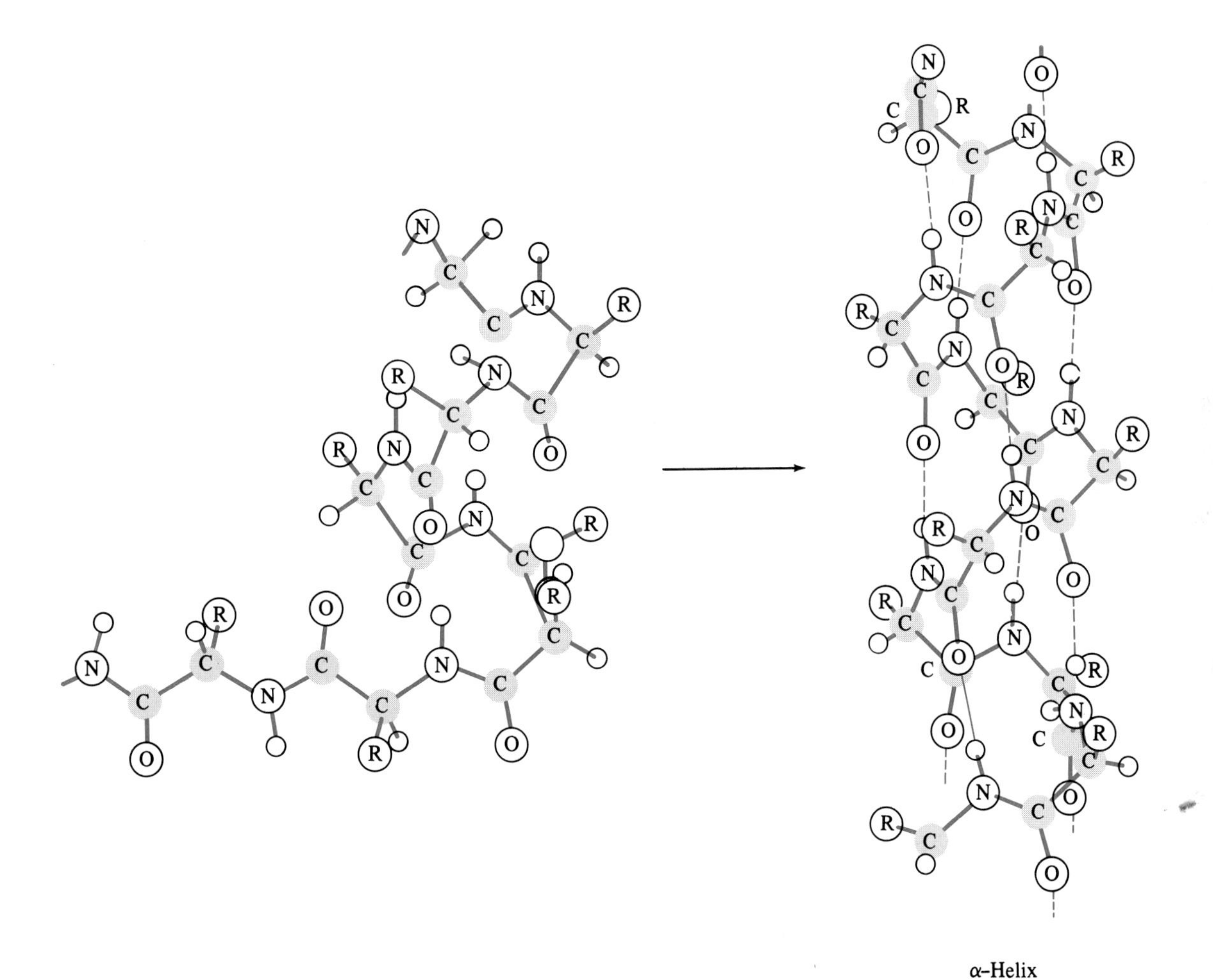

FIGURE 3–5 The formation of a right-handed α-helix. The dashed lines denote hydrogen bonds. The R groups extend out from the helix.

THE LEVELS OF PROTEIN STRUCTURE

Level One: Primary Structure

primary structure

The amino acid sequence of the polypeptide chains (and the presence of disulfide bonds) is called the **primary structure**. The polypeptide chain is held together by the *peptide bond.*

Level Two: Secondary Structure

secondary structure

α-helix

β-pleated sheet

The **secondary structure** of a protein molecule is defined as the local spatial organization of the polypeptide chain. Put another way, the local geometric relationship of neighboring amino acid residues within the polypeptide chain defines the secondary structure. The **α-helix** and the **β-pleated sheet** are two important structures that exhibit secondary structure.

The α-helix. Although the atoms associated with the peptide bond are coplanar and, therefore, make the peptide bond a rigid structure, the single bonds associated with the α-carbon atoms can undergo a 360° rotation. Thus, a kinked right-handed helix (α-helix) is possible. Its formation is depicted in Figure 3–5. The helix is held together and stabilized by hydrogen bonds between the carbonyl oxygen and the hydrogens of the amide bond. The R groups of the amino acids associated with the α-helix extend out of the complex cylindrical structure, thus minimizing molecular interactions between these groups. Certain R groups that have bulky structures or that are electrically charged prevent an α-helix from forming because of atomic interactions between the atoms of the R groups. Such amino acids would destabilize and disrupt the α-helix. Proline, an imino acid, cannot form a normal peptide bond. Thus, proline will disrupt the α-helix and cause a kink or bend in the polypeptide chain. This unusual feature is summarized in Figure 3–6.

The β-pleated Sheet. Certain proteins (silk fibroin) are composed of molecules in nearly 100 percent β-pleated sheet form. In this structure, the polypeptide chains are stretched out to their maximum extent, yet still exhibit a "zig-zag" or "kink" pattern. The R groups extend above and below the main polypeptide chain. Separate chains can be arranged and held together via interchain hydrogen bonds according to Figure 3–7. The structure has the appearance of a pleated sheet—hence the name *β-pleated sheet.*

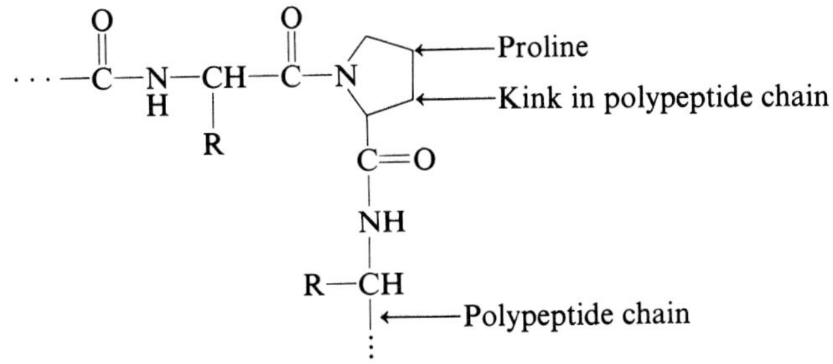

Proline

Kink in polypeptide chain

Polypeptide chain

FIGURE 3–6
A kink in a polypeptide chain caused by proline.

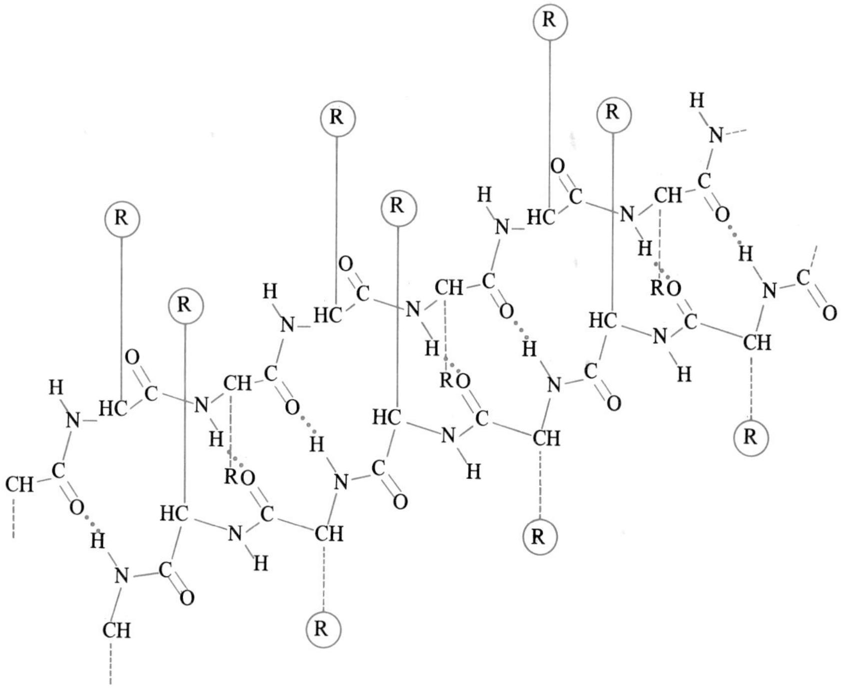

FIGURE 3–7
The β-pleated sheet.

The polypeptide chains of the β-pleated sheet can be arranged in one of two arrangements: (1) *parallel* or (2) *antiparallel* (see Figure 3–8). Most protein are not solely in α-helix or β-pleated sheet organization; instead, they contain regions of one or the other structure, as shown in Figure 3–9.

1. Parallel β–pleated sheet:

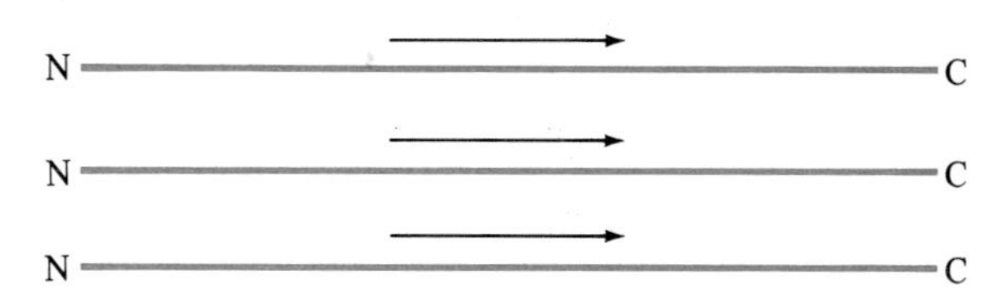

2. Antiparallel β–pleated sheet:

N — C
C — N
N — C

FIGURE 3–8
Parallel and antiparallel arrangements of the β-pleated sheet. "N" refers to the N-terminal; "C" refers to the C-terminal.

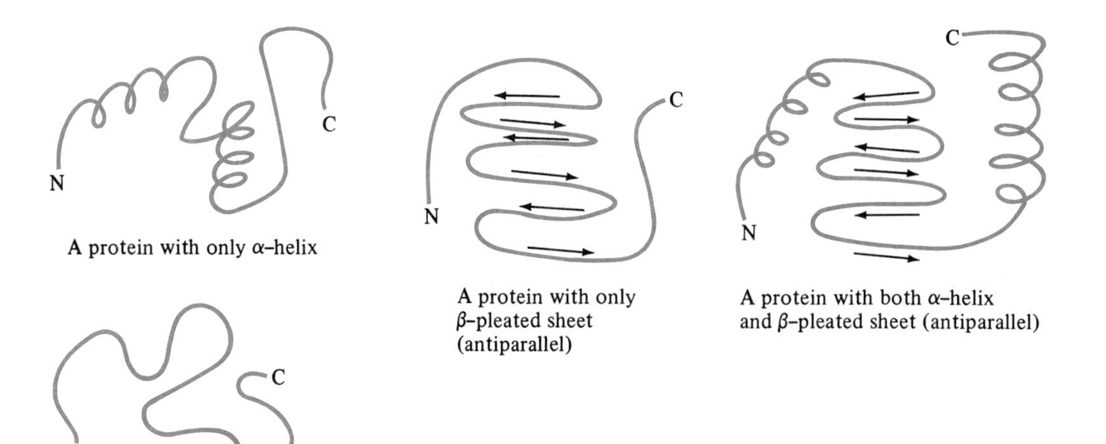

FIGURE 3–9 Types of proteins with secondary structure.

Level Three: Tertiary Structure

The specific three-dimensional structure of the protein in solution is called its **tertiary structure**. This is built up from primary and secondary structures by means of a number of bonding forces:

1. Hydrogen bonds between α-helix or β-pleated sheet structures.
2. Hydrogen bonding between certain amino acids sidechains.

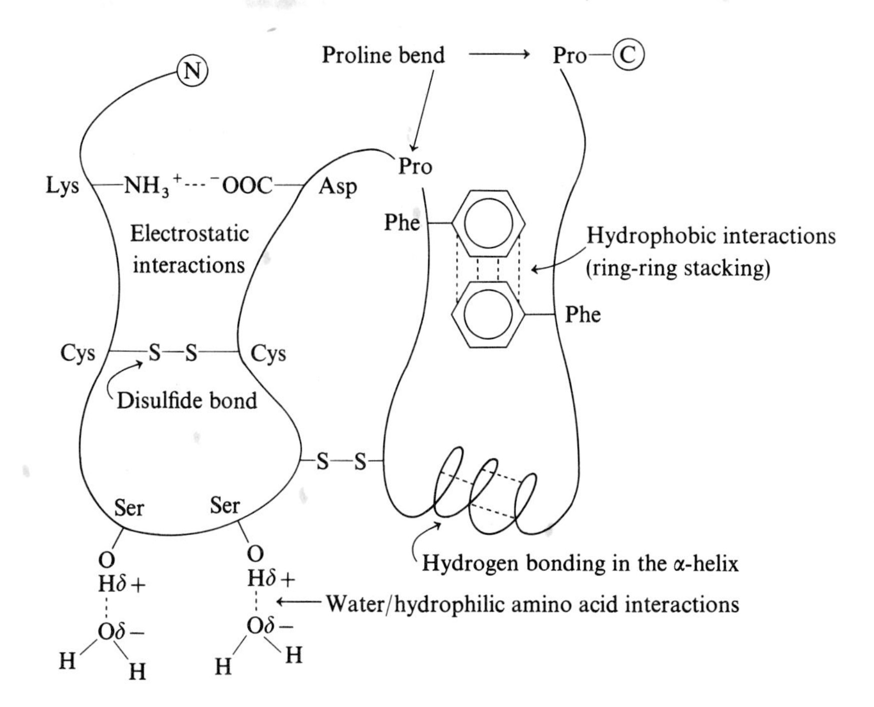

FIGURE 3–10 A hypothetical protein exhibiting tertiary structure.

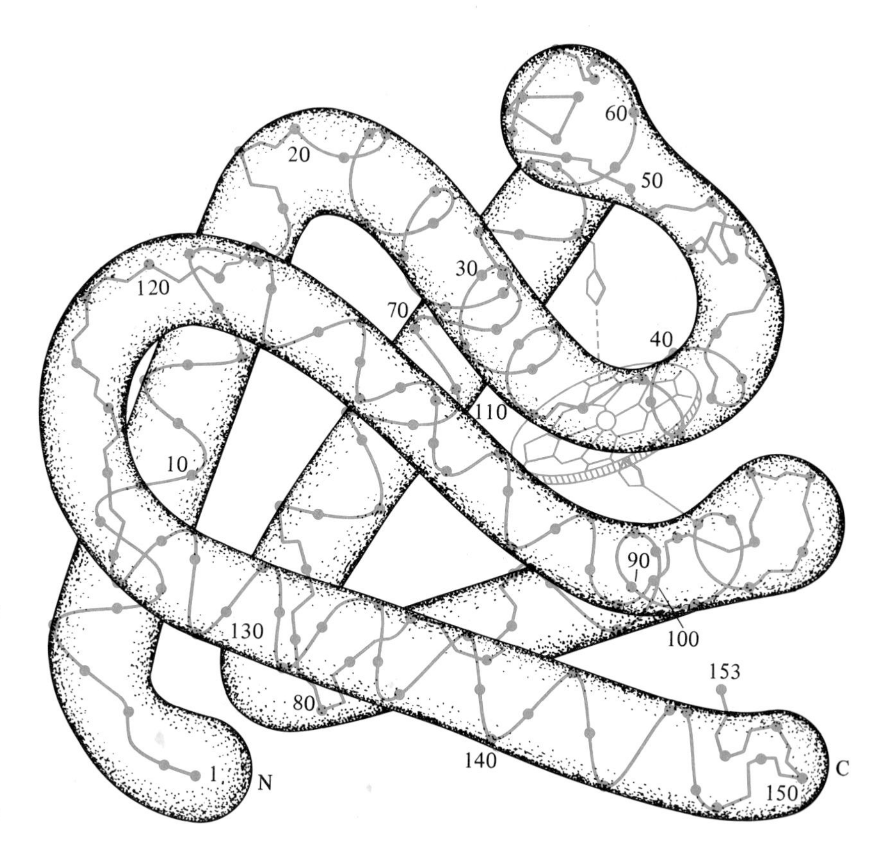

FIGURE 3–11
The 3-dimensional structure of myoglobin. A significant portion of the protein molecule is in α-helix form. The heme group (the part that binds an O_2 molecule) is represented by a disk. (From M. F. Perutz, "The Hemoglobin Molecule," *Scientific American*, November, 1964.)

3. Disulfide bonds between cysteine residues.
4. Electrostatic bonds (salt links).
5. Hydrophobic interactions between hydrophobic R groups.
6. Water/protein interactions.

The tertiary structure of a hypothetical protein held together by these bonding forces is depicted in Figure 3–10. The actual tertiary structure of the protein *myoglobin* is shown in Figure 3–11.

FIGURE 3–12
The quaternary structure of a hypothetical protein molecule (a tetramer).

Level Four: Quaternary Structure

Certain proteins can aggregate into multipolypeptide structures, each chain of which is called a **subunit**. These multisubunit structures are called **oligomers**. The arrangement of the individual subunits into an oligomer is called **quaternary structure**. Figure 3–12 shows a typical oligomeric protein (in this case, a *tetramer*). For reasons not fully understood, most oligomeric proteins are organized with even numbers of subunits (two subunits = a *dimer*, four subunits = a tetramer, and so on).

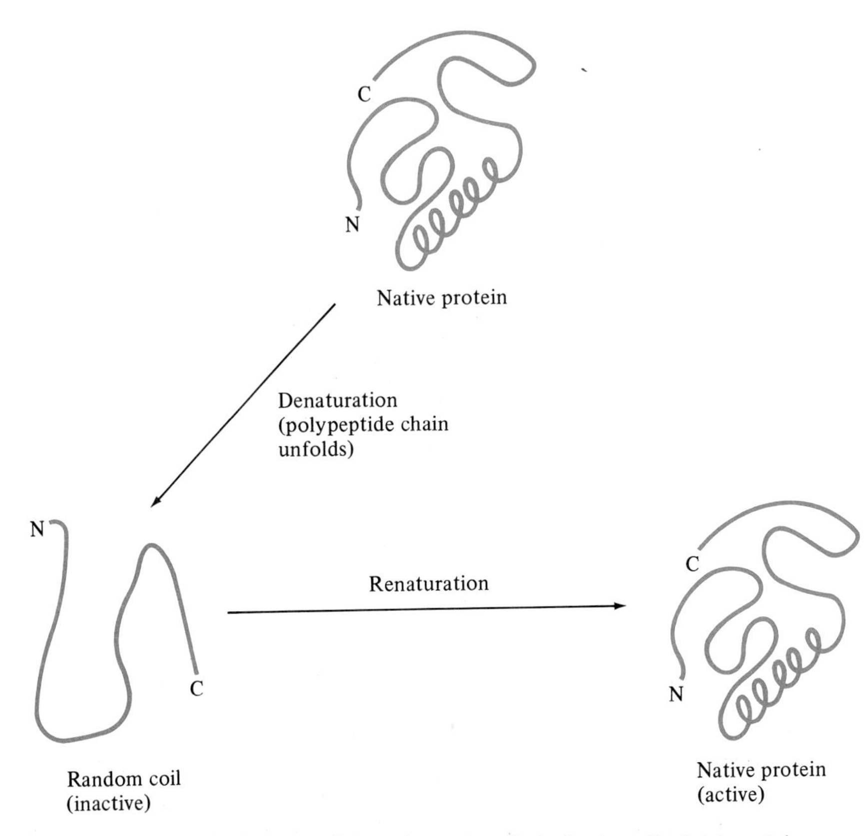

FIGURE 3–13 The denaturation and renaturation of a hypothetical protein. The tertiary structure of the denatured protein is restored upon renaturation.

denaturation

It should be noted that a polypeptide chain can unfold from its specific three-dimensional structure, a process called **denaturation**. A protein can be denatured by treating it with (1) heat, (2) strong acids or bases, (3) organic solvents, or (4) concentrated salt solutions. In essence, these treatments break the bonding forces holding the polypeptide chain(s) together in a specific shape. The peptide bonds, however, are not usually broken, unless drastic and long-term treatment with acids or bases is used. When denaturation occurs, the protein often loses its biological properties and activity, as well as its solubility in water. This is exemplified when one boils an egg. The proteins of the egg white lose their solubility when subjected to heat and become insoluble. Under special conditions, a protein can be renatured by removing the strong denaturing agents and returning the molecule to its previous environment. The polypeptide chains will refold and resume their original three-dimensional shape. Some proteins literally snap back during renaturation; under certain conditions, even their specific biological activity can be restored. These changes are summarized in Figure 3–13.

SUMMARY

Amino acids are the basic building blocks for polypeptides and proteins.

The 20 amino acids commonly found in proteins (Table 3–1) are classified into one of four categories: (1) hydrophobic, (2) hydrophilic (uncharged), (3) positively charged (at pH 7) hydrophilic, and (4) negatively charged (at pH 7) hydrophilic.

Hydrophobic amino acids are minimally soluble in water. Amino acids possess ionizable groups that can behave as either acids or bases. Amino acids ionize to yield the zwitterion form (which has zero net electrical charge). Each amino acid can undergo reversible ionization reactions in which the carboxyl proton is easily lost to the solution, followed by the α-amino proton. Some amino acids possess ionizable R groups and, therefore, can participate in more complex ionization reactions. Most amino acids (except glycine) can exist in either D- or L-enantiomeric forms. All proteins are constructed solely from L-amino acids.

Amino acids can link together to form polypeptide chains. This usually occurs when the carboxyl group of one amino acid condenses with the α-amino group of a second amino acid to form a peptide bond. Small, intermediate, and long polypeptide chains can result from this condensation reaction.

Proteins are polypeptides with at least 50 amino acids. Proteins are extremely important for life and serve many functions, including physical structure, chemical and electron transport, storage, defense, catalysis (enzymes) and contraction. Proteins vary in size from about 6000 to several million daltons and can be globular or fibrous in shape. Conjugated proteins are distinguished by a non–amino acid (prosthetic) group conjugated to the polypeptide chain. Because proteins are constructed from at least 20 different amino acids, an extremely large number of amino acid sequences are possible, each representing a different protein.

The peptide bond results from the condensation of a carboxyl group with an α-amino group. The peptide bond is rigid and its atoms are coplanar.

The four levels of protein structure include: primary structure (the amino acid sequence of the polypeptide chain), secondary structure (local spatial organization within the polypeptide chain), and tertiary structure (overall three-dimensional shape). Proteins are held together by a number of covalent and noncovalent bonding forces. The organization of subunits into multisubunit aggregates (oligomers) is called quaternary structure. Biological activity and function result from the shape of the protein.

REVIEW QUESTIONS

1. We have presented a number of important concepts and terms in this chapter. Define (or briefly explain) the following terms:

a. Amino acid
b. Hydrophobic amino acid
c. Positively charged amino acid
d. Sulfhydryl group
e. Zwitterion
f. Isoelectric pH (pH_I)
g. Asymmetric carbon atom
h. Peptide bond
i. Protein
j. Conjugated protein
k. α-helix
l. β-pleated sheet
m. Tertiary structure
n. Disulfide bond
o. Electrostatic bond
p. Subunit
q. Oligomer

Refer to Table 3–5 when answering Problems 2–5.

2. Write the chemical structure for each of the following amino acids:

a. The fully protonated form of isoleucine.
b. The fully protonated form of lysine.
c. The fully deprotonated form of glutamic acid.

3. In Question 2, give the net electrical charge for each chemical species mentioned

a. At pH 1. **b.** At pH 7. **c.** At pH 12.

* **4.** Starting with the fully protonated form of lysine, write the complete stepwise titration equation

when excess base (OH^-) is added. Indicate the net electrical charge on each species at each step.

5. Consider the following polypeptide:
   N—Gly—Lys—Met—Ser—Tyr—Glu—C
   a. Draw the complete chemical structure, including the peptide bond backbone.
   b. What would be the net electrical charge for the polypeptide molecule at
   (1) pH 1? (2) pH 7? (3) pH 12?
6. Why doesn't glycine exist as D- and L-enantiomeric forms?
7. If a typical amino acid has an average molecular weight of 120 daltons, how many amino acids are found in a protein with a molecular weight of 50 000 daltons? (Assume each amino acid would lose a mass of 18 daltons as it became part of the polypeptide chain; this is because a water molecule is lost during peptide bond formation.)
8. What is peculiar about the amino acid sequence of the polypeptide bradykinin found in Table 3–3?
9. Which of the following pairs of amino acids might be expected to exhibit hydrogen bonding interactions between sidechains?
   a. Tyr—Glu **b.** Phe—Phe **c.** His—Cys
   d. Asp—Ser **e.** Ala—Pro
10. Which of the following pairs of amino acids might be expected to exhibit electrostatic interactions between sidechains:
    a. Leu—Ile **b.** Lys—Asp **c.** Glu—Glu
    d. Asp—Arg **e.** Asn—Glu

TABLE 3–5 The pK_a Values for the Twenty Amino Acids

Amino Acid	pK_a α-COOH	pK_a α-NH_3^+	pK_a R Group
Alanine	2.35	9.69	—
Arginine	2.17	9.04	12.48
Asparagine	2.02	8.8	—
Aspartic acid	2.09	9.82	3.86
Cysteine	1.71	10.78	8.33
Glutamic acid	2.19	9.67	4.25
Glutamine	2.17	9.13	—
Glycine	2.34	9.6	—
Histidine	1.82	9.17	6.0
Isoleucine	2.36	9.68	—
Leucine	2.36	9.60	—
Lysine	2.18	8.95	10.53
Methionine	2.28	9.21	—
Phenylalanine	1.83	9.13	—
Proline	1.99	10.60	—
Serine	2.21	9.15	—
Threonine	2.63	10.43	—
Tryptophan	2.38	9.39	—
Tyrosine	2.20	9.11	10.07
Valine	2.32	9.62	—

SUGGESTED READING

Amino Acid Chemistry

Segel, I. H. *Biochemical Calculations.* 2nd ed., pp. 69–83. New York: John Wiley, 1976.

Peptide and Protein Chemistry

Dickerson, R. E., and Geis, I. *The Structure and Action of Proteins.* Menlo Park, CA: W. A. Benjamin, 1969.

Lehninger, A. L. *Biochemistry*, 2nd ed., Chs. 3, 5–7. New York: Worth, 1975.

Kendrew, J. C. "The Three-Dimensional Structure of a Protein Molecule." *Scientific American* 205 (1961): 97–110.

NOTE

1. Iversen, L. L. "The Chemistry of the Brain," *Scientific American*, 241 (1979): 146.

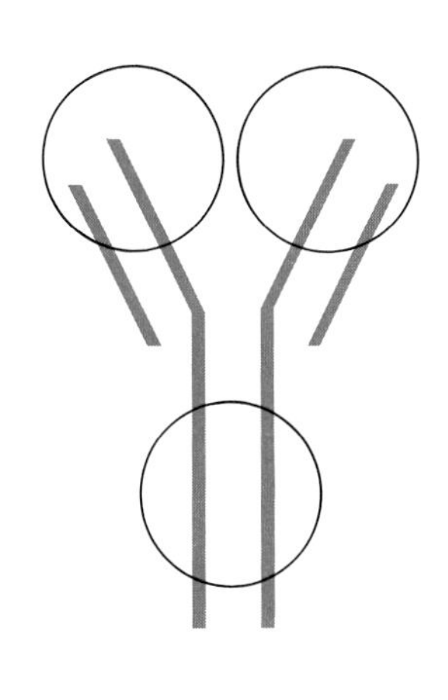

Chapter 4

Functional Proteins: Hemoglobin and Collagen

4.1 INTRODUCTION

hemoglobin

collagen

Having discussed the general structure of proteins, let us now amplify upon those concepts by describing the chemistry of hemoglobin and collagen. **Hemoglobin** (a representative globular protein) has been extensively studied; more is probably known about the structure and function of this molecule than of any other protein. **Collagen** (a representative fibrous protein) will also be discussed, since nearly one-third of all the protein in the human body is collagen. We will also present specific disease states that impair the normal functioning of each of these proteins.

4.2 RED BLOOD CELLS AND THE STRUCTURE OF HEMOGLOBIN

Blood is composed of two main components: (1) *plasma* (a straw-colored solution of proteins, inorganic ions, and other nutrients), and (2) *blood cells*. The average blood volume of a normal adult is approximately 5 liters. Of this, about 3 liters is plasma, and the remaining 2 liters is red blood cells and white blood cells.[1] Each cubic millimeter (mm^3) of blood contains approximately 5×10^6 red blood cells[2]; therefore, each of us has about 2.5×10^{13} red blood cells in our bodies.

The red blood cell is a flat, disk-shaped cell with a depression in the center of each face (called a *biconcave disk*). It does not have a nucleus or other cellular organelles. The red blood cell loses its nucleus during its normal development and maturation. The internal cytoplasm of the red blood cell is approximately 20 percent protein; about 90 percent of this protein is *hemoglobin*. Thus, one can think of the red blood cell as a "bag" of hemoglobin surrounded by a lipid-protein cell membrane. Obviously, the cell is not that simple, since it contains hundreds of other proteins and enzymes that carry out a variety of metabolic reactions. Yet, the primary purpose of the red blood cell is to bind and transport oxygen for distribution to the tissues. Both oxygen and carbon dioxide gases diffuse across the membrane of the red blood cell, as shown in Figure 4–1.

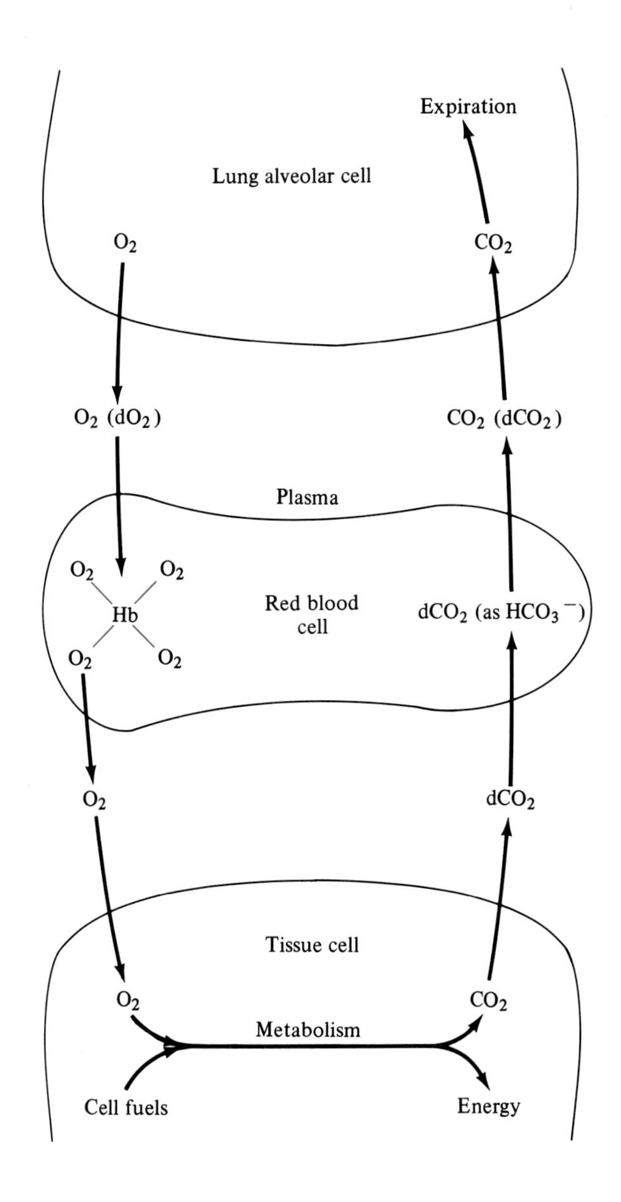

FIGURE 4–1
Schematic representation of oxygen and carbon dioxide transport. O_2 = oxygen; dO_2 = dissolved oxygen; CO_2 = carbon dioxide; dCO_2 = dissolved carbon dioxide; Hb = hemoglobin.

The hemoglobin molecule is the primary oxygen-binding protein of the body. The molecule is composed of four subunits and, thus, is a tetramer. Two of the polypeptide chains (called the α-subunits) contain 141 amino acids; the other two chains (the β-subunits) contain 146 amino acids. Do not be confused by the notation used to distinguish the different chains (the Greek letters α and β). This notation does *not* mean that one type of polypeptide chain is an α-helix while the other type of chain is a β-pleated sheet; actually, both types of chains contain large segments in α-helix form. Each chain has a molecular weight of 16 000 daltons; therefore, the entire hemoglobin tetramer has a molecular weight of about 64 000 daltons. The separate chains are globular in shape. Each chain is bonded to the other chains via specific noncovalent interactions (salt links and hydrogen bonds). The generalized structure of the hemoglobin molecule is shown in Figure 4–2; a more detailed representation is shown in Figure 4–3 on the next page.

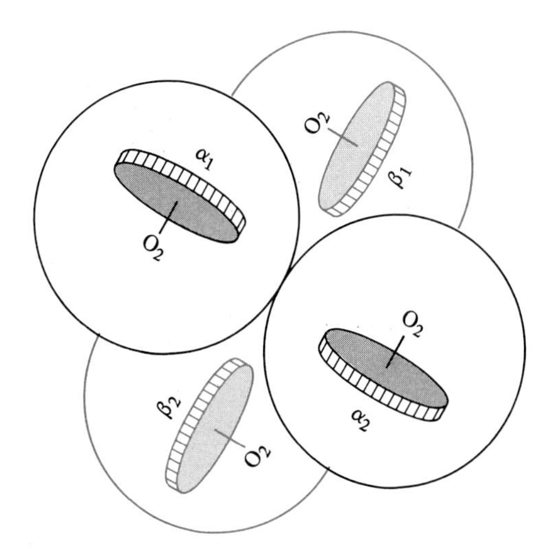

FIGURE 4–2
Generalized structure of the hemoglobin molecule. The molecule contains four subunits (two α-chains and two β-chains). Each chain contains one heme group (the disk-shaped object). This diagram represents the fully oxygenated form of hemoglobin.

heme

Each chain contains one **heme** group (protoporphyrin IX) as a bound prosthetic group. Each heme group can bind one oxygen molecule; therefore, the entire hemoglobin molecule (with a total of four heme groups) can bind four oxygen molecules. The heme group is a complex, disk-shaped organic molecule with an iron atom in the center of a tetrapyrrole ring system. The iron can bind to a total of six atoms. Four of the six bonds are associated with the four nitrogen atoms of the tetrapyrrole ring system; a fifth position binds the nitrogen atom of the histidine R group. An oxygen molecule can occupy the sixth binding position. The entire heme group is actually found in a hydrophobic "pocket" within each polypeptide chain, as shown in Figure 4–4 on page 57.

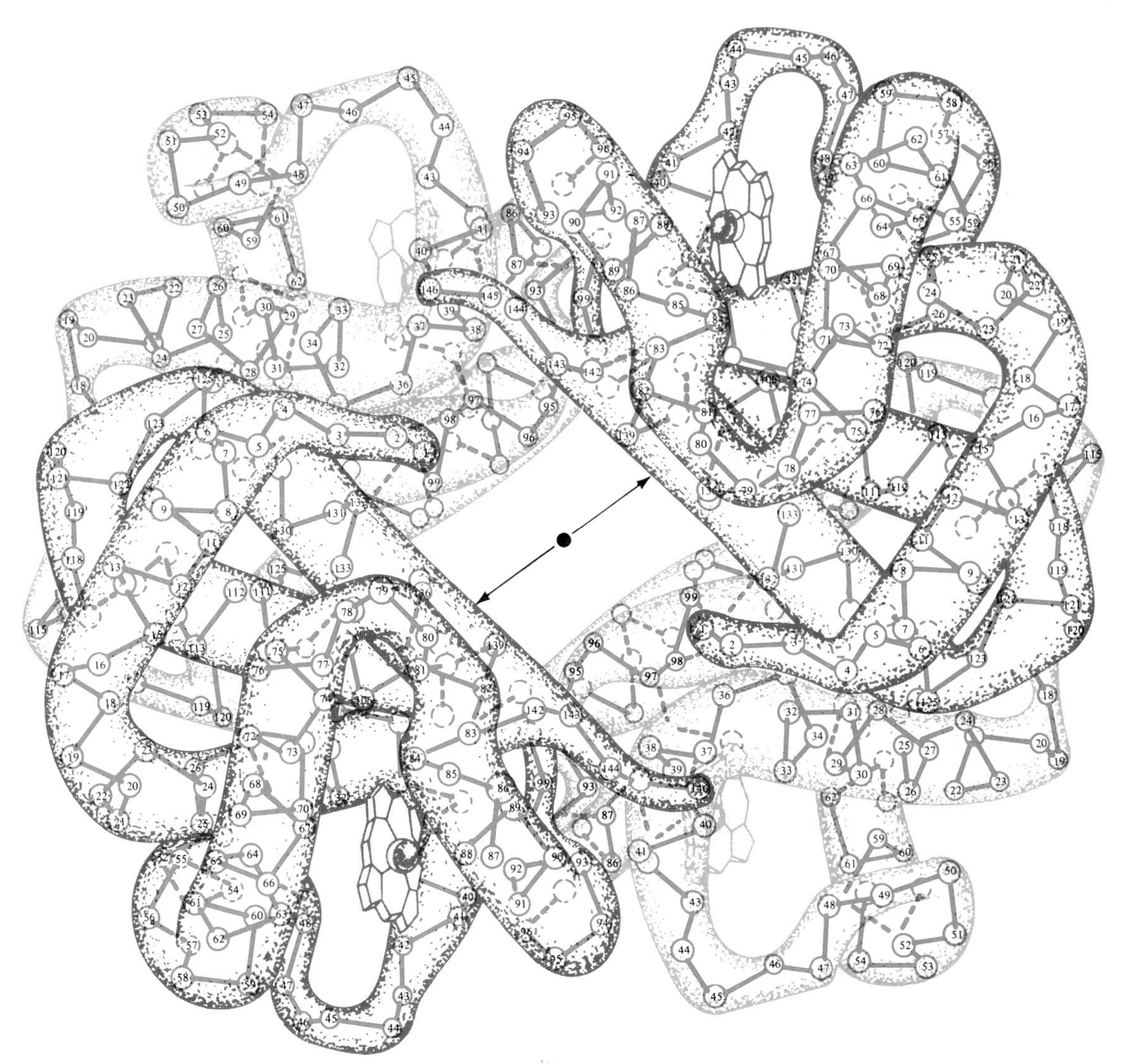

FIGURE 4–3 Top view of oxygenated hemoglobin. The β-chains are shown in dark color, and the α-chains in light color. The chains exhibit a high degree of α-helix structure. Also shown is the central cavity in which 2,3-diphosphoglycerate (DPG) can bind. (Drawing kindly supplied by Irving Geis, copyright © 1979.)

4.3 HOW DOES HEMOGLOBIN BIND OXYGEN?

A number of researchers have proposed various models to explain how the hemoglobin molecule binds and releases oxygen. One model, developed in the 1960s by Dr. Daniel Koshland, Jr., explains nicely how hemoglobin works.[3] Koshland suggested that the hemoglobin tetramer binds oxygen in

CH_2
CH
CH_3 ← (Methyl group)
H_3C
CH ═ CH_2 ← (Vinyl group)
N N
Fe
(Fe–iron atom)
N N
H_3C
CH_3
CH_2 CH_2
CH_2 CH_2
COO^- COO^- ← (Propionic acid group)

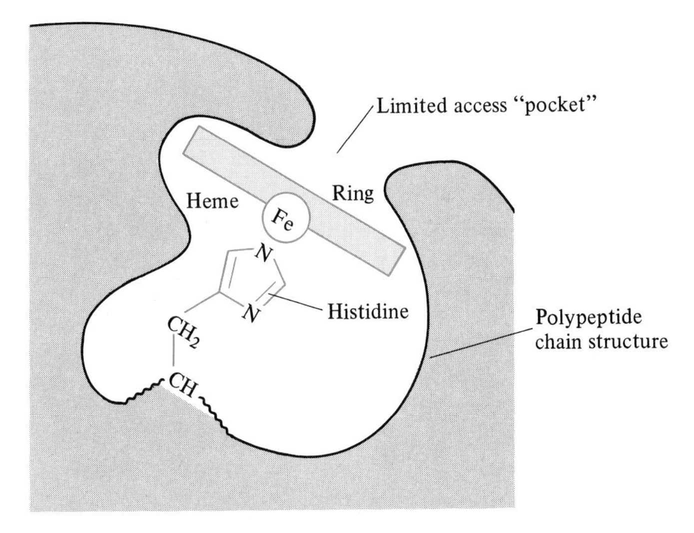

FIGURE 4–4
The chemical structure of the heme group and its orientation in the hydrophobic pocket of one of the hemoglobin subunits.

a sequential and cooperative manner. One oxygen molecule first binds to the heme group of one chain, followed, in sequence, by the binding of the second, third, and fourth oxygen molecules to the other chains. The binding of one oxygen molecule *enhances* the binding of the other oxygen molecules. Thus, the four subunits "**cooperate**" with each other to enhance oxygen binding. Koshland proposed that during the binding of an oxygen molecule to a subunit, the three-dimensional shape of that subunit changes and that this change in shape (**conformational change**) induces changes in the other subunits. The change in the shape of the individual subunits ultimately

cooperate

conformational change

increases the binding affinity of the oxygen for the heme groups. The Koshland sequential binding model is summarized in Figure 4–5. Subunits *without* bound oxygen are symbolized as circles (○); subunits *with* bound oxygen are symbolized as squares (■). Hemoglobin without bound oxygen is called *deoxyhemoglobin*; hemoglobin with bound oxygen is called *oxyhemoglobin*. A change from a circle (○) to a square (■) configuration

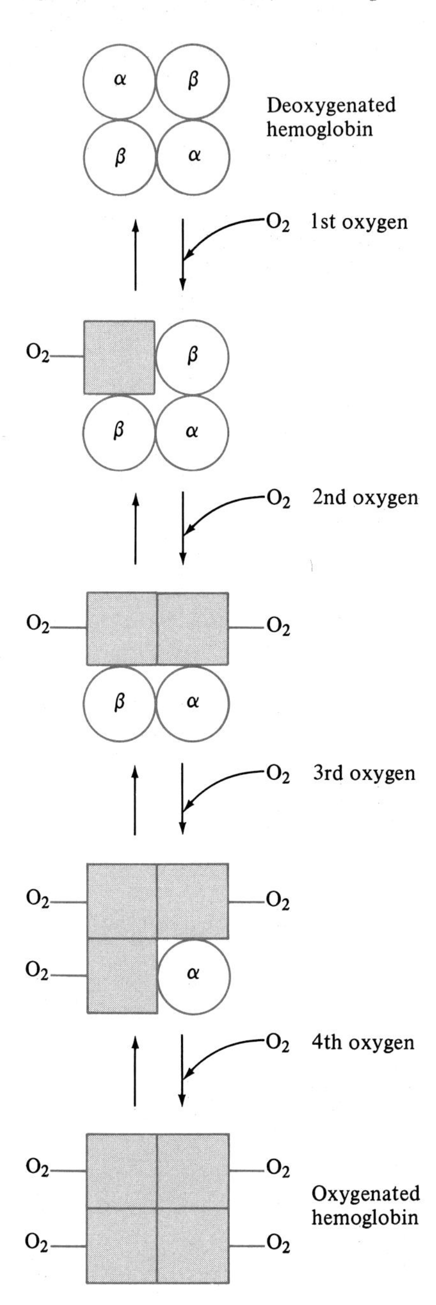

FIGURE 4–5
The Koshland sequential binding model. The chains symbolized by circles (○) have a low affinity for oxygen; those symbolized by squares (■) have a high affinity for oxygen. (Adapted from Lehninger, A. L., *Biochemistry*, 2nd ed. (New York: Worth, 1975), p. 148.)

represents a change in the shape of the specific subunit. The events depicted in Figure 4–5 occur extremely quickly within the capillaries of the lung alveoli. The oxygenated hemoglobin then travels to the tissue capillaries, where the oxygen is released (also, possibly, in a sequential manner) to the extracellular fluid and diffuses into the cells. The depleted hemoglobin (deoxyhemoglobin) then travels from the tissue capillaries back to the lungs for a new cycle.

Although we have presented only one model for how hemoglobin binds oxygen, other models have also been developed and may be equally valid.

The Koshland sequential binding model (and other similar models) is based upon excellent experimental results. We now have a fairly complete picture of what is happening at the molecular level when hemoglobin binds oxygen. As stated previously, each heme group resides in a hydrophobic pocket (see Figure 4–3). This pocket prevents the easy access of oxygen to the heme group. If we look edge-on at the heme group, we see how the histidine residue is arranged and associated with the central iron atom (Figure 4–6). Experimental evidence shows that the iron atom in deoxyhemoglobin has an ionic radius of 2.24 Å (1 Å = 1×10^{-8} cm) and cannot easily fit within the central cavity of the tetrapyrrole ring system. This situation is summarized in Figure 4–6(a). When an oxygen molecule does bind to the iron atom, the ionic radius of the iron atom *decreases* to 1.99 Å. This decrease in size allows the iron to fit easily into the central cavity of the heme ring; thus, the iron pops up into the plane of the heme ring. The movement is small—only 0.75 Å—but it has dramatic consequences. As the iron moves into the heme ring, it carries the histidine residue with it. Because histidine is actually part of the polypeptide chain, the entire polypeptide chain moves along with it. As the polypeptide chain moves, a number of key electrostatic interactions (salt links) holding the separate chains together are broken,

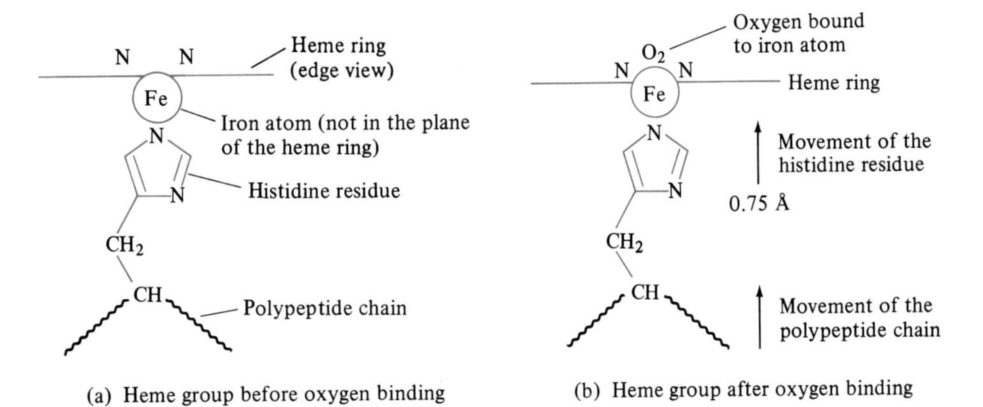

FIGURE 4–6 Iron atom movement caused by the binding of oxygen. As oxygen binds to the iron atom, the iron shrinks and is able to move into the plane of the heme ring. As it does, the histidine residue also moves, triggering the movement of the entire polypeptide chain.

causing very specific conformational changes within the separate chains. In essence, the movement within one chain is translated to the other chains, and all the chains move. The result is that the two β-chains move closer together in a "pincer" movement, as depicted in Figure 4–7. This movement opens up the hydrophobic pockets within the second, third, and fourth subunits, thus allowing easier access to these heme groups and easier binding of oxygen to them (Figure 4–8). In essence, the hemoglobin subunits "co-

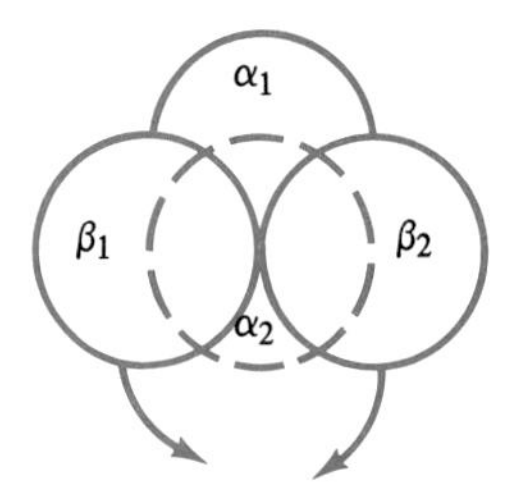

FIGURE 4–7 The "pincer" movement of the β-chains. (Adapted from Bhagavan, N. V., *Biochemistry*, 2nd ed. (Philadelphia: J. B. Lippincott, 1978), p. 716.)

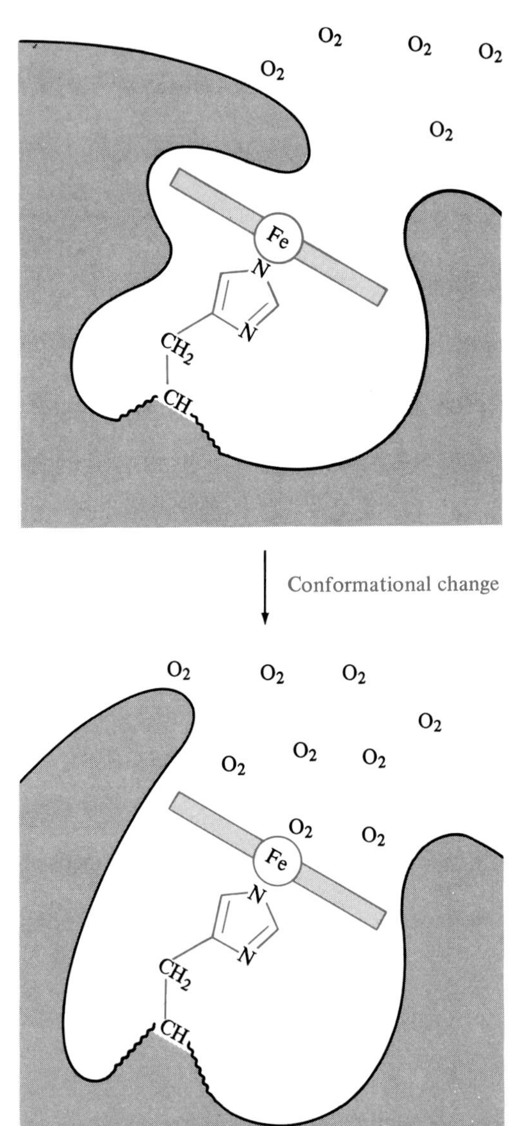

FIGURE 4–8 The opening of the hydrophobic pocket surrounding the heme group.

cooperativity

operate" with each other to efficiently bind and release oxygen molecules. This phenomenon is called **cooperativity**. Multisubunit proteins (oligomers) such as hemoglobin exhibit cooperativity between subunits. Obviously, a single-subunit protein *cannot* exhibit this property. *Myoglobin* (another oxygen binding protein) contains only one subunit and, therefore, does not exhibit cooperativity.

oxygen saturation curve

By now, you should be impressed with the unique design of the hemoglobin molecule. It is a perfect little machine, binding and releasing oxygen with great efficiency. The ability of hemoglobin to bind oxygen is depicted in Figure 4–9. The curve is called an **oxygen saturation curve**. The percentage of hemoglobin saturated (bound) with oxygen is noted on the vertical axis; the partial pressure of oxygen (pO_2) is noted on the horizontal axis in millimeters of mercury (mm Hg, or torr).

sigmoid curve

The oxygen saturation curve simply depicts what percentage of a hemoglobin sample is saturated with oxygen at a certain oxygen pressure (oxygen concentration). For example, at an oxygen concentration (pO_2) of 26 torr, 50 percent of the total number of hemoglobin molecules in the sample are saturated with oxygen. Notice that the saturation curve approximates an S shape; this kind of curve is called a **sigmoid curve**. At low oxygen pressures (less than 20 mm Hg), very little of the hemoglobin is saturated with oxygen. At pressures between 20 and 60 mm Hg, the amount of saturated hemoglobin increases dramatically. At pressures of 80–100 mm Hg, nearly all of the hemoglobin is saturated with oxygen. Interestingly, the

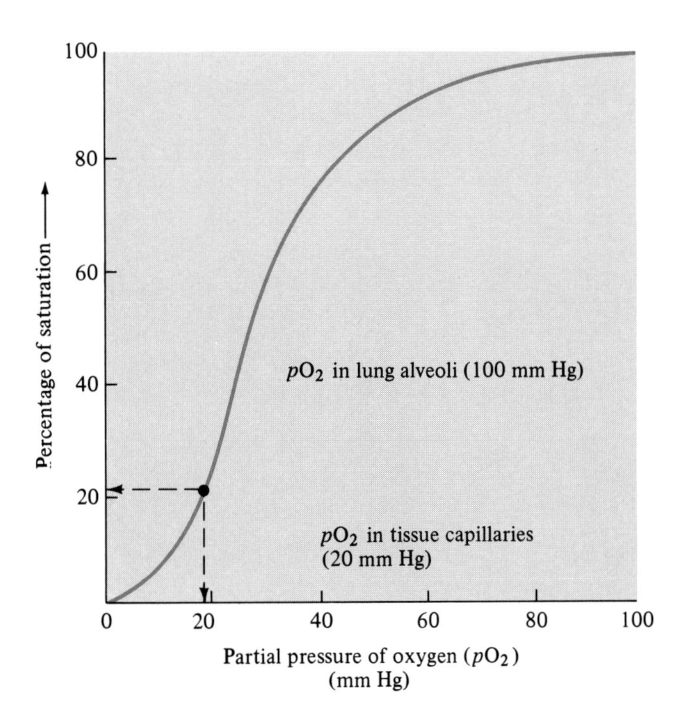

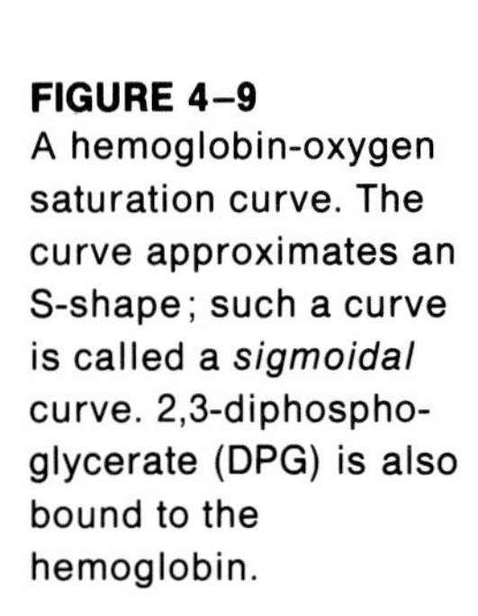

FIGURE 4–9
A hemoglobin-oxygen saturation curve. The curve approximates an S-shape; such a curve is called a *sigmoidal* curve. 2,3-diphosphoglycerate (DPG) is also bound to the hemoglobin.

partial pressure of oxygen within the capillaries of the lung alveoli is around 100 mm Hg. Thus, the hemoglobin within the lung alveoli is fully saturated with oxygen. The partial pressure of oxygen within the tissue capillaries, on the other hand, is about 20 mm Hg. This means that the oxygen will be *easily released* from the hemoglobin molecule to the tissue capillary. The curve is a sigmoid curve because the hemoglobin subunits exhibit cooperativity when binding oxygen molecules. Myoglobin (a single-subunit protein) does not exhibit cooperativity and, therefore, would not be expected to produce a sigmoid oxygen saturation curve. In fact, myoglobin produces a *hyperbolic* oxygen saturation curve instead.

rightward shift

Under certain conditions, oxygen saturation curves will shift to the right as depicted in Figure 4–10. This **rightward shift** has important physiological implications. Examine curve B in Figure 4–10, and compare it to curve A. Because curve B is shifted to the right, more oxygen can be released from the hemoglobin molecule to the tissues, where it is needed. Look at a fixed partial pressure (say, 26 mm Hg). Notice that the nonshifted curve A indicates that the hemoglobin is 50 percent saturated with oxygen. This means that at this partial pressure of oxygen, the hemoglobin does not easily release its oxygen to the environment. Since curve B is shifted to the right, the percentage of hemoglobin that is saturated with oxygen has *decreased* (to 10 percent at the same pO_2, 26 mm Hg). Thus, more oxygen has been released to the tissues.

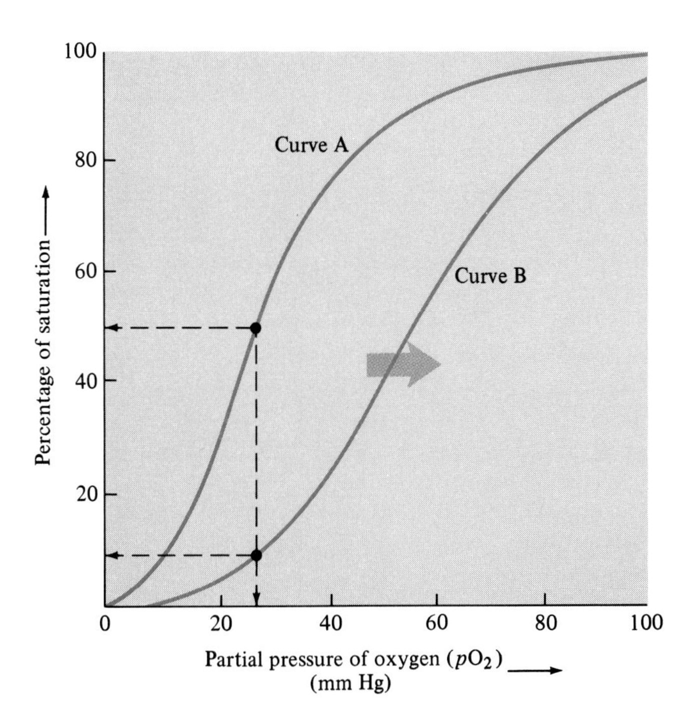

FIGURE 4–10
Hemoglobin-oxygen saturation curves demonstrating the rightward shift. Curve A: without rightward shift; curve B: with rightward shift.

The rightward shift can occur in the tissue capillaries as the result of (1) *decreased* tissue pH, (2) *increased* tissue temperature, (3) *increased* tissue pCO_2, and (4) *increased* 2,3-diphosphoglycerate (DPG) concentrations. These conditions usually occur when the skeletal muscles are involved in strenuous exercise. During periods of exercise, tissue metabolism produces excess acid (from lactic acid), excess carbon dioxide (CO_2), and excess heat as waste products.

2,3-diphosphoglycerate (DPG)

The compound **2,3-diphosphoglycerate (DPG)** is normally present in the red blood cell and regulates the binding and release of oxygen from the hemoglobin molecule.

$$\begin{array}{l} COO^- \\ | \\ HC-O-PO_3^{2-} \\ | \\ H_2C-O-PO_3^{2-} \end{array}$$

2,3-Diphosphoglycerate (DPG)

Specifically, DPG binds into the central cavity of deoxyhemoglobin (see Figure 4–3). Two additional salt links are thus generated between the DPG molecule and the hemoglobin molecule. This situation makes the conformational transition between deoxyhemoglobin and oxyhemoglobin more difficult. In essence, the presence of DPG in the central cavity of the deoxyhemoglobin molecule decreases its affinity for oxygen. Without DPG, hemoglobin would release very little oxygen to the tissue cells. The oxygen saturation curves shown in Figures 4–9 and 4–10 are partly the result of the action of DPG.

THE BOHR EFFECT: OTHER TRANSPORT PROPERTIES OF HEMOGLOBIN

Besides binding oxygen, hemoglobin can reversibly bind both CO_2 and protons (H^+) (see Figure 4–11). For instance, the CO_2 from normal cellular metabolism can bind to the N-terminal of one of the four polypeptide chains of the hemoglobin tetramer, forming a *carbamate derivative*.

Hemoglobin chain

$$C\text{—}NH_2 + CO_2(d) \rightleftharpoons C\text{—}NH\text{—}\boxed{COO^-} + H^+$$

N-terminal

Carbamate derivative (bound CO_2)

Only 3–4 percent of the total CO_2 in the body is bound by hemoglobin. Most is transported from the tissues to the lungs as bicarbonate (HCO_3^-) (see Chapter 2).

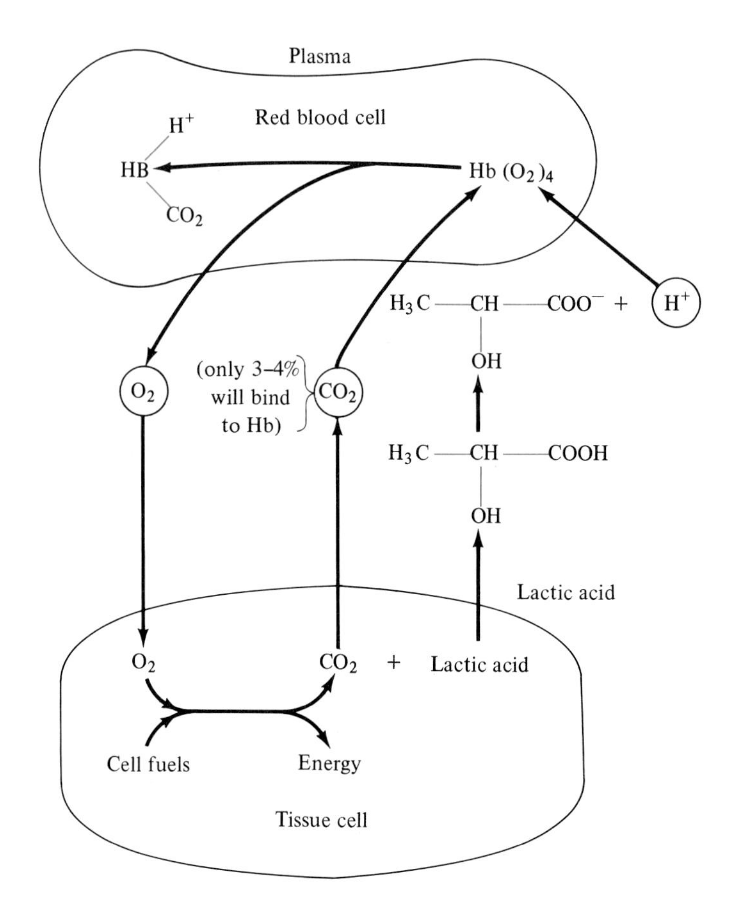

FIGURE 4–11
The Bohr effect.
O_2 = oxygen;
CO_2 = carbon dioxide;
Hb = hemoglobin.

Approximately six histidine residues in each hemoglobin molecule can reversibly bind protons according to the following equation:

H N N CH$_2$ CH $\underset{}{\overset{+H^+}{\rightleftarrows}}$ H N N$^+$ (H) CH$_2$ CH

Polypeptide chain

During strenuous exercise, large amounts of CO_2 and H^+ are generated within the active tissue cells. The H^+ comes from lactic acid, a waste product of cellular metabolism. The CO_2 and H^+ bind to oxyhemoglobin and *promote the release of oxygen to the tissue cells*, where it is needed. The hemoglobin that has released its oxygen and bound H^+ and CO_2 travels to the lungs. Within the lung alveoli, high concentrations of oxygen promote

Bohr effect

the release of the bound H^+ and CO_2, allowing oxygen to bind to the deoxyhemoglobin. These relationships are commonly known as the **Bohr effect**, and are diagrammed in Figure 4–11.

FETAL HEMOGLOBIN

The partial pressure of oxygen within the placenta is only about 30 mm Hg. Normal hemoglobin would not be sufficiently saturated with oxygen at this low pO_2. Fortunately, the fetus contains a different type of hemoglobin that has a greater affinity for oxygen at low pO_2 values. Fetal hemoglobin, therefore, exhibits an oxygen saturation curve different from that of adult hemoglobin. These differences are summarized in Figure 4–12.

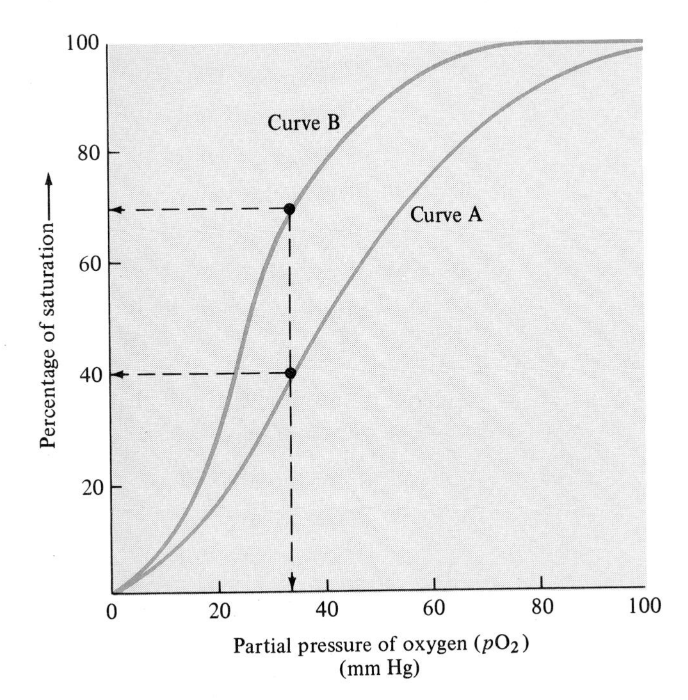

FIGURE 4–12
Hemoglobin-oxygen saturation curves for maternal hemoglobin (curve A) and fetal hemoglobin (curve B).

Fetal hemoglobin actually contains a different type of polypeptide chain than adult hemoglobin. Specifically, it contains two α-chains and two γ-chains (and no β-chains).

$$\text{Hemoglobin F (Fetal)} = \alpha_2^{\text{Adult}}, \gamma_2^{\text{Fetal}}$$
$$\text{Hemoglobin A (Adult)} = \alpha_2^{\text{Adult}}, \beta_2^{\text{Adult}}$$

Figure 4–13 shows the sequence in which these chains are produced during fetal development. Observe that α-chain production begins very soon after fertilization occurs (by the 8th week). Production of a different chain, the

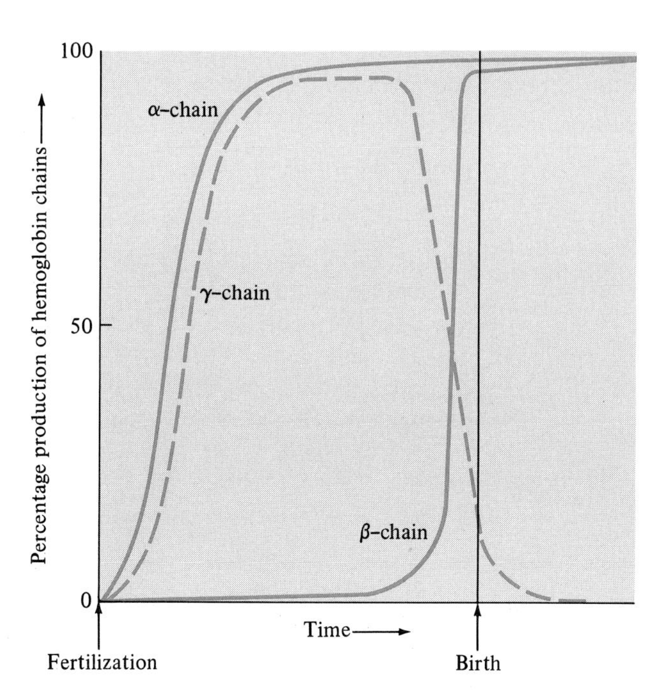

FIGURE 4–13
The production of α-, β-, and γ-chains during fetal development.

γ-chain, begins at the same time and quickly matches α-chain production. The production of the β-chain does not begin until the very end of the nine-month gestation period. At the time of birth, the fetus shuts down γ-chain production and accelerates β-chain production until it matches the rate of α-chain production.

WHY IS BLOOD RED?

absorption spectrum

Simply stated, blood is red because the heme groups of the hemoglobin molecule absorb blue and green wavelengths and transmit red wavelengths. This property is shown in Figure 4–14. Figure 4–14 is called an **absorption spectrum** and is generated using a *spectrophotometer*. A sample of blood is placed in this instrument. A light source within the spectrophotometer generates a light beam that is broken up into the various colors of the spectrum, each with its specific range of wavelengths. These colors are directed through the blood sample. The amount of light *absorbed* by the sample is measured and recorded on a meter. Blue colors have wavelengths of 400–450 nanometers (nm), green colors have wavelengths of 450–550 nm, and red colors have wavelengths above 600 nm. As you can see from Figure 4–14, blood hemoglobin significantly absorbs blue (410 nm) and green (570 nm) wavelengths, but not red wavelengths (those above 600 nm). Interestingly, the oxygenation of hemoglobin significantly changes the absorption spectrum of the molecule, particularly in the green region.

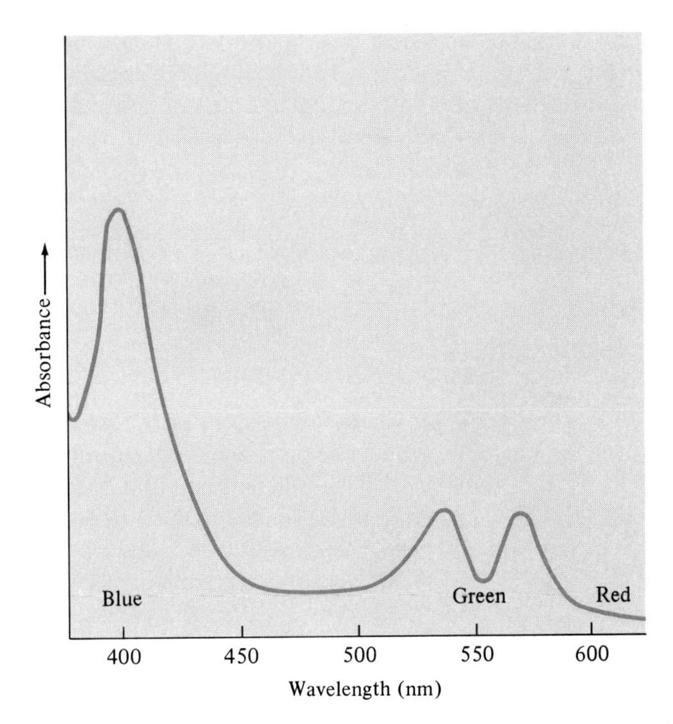

FIGURE 4–14
The absorption spectrum of hemoglobin (oxygenated form). Hemoglobin significantly absorbs blue and green wavelengths, but not red wavelengths. Thus, blood is red.

Fully oxygenated hemoglobin possesses two distinct absorption peaks in the green region; deoxygenated hemoglobin has only one peak. Thus, oxygenated hemoglobin is a brighter red than deoxygenated hemoglobin. (Note, however, that deoxygenated blood is still red and not blue, as is often mistakenly thought.)

Hemoglobin can be measured in the clinical laboratory by taking a blood sample, treating the sample with a reagent containing cyanide (Drabkin's reagent), and determining the absorbance of the cyanide/hemoglobin complex using a spectrophotometer. The cyanide/hemoglobin complex exhibits an absorbance peak at 540 nm. The magnitude of this peak is compared to that of the corresponding peak from a standard containing a known amount of the cyanide/hemoglobin complex. Thus, the hemoglobin concentration of the unknown sample can be calculated. Normal and abnormal hemoglobin concentrations are shown in Table 4–1. Hemoglobin

TABLE 4–1 The Concentration of Hemoglobin in the Blood

	Blood Hemoglobin (g/100 ml)
Adult Males	13–18
Adult Females	11–16
Newborns	14–23

Source: From Sigma Technical Bulletin #525, Sigma Chemical Company P. O. Box 14508, Saint Louis, Missouri 63178.

anemia

polycythemia

concentrations below 12 g/100 ml blood usually indicate **anemia** (low red blood cell count); hemoglobin concentrations much greater than 14 g/100 ml of blood can indicate a condition called **polycythemia** (the production of too many red blood cells). Newborns often have a very high hemoglobin concentration.

4.4 DISEASE STATES ASSOCIATED WITH ABNORMAL HEMOGLOBIN

SICKLE-CELL ANEMIA (HEMOGLOBIN S DISEASE)

sickle-cell anemia

Approximately 9 percent of the American black population are carriers of the genetic disease **sickle-cell anemia**; about 1 in 70 are afflicted by the disease.[4] The afflicted individuals have inherited two copies of an altered gene that codes for the β-chain of hemoglobin, one copy from each parent. They are, therefore, *homozygous* for the genetic trait.

This disease is characterized by severe anemia. As the name implies, individuals with the disease have red blood cells that exhibit a characteristic "sickle" shape. Interestingly, the red blood cells in these individuals have the normal biconcave disk shape when fully oxygenated, but assume the sickle shape when *deoxygenated.* Under these conditions, the sickled cells block small blood vessels and capillaries and, thus, deprive the surrounding tissues of oxygen. This condition aggravates the deoxygenation and induces a very painful *sickling crisis.* Severe anemia occurs because the sickled cells are destroyed by the spleen, thus significantly decreasing the red blood cell count.

A great deal of research has finally demonstrated that the primary genetic defect occurs in the β-chain of the hemoglobin molecule. Specifically, the amino acid valine has been substituted for glutamic acid in position number 6 counting from the N-terminal region of the β-chain. This is shown in Figure 4–15. Notice that valine is a hydrophobic amino acid, whereas glutamic acid is negatively charged at physiological pH values of 7–7.4. Thus, in sickle-cell anemia, the entire hemoglobin has lost two negative charges and acquired two additional hydrophobic amino-acid residues. Hydrophobic regions therefore exist on the hemoglobin surface. These "sticky" hydrophobic regions cause the hemoglobin molecules to *aggregate*, or clump, into very long fibers. Once the hemoglobin aggregates into these fibers (this happens in deoxygenated blood), the osmotic pressure in the red blood cell decreases, causing the cell to collapse into the characteristic "sickle cell" shape. The distorted cells often aggregate in the junctions of the small capillaries, causing a painful sickling crisis. This sequence of events is shown in Figure 4–16.

Research efforts are underway to develop drugs or specific therapies to prevent the hemoglobin from aggregating. Test-tube trials have shown that *potassium cyanate* can prevent hemoglobin from aggregating. Clinical

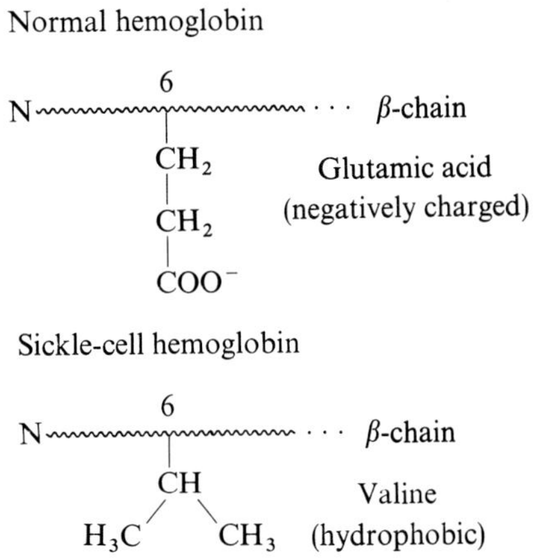

Normal hemoglobin

N ~~~ 6 ~~~ ... β-chain

CH_2

CH_2 — Glutamic acid (negatively charged)

COO^-

Sickle-cell hemoglobin

N ~~~ 6 ~~~ ... β-chain

CH — Valine

H_3C CH_3 (hydrophobic)

FIGURE 4–15
The substitution of valine for glutamic acid in sickle-cell anemia.

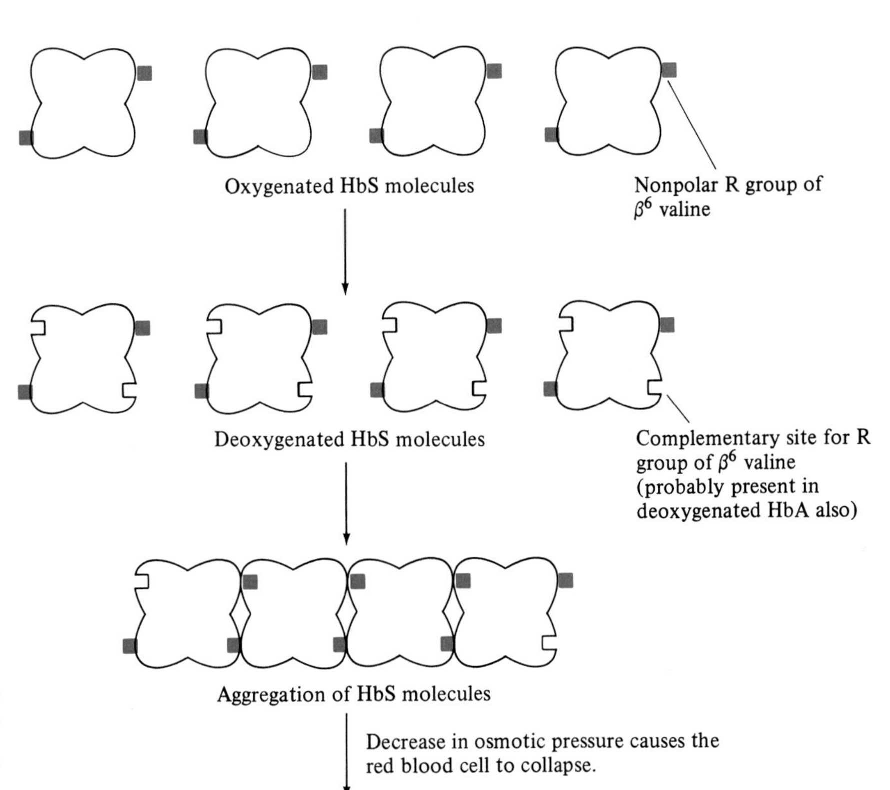

FIGURE 4–16
The aggregation of sickle-cell hemoglobin into long fibers. Hemoglobin aggregation occurs at low pO_2; it is caused by the hydrophobic ("sticky") valine residues on the exterior surface of the molecule. (Adapted from Armstrong, F. B., and Bennett, T. P. *Biochemistry* (New York: Oxford University Press, 1979), p. 110.)

trials using this drug are presently underway. Unfortunately, toxic side effects (neurological disorders) are associated with the use of potassium cyanate.

One might ask how this disease came into existence, and why individuals of specifically Negroid descent are afflicted with it. It is presently believed that a specific mutation occurred in the portion of the DNA molecule that codes for the amino-acid sequence of the β-chain. As a result of this mutation, valine was inserted into the β-chain position normally occupied by glutamic acid. Although the disease is very serious, the mutation may have been beneficial for individuals living in malaria-infested regions of Central Africa. The protozoan that causes malaria (*Plasmodium vivax*) requires red blood cells for its life cycle. Because sickle cells are so abnormal in shape and so fragile, they are rapidly destroyed by the body, along with the infecting protozoan. Thus, the disease confers resistance to malaria by sacrificing a certain percentage of red blood cells that may have been infected with the protozoan.

HEMOGLOBINOPATHIES: DISORDERS INVOLVING HEMOGLOBIN VARIANTS

A number of other types of human hemoglobins have been identified. These hemoglobin molecules differ from normal hemoglobin because a specific mutation in the genes coding for either the α- or β-chains caused the wrong amino acid to be inserted during polypeptide chain synthesis. Such hemoglobins usually contain one wrong amino acid in a specific position, much as in the hemoglobin S of sickle-cell anemia. Although over 250 different types of abnormal hemoglobin molecules have been identified, many do not produce any abnormal clinical symptoms. Unfortunately, some of the others are associated with serious diseases, as summarized in Table 4–2.

TABLE 4–2 Abnormal Hemoglobins

Hemoglobin Type[a]	Substitution[b]	Clinical Symptoms
Hemoglobin S	β^{6}Glu → Val	Hemolytic anemia, sickling
Hemoglobin Köln	β^{98}Val → Met	Unstable hemoglobin
Hemoglobin Zurich	β^{63}His → Arg	Unstable hemoglobin
Hemoglobin M (Boston)	α^{58}His → Tyr	Cyanosis
Hemoglobin Yakima	β^{99}Asp → His	Leftward shift of O_2 saturation curve
Hemoglobin Bethesda	β^{145}Tyr → His	Leftward shift of O_2 saturation curve

Source: From Fairbanks, V. F., "Hemoglobin Derivatives, and Myoglobin," in N. Tietz (ed.), *Fundamental of Clinical Chemistry*, 2nd ed., p. 406 (Philadelphia: W. B. Saunders, 1976).

[a] The different types of hemoglobin are often named after the city in which they were first identified and studied.

[b] The notation describes the type of chain in which the substitution occurred, the position of the substitution, and the amino acid involved in the substitution.

THE THALASSEMIAS

thalassemias

The **thalassemias** are a group of diseases resulting from the decreased production of one of the hemoglobin chains. It should be stressed that decreased chain *production* occurs in this disease, *not* the synthesis of an abnormal hemoglobin chain with an abnormal amino-acid sequence. There are two main types of thalassemia, (1) α-thalassemia and (2) β-thalassemia.

α-Thalassemia. This disease results from decreased α-chain production. A rare, but deadly, subcategory of this disease is called *hydrops fetalis*. Fetuses afflicted with *hydrops fetalis* usually die before term or shortly after birth.

β-Thalassemia. This disease results from decreased β-chain production. Two main varieties of the disease have been studied: (1) β^+-thalassemia, in which a partial loss of β-chain production has occurred, and (2) β^0-thalassemia, in which a complete loss has occurred. The molecular basis for β-thalassemia has not been completely elucidated. Apparently, the gene coding for the β-chain is still intact, but its expression has somehow been inhibited, resulting in either partial or complete loss of β-chain production. β-Thalassemia is common in the regions surrounding the Mediterranean sea and throughout central and southeast Asia, usually in malaria-infested regions. Both the α- and β-forms of the disease produce a mild anemia.

4.5 THE STRUCTURE AND FUNCTION OF COLLAGEN— A FIBROUS PROTEIN

collagen

Approximately 33 percent of the total protein in our bodies is **collagen**. From collagen is constructed skin, tendons, ligaments, and the structural framework to which cells adhere. A related protein, *elastin*, is found in the lungs, skin, and arteries, where great elasticity is required along with physical strength.

tropocollagen

Collagen is constructed from **tropocollagen**. Tropocollagen is a protein with a molecular weight of around 290 000 daltons. It is constructed from three polypeptide chains wrapped around each other in a helix, much like the individual strands of a rope. Two of these chains, the α_1-chains, are identical; the third chain, the α_2-chain, differs from the α_1-chain with respect to amino acid sequence and composition. Each chain is about 1000 amino acids in length. The structure of the tropocollagen molecule is shown in Figure 4–17.

The amino acid sequence of the α_1- and α_2-chains has been determined. It is

$$\cdots(\text{Gly—X—Y})_n\cdots$$

Tropocollagen molecule with nonhelical N- and C-terminal regions; these regions are important in cross-link formation.

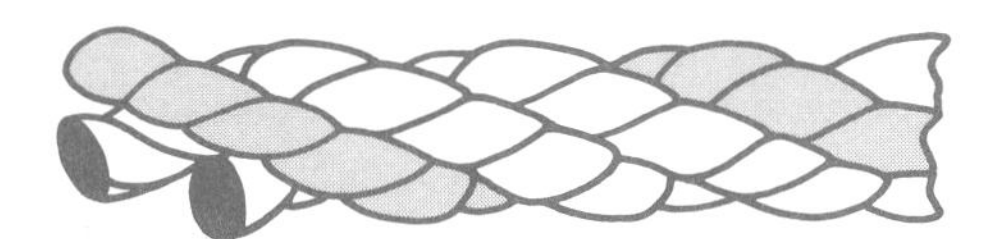

A portion of the tropocollagen triple helix

FIGURE 4–17
A structural diagram of the tropocollagen molecule showing how the three polypeptide chains are coiled into a right-handed helix.

in which the (Gly—X—Y) sequence is repeated many times. The X and Y amino acid positions are variable, but they are usually occupied by alanine and proline, respectively. Interspersed within the (Gly—X—Y) sequence can be found an occasional lysine residue or hydroxyproline or hydroxylysine derivative. The (Gly—X—Y) sequence is not found in the ends of the three strands of the tropocollagen molecule. These unique ends are about 20 amino acids in length.

There are four main subclasses of tropocollagen, as shown in Table 4–3. Each subclass requires a slightly different type of polypeptide chain.

Once tropocollagen has been synthesized within cells capable of making this protein, the individual tropocollagen molecules are excreted out of the cell and are *cross-linked* by the enzyme *lysyl oxidase.* This enzyme catalyzes the condensation of lysine residues from two separate tropocollagen molecules, thus linking the tropocollagen helices together. The lysine residues

TABLE 4–3 Types of Tropocollagen

Polypeptide Chain Composition	Found in the Following Body Tissues
Type I (α_1-Type I)$_2\alpha_2$	The most abundant form of tropocollagen in the body—found in skin, bone, tendons, blood vessels, and lung tissue (high tensile strength)
Type II (α_1-Type II)$_3$	Present in cartilaginous tissue (relatively low tensile strength)
Type III (α_1-Type III)$_3$	Major constituent of fetal skin (not found in adult skin)
Type IV (α_1-Type IV)$_3$	Basement membrane collagen (often associated with carbohydrates)

Note: The α_1 chain exists in four distinct forms, α_1-Types I–IV. Each chain varies slightly in amino acid sequence and composition.

Lysine — Chain 1: $\text{CH—(CH}_2)_4\text{—NH}_3^+$

Lysine — Chain 2: $^+\text{H}_3\text{N—(CH}_2)_4\text{—CH}$

↓ Lysyl oxidase (cross-link formation), H_2O released

$\text{CH—(CH}_2)_2\text{—CH=C—(CH}_2)_2\text{—CH}$ with CH=O on C

Cross-link

that participate in cross-link formation are in specific regions of the triple helix.

Thus, many tropocollagen molecules are organized into a "staggered" fiber, as shown in Figure 4–18.

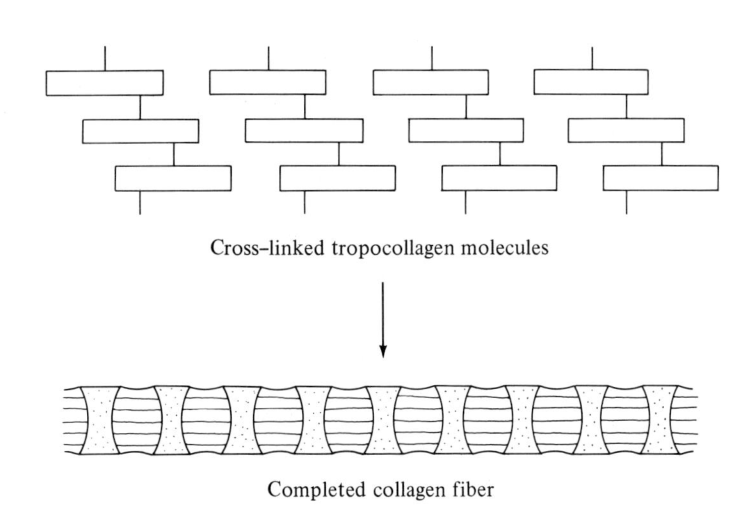

FIGURE 4–18
The organization of tropocollagen molecules into a completed collagen fiber. The individual tropocollagen molecules (actually 3-stranded helices) are cross-linked in a staggered manner, as shown in the upper portion of the figure.

4.6 DISEASE STATES ASSOCIATED WITH ABNORMAL COLLAGEN STRUCTURE AND METABOLISM

Recent research has indicated that a number of very serious disease states are related to *abnormal cross-linking* of individual tropocollagen molecules. Two subcategories of abnormal cross-linking are evident: (1) *low cross-linking* and (2) *high cross-linking.* As the name implies, low cross-linking means that too few lysine cross-links are present within the collagen fiber. This condition results from an abnormal (inactive) lysyl oxidase or from abnormal tropocollagen end regions. A low percentage of cross-links produces a weak and fragile collagen fiber. High cross-linking means that too many cross-links are present. High cross-linking produces a rigid collagen fiber, unable to withstand the normal elastic movements required of the molecule. High cross-link formation results from overactive lysyl oxidase or from an abnormal tropocollagen molecule containing too many lysine residues. A variety of disease states result from high and low cross-linking, as summarized in Table 4–4.

TABLE 4–4 Disease States Associated with Abnormal Collagen Cross-Linking

A. *Low Cross-Linking* (therefore, the collagen has low tensile strength)
   1. Spinal deformities (defective formation of bone from cartilage)
   2. Dislocation of joints (weak connective tissue, subject to tearing)
   3. Aortic aneurysms (weak blood-vessel walls)
   4. Dermal (skin) lesions

B. *High Cross-Linking* (therefore, the collagen is rigid)
   1. Arterial wall fibrosis (increased deposition of highly cross-linked (?) collagen, leading to arterial fibrotic lesions and plaques)
   2. Pulmonary fibrosis (increased deposition of highly cross-linked (?) collagen; decreased lung elasticity; abnormal pulmonary function)
   3. Related aging problems (age-related abnormalities such as decreased skin elasticity, stiffness in joints, and so on)

SUMMARY

Hemoglobin is a representative globular protein. Collagen is a representative fibrous protein.

The red blood cell contains a high percentage of hemoglobin (the oxygen binding protein). The hemoglobin molecule is composed of two α-chains and two β-chains. A heme group is located in a hydrophobic pocket in each chain.

Research suggests that the hemoglobin molecule binds oxygen in a sequential manner. One oxygen molecule binds to the first heme group of one chain, followed in sequence by the binding of oxygen molecules to the second, third, and fourth subunits. Increased affinity for oxygen is produced as a consequence of this sequential binding. The mechanism can be explained on a molecular level. Hemoglobin exhibits a sigmoidal oxygen saturation curve. At pO_2 values of 90–100 mm Hg (occurring in the lung alveoli), hemoglobin is fully saturated with oxygen. At pO_2 values less than 20 mm Hg (occurring in the body tissues), very little oxygen is bound to the hemoglobin. The hemoglobin-oxygen saturation curve shifts to the right under certain physiological conditions, such as increased tissue temperature, decreased tissue pH, increased tissue pCO_2, and increased DPG concentrations. Hemoglobin can also bind CO_2 and protons, which promote the release of oxygen to the tissues (Bohr effect). Fetal hemoglobin is constructed from two α-chains and two γ-chains. Prior to birth, fetal γ-chain production stops and β-chain production increases.

Blood is red because the heme groups of hemoglobin absorb blue and green wavelengths and transmit red wavelengths. Normal blood hemoglobin concentrations are 13–18 g/100 ml (males) and 11–16 g/100 ml (females).

Approximately 9 percent of the American black population are carriers of the genetic disease sickle-cell anemia (a severe anemia in which red blood cells assume a sickle shape under deoxygenated conditions). The glutamic acid residues at position 6 of the β-chains are replaced by valine residues. Because of this defect, deoxyhemoglobin tends to aggregate into long fibers. α-Thalassemia results from decreased α-chain production; β-thalassemia results from decreased β-chain production.

From collagen is constructed skin, tendons, ligaments, and the structural framework to which cells adhere. Collagen, in turn, is constructed from tropocollagen, a molecule composed of three strands wrapped around each other in a triple helix. Nonhelical end regions are important for cross-linking individual tropocollagen molecules by means of lysyl oxidase into a giant collagen fiber.

Low collagen cross-linking produces a weak and fragile collagen fiber, which causes such disorders as aortic aneurisms, skin lesions, and dislocation of joints. High collagen cross-linking produces a rigid collagen molecule, which may cause such disorders as arterial and pulmonary fibrosis as well as a number of defects associated with old age.

REVIEW QUESTIONS

1. We have presented a number of important concepts and terms in this chapter. Define (or briefly explain) each of the following terms:

a. Red blood cell
b. Hemoglobin molecule
c. α-Chain
d. Sequential binding model
e. Conformational change
f. Cooperativity
g. pO_2
h. Sigmoid saturation curve
i. Rightward shift
j. 2,3-Diphosphoglycerate
k. Fetal hemoglobin
l. Carbamate derivative
m. Bohr effect
n. Sickle-cell anemia
o. Thalassemias
p. Tropocollagen
q. Collagen fiber
r. Cross-link
s. Lysyl oxidase

2. Using the data found on page 53, describe how the author arrived at the value of 2.5×10^{13} as the total number of red blood cells in the body. Defend your answer.

***3.** Assume that the average weight of a red blood cell is 1.5×10^{-10} g.

a. Using the data on page 53 and using the value of 64 000 g/mole for the molecular weight of hemoglobin, calculate how many hemoglobin molecules are found in one red blood cell. (Avogadro's number is 6.02×10^{23} particles/mole.)

b. How many O_2 molecules can this red blood cell carry at any one time?

***4.** Suggest plausible reasons why the red blood cell assumes the shape of a biconcave disk.

5. What amino acids are closest to the heme groups of hemoglobin?

6. Would hemoglobin exhibit cooperativity if it consisted of only one polypeptide chain? Why or why not?

7. Suggest what the shape of the hemoglobin-oxygen saturation curve would be if hemoglobin did not bind oxygen in a cooperative manner.

***8.** Suppose life was found on Mars and that Martians required oxygen, had lungs, and had a circulatory system much like ours. Also suppose that Martians had an oxygen-binding protein much like our hemoglobin (call it *marglobin*). Draw an oxygen saturation curve for marglobin, assuming that marglobin is a tetramer and that it binds oxygen in a cooperative manner. The pO_2 value for the Martian atmosphere is only 10 mm Hg.

9. A person is at rest, but hyperventilating. Blood gas analysis reveals that this individual is in respiratory alkalosis (see Chapter 2). Which of the following statements is correct?

a. A rightward shift of the hemoglobin-oxygen saturation curve would be observed.

b. A rightward shift would not be observed.

c. A slight rightward shift would be observed.

d. None of the above are correct.

Explain your answer.

10. Chlorophyll, the green pigment in plants, has a chemical structure very similar to heme except that a magnesium ion, not an iron ion, is in the center of the ring system and the ring system has different side-chains.

a. What portions of the visible spectrum does chlorophyll absorb? What portions does it transmit?

b. Draw an absorption spectrum for chlorophyll.

11. We have stated that the amino acid sequence for the individual chains of tropocollagen is (Gly—X—Y), and that alanine and proline occupy the X and Y positions, respectively. What would you conclude about the secondary structure of the α_1- and α_2-chains, which both have this amino acid sequence and composition?

12. Why are the α_1- and α_2-chains of tropocollagen wrapped around each other like the strands of a rope?

SUGGESTED READING

Hemoglobin Structure and Function

Bhagavan, N. V. *Biochemistry*, 2nd ed., pp. 700–66. Philadelphia: J. B. Lippincott, 1978.

Dickerson, R. E., and Geis, I. *The Structure and Action of Proteins*, pp. 44–59. Menlo Park, CA: W. A. Benjamin, 1969.

Guyton, A. C. *Textbook of Medical Physiology*, 5th ed., pp. 56–66, 543–56. Philadelphia: W. B. Saunders, 1976.

McGilvery, R. W., and Goldstein, G. *Biochemistry—A Functional Approach*, pp. 228–52. Philadelphia: W. B. Saunders, 1979.

Tietz, N., ed. *Fundamentals of Clinical Chemistry*, 2nd ed., pp. 401–14. Philadelphia: W. B. Saunders, 1976.

Collagen Structure and Function

Dickerson, R. E., and Geis, I. *The Structure and Action of Proteins*, pp. 40–42. Menlo Park, CA: W. A. Benjamin, 1969.

McGilvery, R. W., and Goldstein, G. *Biochemistry—A Functional Approach*, pp. 150–65. Philadelphia: W. B. Saunders, 1979.

NOTES

1. Guyton, A. C., *Textbook of Medical Physiology*, 5th ed. (Philadelphia: W. B. Saunders, 1976), p. 56.
2. Guyton, *Textbook of Medical Physiology*, 5th ed., p. 56.
3. Koshland, D. E., Nemethy, G., and Filmer, P., *Biochemistry* 5 (1966): 365.
4. Montgomery, R., et al., *Biochemistry—A Case-Oriented Approach*. 3rd ed. (St. Louis: C. V. Mosby, 1980), p. 81.

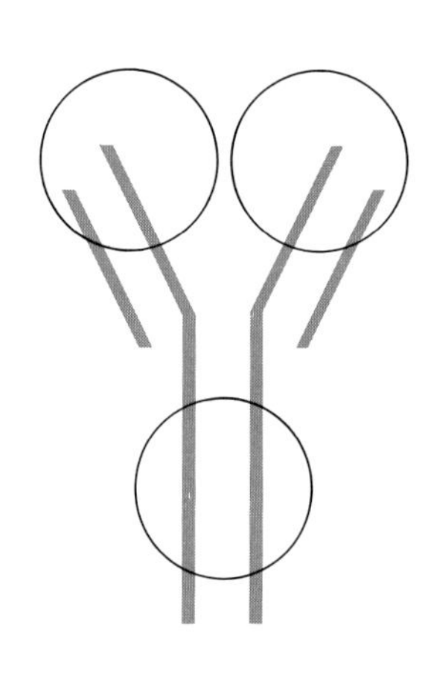

Chapter 5

Enzymes I: A General Description

5.1 INTRODUCTION

enzymes

Enzymes are proteins that are capable of speeding up chemical reactions (a process called *catalysis*). The enzyme is neither consumed nor irreversibly altered during this process. The term *enzyme* was first used by W. Kühne in 1876. It means (in Greek) "in leaven (yeast)." He, like other chemists of his day, realized that yeast contained substances that caused various processes to occur faster.

Originally, enzymes were not thought to be proteins. It was not until the 1920s and 1930s that the protein nature of an enzyme was firmly established. In 1926, J. B. Sumner isolated and purified an enzyme called *urease* and proved that his purified urease preparation was indeed a protein.

In this chapter, we will discuss what enzymes are, how they are classified, and how they speed up chemical reactions.

5.2 A GENERAL DESCRIPTION OF ENZYMES

catalyst

As stated in the introduction, enzymes are protein molecules that possess *catalytic activity*. A **catalyst** is any substance that increases the rate of a chemical reaction, but is not itself consumed or irreversibly altered during the process. Although a catalyst actually participates in the reaction, it is later regenerated.

The cells in your body (and in all living organisms) carry out thousands of different chemical reactions. These reactions constitute the *metabolism* of the cell, and are required for (1) extracting matter and energy from the environment; (2) transporting various chemicals, (3) synthesizing new biomolecules from precursor building blocks; (4) moving the cell or organism within its environment; and (5) replicating the parent cell to yield a new generation. Most of these reactions would normally proceed very slowly, or not at all, without enzymes. Enzymes speed up these necessary reactions and thus make life processes possible. Each chemical reaction found within the cell is catalyzed by a separate enzyme. Thousands of different enzymes therefore exist in the cell, each responsible for the catalysis of a specific chemical reaction.

Since enzymes are protein molecules, they possess tertiary structure and, therefore, have a unique three-dimensional shape (see Chapter 3). Some enzymes are oligomers and possess quaternary structure as well. The shape of the enzyme is critical to the proper binding of the *reactant* molecule and significantly contributes to the catalytic activity of the enzyme. The molecule on which the enzyme acts is called the **substrate**. Chemical bonds are either made or broken within the substrate. The polypeptide chain of the enzyme is folded in such a way that a small three-dimensional "pocket" or "cleft" is located somewhere on the surface of the enzyme. At this pocket, called the **active site**, the substrate binds to the enzyme to form an enzyme/substrate complex (ES complex). This complex undergoes further reactions to yield the product along with the regenerated enzyme. One can think of an enzyme molecule as a lock, the substrate as a key, and the active site as the keyhole of the lock. This description is called the lock-and-key theory of enzyme structure.

substrate

active site

One should note that the active site—the keyhole—has a specific three-dimensional shape that accommodates only one type of substrate molecule. Substrate molecules with other shapes do not fit into the active site of the enzyme molecule. Thus, enzymes exhibit **specificity**—they only bind a very specific type of substrate molecule, one capable of fitting into the active site. Actually, there are four main classes of specificity:[1]

specificity

1. **Absolute Specificity**. An enzyme exhibits absolute specificity when it acts on only one substrate and catalyzes only one reaction.

2. **Group Specificity**. Certain enzymes can act on a class of substrates that have a *common functional group*. For example, the phosphatases can act on a variety of different substrates, but each must have a phosphate functional group.

$$R{-}O{-}\overset{\overset{\displaystyle O}{\|}}{\underset{\underset{\displaystyle O^-}{|}}{P}}{-}O^-$$

3. **Linkage Specificity**. Certain enzymes are specific for a particular type of chemical bond (or linkage). For example, the esterases catalyze the hydrolysis of esters:

$$R_1-O-\overset{\overset{O}{\|}}{C}-R_2 \xrightarrow[H_2O]{\text{Esterase}} R_1-OH + HO-\overset{\overset{O}{\|}}{C}-R_2$$

(Ester link)

4. **Stereospecificity**. Finally, certain enzymes can discriminate between D-stereoisomers and L-stereoisomers and catalyze a reaction using only one of them. This unique ability is explained in Figure 5–1.

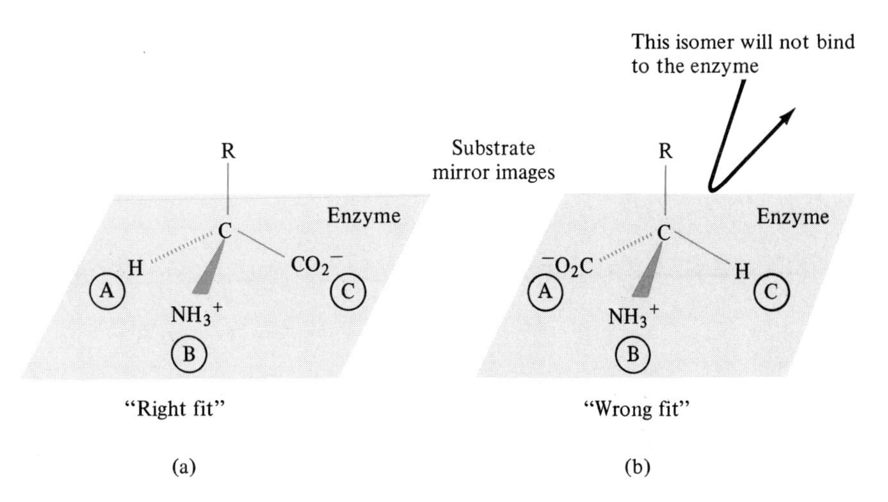

FIGURE 5–1 A stereospecific enzyme that can only bind one of two possible stereoisomers. (Adapted from Page, D. S., *Principles of Biological Chemistry* (Boston: Willard Grant, 1976), p. 103.)

Here we see a substrate binding to the enzyme active site at three points, A, B, and C. Notice that the structure in Figure 5–1(a) produces the "right fit," in which the hydrogen atom is associated with binding site A, the amino group with binding site B, and the carboxyl group with binding site C. The mirror image of this structure (the stereoisomer) cannot align with the correct binding sites (see Figure 5–1(b)). The carboxyl group is now associated with binding site A and the hydrogen atom with binding site C.

The active site is surrounded by amino acid R groups that (1) bind the substrate into the active site and (2) may participate in the catalytic event or process. Figure 5–2 (page 80) demonstrates how the polypeptide chain folds, bringing together certain amino acids that ultimately make up the active site.

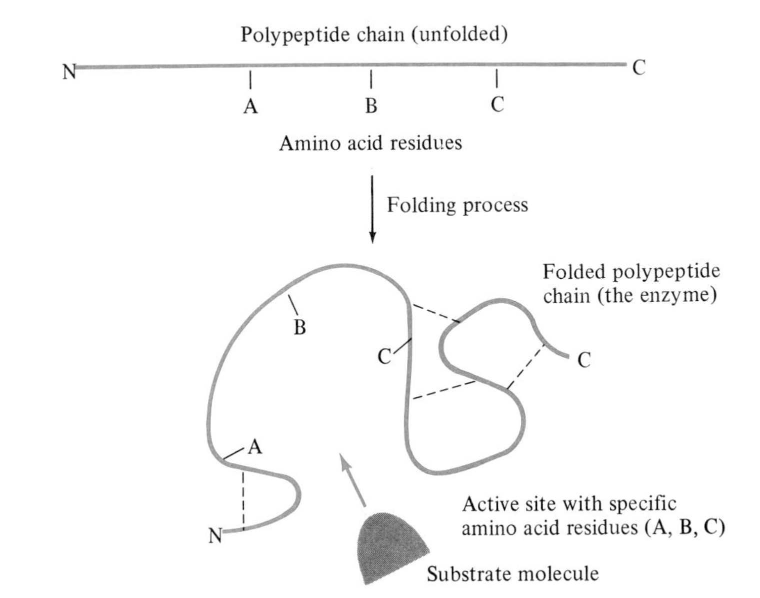

FIGURE 5–2 Folding the polypeptide chain to yield the complete enzyme molecule with active site.

FIGURE 5–3 A 3-dimensional representation of the α-chymotrypsin molecule. The polypeptide chain backbone is shown. Amino acid residues 195, 57, and 102 are part of the active site and participate in the catalytic event. Amino acid residue 16 is also essential for activity.

Figure 5–3 shows a detailed structure of an enzyme molecule.

The chemical and physical properties of the different amino-acid R groups aid in binding the substrate to the active site. For example, negatively charged amino acids such as aspartic and glutamic acid can bind a positively charged substrate via electrostatic interactions. Likewise, positively charged amino acids such as histidine, lysine, and arginine can bind negatively charged substrate molecules. Hydrophobic amino acids can bind hydrophobic substrates via hydrophobic/hydrophobic attractions; hydrophilic amino acids can bind polar substrates via hydrogen bonding. A few of these interactions are summarized in Figure 5–4.

Certain amino-acid R groups in the active site aid in the catalytic event of the enzyme. For example, some amino acids act as either acids or bases and either donate protons to the bound substrate or accept protons from the substrate. Other amino-acid R groups attract electrons from the substrate (are electrophilic) or bring electrons to the substrate (are nucleophilic). These processes are often critical for making or breaking chemical

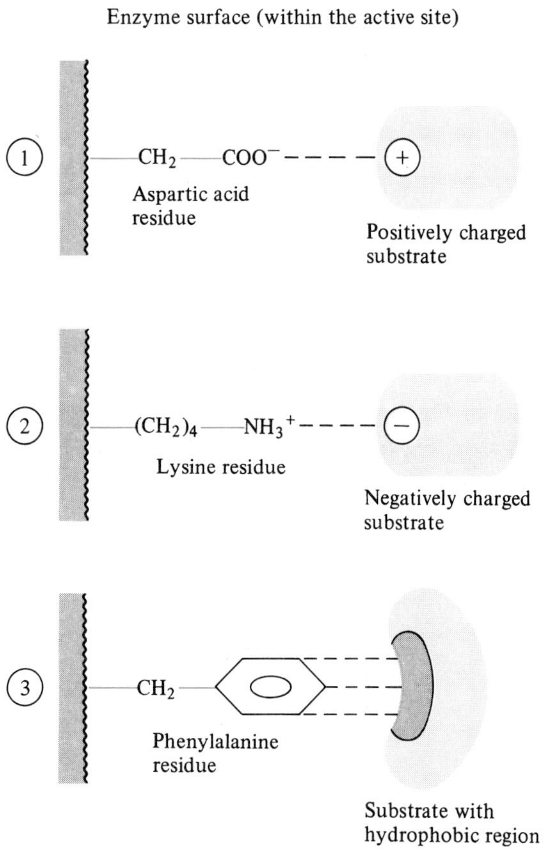

Enzyme surface (within the active site)

① CH_2—COO^- - - - - ⊕
Aspartic acid residue
Positively charged substrate

② $(CH_2)_4$—NH_3^+ - - - - ⊖
Lysine residue
Negatively charged substrate

③ CH_2
Phenylalanine residue
Substrate with hydrophobic region

FIGURE 5–4
Representative substrate/amino acid binding interactions within the active site.

bonds as required by the catalytic event. Figure 5–5 summarizes a few amino acids that participate in the catalytic event within the active site.

Enzyme activity sometimes results only from the three-dimensional shape of the enzyme molecule. In other cases, it results from the protein's three-dimensional shape together with a *bound* **cofactor**. This type of enzyme is called a *conjugated enzyme*, or *holoenzyme*. If the cofactor is tightly bound to the protein molecule (usually by covalent bonds), the cofactor is referred to as a *prosthetic group*. If the cofactor is not tightly bound to the protein, then it is called a **coenzyme**. Most coenzymes are bound to the protein via electrostatic interactions, hydrogen bonds, or other noncovalent associations. Various metal ions (Mg^{2+}, Mn^{2+}, Zn^{2+}, Cu^{2+}, and so on) act as cofactors and are required for enzyme activity. Enzymes that require metals for activity are called metalloenzymes. Other rather complex organic molecules (vitamin derivatives) act as coenzymes. Some metal ion cofactors and some vitamin coenzymes are summarized in Table 5–1. Cofactors are bound into the active site along with the substrate. They participate in the catalytic event by acting as an acid, a base, or an electrophile. Chemical bonds are made or broken (by electron-pair shifts) as a result of the chemical properties of the cofactor.

cofactor

coenzyme

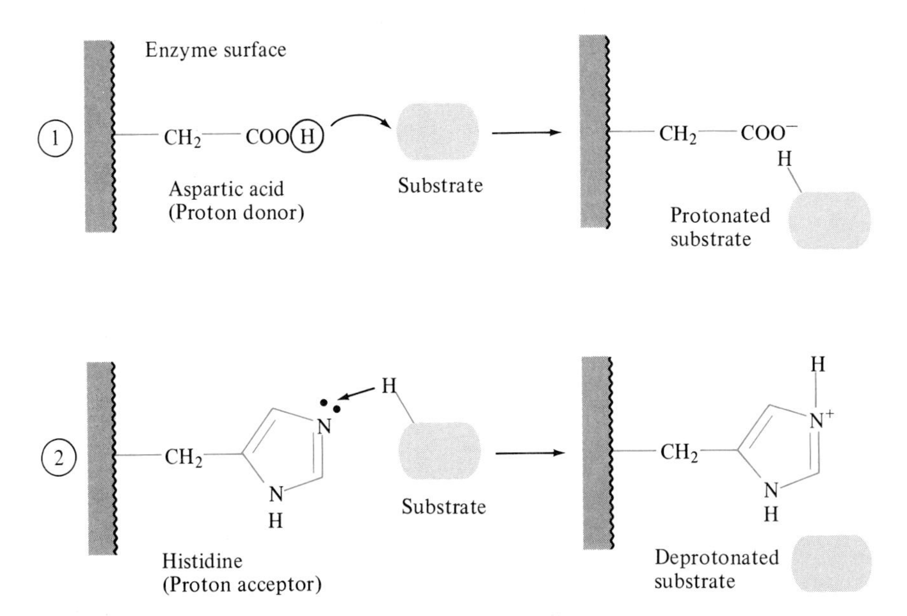

FIGURE 5–5 Amino acid R groups acting as acids or bases during enzyme catalysis. Diagram 1 shows how aspartic acid behaves as an acid (proton donor) in protonating the substrate. Diagram 2 shows how histidine behaves as a base (proton acceptor) in deprotonating the substrate.

TABLE 5–1 Metal Ion Cofactors and Vitamin Coenzymes

Cofactor/Coenzyme	Precursor	Function	Enzyme(s)
Cu^{1+}/Cu^{2+}	——	Enzyme catalysis	Cytochrome oxidase of the electron transport system
Fe^{2+}/Fe^{3+}	——	Enzyme catalysis	Cytochrome enzymes of the electron transport system; peroxidases
Mg^{2+}	——	Enzyme catalysis	Enzymes requiring *ATP*
Zn^{2+}	——	Enzyme catalysis	Alcohol and lactate dehydrogenases
*F*lavin *A*denine *D*inucleotide (*FAD*)	Riboflavin; Vitamin B_2	Electron/hydrogen transfer	Succinate dehydrogenase of Krebs cycle
*F*lavin *M*ono*n*ucleotide (*FMN*)	Riboflavin Vitamin B_2	Enzyme catalysis	L-Amino acid oxidase
Lipoic acid	——	Transfer of hydrogens and acetyl groups	Serves as the "swinging arm" in the pyruvate dehydrogenase complex
*N*icotinamide *A*denine *D*inucleotide (NAD^+)	Niacin	Oxidation-reduction reaction (hydrogen transfer)	Lactate dehydrogenase
*N*icotinamide *A*denine *D*inucleotide *p*hosphate (*$NADP^+$*)	Niacin	Reduction reactions	Fatty acid synthetase complex
Biotin	Biotin	Transfer of carbon dioxide (CO_2)	Acetyl-CoA-Carboxylase
Cobamide	Vitamin B_{12}	Transfer of methyl groups	Methylmalonyl-CoA-mutase
Coenzyme A (CoA)	Pantothenic acid	Transfer of acetyl groups	Pyruvate dehydrogenase complex
Pyridoxal phosphate	Vitamin B_6	Transfer of amino groups	Transaminases
Tetrahydrofolic acid (FH_4)	Folic acid	Transfer of single carbon groups	Glycine synthase
*T*hiamine *P*yro*p*hosphate (TPP)	Thiamine	Removal of CO_2 group	Pyruvate dehydrogenase complex

To summarize, an enzyme is a complex polypeptide chain folded into a unique three-dimensional shape. The manner in which the chain folds determines which amino acids will be in close proximity with each other within the active site. These amino acids bind the substrate into the active site (often along with cofactors) and engage in the catalytic process by which chemical bonds are made or broken in the substrate.

5.3 HOW DOES AN ENZYME WORK?

One must first appreciate the fact that enzymes cause chemical reactions to proceed thousands—even millions—of times faster than normal. An uncatalyzed chemical reaction may take days or weeks to reach *equilibrium* (the point at which the rate of formation of products equals the rate of reformation of reactants). With an enzyme present, this same reaction can attain equilibrium in only a matter of seconds. How does an enzyme accomplish this amazing feat?

In order to answer this question, we must first develop the concept of the "energy hill" and the transition state. Suppose we have two chemicals A and B (reactants) that can react to form the products C and D according to Equation 5–1.

$$\underset{\text{Reactants}}{A + B} \longrightarrow \underset{\text{Products}}{C + D} \tag{5–1}$$

activation energy
transition state

If reactant molecules A and B are to react, they must collide with enough *kinetic energy* (energy of movement) to make (or break) a chemical bond during the collision. Thus, reactants A and B must have enough energy (**activation energy**, E_a) to reach what is known as the **transition state** ($[A–B]^*$). This transition state is a high-energy intermediate in which A and B have enough energy to be converted into the products C and D. Once the transition state has been reached, product formation is "downhill" in terms of energy. Reactants A and B thus experience an energy hill, as depicted in Figure 5–6. Do not, however, be confused by Figure 5–6. The reactants do not move over a literal hill. Rather, this figure is an *abstract representation* of the energy required to achieve the transition state. If most of the reactant molecules do not possess sufficient kinetic energy, few will reach the energy of the transition state and, therefore, few product molecules will be formed. If, however, sufficient energy is supplied to the system (often in the form of heat), a number of reactant molecules will have sufficient energy to reach the transition state and more product molecules will be formed.

In an *enzyme-catalyzed reaction*, on the other hand, reactants A and B enter the active site of the enzyme, bind to it, and react to form products C and D. In this situation, the energy hill is much *lower* than in the uncatalyzed reaction (Figure 5–6). This means that reactants with lower energy can form the new transition state, which has a lower activation energy. Thus, enzymes catalyze unfavorable chemical reactions by significantly decreasing the energy of activation. In essence, an enzyme provides the reactants with an alternative route that has a lower energy hill. This concept can be explained by considering the following analogy (Figure 5–7). Consider two drivers, A and B. Driver A chooses route #1 (the uncatalyzed route) over the mountain (the energy hill). This route is very long, very steep, and requires a great deal of energy. Driver B chooses route #2 (the enzyme-catalyzed route). This route goes through a mountain pass and does not require large amounts of energy; thus, driver B gets to his destination quicker.

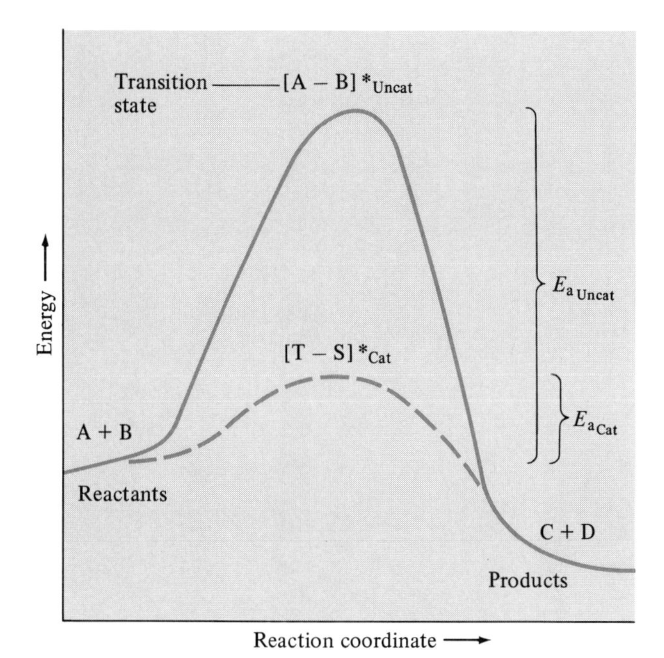

FIGURE 5–6 Energy diagram for the reaction A + B → C + D. The solid line indicates the energy hill for the uncatalyzed reaction; the dashed line indicates the energy hill for the enzyme-catalyzed reaction. The transition state is noted by the symbols [A—B]* and [T—S]*. E_a represents the activation energy.

Let us summarize the concepts discussed in the previous two sections. Consider enzyme E, which is a protein folded into a unique three-dimensional shape that includes an active site. Substrate S can bind into the active site to form an enzyme/substrate complex ES. The amino acids that surround the active site bind the substrate (and also any necessary cofactors). Specific amino acid residues (together with a cofactor, if required) induce the catalytic event, during which chemical bonds are made or broken. During this process, the ES complex achieves a transition state [ES]* at a significantly lower activation energy than that of the uncatalyzed reaction. The ES transition state breaks down into an enzyme/product complex EP. This EP complex finally yields products from the active site, regenerating enzyme E at the same time. These steps are summarized in Equation 5–2.

$$E + S \rightleftharpoons ES \rightleftharpoons [ES]^* \rightleftharpoons EP \rightleftharpoons E + P \qquad (5\text{–}2)$$

One should note that the binding of the substrate to the enzyme is *reversible*. Once the [ES]* transition state has been achieved, it will yield the EP complex and, ultimately, free enzyme and free products.

Finally, we should note that few enzymes work on only a single substrate. Rather, two or more different substrates must bind into the active

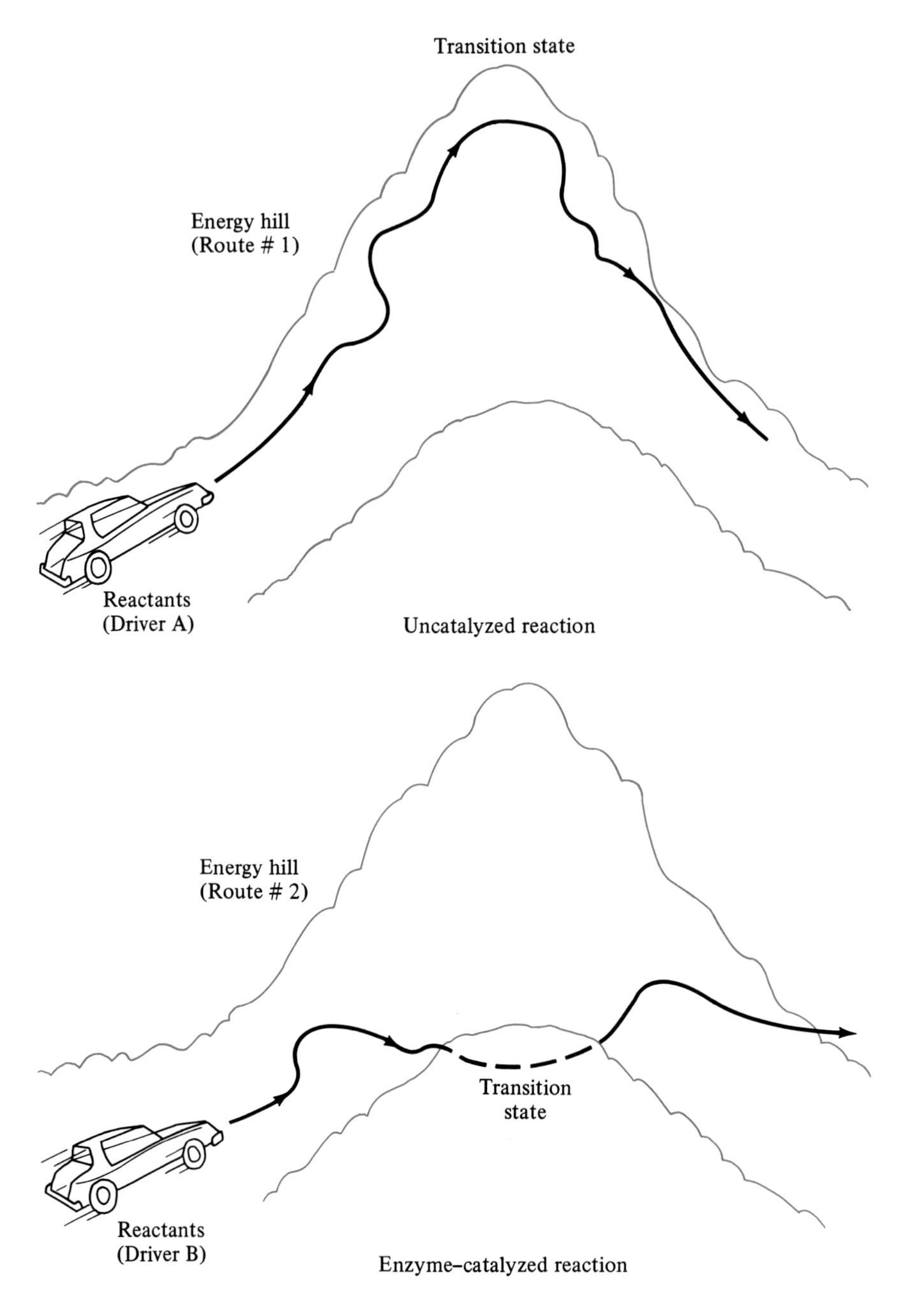

FIGURE 5–7 A comparison between the energy hills of uncatalyzed and enzyme-catalyzed reactions.

site of most enzymes before catalysis occurs. For example, the enzyme *hexokinase* catalyzes the following reaction:

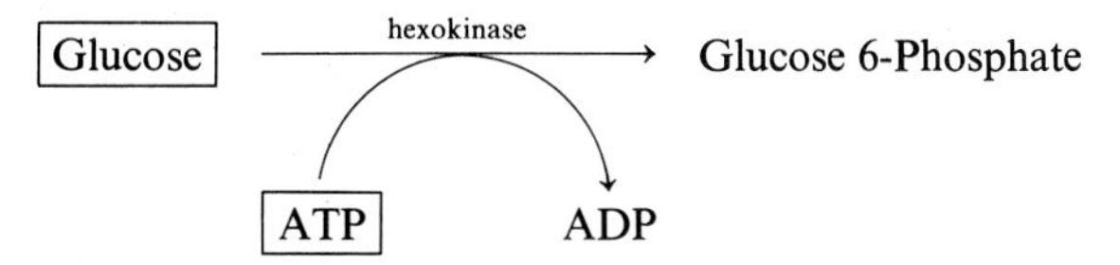

Glucose and *ATP* (*a*denosine *tri*phosphate) are the two substrates of hexokinase. When two or more substrates bind into the active site, the *sequence* of binding is often important. The ordered and compulsory binding sequence is an example. This type of binding demands that one substrate (A) bind to the enzyme before the second substrate (B) can do so.

$$E + A \rightleftharpoons EA$$
$$EA + B \rightleftharpoons EAB$$

Other binding sequences are possible.

5.4 CLASSIFICATION OF ENZYMES

Literally thousands of different enzymes have been isolated and studied. This situation has produced a problem. How do we name and classify all of these enzymes? Unfortunately, because investigators in the early part of this century did not appreciate this problem, they named newly discovered enzymes in a rather haphazard manner. Many of these "common" names are still with us today, firmly entrenched in the scientific literature and in textbooks. At first, investigators adopted a rather simple naming system: To the substrate on which an enzyme acted was added the suffix "-ase." For example, the enzyme name *arginase* was built up by adding "-ase" to the (slightly shortened) substrate name "arginine." Other enzymes, however, were named according to the substrate used and the type of reaction catalyzed. For example, *glucose oxidase* catalyzes the oxidation of glucose (that is, the removal of electrons from glucose).

In order to systematize the naming of enzymes, the International Enzyme Commission (IEC) adopted a system based on the *type of reaction* that the enzyme catalyzes. Six major categories have been described; these are summarized in Table 5–2. The **oxidoreductases** catalyze oxidation-reduction reactions (reactions that remove electrons from or add electrons to the substrate). The **transferases** catalyze the transfer of various functional groups from one substrate molecule to another. The **hydrolases** catalyze the addition of a water molecule across a bond, a process called *hydration* that cleaves the bond. The **lyases** catalyze the removal of various groups from substrates,

oxidoreductases

transferases

hydrolases

lyases

TABLE 5–2 The Classification of the Enzymes

1. Oxidoreductases (catalyze oxidation-reduction reactions):
   - 1.1 Acting on >CH—OH
   - 1.2 Acting on >C=O
   - 1.3 Acting on —CH=CH—
2. Transferases (catalyze the transfer of functional groups):
   - 2.1 One-carbon groups

   - 2.2 Aldehydic or ketonic groups (—CH(=O) or R—C(=O)—R)
   - 2.3 Acyl groups (hydrocarbon chains)
   - 2.4 Glycosyl groups (sugars)
3. Hydrolases (catalyze hydrolysis reactions—the breaking of chemical bonds by adding water):
   - 3.1 Esters (R—O—C(=O)—R)
   - 3.2 Glycosidic bonds (between sugar molecules)
   - 3.4 Peptide bonds (in proteins)
4. Lyases (catalyze the removal of chemical groups, with double-bond formation):
   - 4.1 CO_2 removal (decarboxylase)
   - 4.2 NH_3 removal (deaminase)
5. Isomerases (catalyze isomerization reactions):
   - 5.1 Racemases (catalyze the interconversion of D and L forms).
6. Ligases (catalyze the formation of chemical bonds, with ATP cleavage):
   - 6.1 C—O bonds
   - 6.2 C—S bonds
   - 6.3 C—N bonds
   - 6.4 C—C bonds

Note: This table only lists a few of the subcategories.

isomerases · ligases

in the process forming double bonds. The **isomerases** catalyze *intra*molecular rearrangements (for example, the interconversion of D- and L-isomeric forms). Finally, the **ligases** catalyze the formation of chemical bonds. This last class of enzymes requires the input of chemical energy (in the form of adenosine triphosphate, ATP) if new bonds are to be formed. Each major category contains subcategories, and every conceivable type of enzyme can be classified in an accurate and systematic approach.

In this system, each enzyme is given its own systematic name and classification number. For example, the enzyme *lactate dehydrogenase* (*LDH*) has the following Enzyme Commission (EC) name and number; EC 1.1.1.27 L-lactate: NAD oxidoreductase. This means that LDH belongs to the first major class (1. = oxidoreductases), removes electrons from the substrate (lactic acid) and utilizes the enzyme cofactor NAD^+ (*n*icotinamide *a*denine *d*inucleotide).

SUMMARY

Enzymes are protein catalysts that increase the rate of a chemical reaction, but are not consumed or altered in the process. The polypeptide chain of an enzyme is folded in a manner that produces an "active site." The substrate binds to the active site in a very specific manner to form an enzyme/substrate (ES) complex. Once the enzyme/substrate complex has formed, the catalytic event occurs, forming reaction products. Enzymes exhibit four kinds of specificity: absolute specificity, group specificity, linkage specificity, and stereospecificity. The active site is surrounded by a number of amino acid residues that bind the substrate molecule to the active site and can also be responsible for the catalytic process. Enzyme cofactors (metal ions and vitamin coenzymes) can also bind into the active site, thus aiding the enzyme in forming or breaking chemical bonds.

If molecules are to react with each other, they must possess sufficient kinetic energy (energy of movement) to make or break chemical bonds after they have reached the transition state. Enzymes accelerate chemical reactions by providing an alternative reaction route that has a lower activation energy. They do this by bringing reactant molecules together within the active site (which also may contain bound cofactors).

In the past, enzymes were named by adding the suffix "-ase" to the name of the substrate or the name of the reaction being catalyzed. Enzymes are now classified by a more systematic scheme. There are six main classes of enzymes: oxidoreductases, transferases, hydrolases, lyases, isomerases, and ligases.

REVIEW QUESTIONS

1. We have presented a number of concepts and terms in this chapter. Please define (or briefly explain) each of the following terms:
   a. Enzyme
   b. Catalyst
   c. Substrate
   d. Active site
   e. Absolute specificity
   f. Coenzyme
   g. Cofactor
   h. Transition state
   i. Activation energy (E_a)
   j. Oxidoreductase
   k. Ligase
2. What is the principal difference between
   a. Absolute and group specificity?
   b. Group and linkage specificity?
3. Referring to Figure 5–1, suggest what amino acid residues might be involved in binding the substrate to the stereospecific active site shown in the diagram. Specifically, what amino acid residues might occupy positions A, B, and C?
4. Is it likely that an enzyme would bind more than one type of cofactor for activity? Why or why not?
5. *Draw an energy diagram similar to that shown in Figure 5–6 for an uncatalyzed reaction in which the products C and D have a *greater* final energy level than the reactants A and B. What does this tell you about the direction of the reaction?
6. A *random* binding sequence exists for certain enzymes. How would this binding sequence differ from the ordered and compulsory binding sequence mentioned in the text?
7. Naming enzymes and reactions:
   a. Give a trivial name for enzymes that act on the following substrates:
   (1) Urea (2) Glucose 6-phosphate
   (3) Adenosine triphosphate (ATP)

b. Suggest possible trivial names for enzymes that catalyze the following hydrolysis reactions:

(1) $^{+}H_3N{-}CH(R_1){-}C(=O){-}NH{-}CH(R_2){-}COO^{-} \longrightarrow 2\ R_X{-}CH(NH_3^{+}){-}COO^{-}$

Free amino acids

(2) Sugar 1—Sugar 2 ⟶ sugar 1 + sugar 2

c. Suggest what class of enzymes would catalyze each of the following reactions:

(1) $R{-}O{-}PO_3^{2-} \xrightarrow{H_2O} R{-}OH + {}^{-}O{-}P(=O)(O^{-}){-}O^{-}$

(2) $R_1R_2C{=}O \longrightarrow R_1R_2CH{-}OH$ (NADH + H^{+} → NAD^{+})

(3) Glucose + ATP ⟶ Glucose 6-phosphate + ADP

SUGGESTED READING

Enzyme Structure and Function

Bernhard, S. A. *The Structure and Function of Enzymes*, New York: W. A. Benjamin, 1968.

Bhagavan, N. V. *Biochemistry*, 2nd ed. Ch. 3. Philadelphia: J. B. Lippincott, 1978.

Lehninger, A. L. *Biochemistry*, 2nd ed., Chs. 8 and 9. New York: Worth, 1975.

Tietz, N, ed. *Fundamentals of Clinical Chemistry*, 2nd ed., Ch. 12. Philadelphia: W. B. Saunders, 1976.

Special Articles and Topics

Koshland, D. E. "Protein Shape and Biological Control." *Scientific American* 229 (1973): 52.

Phillips, D. C. "The Three-Dimensional Structure of an Enzyme Molecule." *Scientific American* 215 (1966): 78.

NOTE

1. Henrickson, C., and Byrd, L. *Chemistry for the Health Professions*, p. 718, Belmont, CA: Wadsworth Publishing Co., 1980.

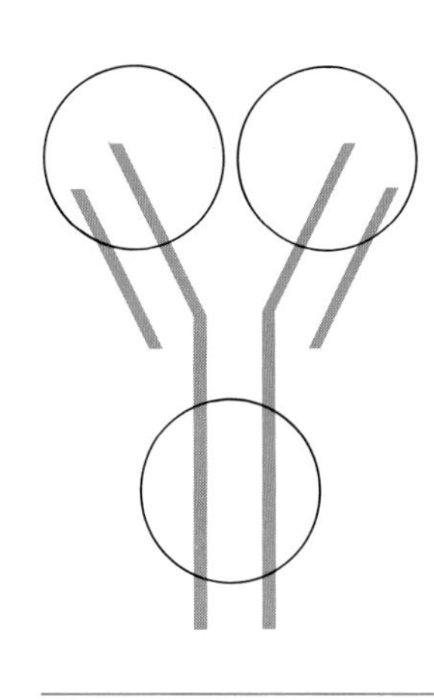

Chapter 6

Enzymes II: Factors That Influence Enzyme Activity

6.1 INTRODUCTION

Enzyme structure and function was discussed in the previous chapter. In this chapter, we will describe various factors that influence the activity of enzymes. These factors include temperature, pH, substrate concentration, enzyme inhibitors, covalent modification of the enzyme molecule, cellular enzyme concentration, and allosteric effectors. We will then discuss the importance of enzyme activity assays in clinical medicine.

6.2 ENZYME ACTIVITY AND ENZYME ASSAYS

To measure the catalytic activity of an enzyme, a biochemist could measure either the rate of disappearance of the substrate or the rate of appearance of the product. The rate of appearance of the product is the more convenient of the two choices, especially if the product is colored and, therefore, absorbs light in the visible region of the spectrum. The enzyme and substrate are combined in a test tube, placed in a spectrophotometer, and the absorbance of the mixture is measured at a specific wavelength. As more product is formed during the reaction, more color is produced and, thus, the absorbance of the mixture increases. A typical enzyme assay is shown in Figure 6–1. The increase in absorbance is usually linear over a certain period of time. As the substrate in the test tube is depleted, the reaction, and with it the rate of increase in absorbance, begins to slow down. Most enzyme assays

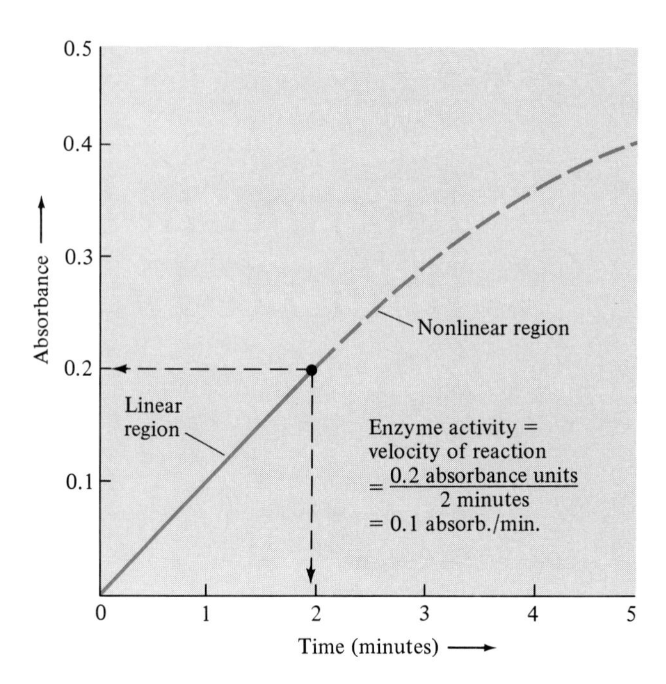

FIGURE 6–1
An enzyme activity assay. An enzyme is changing S into P. Product P is colored and absorbs light in the visible region of the spectrum. As the concentration of P increases, the absorbance also increases. The increase in absorbance is measured with a spectrophotometer and used to calculate the activity of the enzyme.

are run only during the time in which the increase in absorbance is linear with time.

enzyme activity

The **enzyme activity** is defined as the amount of enzyme that will convert a certain amount of substrate into products in a specified period of time under conditions of constant temperature and pH. Since each research investigator has developed his or her own assay for a particular enzyme, a great deal of confusion has existed in the scientific literature regarding the measurement of enzyme activity. For example, consider the enzyme *alkaline phosphatase*. A number of investigators have studied this enzyme using different assays, and have expressed the activity in a variety of ways (see Table 6–1).

International Unit (I.U.)

To eliminate this source of confusion, the International Enzyme Commission adopted a standard unit of enzyme activity called the **International Unit (I.U.)**. One I.U. is defined as the amount of enzyme that can convert 1 micromole (μmol) of substrate into product per min at 25 °C. One μmol $= 1 \times 10^{-6}$ mole. The international unit has been recently redefined; it will be replaced by the *Katal*, in which activity is based upon the number of moles of substrate transformed into product per second at 25 °C.

$$1 \text{ I.U.} = 10^{-6} \text{ mol/60 seconds} = 16.7 \times 10^{-9} \text{ mol/second}$$

As we shall see in a moment, these standards are important because the measurement of enzyme activity is critical in the diagnosis of disease states.

specific activity (S.A.)

Another term that is often used is **specific activity (S.A.)**. Specific activity is simply the number of International Units of enzyme per milligram (mg) of protein. This term indicates the *purity* of the enzyme preparation.

TABLE 6–1 Different Assay Methods for Measuring Alkaline Phosphatase Activity

Investigator(s)	Assay and Definition of Activity Unit
Bodansky	1 unit = the amount of enzyme that will hydrolyze 1 mg of phosphate in 60 minutes (pH 8.6, 37 °C)
Kind and King	1 unit = the amount of enzyme that will liberate 1 mg of phenol in 15 minutes (pH 10, 37 °C)
Bessey-Lowry-Brock	1 unit = the amount of enzyme that will hydrolyze 1 millimole of substrate in 60 min (pH 10.2)
Bowers-McComb	1 unit = the amount of enzyme that will hydrolyze 1 millimole of substrate in 1 min (pH 10.2, 30 °C)

Source: From Tietz, N. (ed.), *Fundamentals of Clinical Chemistry*, 2nd ed. (Philadelphia: W. B. Saunders, 1976), p. 591.

Some relatively crude enzyme preparations are contaminated with many other proteins. Thus, 1 mg of a crude enzyme preparation may contain only 0.1 mg of the enzyme under study and 0.9 mg of contaminating proteins. Preparations that contain more of the enzyme under study will have a higher specific activity. Consider preparation 1, in which 1 mg of the preparation has a specific activity of 1 (S.A. = 1 = 1 International Unit/1 mg). A second, more highly purified preparation has a specific activity of 1000 (S.A. = 1000 = 1000 International Units/1 mg). This means that the second preparation contains a larger amount of the enzyme under analysis and, therefore, a smaller amount of contaminating proteins. Specific activity is important when purifying enzymes from crude preparations of starting material.

turnover number

A term related to specific activity is **turnover number**. Turnover number (T.N.) is the number of substrate molecules converted into product per unit time when the enzyme is fully saturated with substrate. Put another way, turnover number tells how many substrate molecules are converted into products by each enzyme molecule. It tells us exactly how fast an enzyme can work or "turn over" substrate into products. Some relatively slow enzymes have turnover numbers of only a few hundred or thousand substrate molecules per second per enzyme molecule. Other enzymes (such as *carbonic anhydrase*, see Chapter 2) operate much faster. Carbonic anhydrase can turn over 600 000 substrate molecules (CO_2 and H_2O) per second to form carbonic acid (H_2CO_3). Thus, one carbonic anhydrase molecule can combine one CO_2 and one H_2O molecule every 1.7 microseconds (μsec). One (μsec = 1×10^{-6} seconds. Clearly this enzyme operates very, very quickly!

ENZYMES AND THE DIAGNOSIS OF DISEASE

A number of different enzymes are normally present in blood plasma. There are two main classes of plasma enzymes, (1) the *plasma specific* enzymes and (2) the *non–plasma specific* enzymes. Plasma specific enzymes are enzymes that are normally present in the blood and are responsible for catalyzing various reactions that normally occur in the blood. *Plasmin* and *thrombin* are examples of plasma specific enzymes; they are involved in the clotting of blood. The non–plasma specific enzymes are enzymes that have been secreted from cells directly into the blood or that have "leaked" into the blood from damaged tissue cells. These enzymes do not have any physiological or biochemical function in the blood; they are simply circulating there as a result of their secretion from the tissues. *Lactate dehydrogenase* (*LDH*), *creatine phosphokinase* (*CPK*), and *alkaline phosphatase* are examples of non-plasma specific enzymes. Generally, non–plasma specific enzymes are present at considerably lower concentrations in plasma than in tissues. However, the plasma concentration of these enzymes significantly increases during the course of various disease states. A disease might significantly alter the permeability of the cell membrane or may cause the actual destruction of the cell membrane. If destruction does occur, large amounts of *intracellular* enzymes (enzymes normally confined in the cell) will leak from the cells into the blood. Thus, increased amounts of intracellular enzymes would be present in the blood. Specific enzyme assays performed on blood can easily detect increased amounts of a particular enzyme.

Many enzymes are normally present at higher concentrations in a specific organ. If the organ is diseased, then increased amounts of certain enzymes will be present in the blood and can be identified by specific enzyme assays. For example, an abnormally high level of serum lactate dehydrogenase isoenzymes 1 and 2 indicates myocardial infarction (a heart attack);

TABLE 6–2 Increased Serum Enzyme Levels Associated with Various Disease States

Organ	Enzyme	Disease State(s)
Bone	Alkaline phosphatase	Paget's disease (increased osteoblastic cellular activity) Rickets Metastatic bone tumors
Heart	Lactate dehydrogenase (isoenzymes 1 and 2)	Myocardial infarction
	Aspartate transaminase	Myocardial infarction
	Creatine phosphokinase (CPK)	Myocardial infarction
Liver	Aspartate transaminase (AST)	Hepatocellular disease
	Lactate dehydrogenase (isoenzyme 5)	Hepatocellular disease
Prostate	Acid phosphatase	Prostatic carcinoma
Skeletal muscle	Creatine kinase	Muscular dystrophy
	Aldolase	Diseases involving muscle disintegration

an abnormally high level of lactate dehydrogenase isoenzyme 5 indicates liver disease. (The biochemistry of lactate dehydrogenase isoenzymes will be discussed in greater detail in Section 6.5.) High serum acid phosphatase can indicate cancer of the prostate; high creatine kinase can indicate abnormalities in the skeletal muscles. Table 6–2 summarizes a variety of disease states in which specific enzymes are elevated.

6.3 WHAT FACTORS INFLUENCE ENZYME ACTIVITY?

A number of environmental and chemical factors influence and regulate the activity of enzymes.

THE EFFECT OF TEMPERATURE

The rate of an enzyme reaction is considerably affected by temperature, as shown in Figure 6–2. Generally, enzymes operate very slowly at temperatures at or near 0 °C, because molecular motion is significantly decreased at this temperature. At temperatures above 10 °C, enzymatic rates rise significantly. An enzymatic rate usually doubles for every 10 °C increase in temperature up to what is known as the **optimum temperature** range of the enzyme. Since the temperature of our bodies is closely regulated to 37 °C,

optimum temperature

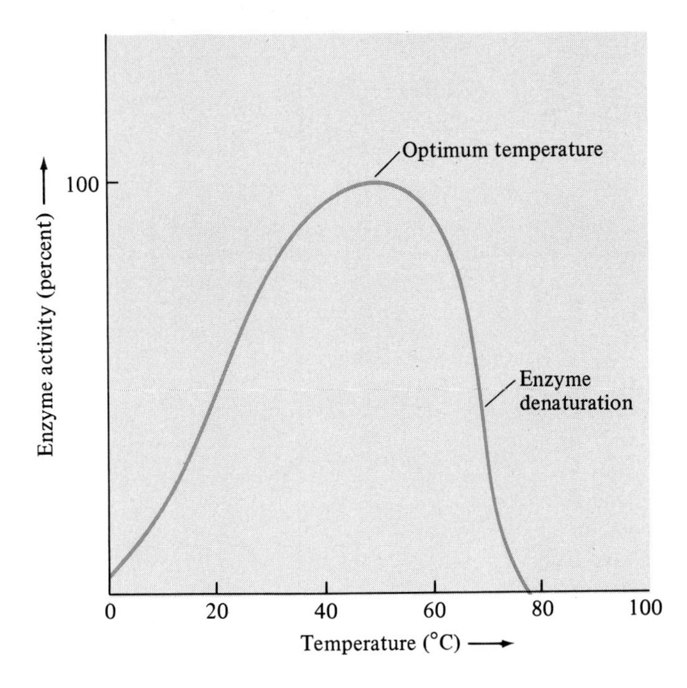

FIGURE 6–2
The effect of temperature on enzyme activity. All enzymes exhibit an optimum range of temperature (temperatures that allow the enzyme to attain 100% activity). Enzyme denaturation begins at temperatures exceeding 60° to 70° C.

one would expect that the enzymes within the cells of our bodies would have an optimum temperature of 37 °C, as, in fact, they do. At temperatures greater than 60 to 70 °C, most enzymes begin to denature and lose their activity. Rarely do enzymes survive temperatures greater than 100 °C.

Referring to Chapter 3, you will recall that extremes in temperature disrupt the specific three-dimensional shape of the enzyme.

THE EFFECT OF pH

optimum pH

Enzyme activity is also regulated by changes in the pH of the environment. Figure 6–3 depicts a typical *optimum pH curve* in which enzyme activity is significantly less on either side of the **optimum pH**. The curve usually approximates a bell shape. Each enzyme has its own optimum pH. Some enzymes exhibit a rather broad pH optimum, covering a wide range of pH; others are very narrow, covering only a few tenths of a pH unit. The proteolytic enzyme, pepsin, has a optimum pH of 1–2 and is, therefore, active in the acidic stomach. Most other enzymes function in a more neutral or alkaline environment (trypsin is active in the small intestine, pH = 7–8). It should be stressed that most cells and organ systems have relatively constant pH values.

One might ask, why does pH affect the activity of an enzyme? We can answer this question by referring to the chemical and physical properties of the amino acid R groups (Chapter 3). You may recall that pH significantly affects the protonation or deprotonation of the amino acid. Let us assume

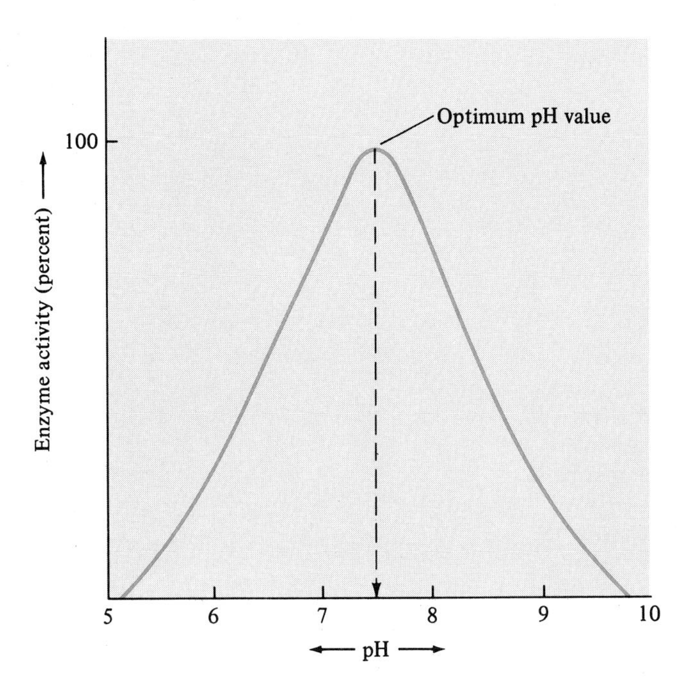

FIGURE 6–3
The effect of pH on enzyme activity. All enzymes have an optimum pH (a pH at which the enzyme exhibits maximum activity).

that aspartic acid is present in the active site, deprotonated, and that the negatively charged carboxyl group of aspartic acid is required to bind a positively charged substrate into the active site. Acidification of the enzyme would protonate the aspartic acid carboxyl group and, thus, prevent the substrate from binding to the aspartic acid residue in the active site. Since the substrate cannot bind to the active site, it cannot be converted into products and, thus, we would not observe appreciable enzyme activity. This situation is depicted in Figure 6–4. In addition, various substrates can possess acidic or basic functional groups that also are affected by the pH of the solution. Finally, the pH significantly influences the tertiary structure of the enzyme molecule and, thus, the activity of the enzyme.

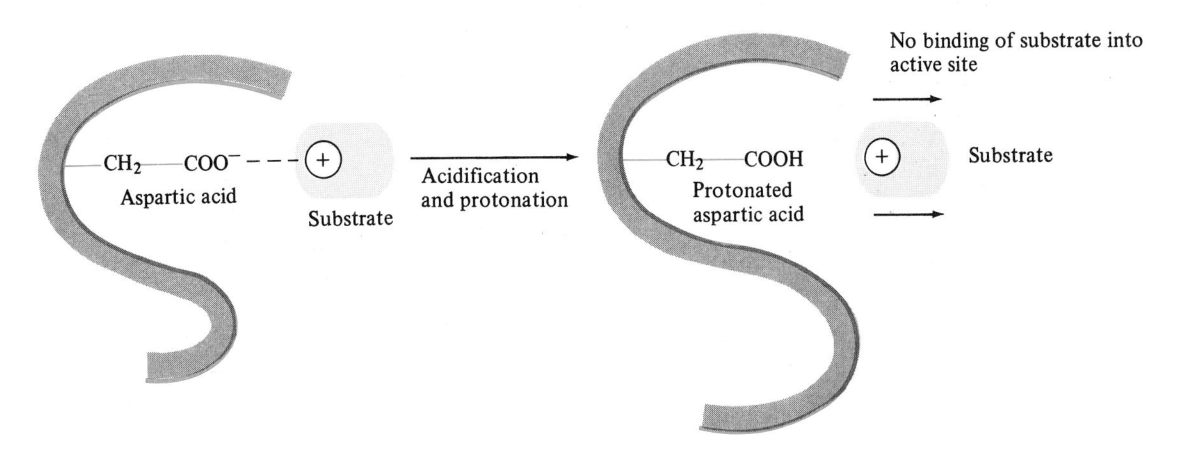

FIGURE 6–4 The effect of changing pH (acidification) on substrate binding.

THE EFFECT OF SUBSTRATE CONCENTRATION

Enzyme activity is also significantly affected by the *amount of substrate* present within the cell and available to the enzyme. This situation is depicted in Figure 6–5 (page 98). Figure 6–5 shows a **substrate saturation curve**. The curve is *hyperbolic*, and shows that enzyme activity (velocity) increases dramatically at low concentrations of substrate. At higher concentrations of substrate, the increase in velocity is not as great. A point is reached in which the velocity will no longer increase no matter how much substrate is present and available to the enzyme. The enzyme is then said to be saturated and cannot convert substrate into products any faster. The enzyme has reached its **"maximum velocity"** (**V_{max}**). This situation is diagrammed in Figure 6–6 (page 99).

substrate saturation curve

maximum velocity (V_{max})

To explain the substrate saturation curve depicted in Figure 6–5, let us digress a moment to 1913, when L. Michaelis and M. L. Menten proposed a theory of enzyme action. Enzyme E can bind to substrate S, forming an

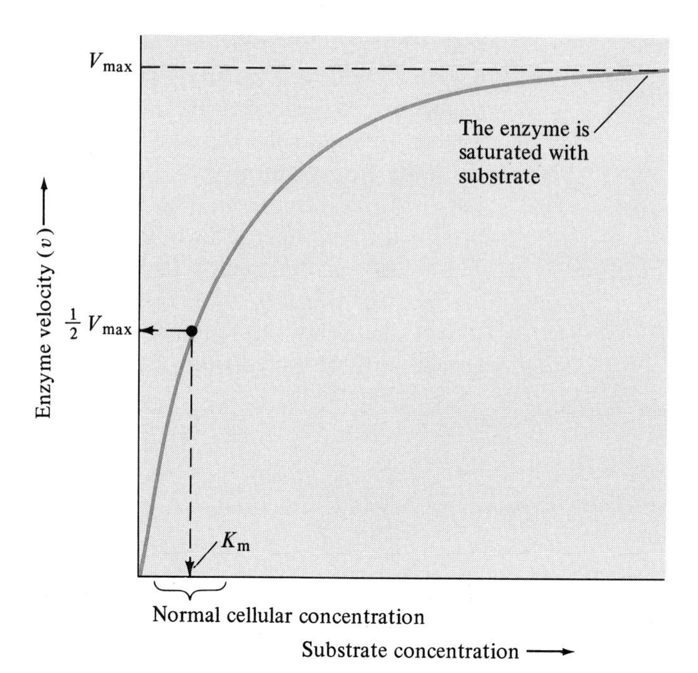

FIGURE 6–5
The effect of substrate concentration on enzyme activity (velocity). The enzyme reaches its maximum velocity (V_{max}) when it is saturated with substrate. The substrate concentration that produces a velocity of $\frac{1}{2}V_{max}$ is called the Michaelis-Menten constant, K_m, for the enzyme.

enzyme/substrate complex (ES complex)

enzyme/substrate complex (ES complex) according to Equation 6–1. This ES complex breaks down to yield free enzyme and product (P).

$$E + S \underset{k_{-1}}{\overset{k_1}{\rightleftharpoons}} ES \xrightarrow{k_2} E + P \qquad (6\text{–}1)$$

The k_1, k_{-1}, and k_2 are rate constants that define how fast the individual reactions proceed.

Michaelis and Menten assumed that the rate of formation of ES complex from free enzyme and products is very slow (the K-2 rate is small and therefore can be neglected). Importantly, they also assumed that the rate of formation of the ES complex (the k_1 rate) was equal to its rate of degradation the k_{-1} and k_2 rates—that is, that the formation of the ES complex has reached a *steady state*. In that case, the concentration of the ES complex will be constant. Using these assumptions, Michaelis and Menten derived Equation 6–2 (the *Michaelis-Menten equation*).

$$v = \frac{V_{max}[S]}{K_m + [S]} \qquad (6\text{–}2)$$

K_m = the Michaelis-Menten constant

where v = velocity of the enzyme reaction; V_{max} = the maximum velocity attainable at very high—*saturating*—concentrations of substrate; [S] = substrate concentration; and **K_m = the Michaelis-Menten constant** = $(k_2 + k_{-1})/k_1$. This equation defines a hyperbolic curve like the one depicted in Figure 6–5. When the substrate concentration is very high (that is, when the enzyme is saturated with substrate), the Michaelis-Menten equation

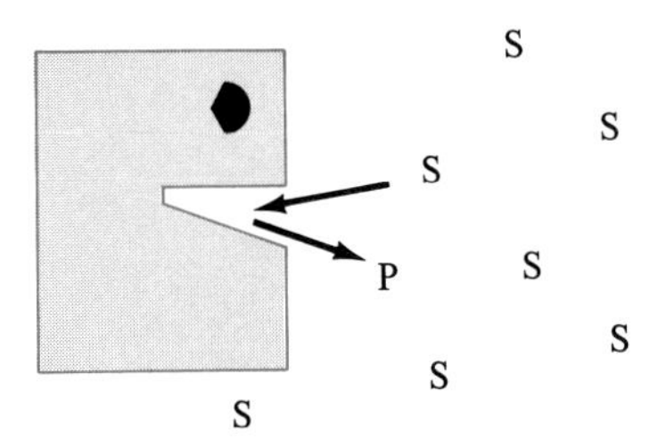

Case 1: Enzyme operating at *low* substrate concentrations

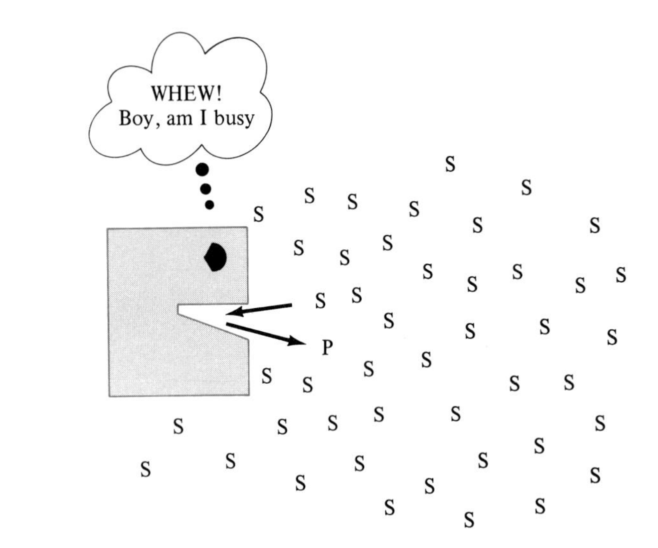

Case 2: Enzyme operating at *high* (saturating) substrate concentrations

FIGURE 6–6
Enzyme saturation.

reduces to $v \approx V_{max}$. Increasing the substrate concentration *cannot* increase the velocity (v) of the reaction. Finally, it should be noted that the K_m constant can also be defined as the *substrate concentration that produces a velocity* (v) *equal to* $\frac{1}{2}$ V_{max}. This can be explained using the Michaelis-Menten equation, and is shown graphically on Figure 6–5.

$$v = \frac{V_{max}[S]}{K_m + [S]}$$

Now, substitute $V_{max}/2$ for v into Equation 6–2.

$$\frac{V_{max}}{2} = \frac{V_{max}[S]}{K_m + [S]}$$

Divide both sides by V_{max};

$$\frac{1}{2} = \frac{[S]}{K_m + [S]}$$

Therefore $$\tfrac{1}{2}(K_m + [S]) = [S]$$

And $$K_m + [S] = 2[S]$$

Therefore $$K_m = [S]$$

A second definition for K_m also exists. In this definition, K_m indicates the *affinity* of the substrate for the enzyme active site. A small K_m value (that is, one for which a small substrate concentration will give $\frac{1}{2}V_{max}$)

indicates that the substrate has a *strong affinity* for the active site and, thus, is rapidly turned into products. A large K_m value (that is, one for which a large substration concentration will give $\frac{1}{2}V_{max}$) indicates that the substrate has relatively *little affinity* for the active site. Thus, fewer substrate molecules will be converted to products at a given substrate concentration.

The concentration of substrate within the cell is usually kept at, or very close to, the K_m of the enzyme. A rapid change in the velocity of the enzyme can, thus, be achieved with only small changes in the cellular concentration of the substrate (see Figure 6–5).

THE ORGANIZATION OF ENZYMES WITHIN THE CELL

Enzyme function is often affected by the *location* of the enzyme within the cell (and within various subcellular organelles). There are two main classes of enzymes; the *soluble enzymes* and the *membrane-bound enzymes*. This classification is based upon the location of an enzyme within a particular physical environment.

soluble enzymes

The **soluble enzymes** are physically dissolved in the aqueous media of the cell (the cytoplasm, and the interior regions of the various cellular organelles). Soluble enzymes can travel throughout the cell media by diffusion; substrates travel to the active site of such enzymes by diffusion as well. Until recently, soluble enzymes were thought to be randomly distributed throughout the cytoplasm and not organized into any meaningful association with other soluble enzymes. Now, scientists speculate that certain soluble enzymes might be loosely associated with each other. This loose association might allow the products of one enzymatic reaction to be immediately utilized as substrates for a second closely associated reaction. Certain soluble enzymes are tightly associated with each other as multienzyme complexes (for instance, the pyruvate dehydrogenase complex).

membrane-bound enzymes

Membrane-bound enzymes constitute about 60 percent of the total number of enzymes within the average cell. As the name implies, these enzymes are firmly bound to the lipid membranes of cells (the plasma membrane and the membranes of the endoplasmic reticulum, mitochondria, nucleus, and other organelles). This facilitates the formation of multienzyme sequences in which the products of one enzymatic reaction can diffuse a short distance to a second stationary enzyme for another reaction. In addition, the activity of membrane-bound enzymes can be regulated by the hydrophobic, lipid membrane. Finally, the fact that enzymes are bound to the membranes of various cellular organelles suggests that certain chemical reactions are localized and "compartmentalized" from each other. One reaction might occur in the cytoplasm, while a *competing* reaction might take place in the mitochondria, away from the cytoplasmic enzyme. This compartmentalization would not be feasible if there were no membranes to separate various regions within the cell. Membranes and membrane-bound enzymes will be discussed in greater detail in Chapter 13.

6.4 INHIBITION OF ENZYME ACTIVITY

A number of substances can bind to enzymes and upon binding, inhibit enzyme activity. These substances are called *inhibitors*. Certain enzyme inhibitors are normally present within the cell; their function is regulating enzyme activity. Others are substances foreign to a particular cell type or organism. These inhibitors are often man-made drugs and toxic agents that happen to bind to enzymes and inhibit their activity.

irreversible
reversible

Enzyme inhibition can be **irreversible** or **reversible**. An irreversible inhibitor forms a *covalent bond* with a specific functional group of the enzyme, thus inhibiting the enzyme. An irreversible inhibitor cannot be removed from the molecule. A reversible inhibitor, on the other hand, can reversibly bind to an enzyme (forms a noncovalent bond with the enzyme) and can, therefore, be removed from the enzyme.

IRREVERSIBLE INHIBITION

A number of man-made toxic agents act as irreversible enzyme inhibitors. For example, *organophosphorus* agents such as *diisopropyl-phosphofluoridate (DIPF)*, *malathion*, and *parathion* bind to a specific serine residue within the active site of the enzyme *acetylcholinesterase*, thus inhibiting the enzyme. This enzyme is responsible for the hydrolysis of acetylcholine into choline and acetate. Acetylcholine participates in nerve impulse transmission; inhibition of acetylcholinesterase, in turn, impairs nerve impulse firing. Acetylcholinesterase inhibitors cause paralysis of striated muscles and spasms of the pulmonary system. These agents are excellent insecticides. Potent nerve gases are also organophosphorus compounds and inhibit acetylcholinesterase. The structure of the insecticide *malathion* is as follows:

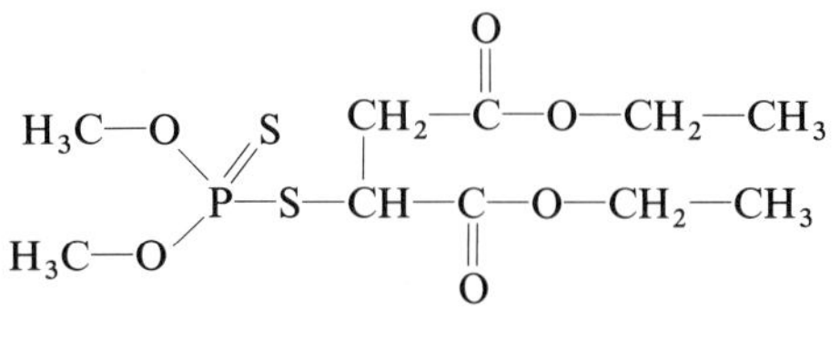

$$(H_3C{-}O)_2P({=}S){-}S{-}CH(CH_2{-}C({=}O){-}O{-}CH_2{-}CH_3){-}C({=}O){-}O{-}CH_2{-}CH_3$$

Malathion

Other examples of irreversible inhibition include the inhibition of the mitochondrial electron-transport system by *cyanide* (CN^-) and the poisoning of a number of enzymes by heavy metal ions. Heavy metals, such as mercury (Hg^{2+}) and lead (Pb^{2+}), bind to key sulfhydryl groups that are required for enzyme activity, thus inhibiting the enzyme. It is for this reason that most heavy metals are so toxic to cells; if ingested in even small doses they can cause serious illness and even death. For example, lead (Pb^{2+}) is found in many paint products. Children often eat paint or surfaces coated with chipped paint and, thus, ingest the dangerous lead. The lead inhibits an enzyme responsible for *heme* biosynthesis. If this enzyme is inhibited for long enough, heme (and, therefore, hemoglobin) synthesis will decrease, causing a serious form of *anemia*.

REVERSIBLE INHIBITION

Substrate Inhibition

Certain enzymes exhibit *substrate inhibition*, as depicted in Figure 6–7. In this situation, very high concentrations of substrate inhibit the enzyme, thus slowing the enzyme velocity considerably.

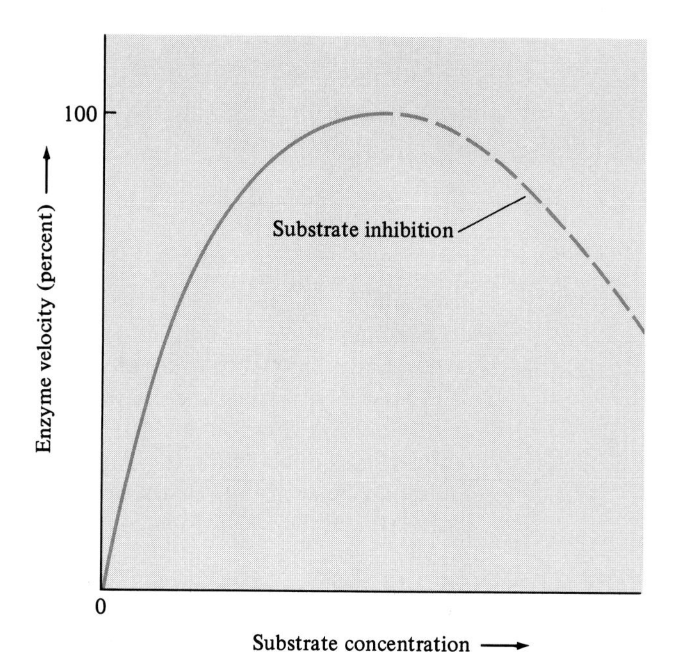

FIGURE 6–7
An enzyme exhibiting substrate inhibition.

Competitive Inhibition

Certain substances inhibit enzyme activity in a competitive manner. These substances are structurally very similar to the substrate molecule and, hence, can bind reversibly to the active site of the enzyme. We say that the inhibitor competes with the actual substrate for the same active site. In essence, a **competitive inhibitor** "looks like" the actual substrate molecule and, thus, can easily fit into the active site. However, although the competitive inhibitor is very similar to the actual substrate, the inhibitor *cannot* be converted into products. The effects of a competitive inhibitor can be reversed by increasing the concentration of the substrate, thus making more substrate molecules available to bind with the enzyme. Competitive inhibition is diagrammed in Figure 6–8.

competitive inhibitor

The competitive inhibition of the enzyme *succinate dehydrogenase* is a classic example. Succinate dehydrogenase catalyzes the oxidation of the

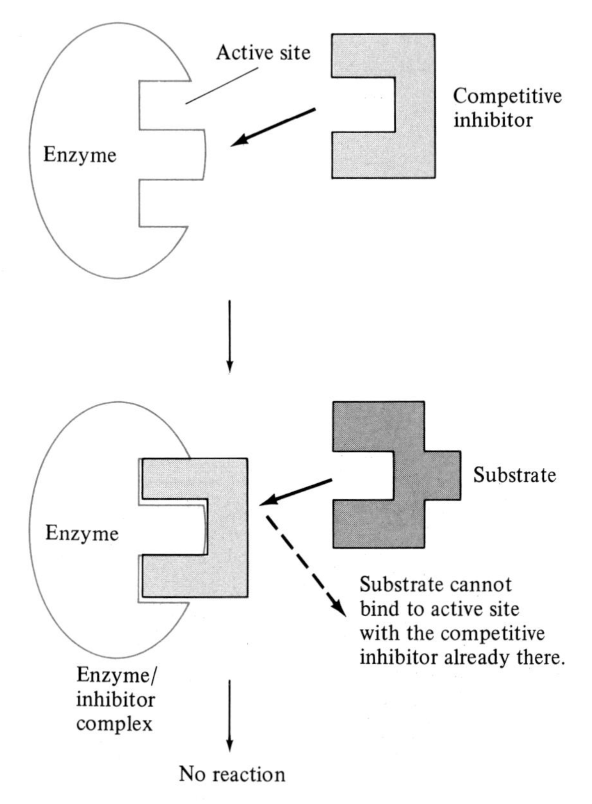

FIGURE 6–8
Competitive inhibition. A competitive inhibitor competes with the substrate for the same active site.

substrate succinate to form fumarate by removing two hydrogens with two electrons.

$$\begin{matrix} COO^{\ominus} \\ | \\ CH_2 \\ | \\ CH_2 \\ | \\ COO^{\ominus} \end{matrix} \xrightarrow[\searrow 2H]{\text{Succinate dehydrogenase}} \begin{matrix} COO^{\ominus} \\ | \\ CH \\ \| \\ HC \\ | \\ COO^{\ominus} \end{matrix}$$

Succinate Fumarate

Malonate is a competitive inhibitor of succinate dehydrogenase. The structure of malonate is shown and compared to the structure of succinate. Notice how similar the two molecules are.

$$\begin{matrix} COO^{\ominus} \\ | \\ CH_2 \\ | \\ CH_2 \\ | \\ COO^{\ominus} \end{matrix} \qquad \begin{matrix} COO^{\ominus} \\ | \\ CH_2 \\ | \\ COO^{\ominus} \end{matrix}$$

Succinate Malonate

Noncompetitive Inhibition

noncompetitive inhibitors

Another class of inhibitors is the **noncompetitive inhibitors**. These substances bind reversibly to a site on the enzyme surface that is *separate* from the active site. This changes the structure of the enzyme so that it can no longer convert substrates into products (see Figure 6–9). Some noncompetitive inhibitors bind to the active site but cannot be displaced by increasing the concentration of substrate (as was the case for competitive inhibition). Some heavy metal ions act as noncompetitive inhibitors, since they reversibly bind to key sulfhydryl groups that are required for enzyme activity.

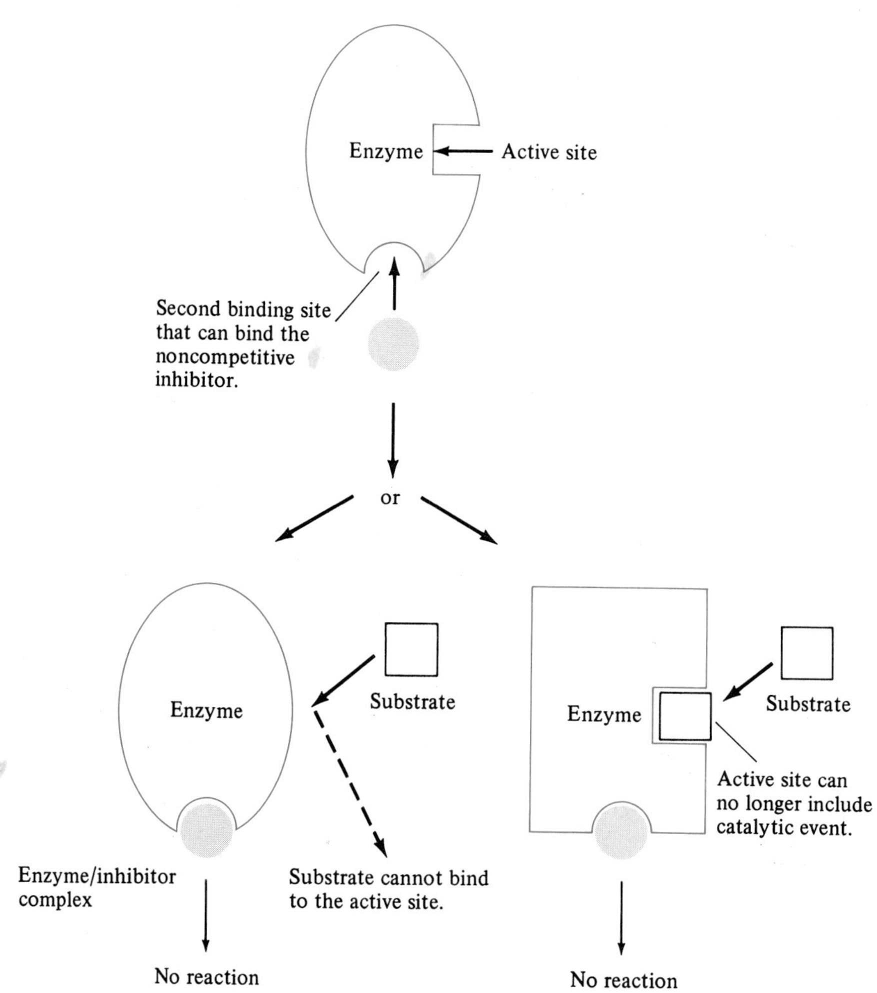

FIGURE 6–9 Noncompetitive inhibition. A noncompetitive inhibitor binds to a second binding site and either prevents the substrate from binding to the active site or renders the enzyme/substrate complex incapable of breaking down into products.

6.5 MULTIPLE-MOLECULAR ENZYME FORMS (ISOENZYMES)

oligomers

Some enzymes are **oligomers** (that is, proteins constructed from more than one subunit). Some oligomeric enzymes are constructed from multiples of the same type of subunit; others are constructed from multiples of different subunits. The enzyme *lactate dehydrogenase* (LDH), a tetramer, is constructed from two different subunits—a *heart (H) subunit* and a *muscle (M) subunit.* LDH isolated from heart tissue is composed primarily of the heart subunits; LDH isolated from skeletal muscle is constructed from muscle subunits. Intermediate forms of the enzyme can exist and are also found in other organs of the body. There are a total of five distinct forms of lactate dehydrogenase. Such forms are called **isoenzymes**. Each isoenzyme is a tetramer and catalyzes the same reaction, but is constructed from a different ratio of the two types of subunits, as shown in Figure 6–10. Interestingly, the five distinct forms of LDH have different levels of catalytic activity and, thus, control the rate of a particular metabolic pathway in the different organs of the body. This unique situation will be explained in greater detail in Chapter 9.

isoenzymes

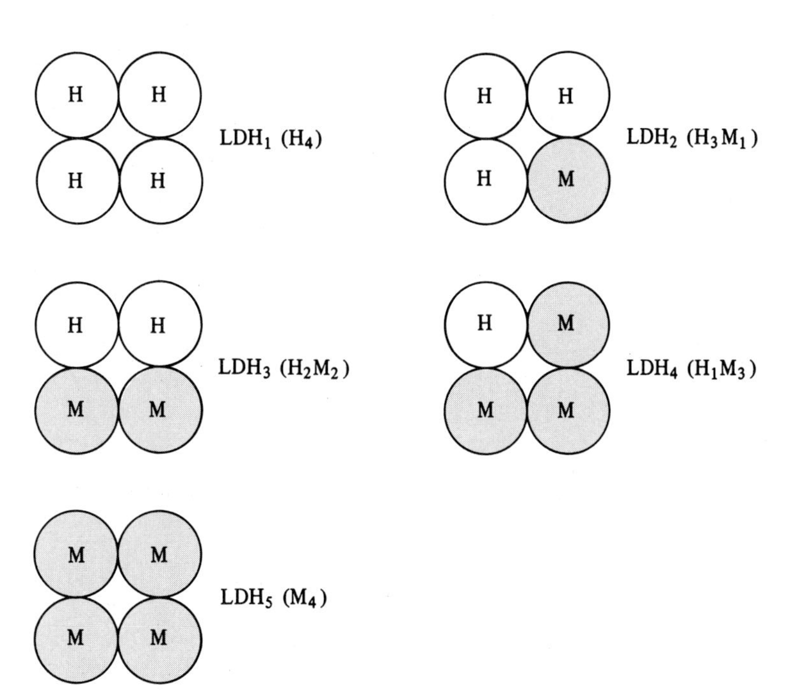

FIGURE 6–10
The multiple-molecular forms of lactate dehydrogenase (isoenzymes).

6.6 COVALENT MODIFICATION OF ENZYMES

Some enzymes can be chemically or enzymatically modified within the cell. In this way, their activity can be controlled. For example, the enzyme *glycogen phosphorylase* (see Chapter 10) exists in two forms, an *active* form and an *inactive* form. The inactive form can be converted into the active form via a specific modification of key serine residues, as shown in Figure 6–11. In this example, the serine residues are *phosphorylated* using the phosphate donor *adenosine triphosphate* (*ATP*). The phosphorylated dimer then aggregates into the active tetramer. Many other examples of **covalent modification** are known to occur in key enzyme systems within the cell.

covalent modification

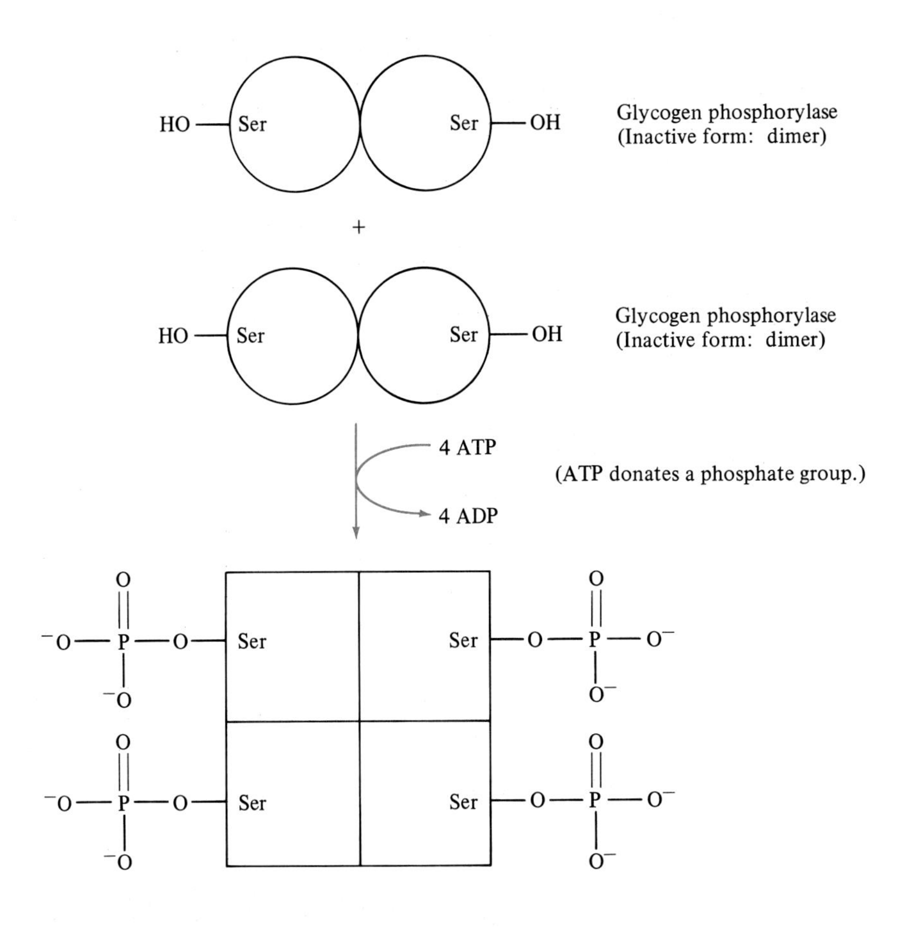

FIGURE 6–11 Covalent modification of glycogen phosphorylase. Phosphorylation of four serine residues (by ATP) causes dimer aggregation and enzyme activation.

ZYMOGENS AND ZYMOGEN ACTIVATION

Often, certain cells within an organ must store enzymes until they are needed. Some enzymes in their active form would digest the internal structures of the cells and, ultimately, the cells themselves. Such enzymes must be stored

zymogens

in an inactive form (called **zymogens**). When zymogens are needed to catalyze specific chemical reactions, they are released from the cells that originally produced and stored them, then activated.

Zymogens are usually activated when the inactive polypeptide chain splits into two or more fragments, one or more of which refolds into an active conformation. This process is summarized in Figure 6–12.

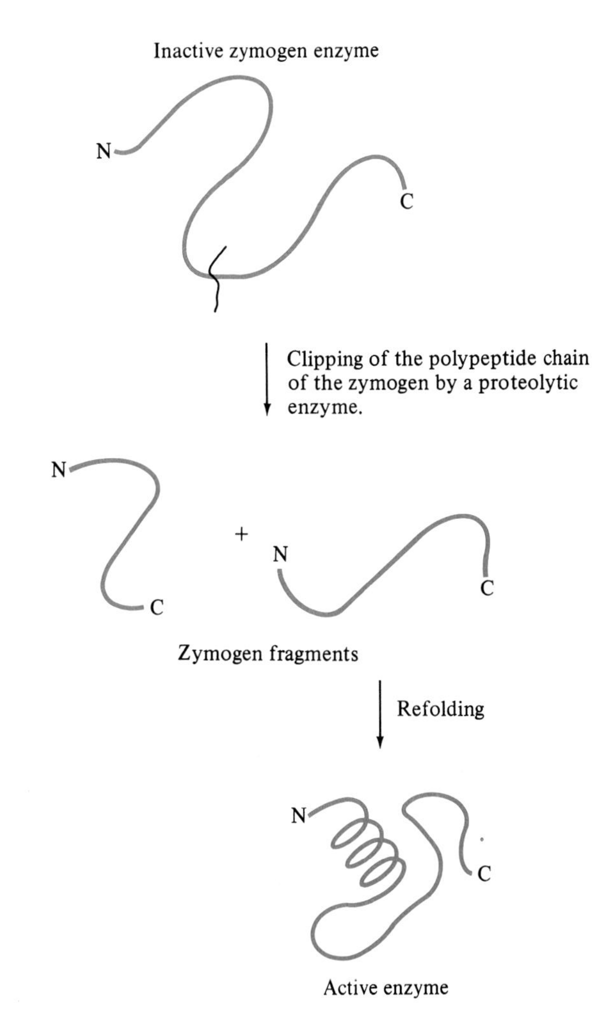

FIGURE 6–12
Zymogen activation.

proteolytic enzymes

To appreciate zymogens and zymogen activation in greater detail, let us consider the activation of the **proteolytic enzymes** (enzymes that digest proteins). A complex activation sequence occurs in our bodies every time we consume a meal that contains proteins to be digested. This sequence is summarized in Figure 6–13.

1. Food enters the stomach.

2. Presence of food stimulates the release of gastric juice, causing increased acidity (H^+).

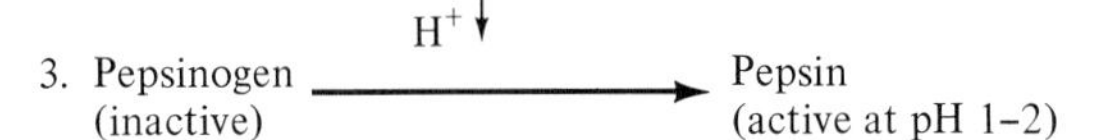

4. Active pepsin begins to digest some proteins in stomach.

5. Acidic contents of stomach enter the small intestine and are neutralized to pH 7–8.

6. Stomach contents entering the small intestine stimulate the release of pancreatic juice, which contains zymogen forms of the proteolytic enzymes, along with intestinal enterokinase (an enzyme).

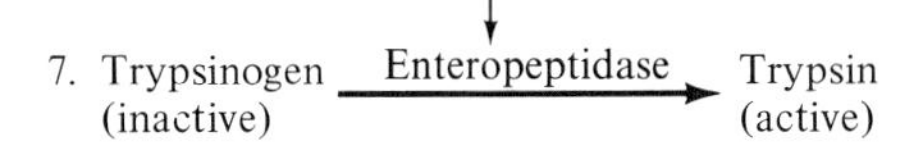

8. Once trypsin has been activated, it activates other zymogens:

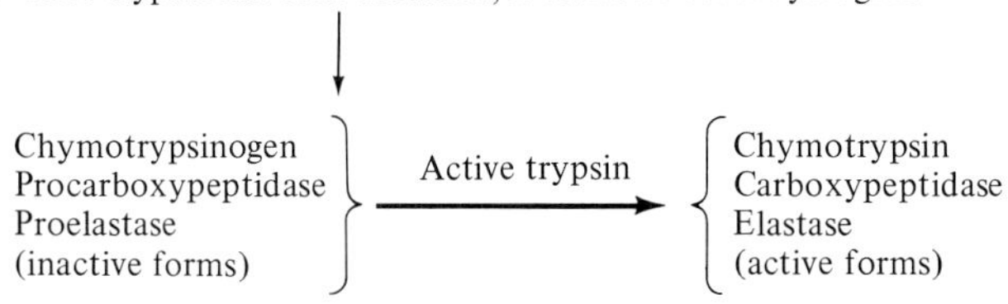

FIGURE 6–13
Proteolytic enzyme activation sequence. Most proteolytic enzymes (enzymes that digest proteins) exist as inactive zymogens. They are activated when the polypeptide chains are clipped at specific points to yield active enzyme fragments.

hemorrhagic pancreatitis

Hemorrhagic pancreatitis is a serious disease state associated with the abnormal activation of proteolytic enzymes. The pancreas is a very important organ; it supplies a number of digestive enzymes, as well as such hormones as insulin and glucagon. The digestive (proteolytic) enzymes are stored as zymogens in the pancreatic cells. When they are needed, they are secreted from the pancreas into the small intestine and activated. Hemorrhagic pancreatitis occurs when the digestive enzymes are prematurely activated while still in the pancreatic cells. Pancreatic cells are destroyed; in serious cases, the pancreas itself is destroyed as well. In essence, the pancreas undergoes autodigestion, accompanied by the leakage of blood and fluids from the pancreas cells. Severe abdominal pain results, with nausea and vomiting. The disease can be clinically evaluated by monitoring the increase in blood serum amylase and lipase enzyme activity. The causes of hemorrhagic pancreatitis include alcoholism, gallbladder disease, infections, and trauma to the pancreas. Alcoholism is probably the major factor. Exactly how pancreatitis develops is still largely unknown.

6.7 REGULATION OF ENZYME SYNTHESIS

Cells do not need all of their enzymes working all of the time. The cell will "turn off" or "turn on" the production of a particular enzyme depending on the need for that enzyme at a given time. Breast tissue *lactose synthetase* will serve as an example.

Lactose synthetase makes *lactose* (milk sugar) according to the following reaction:

$$UDP\text{-Galactose} + \text{Glucose} \xrightarrow{\text{Lactose synthetase}} \underset{\text{(milk sugar)}}{\text{Lactose}} + \text{UDP}$$

Note: UDP-galactose is the uridine diphosphate derivative of galactose. The enzyme is composed of two different subunits (and, thus, is a dimer). When separated from each other, the subunits are inactive; but when they combine, they produce an active lactose synthetase molecule. The production of the two subunits during pregnancy is shown in Figure 6–14. Notice that after fertilization, the breast tissue begins the immediate production of subunit A (this subunit has another function at this time). Subunit B production is inhibited during the nine month gestation period. Obviously, milk sugar is not needed until the baby is born and begins suckling. After the baby is born, the breast begins the rapid production of subunit B. An *active* lactose synthetase is made at this time, and is responsible for the production of lactose needed by the baby. One might ask, how is this system regulated?

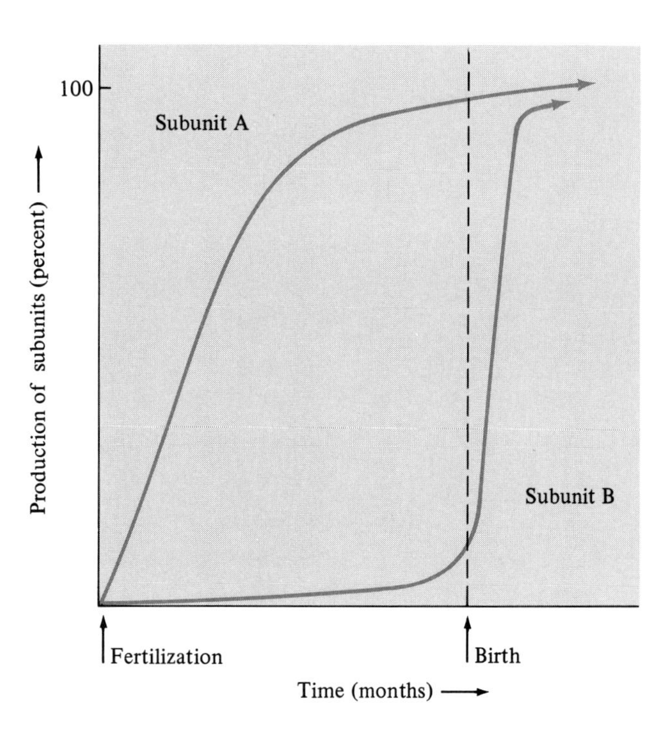

FIGURE 6–14
The production of lactose synthetase subunits by the breast during pregnancy. Subunit B production is inhibited by the placental hormone *progesterone*.

How is subunit B production inhibited during fetal development and then suddenly accelerated after the baby is born? Research indicates that subunit B production is inhibited by the steroid hormone progesterone, which is produced by the placenta. At birth, the placenta is expelled from the body; thus, the source of the inhibitor is removed and subunit B production increases. By reviewing this example, one can see that the body can control various physiological processes by controlling the production of enzymes and their activity.

6.8 ALLOSTERIC ENZYMES AND THEIR REGULATION

allosteric enzymes

allosteric effector

Allosteric ("other space") **enzymes** are oligomers that have not only an active site that binds the substrate, but one or more additional sites for **allosteric effector** molecules. Allosteric effectors (regulators) are substances of small molecular weight that bind to the allosteric sites and control the activity of the enzyme. There are two main classes of allosteric effectors: *positive effectors* that activate the enzyme and *negative effectors* that inhibit it. A simple allosteric enzyme (in this case, a dimer) and its response to a negative allosteric effector is shown in Figure 6–15. In the example cited, the dimeric enzyme has two subunits, a catalytic (C) subunit that contains the active site and, a regulatory (R) subunit that contains the allosteric-effector binding

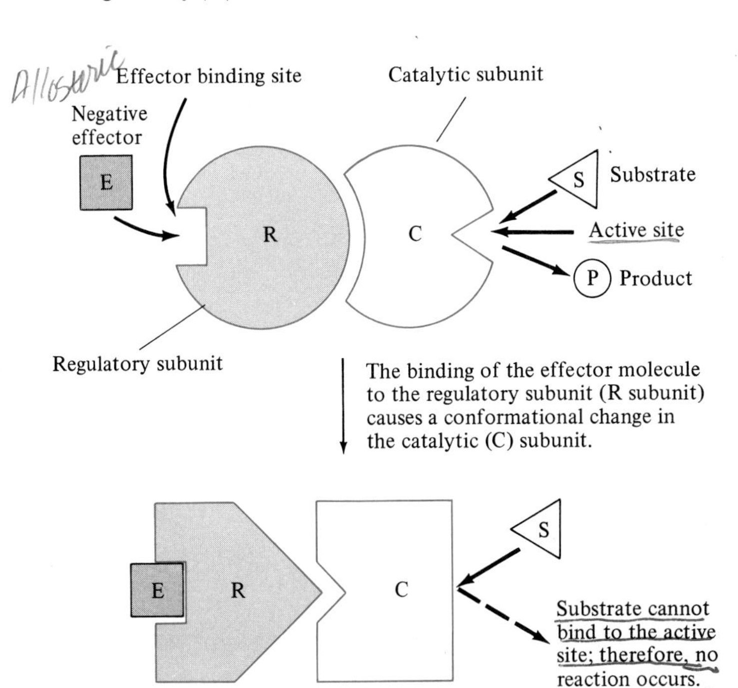

FIGURE 6–15
How an allosteric enzyme works. The binding of a negative effector to an allosteric enzyme shuts down the enzyme.

site. Suppose the negative effector is present and binds to the effector binding site. Upon binding, the negative effector causes a *conformational change* in the entire enzyme molecule. This change in conformation changes the affinity of the active site for the substrate molecule, thus preventing the binding of the substrate. Enzyme activity would *decrease* (no substrate binding, no activity) and the enzyme would be "turned off." The reverse argument holds for a positive effector. In this case, the binding of the positive effector to the binding site would *increase* the affinity of the active site for the substrate and, thus, *stimulate* (turn on) the enzyme.

Why do allosteric enzymes exist in the cell and why are they important? Simply stated, allosteric enzymes control the rate of conversion of substrates into pathway products in a multienzyme sequence. It is important to realize that the metabolic pathways and reactions that occur in the cells of your body are under constant moment-by-moment control. If this was not the case, the reactions would "run wild." A product might be made in excess (and thus wasted), or a precursor substance might be depleted by an enzymatic process that wasted it.

FEEDBACK INHIBITION

feedback inhibition

Feedback inhibition is depicted in Figure 6–16. In feedback inhibition, the end product of a complex metabolic pathway acts as the negative allosteric effector on the first enzyme of the pathway, thus controlling the pathway. As the end product builds up in concentration, it diffuses to the allosteric enzyme, shutting down the enzyme and stopping production of the end product. One can think of pathway regulation as similar to the regulation of city traffic by stoplights, which allow the even and systematic flow of cars through intersections.

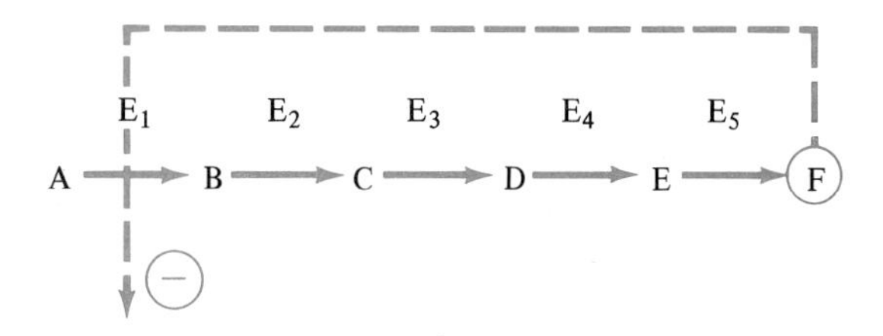

FIGURE 6–16
Feedback inhibition.

Allosteric enzymes are important for a second reason. They provide a means of *controlling enzymatic rates without significantly changing cellular substrate concentrations*. Consider the hyperbolic shape of a substrate saturation curve for a *nonallosteric enzyme* (Figure 6–17) (page 112). To change the enzyme velocity from 10 percent to 90 percent of V_{max}, an 18-fold increase in substrate concentration would be required. This means that the cell must somehow supply 18 times the normal amount of substrate present in the cell.

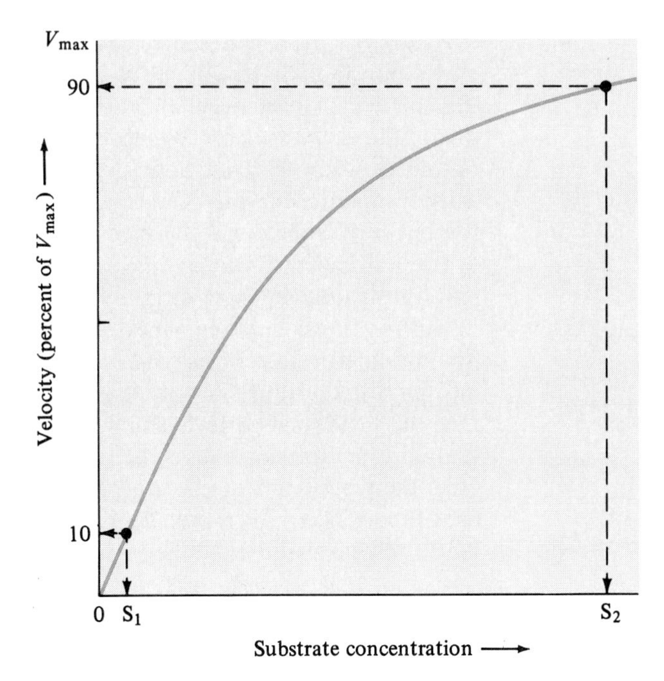

FIGURE 6–17
Substrate saturation curve for a nonallosteric enzyme. The cell would have to change the concentration of the substrate approximately 18-fold to change the enzyme velocity from 10% to 90% of V_{max}.

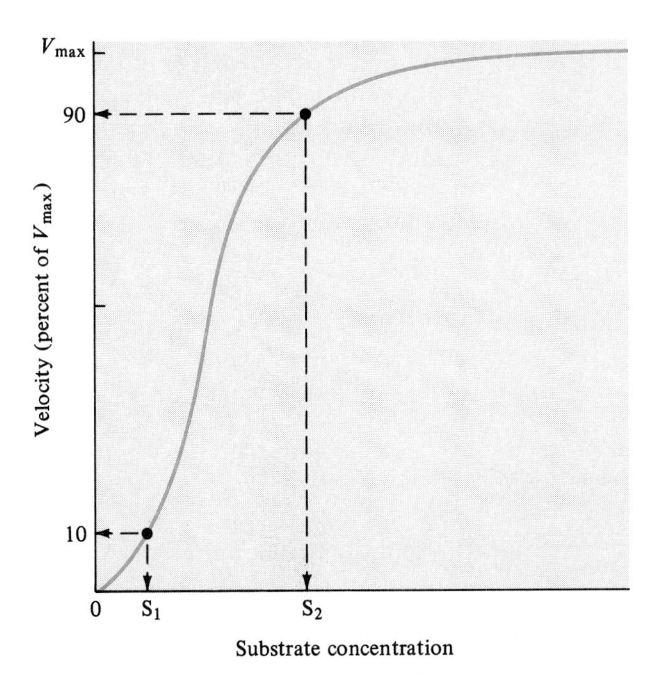

FIGURE 6–18
Substrate saturation curve for an allosteric enzyme (sigmoidal). This cell would have to change the concentration of the substrate by only a factor of 3 to change the enzyme velocity from 10% to 90% of V_{max}.

An allosteric enzyme, on the other hand, exhibits a *sigmoid substrate-saturation curve* (Figure 6–18). In this case, only a 3-fold increase in substrate concentration would be required to increase the velocity from 10 percent to 90 percent of V_{max}. Thus, allosteric enzymes prevent the cell from using up large amounts of substrate unnecessarily.

Positive and negative effectors significantly affect the shape of the sigmoid saturation curve exhibited by allosteric enzymes. This is demonstrated in Figure 6–19. When a positive effector is present, the sigmoid curve is much steeper than the corresponding curve without the effector. This means that at a normal and fixed cellular substrate concentration, the enzyme will exhibit a velocity very close to V_{max}. The enzyme is *activated* by the positive effector. When a negative effector is present, the sigmoid curve is significantly less steep than the corresponding curve without the effector. This means that at the normal cellular substrate concentration, the enzyme will exhibit a very low velocity (possibly only 10–20 percent of V_{max}). The enzyme is *inhibited* by the negative effector. Thus, the velocity of an allosteric enzyme changes in response to allosteric effectors (positive or negative) while the concentration of substrate(s) is held at a relatively constant level.

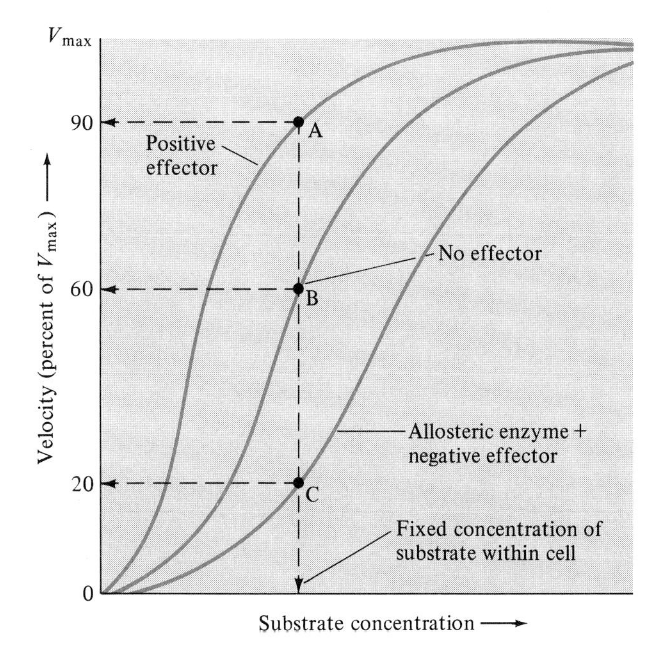

FIGURE 6–19
Substrate saturation curve for an allosteric enzyme in the presence of positive and negative effectors. Curve A: Positive effector. Velocity is 90% of V_{max} (the enzyme is on). Curve B: No effector. Curve C: Negative effector. Velocity is only 20% of V_{max} (the enzyme is off).

SUMMARY

Enzyme activity can be measured by determining the rate of product appearance. The International Unit (I.U.) is defined as the amount of enzyme required to convert one μmole of substrate into product per minute at 25 °C. The specific activity (S.A.) specifies the purity of an enzyme preparation and is defined as the number of I.U.'s per milligram of protein. Turnover number is defined as the amount of enzyme activity per mole of enzyme.

Two classes of enzymes exist in blood plasma,

plasma specific enzymes and non–plasma specific enzymes. Non–plasma specific enzymes are enzymes that have been secreted from tissue cells or lost by leakage. Increased amounts of non–plasma specific enzymes are found in various disease states.

Enzymes exhibit an optimum temperature range (usually 25 to 40 °C) as well as an optimum range of pH. The amount of substrate present and available to the enzyme significantly influences the activity of the enzyme. At V_{max}, the enzyme is saturated with substrate and cannot convert substrate into products any faster. The substrate concentration that gives a velocity of $\frac{1}{2}$ V_{max} is called the Michaelis-Menten constant (K_m).

Enzymes are organized within the cell. There are two main classes of enzymes, soluble enzymes and membrane-bound enzymes. The localization of enzymes within the cell and within various cellular organelles segregates competing biochemical reactions.

Enzyme activity can be inhibited by irreversible and reversible inhibitors. Irreversible inhibitors (malathion, CN^-, and heavy metals) form covalent bonds with a specific functional group of the enzyme. Reversible inhibitors bind reversibly (noncovalently) with the enzyme. Certain enzymes are inhibited by high concentrations of their own substrates (substrate inhibition); others are inhibited by substances that "look like" the actual substrate and bind into the active site, but are not converted into products (competitive inhibitors). Noncompetitive inhibitors bind to a nonactive site on the enzyme surface, inducing structural changes that inactivate the enzyme.

Enzymes can exist as isoenzymes (oliogomeric proteins constructed from different subunits). There are five lactate dehydrogenase (LDH) isoenzymes (H_4, H_3M, H_2M_2, HM_3, and M_4). These isoenzymes catalyze the same chemical reaction, but have different levels of catalytic activity (that is, different V_{max} and K_m values) and, thus, control a metabolic pathway.

Enzyme activity can be changed by covalent modification of the enzyme. For instance, the phosphorylation of specific serine residues causes glycogen phosphorylase dimers to aggregate into an active tetramer.

Zymogens are inactive forms of enzymes that are stored in cells until needed. Zymogens are usually activated by clipping the inactive polypeptide chain into smaller fragments. These fragments then refold into active enzymes. The proteolytic enzymes (enzymes that digest proteins) are stored in the pancreas in zymogen form until they are activated.

The amount of enzyme that is produced within the cell is regulated. Cells do not need all of their enzymes working all of the time. The cell will turn off and turn on the production of a specific enzyme, depending on the needs of the cell at a particular time. The lactose synthetase system in human breast tissue is an example of this type of regulation.

Allosteric enzymes are oligomers with (1) an active site that binds the substrate and (2) one or more additional sites that bind allosteric effector molecules. The allosteric effectors are small-molecular-weight substances that can turn off an allosteric enzyme (negative effectors) or turn one on (positive effectors). Allosteric enzymes control metabolic pathways by feedback inhibition.

REVIEW QUESTIONS

1. A number of concepts and terms are presented in this chapter. Please define (or briefly explain) each of the following terms:
   - **a.** International Unit
   - **b.** Turnover number
   - **c.** Optimum pH
   - **d.** Substrate saturation curve
   - **e.** V_{max}
   - **f.** K_m constant
   - **g.** Enzyme/substrate (ES) complex
   - **h.** Irreversible inhibition
   - **i.** Competitive inhibition
   - **j.** Noncompetitive inhibition
   - **k.** Isoenzyme
   - **l.** Covalent modification
   - **m.** Zymogen
   - **n.** Allosteric enzyme
   - **o.** Negative allosteric effector
   - **p.** Feedback inhibition
   - **s.** Sigmoid substrate-saturation curve

*2. An enzyme preparation produced 1×10^{-2} moles of product in 15 minutes at 25°C.
   a. How many International Units of enzyme activity were present in this preparation?
   b. Assuming that the preparation contained 3 milligrams of protein, calculate the specific activity of the preparation.

*3. How many International Units of enzyme activity are represented by one (1) Bodansky Unit in Table 6–1? Assume that the molecular weight of phosphate is 97 gm/mol.

4. Draw enzyme activity versus temperature curves for the following:
   a. Human alkaline phosphatase at 37°C.
   b. Enzymes of an Arctic fish at 2°C.
   c. Thermophilic (heat-loving) bacteria at 90°C.

5. Draw optimum pH curves for:
   a. Pepsin
   b. Trypsin
   c. Acid and Alkaline phosphatases

6. The following data was obtained for an enzyme-catalyzed reaction:

Substrate (mM)	Velocity (μmol/min)
2.3	0.95
3.0	1.13
4.5	1.42
12.7	2.14
38.5	2.64
100.0	2.94

   a. Graph the data (enzyme velocity versus substrate concentration).
   b. Obtain V_{max} and K_m values for your graph.
   *c. *Bonus Question*: Calculate the turnover number of the enzyme if the molecular weight of the enzyme is 30 000 g/mol and if 1.5 micrograms of enzyme were used in the assay ($1\ \mu g = 1 \times 10^{-6}$ g).

7. How many isoenzyme forms can exist if an enzyme is:
   a. A dimer b. A trimer
   Assume that the dimer and trimer can be constructed from combinations of A and B subunits.

SUGGESTED READING

Bernhard, S. A. *The Structure and Function of Enzymes.* New York: W. A. Benjamin, 1968.

Bhagavan, N. V. *Biochemistry*, 2nd ed., Ch. 3. Philadelphia: J. B. Lippincott 1978.

Koshland, D. E. "Protein Shape and Biological Control." *Scientific American* 229 (1973): 52.

Lehninger, A. L. *Biochemistry*, 2nd ed., Chs. 8 and 9. New York: 1975.

Phillips, D. C. "The Three-Dimensional Structure of an Enzyme Molecule." *Scientific American* 215 (1966): 78.

Tietz, N., ed. *Fundamentals of Clinical Chemistry*, 2nd ed., Ch. 12. Philadelphia: W. B. Saunders, 1976.

NOTE

1. Kachmar, J. F., and Moss, D. W., "Enzymes," in N. Tietz (ed.), *Fundamentals of Clinical Chemistry*, 2nd ed. (Philadelphia: W. B. Saunders, 1976), p. 591.

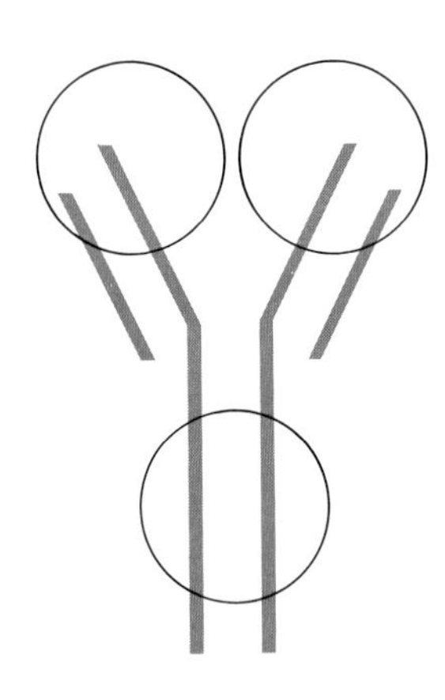

Chapter 7

Carbohydrates

7.1 INTRODUCTION

carbohydrates

Carbohydrates (sugars) are found throughout nature. Some carbohydrates, such as cellulose, serve a *structural role* in plants (including trees); others, such as starch and glucose, are used by both plants and animals as a type of cellular fuel. In fact, over 50 percent of all the carbon in the total biomass takes the form of carbohydrates.

Most carbohydrates are *polyhydroxy aldehydes or ketones*. This means that a carbohydrate can be either an *aldehyde* ($R{-}\overset{O}{\overset{\|}{C}}H$) or a *ketone* ($R{-}\overset{O}{\overset{\|}{C}}{-}R$), and that many hydroxyl groups can be associated with the carbohydrate molecule. Carbohydrates have the general formula $(CH_2O)_n$ and, thus, were originally called hydrates of carbon. Because carbohydrates have so many hydroxyl groups, they are extremely polar molecules and are usually very soluble in water.

monosaccharides
oligosaccharides
polysaccharides

Carbohydrates are divided into different classes based upon their degree of polymerization, as follows: (1) **monosaccharides**, (2) **oligosaccharides**, and (3) **polysaccharides**. The *mono*saccharides contain only one carbohydrate molecule. (mono means "one"). The oligosaccharides contain between two and ten monosaccharide units linked together in small chains (oligo means "a few"). The polysaccharides are constructed from many monosaccharide units linked together in long chains (poly means "many").

In this chapter, we will discuss the chemistry and properties of each of these classes of carbohydrates.

7.2 MONOSACCHARIDES

CLASSIFICATION

Monosaccharides, the simplest carbohydrates, are classified and named according to the number of carbon atoms in the molecule and according to the presence of specific functional groups. Table 7–1 lists a few representative monosaccharides. The smallest contains three carbon atoms and is, therefore, called a *tri*ose.

TABLE 7–1 Monosaccharides Classified According to the Size of the Carbon Chain

Number of Carbon Atoms	Name of Sugar
3	Triose
4	Tetrose
5	Pentose
6	Hexose

aldose

ketose

Monosaccharides can be either an aldehyde or a ketone. If a monosaccharide contains an aldehyde group, it is called an **aldose**. (The suffix "*-ose*" signifies a sugar.) A monosaccharide that contains a carbonyl oxygen bonded to an internal carbon atom (a ketone) is called a **ketose**. A three-carbon aldose would be called an *aldotriose*; a three-carbon ketose would be called a *ketotriose*. Glycer*aldehyde* and dihydroxy*acetone* represent aldoses and ketoses respectively.

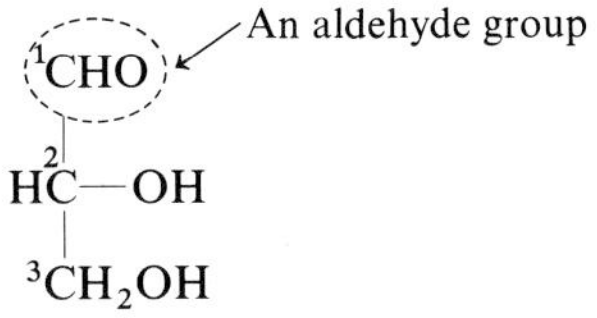

^{1}CHO ← An aldehyde group
|
H^{2}C—OH
|
$^{3}CH_2OH$

Glyceraldehyde
(an aldose)

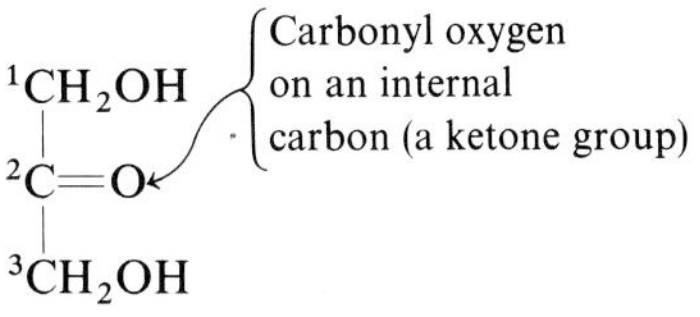

$^{1}CH_2OH$
|
^{2}C=O ← Carbonyl oxygen on an internal carbon (a ketone group)
|
$^{3}CH_2OH$

Dihydroxyacetone
(a ketose)

STEREOCHEMISTRY

asymmetric

All monosaccharides (except dihydroxyacetone) have at least one **asymmetric** carbon atom. One should remember that an asymmetric carbon atom has *four different groups* bound to it. Thus, D- and L-stereoisomers (mirror images) are possible. For example, glyceraldehyde can exist as either D-glyceraldehyde or L-glyceraldehyde, as shown in Figure 7–1 (page 118).

The structures shown in Figure 7–1 are called *Fischer projections*. A Fischer projection is a planar projection derived from the three-dimensional

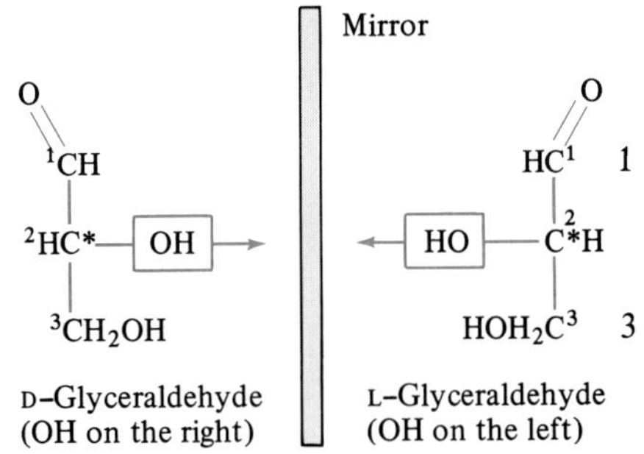

Mirror

$$\begin{array}{c} O \\ \| \\ {}^{1}CH \\ | \\ {}^{2}HC^{*}-\boxed{OH} \rightarrow \\ | \\ {}^{3}CH_2OH \end{array} \qquad \begin{array}{c} O \\ \| \\ HC^{1} \quad 1 \\ | \\ \leftarrow \boxed{HO}-\overset{2}{C}{}^{*}H \\ | \\ HOH_2C^{3} \quad 3 \end{array}$$

D-Glyceraldehyde (OH on the right) L-Glyceraldehyde (OH on the left)

FIGURE 7–1
Fischer projections of D- and L-glyceraldehyde. The asymmetric carbon is identified as C*.

structure of the molecule. In Figure 7–1, the main carbon chain is written in a vertical manner on the page and the functional group is located at the top of the structure. The carbon atoms are numbered from the top to the bottom. Carbon #2 is the asymmetric carbon atom and is noted with an asterisk (*). Carbon atoms #1 and #3 are not asymmetric. (Why?) When glyceraldehyde is written with the OH group on carbon #2 pointing to the right, the structure is called D-glyceraldehyde. The mirror image, with the OH group pointing to the *left*, is called L-glyceraldehyde. Since glyceraldehyde (as well as other monosaccharides) possesses an asymmetric carbon atom, it is *optically active* and can rotate the plane of polarized light in either a clockwise (+) or counterclockwise (−) direction.

Larger monosaccharides can have more than one asymmetric carbon atom; thus, a large number of different stereoisomers can exist. For example, a six-carbon aldose has four asymmetric carbon atoms. The total number of possible stereoisomers (each with six carbon atoms) is therefore $(2)^4 =$ 16 isomers.

Glucose has four asymmetric carbon atoms, as shown in the following structure:

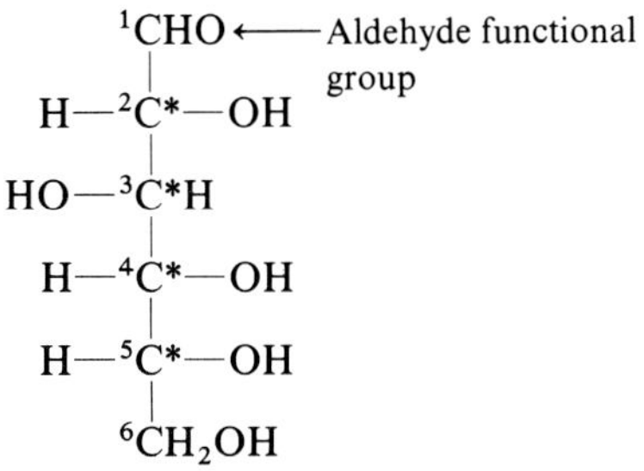

$$\begin{array}{c} {}^{1}CHO \leftarrow \text{Aldehyde functional group} \\ | \\ H-{}^{2}C^{*}-OH \\ | \\ HO-{}^{3}C^{*}H \\ | \\ H-{}^{4}C^{*}-OH \\ | \\ H-{}^{5}C^{*}-OH \\ | \\ {}^{6}CH_2OH \end{array}$$

The asymmetric carbon atom that is farthest from the carbonyl carbon of the aldehyde functional group will define the D or L assignment. If the OH group on carbon #5 pointed to the right, we would call this structure D-glucose; if the OH group on carbon #5 pointed to the left, we would call it L-glucose. Notice that L-glucose is an exact mirror image of D-glucose—all of the OH groups point in the opposite directions as compared to the corresponding D-glucose groups. We call D- and L-glucose **enantiomers**.

enantiomers

D-Glucose can be related back to D-glyceraldehyde; L-glucose can be related back to L-glyceraldehyde.

```
        CHO                    CHO
         |                      |
        HC—OH              HO—CH
         |                      |
    HO—CH                      HC—OH
         |                      |
        HC—OH              HO—CH
         |                      |
    H—C*—(OH)→          ←(HO)—C*—H
         |                      |
        CH2OH                  CH2OH

      D-Glucose              L-Glucose
```

epimers

Sugars that differ in the configuration of the OH group around one specific asymmetric carbon atom are called **epimers**. For example, D-glucose and D-mannose are epimers, because they differ in the configuration of the OH group on carbon #2. Notice, however, that both D-glucose and D-mannose are still D-sugars.

```
1        CHO                    CHO       1
          |                      |
2        HC—(OH)→        ←(HO)—CH         2
          |                      |
     HO—CH                  HO—CH
          |                      |
         HC—OH                  HC—OH
          |                      |
      H—C*—OH                H—C*—OH
          |                      |
         CH2OH                  CH2OH

       D-Glucose              D-Mannose
```

Many different epimeric pairs naturally exist in nature. Figure 7–2 (page 120) shows the family of D-aldose sugars (trioses to hexoses). The aldopentoses and aldohexoses are the most abundant in nature. Figure 7–3 (page 121) shows the family of D-ketose sugars (trioses to hexoses). In both cases, the D-sugars are found throughout nature, whereas the L-sugars are rare.

CYCLIC SUGARS

Although we have written sugars in the open-chain form, only trioses and tetroses actually exist in solution as open chains. The larger sugars (pentoses and hexoses) are in what is known as *cyclic form* in solution.

Consider ring formation for the six-carbon sugar D-glucose. The single bonds connecting the carbon atoms of this sugar can undergo rotation, thus bringing the OH group on carbon #5 close to the aldehyde group of carbon #1. When this occurs, a reaction between the alcohol group (OH group) and the aldehyde group will occur, forming a **hemiacetal**. The gener-

hemiacetal

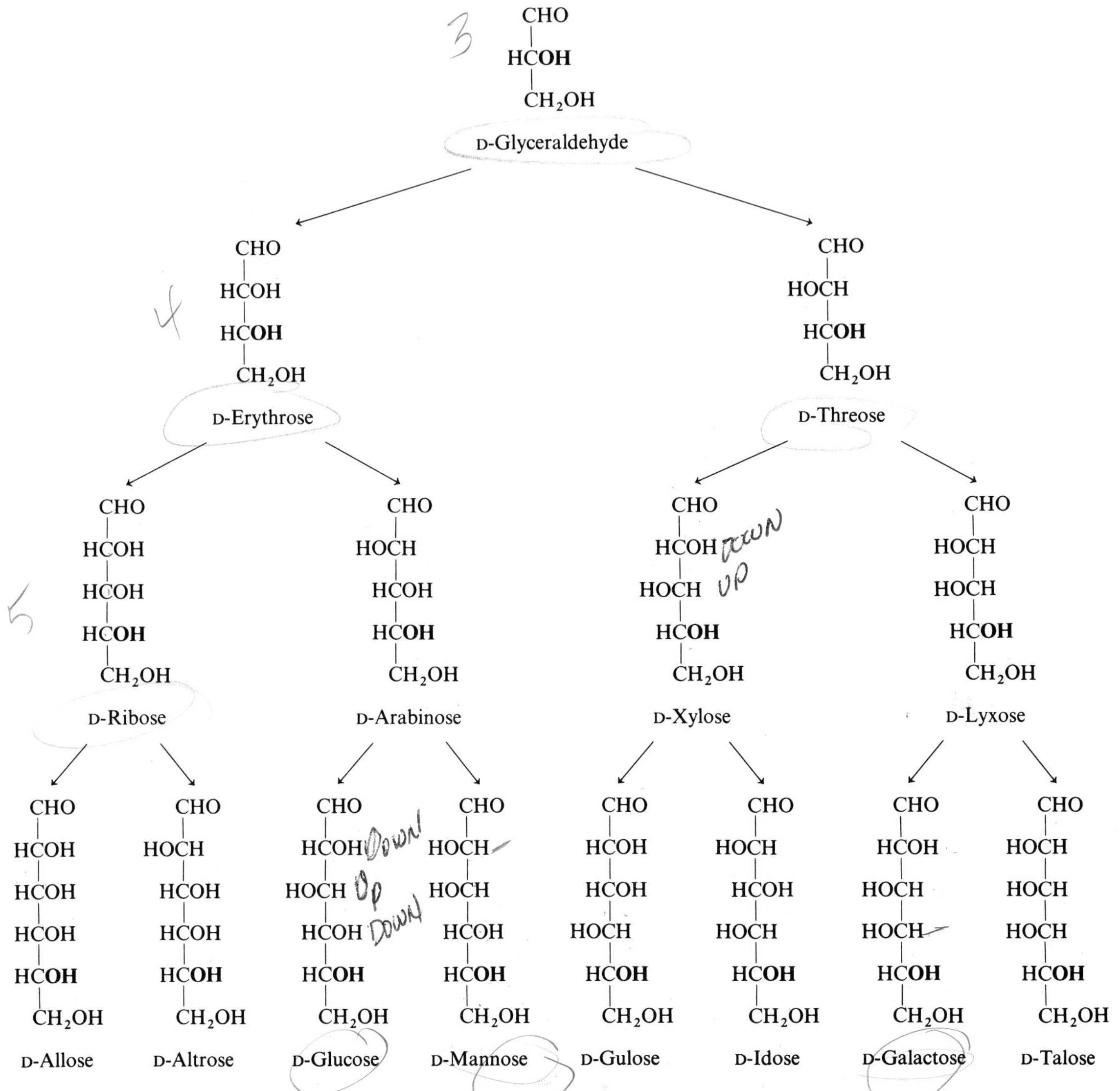

FIGURE 7–2 The "family" of D-aldose sugars, from trioses to hexoses. The OH groups in boldface define the D-isomer assignment.

alized reaction for hemiacetal formation can be noted as:

$$R_2{-}\overset{\displaystyle O}{\overset{\|}{C}}{-}H + R_1{-}OH \rightleftharpoons R_1{-}O{-}\underset{\displaystyle R_2}{\underset{|}{\overset{\displaystyle OH}{\overset{|}{C}H}}}$$

Hemiacetal

Hemiacetal formation causes the sugar to lock into a ring, as shown in

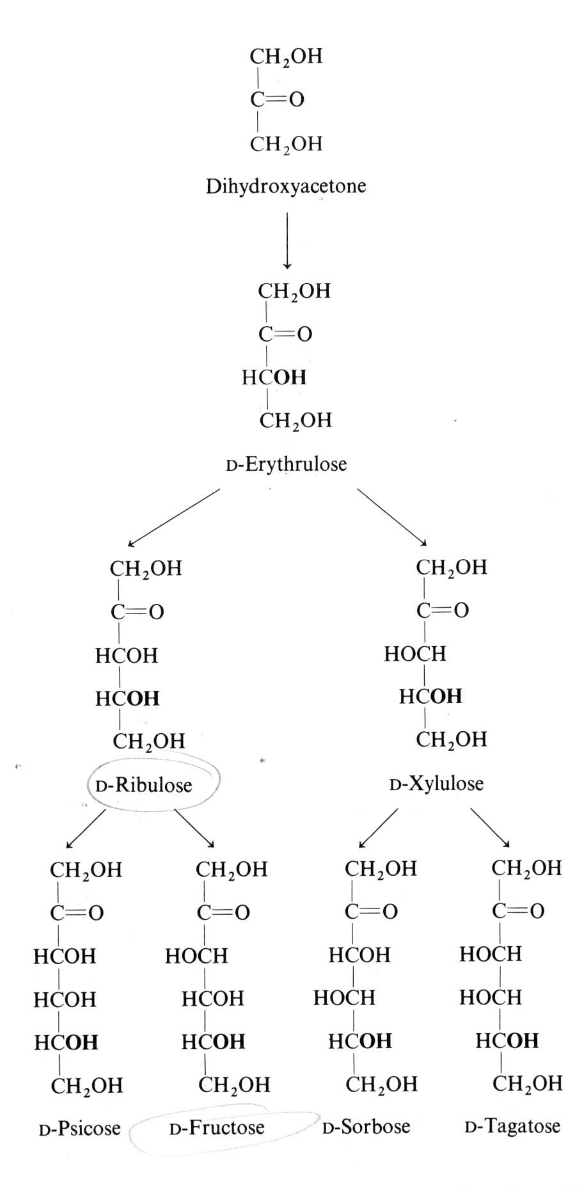

FIGURE 7–3
The "family" of D-ketose sugars, from trioses to hexoses. The OH groups in boldface define the D-isomer assignment.

Figure 7–4. One should note that the OH group of carbon #5 contributes the oxygen atom that bridges the #1 and #5 carbon atoms within the cyclic structure. A new OH group is also produced as a result of hemiacetal formation; this group is located on carbon #1 (the *anomeric* carbon atom). The six-membered structure is called a **pyranose ring**; thus, D-glucose is actually called D-*glucopyranose*.

pyranose ring

When the OH group on the anomeric carbon (#1) is in the "up" position, we refer to this structure as the *β-form* (β-D-glucopyranose, in this example). If the OH group on the anomeric carbon is in the "down" position, then the structure is called the α-form (α-D-glucopyranose). The α and β

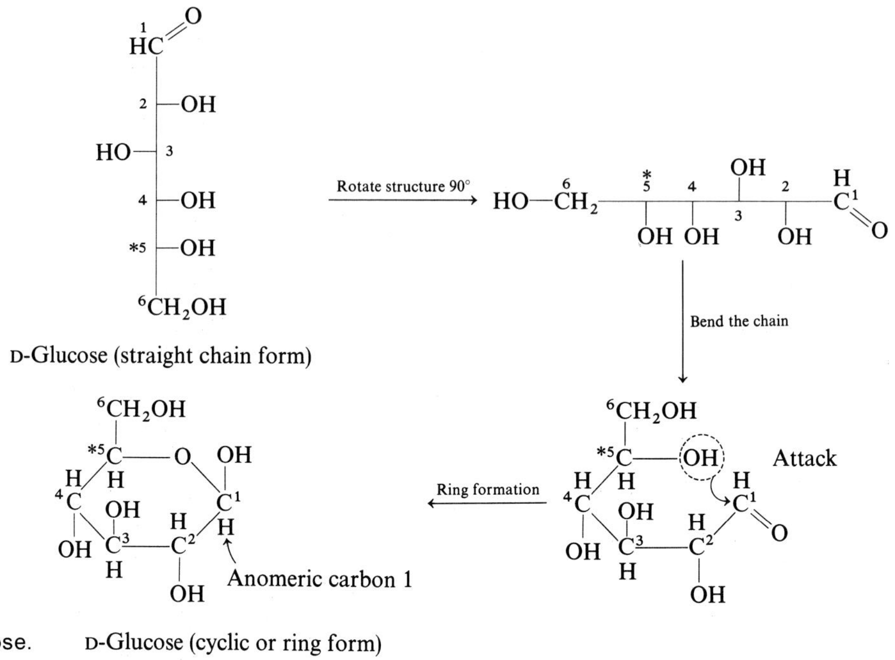

D-Glucose (straight chain form)

Rotate structure 90°

Bend the chain

Attack

Ring formation

Anomeric carbon 1

D-Glucose (cyclic or ring form)

FIGURE 7–4
Cyclization of D-glucose.

anomers

forms of a sugar are called **anomers**, because the two forms differ only in the orientation of the OH group on the *anomeric carbon atom*.

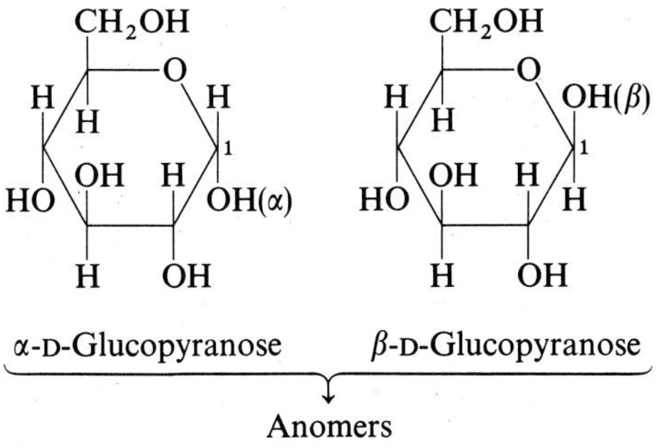

α-D-Glucopyranose β-D-Glucopyranose

Anomers

mutarotation

Anomeric forms of a sugar in solution can undergo very rapid interconversion via a process called **mutarotation**. In mutarotation, an anomeric form of a sugar reverts back to the open-chain form, which then locks into the second anomeric form. Eventually, an equilibrium mixture results that contains a certain concentration of the α form, a certain concentration of the β form, and a small amount of the open-chain form. For example, if one dissolves glucose in water, the solution will contain approximately 33 percent α-D-glucose, 67 percent β-D-glucose, and a trace of glucose in the open-chain form.

$$\underset{(33\%)}{\alpha\text{-D-glucose}} \rightleftharpoons \underset{(\text{trace})}{[\text{open-chain}]} \rightleftharpoons \underset{(67\%)}{\beta\text{-D-glucose}}$$

One might ask, why is the β form favored over the α form? The reason is that the OH groups of the β form cannot interact (and thus interfere) with each other and with the carbon skeleton of the sugar, as those of the α anomeric form do.

It should be noted that the ring structures of sugars do not form planar molecules in which the carbon atoms all reside in a single plane. Instead, six-membered ring structures assume a puckered three-dimensional "chair" configuration, as shown in the following diagram:

HO, CH_2OH, O, HO, OH, $OH(\beta)$, 1

β-D-Glucopyranose
(chair form)

hemiketal
furanose

Ketoses can undergo ring formation in which an alcohol group reacts with the carbonyl oxygen of the ketone to form a **hemiketal** structure. Aldopentoses and ketohexoses form five-membered ring structures called **furanose** rings; most aldohexoses form the more stable *aldopyranose* ring.

$^1CH_2OH$
$^2C{=}O$
$HO{-}^3CH$
$H^4C{-}OH$
$H^5C{-}OH$
$^6CH_2OH$

D-Fructose

Hemiketal formation →

$HO{-}\overset{6}{C}H_2$, O, $OH(\beta)$, 5, H, OH, 2, H, 4, 3, CH_2OH, 1, OH, H

β-D-Fructo*furanose*
(five-membered ring)

REDUCING SUGARS

reducing sugars

It is well known that the aldehyde group of an aldose can *reduce* a number of substances (that is, donate one or more electrons to these substances). Ketoses can also reduce substances, but not as easily as aldoses. This happens because the ring forms of most aldoses and ketoses can spontaneously open up and yield the open-chain form of the sugar, thus exposing the free carbonyl group for participation in a reduction process. Sugars that can reduce divalent copper (Cu^{2+}) to the monovalent species (Cu^+) are called **reducing sugars**. *Nonreducing* sugars, such as sucrose, cannot reduce Cu^{2+} to Cu^+. The *Benedict's test* can be used to identify which sugars are reducing. In this test, a sugar is heated in a copper-citrate solution (Benedict's reagent, containing Cu^{2+}). If a reducing sugar is present, it will reduce the Cu^{2+} to Cu^+, causing a brick-red precipitate of copper oxide (Cu_2O) to form.

OTHER MONOSACCHARIDES

A number of monosaccharides are especially important to living systems. We have already described the structure and chemical properties of D-glucose. This sugar is extremely important because (among other things) it is used by cells as a fuel. Specifically, the hydrogen atoms associated with the carbon and oxygen atoms of glucose are removed from the molecule and used in producing a high-energy compound called *adenosine triphosphate* (*ATP*).

Derivatives of glucose (chemical modifications of glucose) are also important. For example, *phosphorylated* forms of glucose are important in the energy production steps preceeding ATP formation. α-D-*Glucose 1-phosphate* and α-D-*glucose 6-phosphate* are examples of phosphorylated forms of glucose.

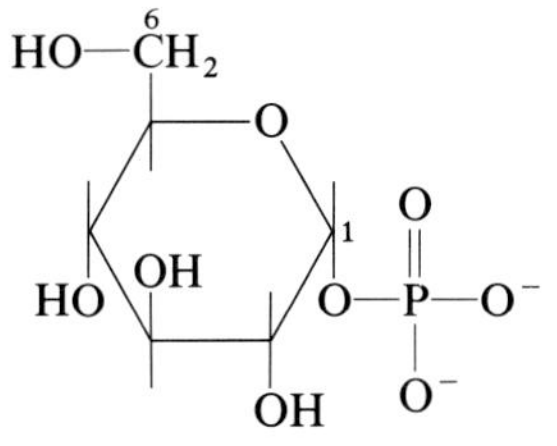

α-D-Glucose 1-phosphate 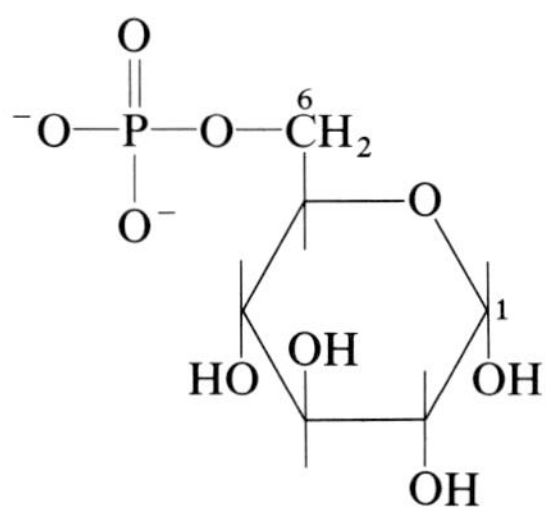 α-D-Glucose 6-phosphate

In another derivative of glucose, called D-*glucosamine*, an amino group (—NH_2) is substituted on the #2 carbon atom in place of the OH group. This amino group, in turn, can be *acetylated* to form *N-acetyl-*D-*glucosamine*. Both D-glucosamine and N-acetyl-D-glucosamine are used in constructing structural components within cells. Long chains of N-acetyl-D-glucosamine form a substance called *chitin*, which makes up the exoskeleton (hard outer shell) of crabs and lobsters.

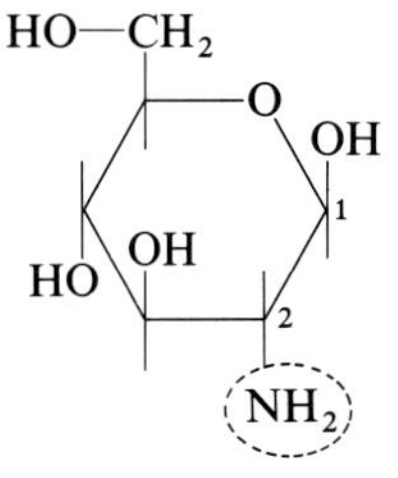

N-acetyl group

D-Glucosamine (an amino sugar) N-Acetyl-D-glucosamine

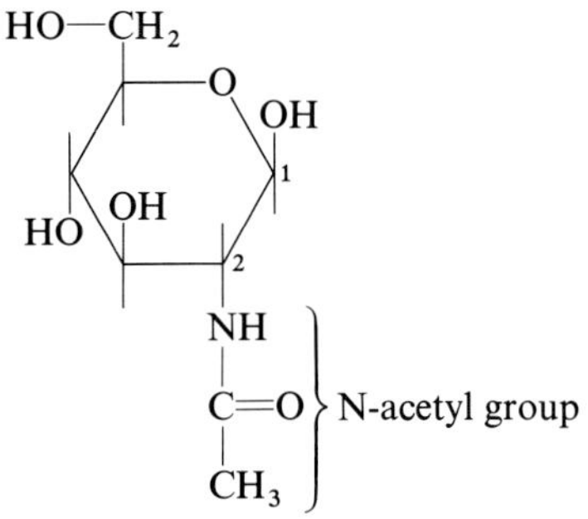

Bone, cartilage, and skin contain another derivative of glucose, *glucuronic acid.* This sugar possesses a carboxylic acid group on the #5 position of the ring.

^{-}OOC — 5 — O — OH — 1 — HO — OH — OH

D-Glucuronic acid

ribose
deoxyribose

Finally, two pentoses, **ribose** and **deoxyribose**, are extremely important because they are used in constructing *nucleic acids*. Ribose (along with phosphate groups) forms the backbone of the nucleic acid *ribonucleic acid (RNA)*. Deoxyribose (along with phosphate groups) forms the backbone of *deoxyribonucleic acid (DNA)*. Deoxyribose differs from ribose in that the OH group on carbon #2 has been replaced by a hydrogen atom (deoxy form).

HO—CH_2 (5) O OH (1) 2 OH OH — Ribose

HO—CH_2 (5) O OH (1) 2 OH H — *Deoxy*ribose

7.3 OLIGOSACCHARIDES

Oligosaccharides are small carbohydrate chains that contain between two and ten monosaccharide units. The disaccharides, which are constructed from two sugars held together by the **glycosidic link**, are the most abundant oligosaccharides in nature. The glycosidic link connects the OH group of the anomeric carbon with a carbon atom of the second sugar. Maltose, lactose, and sucrose are three common disaccharides.

glycosidic link

maltose

Maltose is a disaccharide constructed from two D-glucose residues. The glycosidic link connects the OH group on the anomeric carbon atom of the first glucose residue with the #4 carbon atom of the second glucose residue. The anomeric OH group is in the α position, and the glycosidic link is symbolized as $\alpha(1 \rightarrow 4)$. Since the anomeric carbon atom of the first glucose residue is involved in the glycosidic bond, this glucose residue cannot unloop to yield the open chain. However, the second glucose residue does have such a free anomeric carbon atom and thus can yield a free aldehyde group. This end of the molecule is called the *reducing end*, because it can reduce Benedict's reagent. The structure of maltose is shown in the following diagram:

HO—CH$_2$ HO—CH$_2$ O O OH(β) Reducing end 1 4 1 HO OH O OH OH OH

$\alpha(1 \rightarrow 4)$ Glycosidic link

Maltose

lactose

Lactose (milk sugar), another reducing disaccharide, is constructed from a galactose residue and a glucose residue. The glycosidic link is $\beta(1 \rightarrow 4)$, which means that the OH group on the anomeric carbon atom of galactose is in the β position. The glycosidic link is noted with an unusual chemical bonding symbol.

O

One can think of the lactose molecule as consisting of a galactose residue end-linked to a glucose residue. Because of the $\beta(1 \rightarrow 4)$ glycosidic bond, the glucose residue extends slightly below the galactose residue as shown in the structure to the right.

HO—CH$_2$ HO—CH$_2$ O O HO OH(β) 1 O 4 OH OH OH OH

$\beta(1 \rightarrow 4)$

D-Galactose D-Glucose

Lactose

Another view of lactose:

O (β) O 4 O 1

Galactose

Glucose

sucrose

Sucrose (cane sugar, table sugar) is constructed from a glucose residue and a fructose residue. The glycosidic link is $\alpha(1 \rightarrow 2)$, which means that the OH group on the anomeric carbon atom of glucose is in the α position and is bonded to the #2 carbon atom of the fructose in an unusual *inverted* position. Because the anomeric carbon atoms of both the glucose and the fructose are involved in the glycosidic link, neither sugar can yield a carbonyl oxygen. Thus, sucrose is a *nonreducing sugar*.

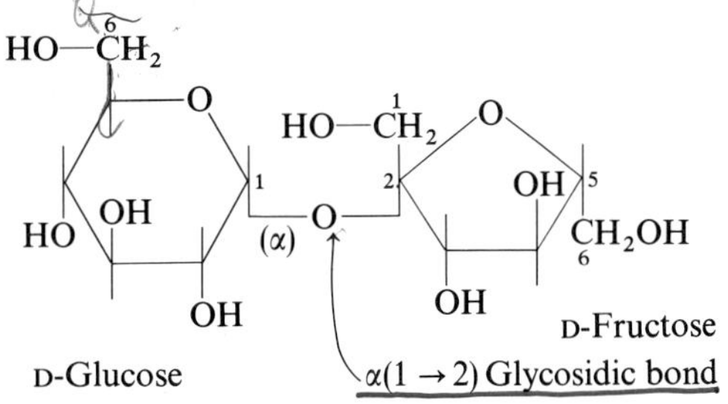

Sucrose
(a nonreducing sugar)

7.4 POLYSACCHARIDES

Most of the carbohydrates found in nature are in the form of very long and complex chains, called polysaccharides. A polysaccharide is constructed from many individual monosaccharide units. There are two main classes of polysaccharides (1) the **homopolysaccharides** and (2) the **heteropolysaccharides**. The homopolysaccharides are constructed from only one kind of monosaccharide unit (*homo* means "the same"). For example, the polysaccharide *cellulose* is constructed from repeating units of D-glucose. Heteropolysaccharides (*hetero* means "different") are constructed from more than one type of monosaccharide unit. For example, *hyaluronic acid* is constructed from alternating residues of *glucuronic acid* and *N-acetyl*-D-*glucosamine*.

homopolysaccharides
heteropolysaccharides

HOMOPOLYSACCHARIDES

A number of homopolysaccharides serve as either a storage form of cellular fuel (for example, starch) or as a cellular structural component (for example, cellulose). We will discuss the starches first, then the chemistry and structural features of cellulose.

starch
amylose

Starch consists of two main components, *amylose* and *amylopectin*. **Amylose** is an unbranched chain hundreds of units long consisting of glucose residues held together by α(1 → 4) glycosidic bonds. One end of the molecule is called the nonreducing end, because the anomeric carbon atom of this terminal glucose residue is involved in the glycosidic link. The other end is called the reducing end, because the anomeric carbon atom can participate in forming the free aldehyde group. The structure of amylose is shown in the following diagram:

Nonreducing end

Reducing end

Amylose

The amylose molecule actually exists as a long helical coil, as shown in Figure 7–5. Because amylose can form long helices, it can react with iodine molecules to yield an intensely, blue starch/iodine complex. This starch/iodine reaction can be used as a test for the presence of starch.

Amylose can be enzymatically digested by two different types of enyzmes, α-amylase and β-amylase. *α-Amylase*, which is present in saliva and pancreatic juice, randomly hydrolyzes the $\alpha(1 \rightarrow 4)$ links and, thus, produce a mixture of free glucose and maltose. *β-Amylase* (present in malt) also hydrolyzes the $\alpha(1 \rightarrow 4)$ glycosidic links, but in a very ordered fashion. Beginning at the nonreducing end, β-amylase clips alternating $\alpha(1 \rightarrow 4)$ links, yielding free maltose (two glucose residues/maltose).

amylopectin

Amylopectin, the second major component of starch, is a highly branched polysaccharide of repeating glucose residues connected together by both $\alpha(1 \rightarrow 4)$ and $\alpha(1 \rightarrow 6)$ glycosidic links. $\alpha(1 \rightarrow 6)$ links occur every 24 to 30 glucose residues, producing what is known as a *branch point*. A portion of an amylopectin molecule is shown in the following diagram[1]:

Branch point $\alpha(1 \rightarrow 6)$ linkage

Main $\alpha(1 \rightarrow 4)$ chain

The amylopectin molecule is a spherical, highly branched structure resembling a tumbleweed, as shown in Figure 7–6.

Finally, it should be noted that both α- and β-amylases can digest amylopectin, releasing smaller saccharide chains; however, enzymatic attack stops when an $\alpha(1 \rightarrow 6)$ branch point has been reached. Isomaltase

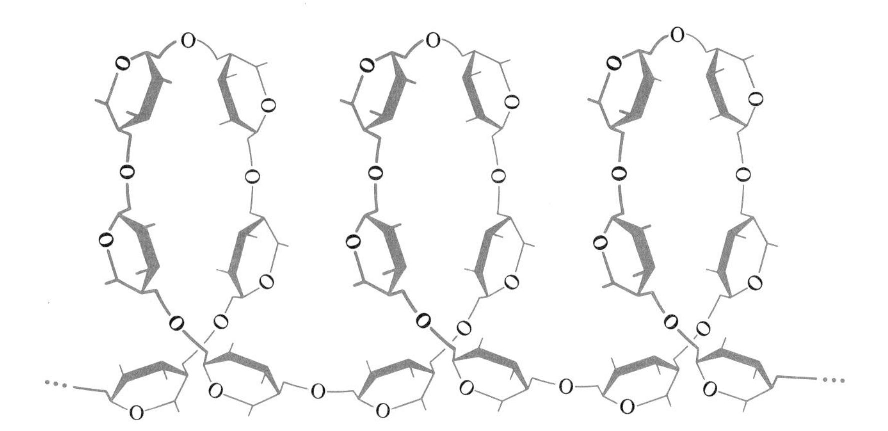

FIGURE 7–5
The helical coil of amylose. (Adapted from Lehninger, A. L., *Biochemistry*, 2nd ed. (New York: Worth, 1975), p. 264.)

then takes over and hydrolyzes the $\alpha(1 \rightarrow 6)$ branch points. Therefore, α- or β-amylase, in combination with the debranching enzyme, can completely digest amylopectin into free maltose and glucose.

glycogen

In animals, **glycogen** is the primary storage form of glucose. Similar in structure to amylopectin, glycogen is more highly branched and thus forms granules in the cytoplasm of liver and skeletal muscle cells. Enzymes associated with these granules phosphorolyze the glycogen molecules and yield free glucose (as glucose 1-phosphate), which can be used to produce energy (ATP). This metabolic process will be discussed in detail in Chapters 9 and 10.

cellulose

Cellulose is the primary structural component of all plant cells and, in fact, is the most abundant carbohydrate on the face of the earth. Just think

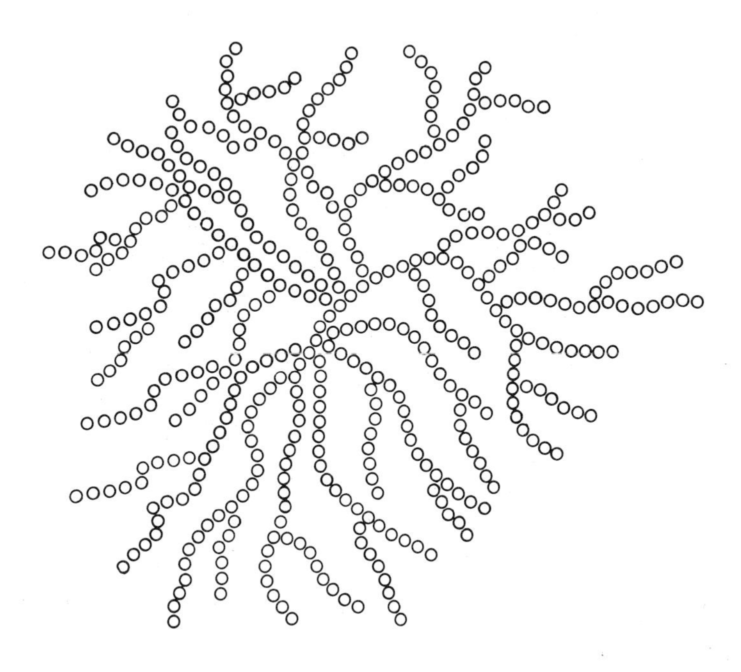

FIGURE 7–6
The highly branched character of amylopectin (and glycogen). Individual glucose units are linked together by $\alpha(1 \rightarrow 4)$ links; the branch points are $\alpha(1 \rightarrow 6)$ links.

of all the plants and trees that populate the earth and you can start to appreciate how much cellulose there actually is. Cellulose is constructed from repeating units of D-glucose, held together by $\beta(1 \rightarrow 4)$ glycosidic bonds as shown in the following diagram:

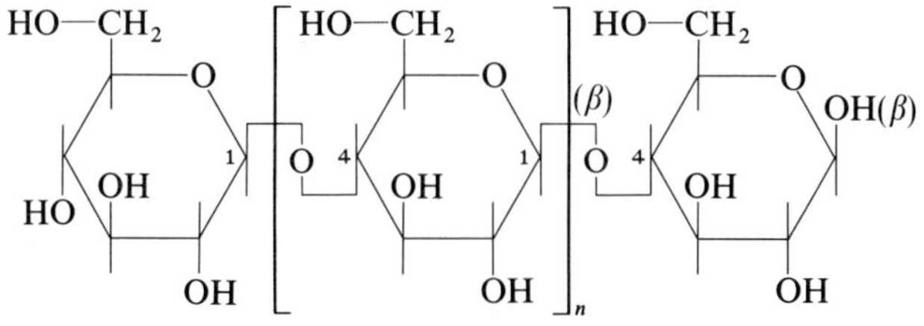

Cellulose (with $\beta(1 \rightarrow 4)$ links)

Interestingly, the primary difference between amylose and cellulose is the presence of the $\beta(1 \rightarrow 4)$ glycosidic link in cellulose. This subtle difference results in significant differences in the physical properties of the two types of polysaccharides. Cellulose forms long fibrils that, in plant cell walls, are cemented together in a matrix of other carbohydrates (pectin and hemicellulose). This fibril/matrix network endows plant cell walls with surprising physical strength. Across the grain, wood can be stronger, weight for weight, than steel.

Most mammals, including humans, cannot hydrolyze the $\beta(1 \rightarrow 4)$ link of cellulose; therefore, they cannot use cellulose as food. Ruminants (cows), on the other hand, have bacteria living in a part of their gastrointestinal tract (called the rumen) that can hydrolyze the $\beta(1 \rightarrow 4)$ link of cellulose into D-glucose using the enzyme *cellulase*.

HETEROPOLYSACCHARIDES

*Hetero*polysaccharides are constructed from more than one type of monosaccharide residue. (The units are also usually modified in some way.) Many examples of heteropolysaccharides exist throughout the biological world; only a few will be discussed here.

chondroitin sulfates

Hyaluronic acid and the **chondroitin sulfates** are typical heteropolysaccharides. Hyaluronic acid is present in the synovial fluid, which lubricates skeletal joints. Chondroitin sulfates are present in cartilage tissue and in arterial walls. Hyaluronic acids are constructed from alternating residues of D-glucuronate and N-acetyl-D-glucosamine held together by alternating $\beta(1 \rightarrow 3)$ (unusual) and $\beta(1 \rightarrow 4)$ glycosidic links, as shown in the diagram at the top of page 131. Negatively charged carboxyl groups are found on the glucuronate residues of hyaluronic acid; these contribute to the overall negative charge of this molecule.

A representative chondroitin sulfate is shown in the next diagram. The chondroitin sulfates differ from hyaluronic acid in that negatively charged

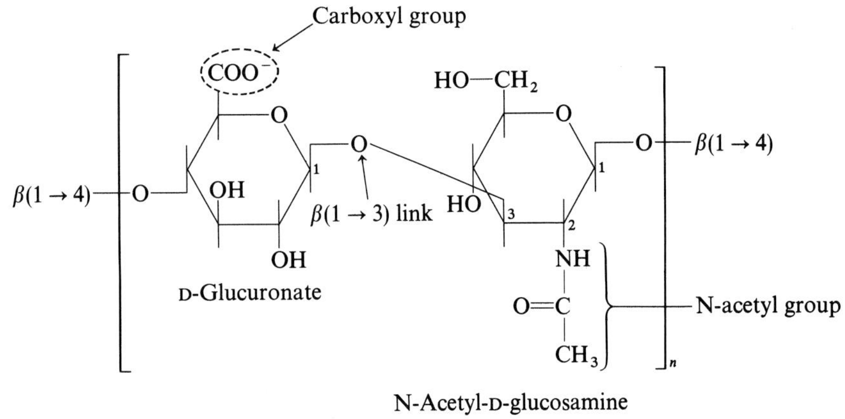

Carboxyl group

$\beta(1 \rightarrow 4)$ · $\beta(1 \rightarrow 3)$ link · $\beta(1 \rightarrow 4)$

D-Glucuronate · N-Acetyl-D-glucosamine · N-acetyl group

Hyaluronic acid

sulfate groups ($SO_3{}^{2-}$) are bound to either the #4 or #6 carbon atoms of the *galactosamine* residue.

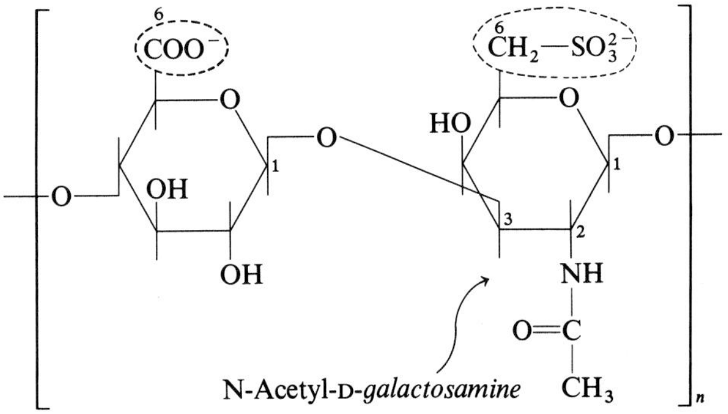

N-Acetyl-D-*galactosamine*

Chondroitin 6-sulfate

Because chondroitin sulfates have a large negative charge, charge-charge repulsion keeps the individual chains apart from each other. This separation causes the molecules to behave as a kind of "molecular sieve" that entraps certain types of molecules within the chain, but excludes others. In addition, chondroitin sulfates can entrap a large number of water molecules, thus forming a gel that can absorb physical stresses and shocks. This physical property is particularly necessary for tissues (like cartilage) that must absorb shocks.

Heparin sulfates are similar in structure to the chondroitin sulfates, but have a more complicated carbohydrate composition. Heparin sulfates contain both glucuronate and iduronate residues associated with alternating glucosamine residues. **Heparin**, the most common of the heparin sulfates, is found primarily in the arterial walls. It inhibits the blood clotting mechanism and helps transport blood serum lipids by liberating an enzyme called *lipoprotein lipase*, which degrades the lipids circulating in the blood. The

heparin

structure of heparin is shown in the following diagram:

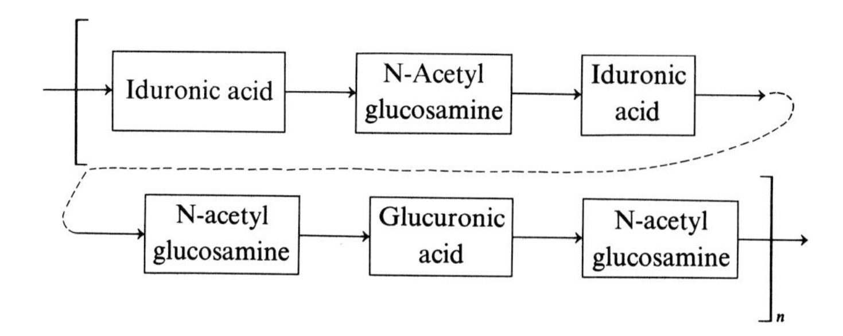

GLYCOPROTEINS

glycoproteins

Glycoproteins are protein molecules that have carbohydrate chains bound to them. The carbohydrates participate in a covalent linkage with the R groups of either serine, threonine, or asparagine, as shown in the following diagram:

HO—CH_2 ... O ... (β) ... O—CH_2—CH Serine

Others sugars ← O ... OH ... OH

D-Glucose (example of a representative sugar)

Polypeptide chain of a protein

A glycoprotein can contain one chain or many. The carbohydrate chains can be very short (one or two sugar residues) or very long. They are often highly branched, and usually do not have a repeating sequence of sugars.

thyroglobulin

Thyroglobulin is a very large glycoprotein (its molecular weight is 660 000 daltons). It is found in the thyroid gland, and is required for thyroid hormone synthesis. This glycoprotein has two types of carbohydrate chains, an A chain and a B chain, composed of mannose, N-acetyl-glucosamine, galactose, and other sugars. There are 9 A chains and 14 B chains in each thyroglobulin molecule.

Glycoproteins have a number of biological functions. Specific enzymes, transport proteins, hormones, and antibodies may be glycoproteins. The glycoprotein is very soluble in water because the attached carbohydrate chains are very polar (the OH groups of the sugar molecules are very hydrophilic). Some scientists speculate that the carbohydrate chains may aid in transporting glycoproteins from one part of the cell to another. Cell/cell contacts may be mediated by interactions between specific glycoproteins residing on the outer surface of cell membranes.

GLYCOLIPIDS

glycolipids

Carbohydrates that are attached to various lipid molecules are called **glycolipids**. The structures and chemical properties of these substances will be discussed in Chapter 13.

SUMMARY

Carbohydrates (sugars) are generally defined as polyhydroxy aldehydes or ketones. There are three main classes of carbohydrates: monosaccharides, oligosaccharides, and polysaccharides.

Monosaccharides are classified according to the number of carbon atoms in the molecule. Monosaccharides are either aldehydes (aldoses) or ketones (ketoses).

All monosaccharides (except dihydroxyacetone) have at least one asymmetric carbon atom. Thus, monosaccharides form D- and L-stereoisomers (enantiomers) that exhibit optical activity. Sugars that differ in the configuration of the OH group around one specific asymmetric carbon atom are called epimers.

The larger monosaccharides (the pentoses and hexoses) assume a cyclic (hemiacetal) form. A hexose forms a six-membered (pyranose) ring structure; a pentose forms a five-membered (furanose) ring structure. The carbon that participates in hemiacetal or hemiketal formation is called the anomeric carbon atom. Two anomeric forms of a D sugar can exist, the α form (in which the OH group is in a "down" position) and the β form (in which the OH group is in the "up" position). Anomeric forms of a sugar can undergo very rapid interconversion (mutarotation).

Glucose is used by many cells as a fuel. Phosphorylated forms of glucose (glucose 1-phosphate and glucose 6-phosphate) are important in the process of energy production. Glucosamine and N-acetyl-glucosamine are important derivatives of glucose.

Ribose is used in constructing ribonucleic acid (RNA). Deoxyribose makes up the backbone of deoxyribonucleic acid (DNA).

Oligosaccharides are small carbohydrate chains that contain between two and ten monosaccharide units. The disaccharides are the most common oligosaccharides. Maltose is constructed from two D-glucose residues connected together by an $\alpha(1 \rightarrow 4)$ glycosidic bond. Lactose is constructed from a galactose residue and a glucose residue connected together by a $\beta(1 \rightarrow 4)$ glycosidic bond. Sucrose is constructed from a glucose residue and a fructose residue connected together by a $\alpha(1 \rightarrow 2)$ glycosidic bond.

There are two main classes of polysaccharides: homopolysaccharides and heteropolysaccharides. Starch (amylose and amylopectin), glycogen, and cellulose are examples of homopolysaccharides. Amylose is a long, unbranched chain of hundreds of glucose residues, each connected to the next by $\alpha(1 \rightarrow 4)$ glycosidic links. Amylopectin is a highly branched polysaccharide constructed from repeating glucose residues connected by both $\alpha(1 \rightarrow 4)$ and $\alpha(1 \rightarrow 6)$ glycosidic links. The $\alpha(1 \rightarrow 6)$ links produce what are known as "branch points." Glycogen, a polysaccharide more highly branched than amylopectin, is the main storage form of glucose within animal skeletal muscle and liver cells. Cellulose is constructed from hundreds of glucose residues held together by $\beta(1 \rightarrow 4)$ glycosidic links. Heteropolysaccharides are constructed from more than one type of monosaccharide residue. Hyaluronic acid and the chondroitin sulfates are examples of heteropolysaccharides.

Glycoproteins are proteins that contain carbohydrates covalently bound to either serine, threonine, or asparagine residues in the polypeptide chain of the protein. The presence of the carbohydrate chains on a protein can significantly influence the physical properties of that protein—for instance, increasing its solubility and altering its electrical charge.

REVIEW QUESTIONS

1. We have presented a number of important concepts and terms in the chapter. Define (or briefly explain) each of the following terms:
   a. Carbohydrate
   b. Oligosaccharide
   c. Polysaccharide
   d. Aldose
   e. Pentose
   f. Fischer projection
   g. D-sugar
   h. Epimeric pair
   i. Hemiacetal
   j. Anomeric carbon atom
   k. Pyranose ring
   l. β-D-glucopyranose
   m. Mutarotation
   n. Reducing sugar
   o. Glycosidic bond
   p. Nonreducing end
   q. Heteropolysaccharide
   r. Branch point
   s. α-Amylase
   t. Debranching enzyme
2. Referring to Figures 7–2 and 7–3, classify each of the following sugars and give its general name:
   a. D-Threose (Example: This is an aldotetrose.)
   b. D-Arabinose **c.** D-Idose
   d. D-Ribulose **e.** D-Fructose
3. **a.** Why are carbon atoms #1 and #3 of glyceraldehyde not considered asymmetric?
   b. Why is dihydroxyacetone not optically active?
4. Draw the mirror image structures (that is, the L-sugars) for each of the following D-sugars:
   a. D-Erythrose **b.** D-Lyxose
   c. D-Galactose ***d.** Dihydroxyacetone
5. Referring to Figure 7–3, p. 120, are D-sorbose and D-tagatose epimers or enantiomers? Why?
6. Draw complete ring structures for:
   a. β-D-Glucose **b.** α-D-Galactose
   c. β-Maltose **d.** Sucrose
   e. Amylose (general structure only)

*7. Referring to Figures 7–2 and 7–3, pp. 120 and 121, draw ring structures for:
   a. Erythrose
   b. D-Mannose (α form)
   c. D-Talose (β form)
   d. D-Fructose (β form)

8. Draw complete ring structures for:
   a. Glucose 4-phosphate
   b. Ribose 5-phosphate
   c. Deoxyribose 1-phosphate
9. Draw the complete ring structure for β-D-galactose-β-D-galactose, including a $\beta(1 \rightarrow 4)$ glycosidic link.
10. **a.** If α-amylase attacks amylose, what would happen to the intense blue color of the amylose/iodine complex if α-amylase were present?
    b. Would iodine produce a blue color if added to amylopectin? Why or why not?
11. Referring back to Chapter 5, Enzymes, what type of specificity do the α- and β-amylases exhibit?

*12. Which amino acids in the active site of α-amylase might be responsible for binding amylose? (Refer back to Chapter 3, Amino Acids and Proteins, and to Chapter 5, Enzymes.)

SUGGESTED READING

Bhagavan, N. V. *Biochemistry*, 2nd ed., Ch. 4. Philadelphia: J. B. Lippincott, 1978.

Lehninger, A. L. *Biochemistry*, 2nd ed., Ch. 10. New York: Worth, 1975.

McGilvery, R. W., and Goldstein, G. *Biochemistry: A Functional Approach*, 2nd ed., Ch. 10. Philadelphia: W. B. Saunders, 1979.

NOTE

1. Amylopectin structure adapted from Lehninger, A. L., *Biochemistry*, 2nd ed. (New York: Worth, 1975), p. 265.

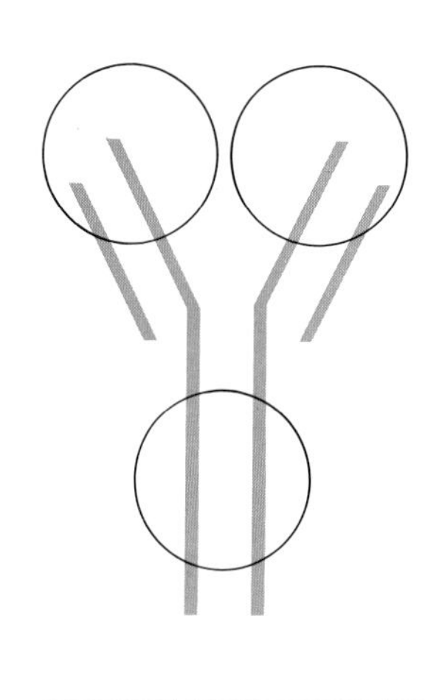

Chapter 8

Bioenergetics and an Introduction to Carbohydrate Metabolism

8.1 INTRODUCTION: GENERAL CONCEPTS OF METABOLISM

If you will recall our discussion of the unusual characteristics of living matter (see Chapter 1), you will remember that living cells can extract matter and energy from the environment. The matter is used by cells in constructing various cellular building blocks and constituents. Some material substances are also used by cells for their chemical energy. That is, the cell (1) transports specific materials (cellular fuels) into itself, (2) extracts the chemical energy in these substances into a useable form, (3) transfers this usable energy to where it is needed in the cell, and (4) uses the energy to drive the various processes within it. Certain cells (photosynthetic plant cells) can directly use the energy of sunlight to drive the photosynthetic process by which carbon dioxide and water are converted into sugars. These sugar molecules are subsequently used as a source of chemical energy by the photosynthetic cells.

metabolism

The chemical processes that occur within the cell constitute the **metabolism** of the cell. Metabolic processes (1) extract chemical energy from the environment; (2) convert various nutrient molecules into the building blocks of larger biomolecules; and (3) assemble these building blocks into proteins, nucleic acids, lipids, and complex carbohydrates for use both by the cell itself and by cells in its environment.

A metabolic process can be more accurately described as a *metabolic pathway*, a simple example of which is summarized in the following diagram:

$$A \xrightarrow{E_1} B \xrightarrow{E_2} C \xrightarrow{E_3} D$$

A metabolic pathway

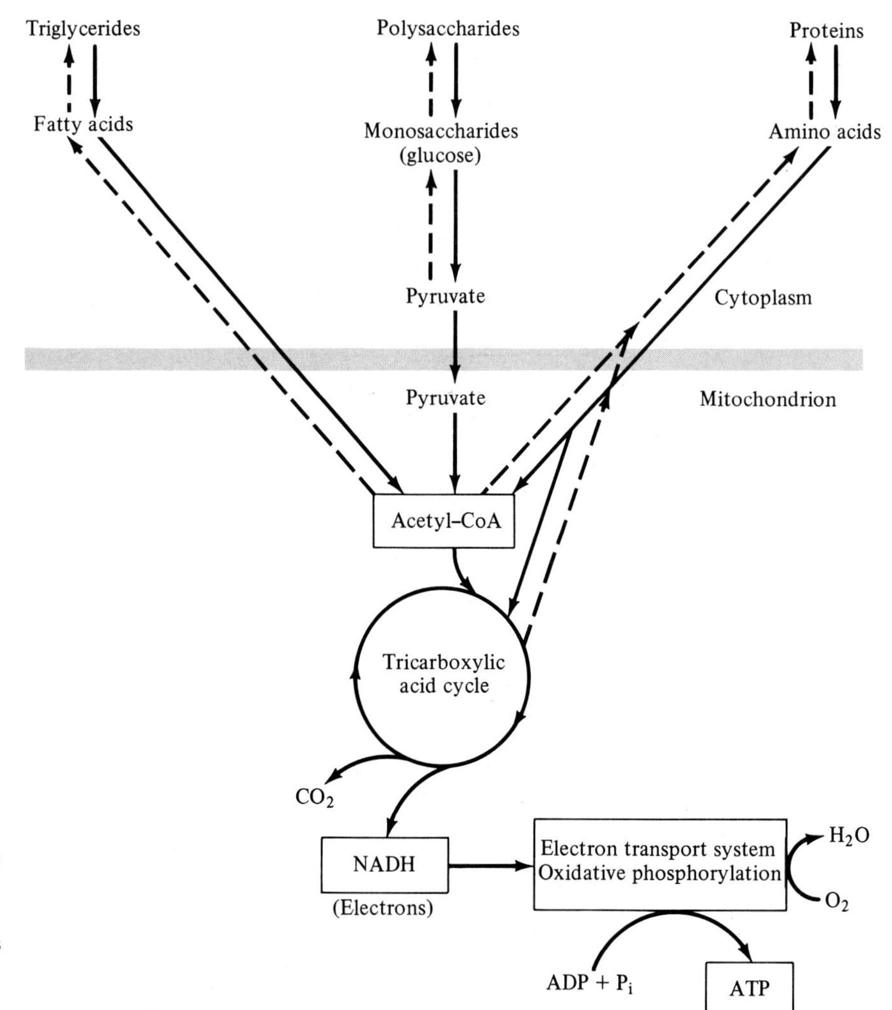

FIGURE 8–1
A summary of the primary anabolic and catabolic metabolic pathways. Catabolic (degradative) pathways that yield ATP: Solid arrows (→). Anabolic (biosynthetic) pathways that require ATP: Dashed arrows (--→).

In this example, substance A is converted into product B, which is converted into product C, which is in turn converted into product D. The individual steps of the pathway are *sequential* and are catalyzed by enzymes E_1, E_2, and E_3. Extremely complex metabolic pathways, often involving between 10 to 20 different enzymatic steps, are known to occur within cells. These processes are under constant moment-by-moment metabolic control (see Chapter 6, Enzymes II) so that the cell can efficiently process the various nutrients and metabolites that it needs. Branched metabolic pathways also exist within the cell.

It should be emphasized that most metabolic pathways are not physically organized within the cell as shown in the previous diagram. That is, the individual enzymes are not generally organized in precise rows as represented in the diagram. The pictorial representation merely aids us in understanding the sequence of the enzymatic steps. Reactant A cannot be directly transformed into product D without the intervening steps catalyzed by enzymes E_2 and E_3. In some pathways, however, enzymes are organized in membranes or in complex aggregates (called multienzymes complexes) in which the product of one enzyme reaction diffuses to the active site of a second, closely associated enzyme for the next step.

catabolic pathways
anabolic pathways

The metabolic processes that occur within a typical cell are actually subdivided into two main categories, (1) the **catabolic pathways** and (2) the **anabolic pathways**. Catabolic pathways are degradative pathways in which rather complex nutrient molecules are degraded into simpler molecules, in the process releasing chemical energy. Anabolic pathways are biosynthetic pathways, in which various cellular constituents are synthesized from simpler precursor molecules. Anabolic pathways require the input of chemical energy to drive their processes to completion. The basic (main line) catabolic and anabolic pathways of the cell are summarized in Figure 8–1. One should note that various types of lipids, carbohydrates, and proteins all feed into a central pathway (called the tricarboxylic acid, or TCA, cycle) via a substance called **acetyl-coenzyme A (acetyl-CoA)**. The importance of the TCA cycle in controlling the metabolism of the cell will be discussed in greater detail in Chapter 11.

acetyl-coenzyme A (acetyl-CoA)

8.2 BIOENERGETICS

As we have already stated, the extraction and transfer of *useable chemical energy* constitutes a significant part of cellular metabolism. Life could not exist without mechanisms for efficiently extracting chemical energy from ingested materials. The study of the processes involved in the extraction and transfer of chemical energy is called **bioenergetics**.

bioenergetics

free energy

To understand bioenergetic principles, we must briefly describe the concept of **free energy**. Free energy can be simply defined as the capacity to perform work at constant temperature and pressure. For example, free energy is required to move a book from point A (a desktop 3 feet from the floor) to point B (a bookshelf 5 feet from the floor). Likewise, free energy is required to make (and break) chemical bonds. These processes require work to insure their completion.

Consider the following reaction:

$$A \longrightarrow B$$

Assume that a chemical bond is broken as A is transformed into B. Heat energy would either be *released* during this process, or would be required to break the bond. The total energy contained in substance A is symbolized by E_{tot}. It is defined as the sum of the *useable energy* and the *unuseable energy*

according to the following equation:

$$E_{tot} = E_{use.} + E_{unuse.} \tag{8–1}$$

The useable energy is called the Gibb's free energy and is symbolized as G. It is energy that can be used to perform useful chemical work. The unuseable energy is energy that cannot be used to perform work and is lost to the environment, usually as waste heat. This waste heat causes the environment to become a little more disordered and chaotic; thus, the amount of total disorder **(entropy)** in the environment increases as a result of the process in which A broke down into B.

entropy

Biochemists use a term called **free-energy change (ΔG)** to measure how much energy is available to perform useful work. The free-energy change can be defined as the *difference* between the energy contents of the products and reactants (G_A and G_B) at constant temperature and pressure.

free-energy change (ΔG)

$$\underset{(G_A)}{\text{A}} \longrightarrow \underset{(G_B)}{\text{B}} \qquad \Delta G = G_B - G_A \tag{8–2}$$

If G_B is less than G_A, ΔG will be negative; if G_B is greater than G_A, then ΔG will be positive.

To understand the implications of the free-energy change, let us examine the following reactions:

$$\text{A} \longrightarrow \text{B}$$

Output energy (ΔG is negative)

Spontaneous reaction

When energy is released to the environment (that is, when ΔG is negative), biochemists call this type of chemical reaction **spontaneous (exergonic)**. In an exergonic reaction, A is easily converted into B, in the process releasing free energy into the environment. This concept is diagrammed in Figure 8–2.

spontaneous (exergonic)

Alternatively, some reactions require the input of free energy to drive the chemical reaction in the direction in which it is written.

$$\text{A} \longrightarrow \text{B}$$

Input energy (ΔG is positive)

Nonspontaneous reaction

This type of reaction is **nonspontaneous (endergonic)**; A will not easily form B without the input of energy.

nonspontaneous (endergonic)

Changes in the free energy of a chemical reaction can be calculated and used to predict whether a chemical reaction is spontaneous or nonspontaneous. The free-energy change for a chemical reaction (ΔG) can be expressed as

$$\Delta G = \Delta G^{\circ\prime} + 2.3\,RT \log \frac{[\text{B}]}{[\text{A}]} \tag{8–3}$$

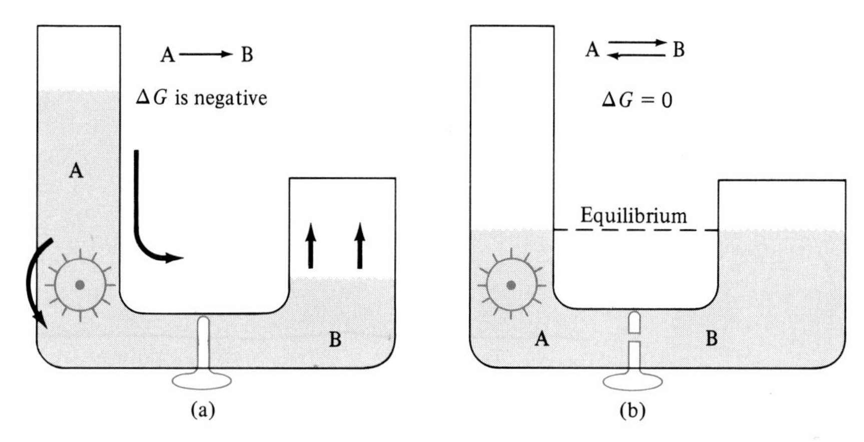

FIGURE 8–2 An analogy to a negative free-energy change during a spontaneous chemical reaction. (a) The diagram depicts the energy levels of A and B prior to reaction. Since the energy content of A is greater than the energy content of B, ΔG will be negative ($\Delta G = G_B - G_A$). (b) The stopcock has been opened, and A flows into the B compartment (A spontaneously reacts to form B). As A is transformed into B, work can be done (represented by the turning water wheel). Eventually, the levels of A and B reach equilibrium, and now $\Delta G = 0$. No work can be performed when $\Delta G = 0$. (Adapted from Segel, I. H., *Biochemical Calculations*, 2nd ed. (New York: John Wiley, 1976), p. 151.)

where ΔG = Actual free-energy change; $\Delta G^{o\prime}$ = Standard free-energy change = $-2.3\,RT\log K_{eq}$ at pH 7; R = Gas constant (1.99 cal/mol·K); T = Absolute temperature (K); [B]/[A] = Ratio of concentrations of B and A.

The "prime" (′) after ΔG in $\Delta G^{o\prime}$ identifies free-energy changes for reactions at pH 7.0.

The actual free-energy change is, thus, the sum of the standard free-energy change and a factor relating the actual concentrations of reactants and products.

When a chemical system reaches equilibrium, $\Delta G = 0$ and no useful work can be done. Thus, Equation 8–3 reduces to:

$$0 = \Delta G^{\circ\prime} + 2.3\,RT\log\frac{[B]}{[A]} \qquad (8\text{–}4)$$

Equation 8–4 can be rearranged in the following manner:

$$-\Delta G^{\circ\prime} = 2.3\,RT\log\frac{[B]}{[A]}$$

$$\therefore \Delta G^{\circ\prime} = -2.3\,RT\log\frac{[B]}{[A]} \qquad (8\text{–}5)$$

A chemical reaction that has reached equilibrium is shown in the following equation:

$$A \overset{K_{eq}}{\rightleftharpoons} B \quad \text{where} \quad K_{eq} = \frac{[B]}{[A]}$$

Equation 8–5 can be rewritten as:

$$\Delta G^{\circ\prime} = -2.3\,RT \log K_{eq} \qquad (8\text{–}6)$$

by substituting K_{eq} for the [B]/[A] ratio in Equation 8–5. The standard free energy change ($\Delta G^{\circ\prime}$) is the free-energy change for a reaction at *standard conditions* with $\Delta G = 0$. (At standard conditions, the actual concentrations of A and B are 1.0 M; the temperature is 25 °C (298 K); the pressure is 1 atmosphere; and the pH is 7.0.) Inserting the appropriate values into Equation 8–6, one can simplify the equation considerably.

$$\begin{aligned} \Delta G^{\circ\prime} &= -2.3\,RT \log K_{eq} \\ &= -(2.3)(1.99\ \text{cal/mol}\cdot\text{K})\,(298\ \text{K}) \log K_{eq} \\ &= \Delta G^{\circ\prime} = -(1363) \log K_{eq} \end{aligned} \qquad (8\text{–}7)$$

Thus, the standard free-energy change is *directly* related to the K_{eq} term. If K_{eq} is large ($K_{eq} \sim 10^3$), then the $\Delta G^{\circ\prime}$ term will be a large negative value as well, indicating a very spontaneous reaction.

$$\underset{\text{(1 molecule)}}{A} \quad \overset{K_{eq}}{\rightleftharpoons} \quad \underset{\text{(1000 molecules)}}{B} \qquad K_{eq} = \frac{[B]}{[A]} = \frac{1000}{1}$$

$$\begin{aligned} \Delta G^{\circ\prime} &= -(1363) \log K_{eq} \\ &= -(1363) \log (10)^3 \\ &= -(1363)(3) \\ &= -4089\ \text{cal/mol} \end{aligned}$$

If the K_{eq} value is small ($K_{eq} \sim 10^{-3}$), then the $\Delta G^{\circ\prime}$ value will be a large, positive value, indicating a *very nonspontaneous* reaction in which very few molecules of A will be converted into B. (Another way of looking at this is that the reaction will proceed spontaneously in the reverse direction.)

$$\underset{\text{(1000 molecules)}}{A} \quad \overset{K_{eq}}{\rightleftharpoons} \quad \underset{\text{(1 molecule)}}{B} \qquad K_{eq} = \frac{[B]}{[A]} = \frac{1}{1000}$$

$$\begin{aligned} \Delta G^{\circ\prime} &= -(1363) \log K_{eq} \\ &= -(1363) \log (10)^{-3} \\ &= -(1363)(-3) \\ &= +4089\ \text{cal/mol} \end{aligned}$$

The relationship between $\Delta G^{\circ\prime}$ and K_{eq} values is summarized in Table 8–1.

TABLE 8–1 Relationship Between $\Delta G^{\circ\prime}$ and K_{eq}

K_{eq}	$\log K_{eq}$	$\Delta G^{\circ\prime}$
1000	3	−4089 cal/mol
100	2	−2726 cal/mol
10	1	−1363 cal/mol
1	0	0
0.1	−1	+1363 cal/mol
0.01	−2	+2726 cal/mol
0.001	−3	+4089 cal/mol

Note: $\Delta G = -(1363)\log K_{eq}$ at 25°C.

actual free-energy change (ΔG)

It must be stressed that most cellular reactions do *not* occur at initial reactant concentrations of 1.0 M. The concentrations of most cellular constituents are significantly less than 1.0 M (0.01–0.001 M, or even less). Therefore, the **actual free-energy change (ΔG)** must be calculated using Equation 8–3 for a chemical reaction occurring within the cell at nonstandard conditions. Representative calculations of the *actual* free-energy change (ΔG) are shown in Table 8–2.

TABLE 8–2 Calculation of *Actual* Free-Energy Change

$$\Delta G = \Delta G^{\circ\prime} + 2.3\,RT\log\frac{[B]}{[A]}$$

(Actual) (Standard)

Case #1: Calculate the *actual* ΔG value when $\Delta G^{\circ\prime} = -4000$ cal/mol (−4.0 kcal/mol); $T = 25$°C(298 K); [A] = 0.1 M; and [B] = 1.0 M.

$$\underset{(0.1\ \text{M})}{\text{A}} \rightleftarrows \underset{(1.0\ \text{M})}{\text{B}}$$

$$\Delta G = -4.0\ \text{kcal/mol} + (2.3)(1.99)(298\ \text{K})\log\frac{(1.0\ \text{M})}{(0.1\ \text{M})}$$

$$= -4.0 + (1363)\log(10)^1$$

$$= -4.0\ \text{kcal/mol} + 1.36\ \text{kilocal/mol}$$

$$\Delta G = -2.64\ \text{kcal/mol}$$

∴ This is a spontaneous reaction.

Case #2: Calculate the *actual* ΔG value when $\Delta G^{\circ\prime} = 1500$ cal/mol (+1.5 kcal/mol); $T = 25$°C(298 K); [A] = 0.6 M; and [B] = 0.4 M.

$$\underset{(0.6\ \text{M})}{\text{A}} \rightleftarrows \underset{(0.4\ \text{M})}{\text{B}}$$

$$\Delta G = +1.5\ \text{kcal/mol} + (2.3)(1.99)(298\ \text{K})\log\frac{(0.4\ \text{M})}{(0.6\ \text{M})}$$

$$= +1.5 + (1363)\log(0.67)$$

$$= +1.5 + (1363)(-0.17)$$

$$= (+1.5\ \text{kcal/mol}) + (-0.23\ \text{kcal/mol})$$

$$\Delta G = +1.27\ \text{kcal/mol}$$

∴ This is a nonspontaneous reaction.

COUPLED REACTIONS

coupled reactions

The chemical energy contained within various nutrient molecules can be extracted by the cell, and *some* of this energy can be used in constructing other biomolecules. This form of chemical work is accomplished by what are known as **coupled reactions,** usually catalyzed by enzymes. In a coupled reaction, the energy contained in a molecule is trapped in the form of an energy-rich intermediate and used to drive a second reaction. This process is depicted in Figure 8–3, in which a circus "teeter board" is used to represent the changes in the free energy for the reactions.

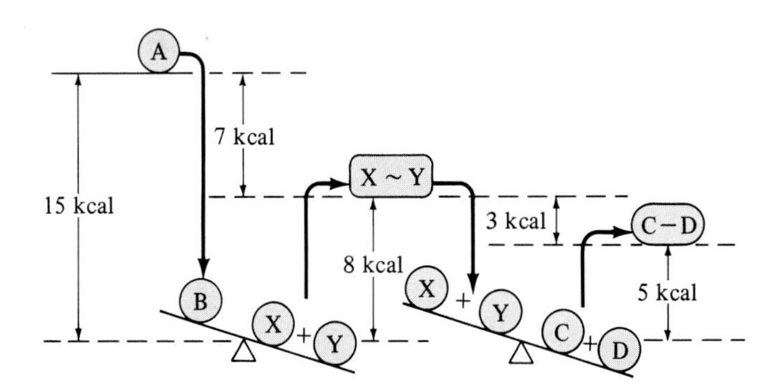

FIGURE 8–3 The free-energy changes associated with coupled reactions (Adapted from Segel, I. H., *Biochemical Calculations*, 2nd ed. (New York: John Wiley, 1976), p. 147.)

Referring to Figure 8–3, as A reacts and forms B, 15 kilocalories/mole (15 kcal) of free energy would be released to the environment. Part of this energy can be trapped in the form of a high-energy intermediate (X ~ Y) (the "squiggle bond" symbolizes its high energy content). Eight kcal of energy are required to synthesize (X ~ Y) from X and Y, and the remaining 7 kcal are wasted as heat during the synthesis. The energy-rich compound X ~ Y is then used to synthesize (C—D) from reactants C and D. Five kcal of energy are required to make (C—D), and the remaining 3 kcal are wasted as heat. In summary, some of the energy contained in substance A is trapped as an energy-rich intermediate. Some of the energy contained in this compound is then used to make compound (C—D). Both reactions waste some heat to the environment. It should be stressed that most cellular reactions transfer energy via what is known as a *common intermediate*, such as B in the following reaction:

$$A \xrightarrow{E_1} \textcircled{B}$$
$$\textcircled{B} \xrightarrow{E_2} C$$

8.3 THE CHEMISTRY OF ADENOSINE TRIPHOSPHATE

adenosine triphosphate (ATP)

During the 1930s and 1940s, a compound called **adenosine triphosphate (ATP)** was isolated from cells, and its role in cellular metabolism was elucidated. The molecule contains a five-membered sugar (ribose), a *purine* ring system (adenine), and a triphosphate "tail" attached to the #5 carbon OH group of the ribose ring. The structure of ATP is shown in the following diagram:

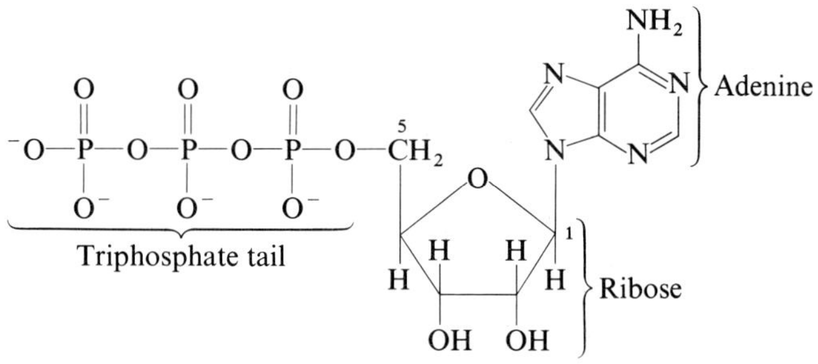

$$^{-}O-\overset{O}{\overset{\|}{\underset{O^-}{\underset{|}{P}}}}-O-\overset{O}{\overset{\|}{\underset{O^-}{\underset{|}{P}}}}-O-\overset{O}{\overset{\|}{\underset{O^-}{\underset{|}{P}}}}-O-\overset{5}{CH_2}-\text{Ribose}-\text{Adenine}$$

Triphosphate tail | Adenine | Ribose

Adenosine triphosphate (ATP)

Divalent magnesium (Mg^{2+}) is usually associated with the negative oxygen atoms of the triphosphate tail, forming an ATP/Mg^{2+} complex. The presence of the Mg^{2+} significantly influences the properties of the ATP molecule.

It has been firmly established that ATP is the primary energy carrier in all life forms. As cellular fuels are broken down, some of the free energy contained in those molecules is conserved in the form of ATP. When cells must synthesize various molecules, free energy is required to make the new chemical bonds; this energy is supplied by the ATP molecule. Specifically, ATP hydrolyzes into **adenosine diphosphate (ADP)** and **inorganic phosphate (P_i)**, releasing a significant amount of free energy.

adenosine diphosphate (ADP)
inorganic phosphate (P_i)
orthophosphate cleavage

The hydrolysis of ATP into ADP and P_i is summarized in the following reaction (called an **orthophosphate cleavage**):

Cleavage point

$$\text{Adenine-Ribose-}CH_2-O-\overset{O}{\overset{\|}{\underset{O^-}{\underset{|}{P}}}}-O-\overset{O}{\overset{\|}{\underset{O^-}{\underset{|}{P}}}}-O-\overset{O}{\overset{\|}{\underset{O^-}{\underset{|}{P}}}}-O^- \xrightarrow{H_2O} \quad (\Delta G^{\circ\prime} = -7.3 \text{ kcal/mol})$$

(ATP)

$$\text{Adenine-Ribose-}CH_2-O-\overset{O}{\overset{\|}{\underset{O^-}{\underset{|}{P}}}}-O-\overset{O}{\overset{\|}{\underset{O^-}{\underset{|}{P}}}}-O^- + {}^{-}O-\overset{O}{\overset{\|}{\underset{O^-}{\underset{|}{P}}}}-OH$$

(ADP) (P_i)

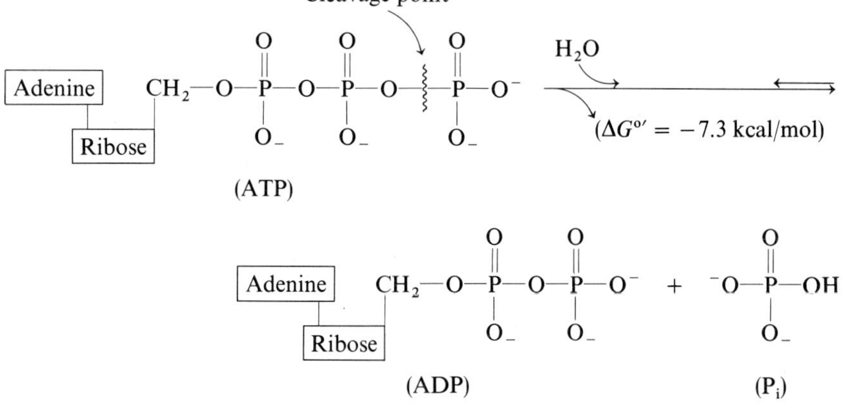

When ATP hydrolyzes under standard conditions into ADP and P_i, approximately 7300 cal/mol (7.3 kcal/mol) of free energy is released. The *actual* free-energy change ($\Delta G'$) within the cell is much greater than the *standard* free-energy change for ATP hydrolysis ($\Delta G' = -14$ to -15 kcal/mol). The actual free-energy change is greater because the intracellular concentrations of ATP, ADP, and P_i are significantly different from 1.0 M and because the pH and the concentration of Mg^{2+} within the cell significantly affect the free energy of hydrolysis. It should also be noted that $+7.3$ kcal/mol ($\Delta G^{o\prime}$) are *required* to synthesize ATP from ADP + P_i.

adenosine monophosphate (AMP)
pyrophosphate (PP_i)

Adenosine triphosphate can undergo a second type of hydrolysis reaction, in which the ATP molecule is hydrolyzed into **adenosine monophosphate (AMP)** and **pyrophosphate (PP_i)**, thereby releasing approximately 7.3 kcal/mol of energy under standard conditions according to the following reaction:

$$\underset{\text{(ATP)}}{\text{Adenine-Ribose-}CH_2{-}O{-}\overset{O}{\overset{\|}{\underset{O^-}{\underset{|}{P}}}}{-}O{-}\overset{O}{\overset{\|}{\underset{O^-}{\underset{|}{P}}}}{-}O{-}\overset{O}{\overset{\|}{\underset{O^-}{\underset{|}{P}}}}{-}O^-} \xrightarrow[(\Delta G^{o\prime} = -7.3\ \text{kcal/mol})]{H_2O}$$

$$\underset{\text{(AMP)}}{\text{Adenine-Ribose-}CH_2{-}O{-}\overset{O}{\overset{\|}{\underset{O^-}{\underset{|}{P}}}}{-}O^-} + \underset{(PP_i)}{{}^-O{-}\overset{O}{\overset{\|}{\underset{O^-}{\underset{|}{P}}}}{-}O{-}\overset{O}{\overset{\|}{\underset{O^-}{\underset{|}{P}}}}{-}O^-}$$

pyrophosphate cleavage

This reaction is called **pyrophosphate cleavage** because it produces *inorganic pyrophosphate* (PP_i). The inorganic pyrophosphate can undergo further hydrolysis, as shown in the following equation:

$$PP_i + H_2O \xrightarrow[\Delta G^{o\prime} = -8.0\ \text{kcal/mol}]{} 2P_i$$

Approximately -8.0 kcal/mol of free energy is released in this reaction. As we shall see in future chapters, the hydrolysis of ATP into AMP and PP_i (with the further hydrolysis of PP_i into 2 molecules of P_i) releases enough free energy to drive certain nonspontaneous chemical reactions.

PHOSPHATE GROUP TRANSFER

A number of different cellular intermediates are *phosphorylated* compounds (that is, they have one or more phosphate groups bound to them). These phosphate groups can undergo hydrolysis in which the phosphate group is removed from the phosphorylated intermediate, thus releasing free energy. Some phosphorylated compounds release a large amount of free energy (phosphoenolpyruvate, $\Delta G^{o\prime} = -14.8$ kcal/mol); other compounds release relatively little (glucose 6-phosphate, $\Delta G^{o\prime} = -3.30$ kcal/mol). Adenosine triphosphate releases an *intermediate amount* of free energy by comparison to phosphoenolpyruvate and glucose 6-phosphate ($\Delta G^{o\prime} = -7.3$ kcal/mol). The high-energy compounds, such as phosphoenolpyruvate, tend to *donate* their phosphate group to a lower-energy phosphate group acceptor, such as ADP. In turn, intermediate-energy compounds, such as ATP, tend to donate their phosphate group to a lower-energy phosphate-group acceptor, such as glucose. This concept can be summarized by the following equations:

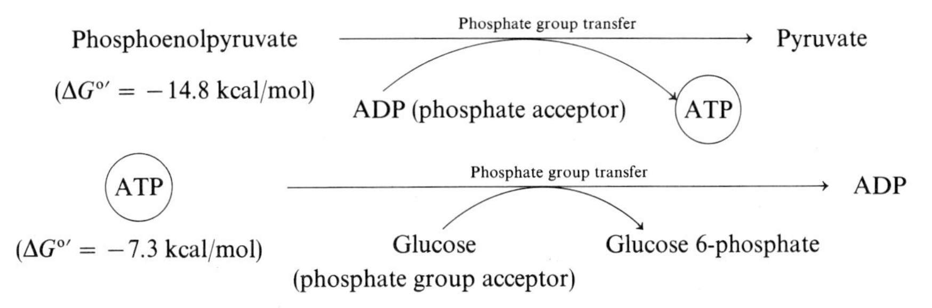

It should be stressed that ADP and ATP assume an *intermediate* position in which ADP acts as a phosphate group acceptor (from high-energy substances) and ATP acts as a phosphate group donor (to low-energy substances). The intermediate role of ADP/ATP is depicted in Figure 8–4.

Finally, we should explain why ATP releases a relatively large amount of free energy upon hydrolysis. Referring to the chemical structure of ATP, we should note that a number of phosphorus-oxygen bonds are present in the triphosphate tail region of the molecule. The last such bond is rather

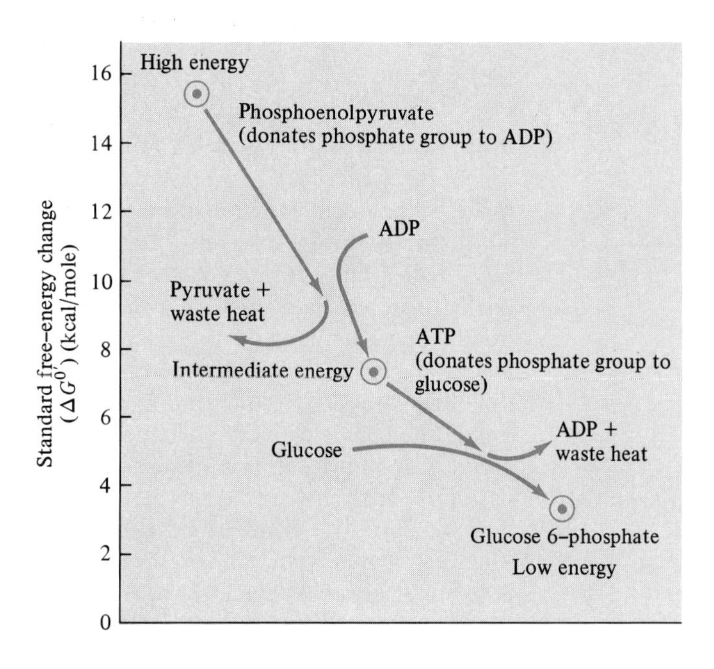

FIGURE 8–4
The intermediate role of adenosine triphosphate (ATP) in phosphate group transfer.

labile (unstable), and when the bond is broken, large amounts of free energy are released.

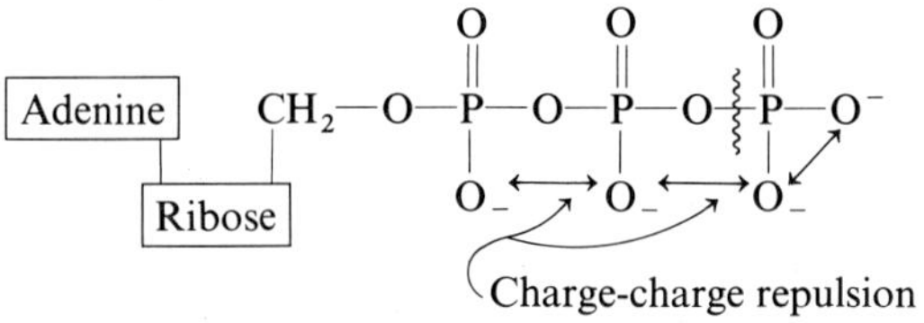

Adenine — Ribose — CH_2—O—P(=O)(O⁻)—O—P(=O)(O⁻)—O~P(=O)(O⁻)—O^-

Charge-charge repulsion

This terminal phosphate group (often symbolized as ~P) is unstable because the entire triphosphate tail contains 4 negatively charged oxygen atoms (at pH 7) bound to the phosphorus atoms. These oxygen atoms experience charge-charge repulsion; this repulsion ultimately splits the terminal phosphate group away from the ATP molecule. The products of ATP hydrolysis (the diphosphate tail of ADP and the inorganic phosphate group P_i) are *more* stable than the original ATP molecule. In addition, they are negatively charged (at pH 7) and, thus, are not likely to reform back into ATP because of charge-charge repulsion. Since the energy content of the products is significantly *less* than the energy content of the reactants, the free-energy change will be a relatively large negative value.

$$\Delta G = G_B - G_A \qquad (8\text{–}2)$$

$$\Delta G_{ATP} = [\underbrace{G_{ADP+P_i}}_{\text{(Products)}} - \underbrace{G_{ATP}}_{\text{(Reactants)}}]$$

8.4 DIGESTION OF CARBOHYDRATES

Carbohydrates supply between 40 and 60 percent of the total calories ingested in our diets. Starch (a homopolysaccharide) and sucrose constitute approximately 90 percent of the **digestible** dietary carbohydrates; glucose, lactose, and other simple sugars constitute the remaining 10 percent.

digestible

indigestible

Indigestible complex carbohydrates, such as cellulose, are also normally ingested each day in the form of vegetables and other fibrous foods. Since humans do not possess enzymes capable of hydrolyzing the $\beta(1 \rightarrow 4)$ glycosidic bonds of cellulose (see Chapter 7, Carbohydrates), cellulose cannot be utilized as cellular fuel. However, the undigestible cellulose fibers provide important bulk and roughage for normal intestinal motility and function. It is speculated that cellulose fibers present within the intestinal contents, or **chyme**, also absorb toxic substances that we either ingest or produce during the digestive process. These toxic substances are, thus, eliminated from our systems before they can be absorbed by the body. The importance of dietary fiber will be discussed in greater detail in Chapter 20 (Vitamins and Nutrition).

chyme

When foods rich in complex and simple carbohydrates are consumed, the food is immediately mixed with saliva, which contains the enzyme *α-amylase*. This enzyme can hydrolyze the $\alpha(1 \rightarrow 4)$ glycosidic bonds of amylose and the outer branches of amylopectin (see Chapter 7, Carbohydrates). Since food is only in contact with saliva for a short time, the salivary amylase is only responsible for the *partial digestion* of the carbohydrates. When the saliva/food mixture is swallowed, it comes in contact with the contents of the stomach, which are acidic and inactivate the salivary amylase. Thereafter, very little enzymatic digestion takes place within the stomach itself.

Once the stomach contents enter the small intestine, the pancreas begins to secrete pancreatic juice, which neutralizes the acidic stomach contents to pH 7–8. The pancreatic juice contains *pancreatic α-amylase*. This enzyme completely degrades the ingested starches—first into much smaller fragments of around 6–7 glucose residues, then into 2-carbon (maltose) and 3-carbon (maltotriose) fragments. These fragments are ultimately converted into individual monosaccharide residues by enzymes in the intestinal **brush border**. This brush border is located on the exterior surface of the intestinal mucosal cells (actually, on the exterior surface of the mucosal cell microvilli), and is composed of a complex network of proteins, glycoproteins, and hydrolytic enzymes. Figure 8–5 is a diagram of the intestinal villus, together with the intestinal mucosal cells and their

brush border

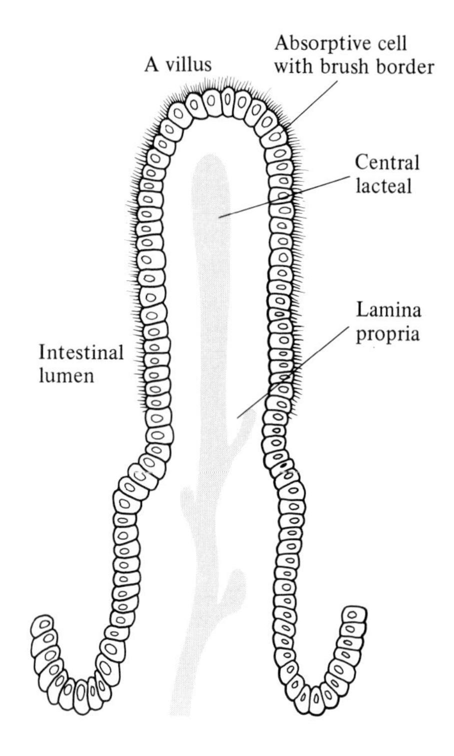

FIGURE 8–5
An intestinal villus showing absorptive mucosal cells with associated brush border. Carbohydrate digestion and transport occur within the brush-border region of the intestinal mucosal cells. (Adapted from Bhagavan, N. V., *Biochemistry*, 2nd ed. (Philadelphia: J. B. Lippincott, 1978), p. 179.)

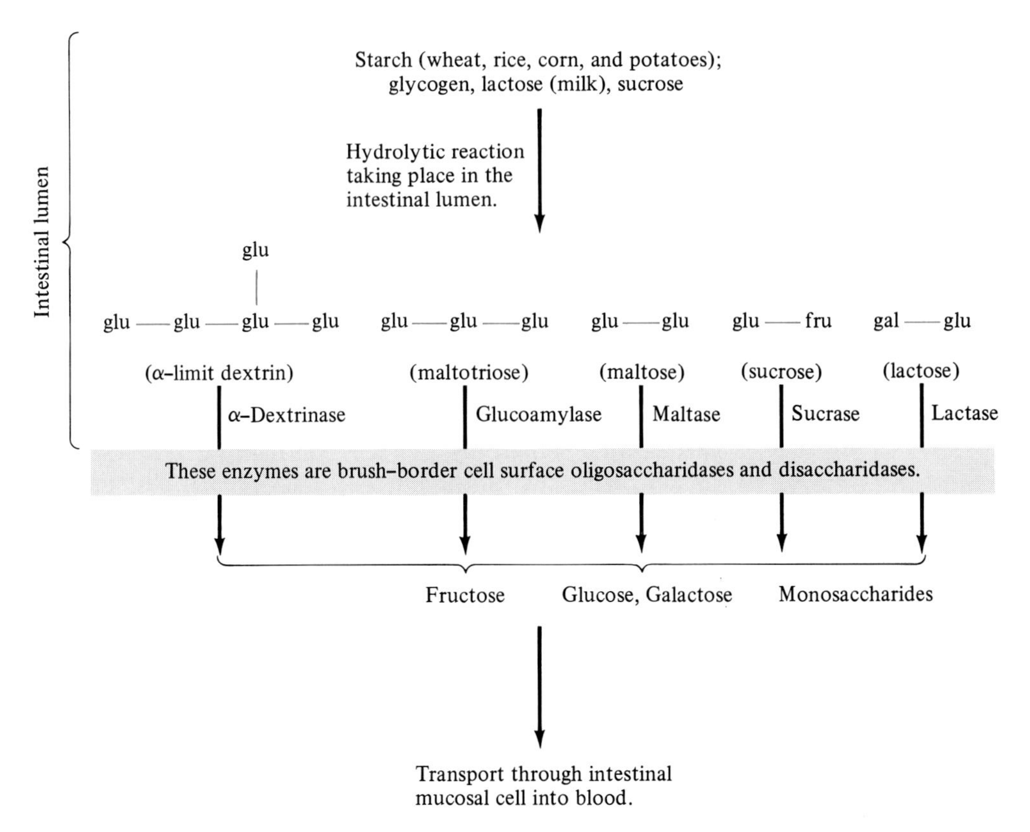

FIGURE 8–6 The digestion and absorption of carbohydrates within the intestinal brush border. (Adapted from Bhagavan, N. V., *Biochemistry*, 2nd ed. (Philadelphia: J. B. Lippincott, 1978), p. 178.)

associated brush border. Figure 8–6 summarizes the different enzymes in the brush border that hydrolyze the various oligosaccharides into monosaccharides.

The monosaccharides thus formed are rapidly absorbed across the membranes of the intestinal mucosal cells and transported into the blood (Figure 8–6). Thus, the concentration of sugars in the blood increases after a carbohydrate meal has been ingested (often in 15 to 30 minutes). Most body cells use monosaccharides as a cellular fuel in which some of the chemical energy contained in the sugar molecule is extracted and conserved in the form of ATP. Some cells use both sugar molecules and fat molecules for energy; others, such as brain cells, depend almost totally on glucose for their energy needs.

Cells must transport sugars from the blood across their cell membranes into their cytoplasm, where the sugars can be broken down and their energy extracted. We will center our attention on the transport of glucose. Since glucose molecules are very hydrophilic, they do *not* readily cross the plasma membrane of the cell. Certain proteins in the cell membrane are specific carriers for glucose; these facilitate the passage of glucose from the blood into the cell. Glucose transport takes place down a concentration gradient—that is, the concentration of glucose outside the cell is *greater* than the concentration of glucose within the cell. This concentration gradient drives the transport of glucose across the membrane. The process is called *facilitated diffusion*, because a carrier protein facilitates the transport of the glucose across the membrane. The transport of glucose into the cells of certain tissues is under strict hormonal regulation. For instance, glucose transport into skeletal muscle and heart muscle cells is regulated by the polypeptide hormone *insulin*. Insulin (as we shall see in the next section) keeps blood glucose levels at about 80 mg glucose/100 ml by promoting the transport of excess glucose from the blood into the tissues. Interestingly, glucose transport into liver cells is apparently not under hormonal control.

8.5 ABNORMAL SUGAR ABSORPTION AND TRANSPORT

BRUSH-BORDER ENZYME DEFICIENCIES

If the intestinal brush border is deficient in one of the enzymes responsible for the hydrolysis of dietary oligo- and disaccharides (see Figure 8–6, p. 148), an intolerance for a particular sugar will develop. For example, if the brush border does not contain a particular **disaccharidase**, then a specific disaccharide will not be hydrolyzed. The presence of this unhydrolyzed disaccharide within the intestinal contents will produce a variety of unpleasant symptoms, such as bloating, flatulence, and loose stools. The symptoms can be relieved by eliminating the offending carbohydrate from the diet.

disaccharidase

lactose intolerance

Lactose intolerance is an example of a common sugar intolerance. In this case, the brush-border enzyme *lactase* is deficient. Lactase is responsible for hydrolyzing the milk sugar lactose into free galactose and glucose, as shown in the following reaction:

$$\boxed{\text{Galactose}} \overset{\beta(1 \rightarrow 4)}{—} \boxed{\text{Glucose}} \xrightarrow{\text{Lactase}} \begin{array}{c}\text{Free galactose}\\ + \\ \text{Free glucose}\end{array}$$

There are two main categories of lactose intolerance, (1) primary (hereditary) lactose intolerance and (2) temporary lactose intolerance. Primary lactose intolerance is very common among certain population groups (Orientals, Indians, and Blacks). Temporary lactose intolerance may develop as a result of temporary damage to the intestinal brush border. Such damage can be produced by gastrointestinal tract infections or by various drugs. For example, infants sometimes develop a severe GI tract infection that destroys the brush-border lactase. Analysis of the loose stools reveals a high lactose content, indicating inadequate degradation of lactose. Analysis of blood samples reveals low blood glucose and galactose levels (which can be brought back to normal levels via intravenous feeding). Large amounts of lactose within the intestine cause the intestine to take up water (hence the loose, watery stools); they also provide an excellent food source for GI tract bacteria, which produce gaseous products (hence the bloating and flatulence). Once the GI tract infection has been brought under control, the brush border will heal and resynthesize new lactase. Therefore, this form of the disease is considered to be temporary.

DISEASE STATES ASSOCIATED WITH ABNORMAL GLUCOSE TRANSPORT

The normal fasting concentration of glucose within the blood is maintained under constant hormonal control at 80 mg glucose/100 ml of blood to insure that various tissues have an adequate supply of glucose. The brain is especially sensitive to variations in glucose level, since its cells depend almost entirely on glucose for chemical energy. When the brain is working hard (for instance, while a person is studying for an exam), the brain consumes over 67 percent of the blood glucose. The remaining 33 percent is consumed by the red blood cells and skeletal muscles. The brain can use substances called *ketone bodies* for energy (see Chapter 14, Lipid Metabolism), but only during periods of severe metabolic stress, such as in starvation. During starvation, when carbohydrate intake is restricted or eliminated, fat reserves are mobilized and yield ketone bodies, which can then be utilized as fuel by the brain cells.

When a carbohydrate meal is consumed, the blood glucose level increases from 80 mg/100 ml to between 125 and 150 mg/100 ml. This increase in the blood glucose concentration signals the *pancreas* to release a small amount of the hormone **insulin** into the blood stream. Insulin enhances the absorption of glucose from the blood into such tissue cells as skeletal and heart muscle cells. Insulin is a small-molecular-weight hormone consisting of two different polypeptide chains (the A chain and the B chain) held together by disulfide bonds (see Chapter 3, Amino Acids and Proteins). Insulin apparently binds to specific receptor sites on the outer surface of cells, thus regulating the transport of glucose into the cell. (Precisely how it regulates glucose transport, however, is still largely unknown.) Since insulin promotes the transport of glucose into the cells, the blood glucose level begins to decrease within about 1 hour after a carbohydrate meal has been ingested, and returns to normal within about 3 hours. The insulin circulating in the blood is quickly degraded (the average lifetime of an insulin molecule in the blood is only 10–15 minutes). A second polypeptide hormone, **glucagon**, is also synthesized and secreted by the pancreas (α-cells). Glucagon acts to *raise* blood glucose levels and, thus, counteracts insulin. At this point, we summarize these concepts in Figure 8–7. Figure 8–7 represents the changes in blood glucose level that follow the ingestion of a carbohydrate meal (or drink); it is called a **glucose tolerance curve**.

insulin

glucagon

glucose tolerance curve

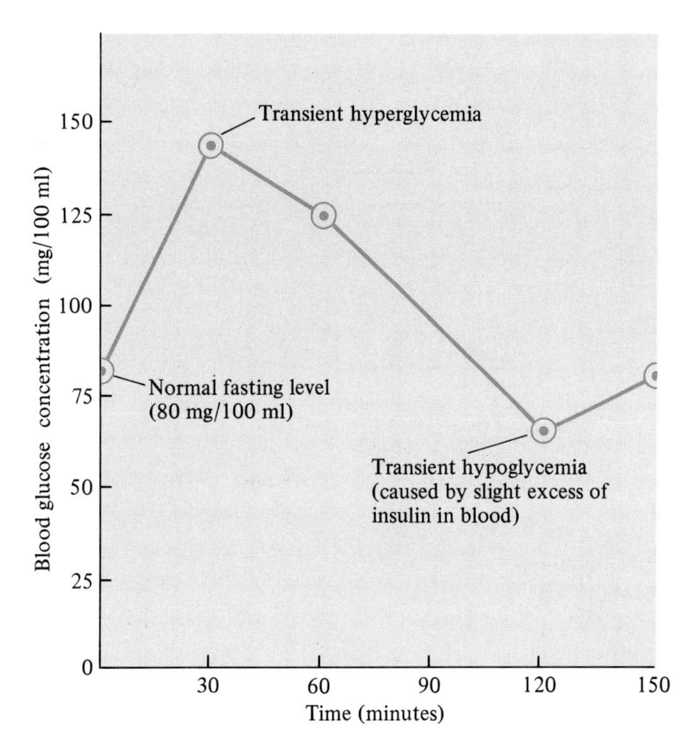

FIGURE 8–7
A normal glucose tolerance curve. The transient hyperglycemic state signals the pancreas to secrete insulin, which promotes the transport of glucose from the blood into the tissue cells. Thus, the blood glucose level falls back to normal.

A glucose tolerance curve can be generated by having a normal individual fast for a period of 12–18 hours. A control blood sample is drawn and analyzed for glucose concentration (the fasting level should be about 70–80 mg/100 ml). A large dose of glucose is then ingested by the test subject in the form of a soda pop solution called Glucola. Blood samples are drawn at 30-minute intervals for 2–3 hours, and the concentration of glucose in these samples is determined. Figure 8–7 represents a normal glucose tolerance test, because the blood glucose levels rise and fall back to a normal level following carbohydrate intake. A normal individual, thus, can tolerate the glucose load on his or her system. We daily subject ourselves to this pattern when we eat or drink carbohydrates in the form of soft drinks, pastries, ice cream, and candy.

hyperglycemia

Abnormal glucose tolerance curves are shown in Figure 8–8. In the case of **hyperglycemia** (*too much* glucose in the blood), fasting blood glucose levels may be as high as 150 mg/100 ml. Hyperglycemia is usually caused by inadequate amounts of insulin in the blood; because of this, very little glucose is actually being transported from the blood into the tissue cells. One can think of this as "cells starving in the midst of plenty."

hypoglycemia

In the case of **hypoglycemia** (*too little* glucose present in the blood), fasting blood glucose levels are usually below 40 mg/100 ml. This clinical situation is usually caused by too much insulin in the blood; thus, too much

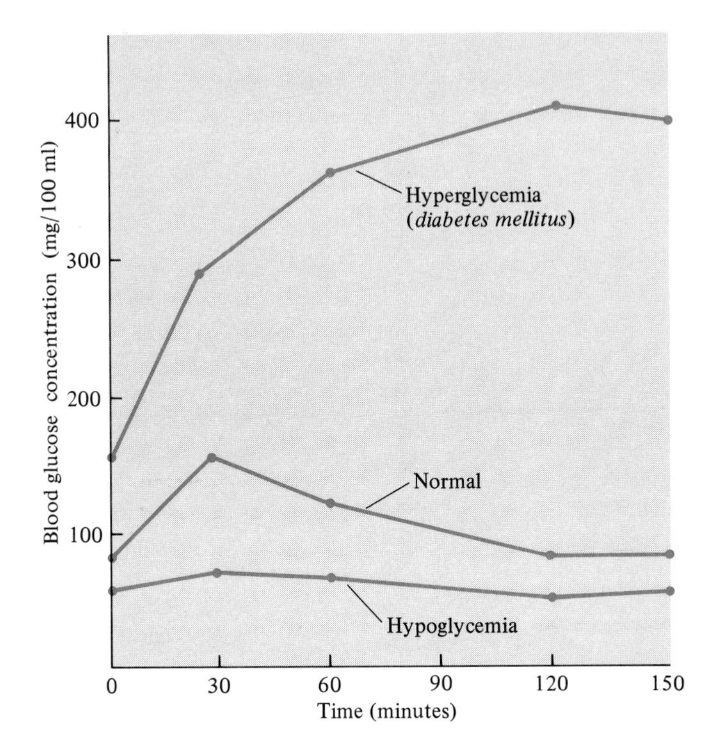

FIGURE 8–8
Abnormal glucose tolerance curves. Hypoglycemia is caused by too much insulin in the blood; hyperglycemia is caused by too little insulin in the blood.

glucose is being transported from the blood into the tissue cells. This condition can be dangerous if the blood glucose level falls to 20–30 mg/100 ml. At these low concentrations, brain cells begin to "starve." Moderate hypoglycemia is characterized by sweating, anxiety, dizziness, trembling, and muscular weakness.

diabetes

polyuria

diabetes mellitus

Having considered the general clinical definitions of hyper- and hypoglycemia, let us consider diabetes in greater detail. The word **diabetes** is used in medicine to indicate any disorder in which an excessive excretion of urine occurs **(polyuria)**. (In fact, the word "diabetes," Greek, means "siphon.") **Diabetes mellitus**, a disorder of carbohydrate, lipid, and protein metabolism, is a specific type of diabetes, characterized by the production of large quantities of urine containing glucose (*mellitus* is the Latin word for "honey," and refers to the fact that the urine is sweet when tasted—a diagnostic test fortunately no longer required of physicians). **Diabetes insipidus**, a pituitary disorder, is another form of diabetes, characterized by the excretion of large quantities of colorless and tasteless (*insipid*) urine. Abnormal glucose metabolism is *not* associated with this disease; thus, large amounts of glucose are not present in the urine.

diabetes insipidus

juvenile-onset diabetes

maturity-onset diabetes

There are two main forms of diabetes mellitus, (1) **juvenile-onset diabetes** and (2) **maturity-onset diabetes**. As the name implies, juvenile-onset diabetes usually becomes manifest during childhood (that is, before the age of 20). In juvenile-onset diabetes, the β-cells of the pancreas are destroyed (possibly by a viral infection); thus, insulin production and secretion is halted. The disease can be controlled through daily injections of insulin and a strict dietary program. Maturity-onset diabetes usually becomes manifest in adulthood. In this form of the disease, a normal amount of insulin is secreted; however, the body does not *release* it soon enough in response to the hyperglycemic state resulting from carbohydrate ingestion. This delay causes blood glucose levels to rise. In addition, afflicted individuals apparently have a decreased sensitivity to insulin (*insulin resistance*). Obesity is associated with maturity-onset diabetes. The disease can often be treated by a careful dietary and exercise regimen.

Diabetes mellitus causes a variety of serious clinical and metabolic disturbances. Blood glucose levels are often very elevated (hyperglycemia) and glucose is ultimately excreted into the urine **(glycosuria)**. Liver glycogen levels fall to very low levels (see Chapter 10, Carbohydrate Metabolism). Thus, an individual has a very low level of reserve carbohydrates. Tissue protein breakdown is stimulated to meet cellular energy needs. Lipid reserves are mobilized (release and secretion of fats from the adipose tissue) and ketone body formation is stimulated (see Chapter 14, Lipid Metabolism). Excessive ketone bodies (such as acetone) are thus produced as an alternative fuel. **Ketone** (or *acetone*) **breath** is often detected when blood ketone body levels reach too high a level. Excretion of large amounts of glucose and ketone bodies into the urine results in dehydration and leads to severe thirst.

glycosuria

ketone breath

Presently, clinical management is largely restricted to daily injections of known doses of commercial insulin (from an animal source) along with a strict dietary regimen. A number of drug companies are involved in producing *human insulin* by recombinant DNA techniques (see Chapter 17). To produce human insulin in this manner, the part of the DNA molecule that codes for the human insulin molecule is isolated and "spliced" into the DNA molecule of a bacterium. The bacteria are grown in culture and synthesize human insulin in large quantities. This insulin is then purified. When FDA approval is obtained, this product will be sold to diabetics as an alternative and better form of insulin (as compared to insulin isolated from animal sources).

Other research efforts are directed towards developing an "artificial pancreas" that more closely reproduces a normal and healthy pancreas. For instance, a functioning "insulin pump" was recently implanted in a patient. Such pumps release a constant amount of insulin into the blood. More sophisticated pumps are being developed to release insulin in response to changing glucose levels. In another approach, pancreatic β-cell tissue was encapsulated and implanted in experimental animals and gave promising results in controlling blood glucose levels over extended periods of time.

SUMMARY

The catabolic (degradative) pathways and the anabolic (biosynthetic) pathways together constitute the metabolism of the cell. Metabolic processes extract chemical energy from the environment; convert various nutrient molecules into the building blocks of larger biomolecules; and assemble these building blocks into proteins, nucleic acids, lipids, and complex carbohydrates.

The free-energy change ($\Delta G'$) defines how much energy is available to perform useful work. The free-energy change can be defined as the difference between the energy content of the products and that of the reactants ($\Delta G' = G_B - G_A$). Free energy is released to the environment ($\Delta G'$ is negative) during a spontaneous (exergonic) reaction. When a reaction is nonspontaneous (endergonic), free energy must be put into the chemical system ($\Delta G'$ is positive) to drive the reaction and form products. No work can be done when a reaction has reached equilibrium ($\Delta G' = 0$). Changes in free energy can be calculated using Equations 8–3 and 8–6. Some of the free energy from a chemical reaction can be trapped in the form of a high-energy intermediate ($X \sim Y$). As ($X \sim Y$) breaks down into X and Y, some of the energy contained in the intermediate can be used to drive the synthesis of (C—D) from reactants C and D.

The structure of adenosine triphosphate (ATP) was presented and its essential role in the transfer of useable chemical energy was described. When ATP undergoes hydrolysis to yield adenosine diphosphate (ADP) and inorganic phosphate (P_i), a significant amount of free energy is released ($\Delta G^{o\prime} = -7.3$ kcal/mol). This energy can be used to drive other chemical reactions. Adenosine triphosphate can undergo a second type of hydrolysis reaction (pyrophosphate cleavage) in which ATP breaks down to AMP and inorganic pyrophosphate (PP_i), releasing -7.3 kcal/mol of free energy. Inorganic pyrophosphate can hydrolyze further, yielding two molecules of P_i along with more free energy. Adenosine triphosphate serves an intermediate role in the acceptance and transfer of phosphate groups. The ATP molecule is called a high-energy compound because the triphosphate "tail" has a high density of negatively charged oxygen atoms bound in close proximity to each other. This results in a very labile phosphorus/oxygen bond that, upon hydrolysis, yields a large amount of free energy.

Carbohydrates supply between 40 and 60 percent of the total calories ingested in our diets. Salivary and pancreatic amylase break down starch (amylose and amylopectin) into small-molecular-weight oligosaccharides. Complete degradation of di- and trisaccharides takes place in the brush border of the intestinal microvilli. The resulting monosaccharides are then transported across the intestinal mucosal cells into the bloodstream.

Deficiencies in one or more of the intestinal brush-border enzymes will impair the transport of the sugars by the intestinal mucosal cells.

The normal concentration of glucose within the blood is maintained at a level of about 80 mg/100 ml. After a carbohydrate meal is consumed, the blood glucose level increases to about 125 to 150 mg/100 ml (a condition called hyperglycemia). This increase stimulates the release of insulin from the pancreas, which in turn stimulates the uptake of glucose from the blood into the tissue cells. The blood glucose level drops back to normal in about 3 hours. Hypoglycemia is characterized by continued low levels of glucose. In diabetes mellitus, an example of hyperglycemia, fasting blood sugar levels may be as high as 150 mg/100 ml. Diabetes mellitus is subdivided into two main categories: (1) juvenile-onset diabetes, in which pancreatic β-cells have been destroyed (therefore, no insulin is produced), and (2) maturity-onset diabetes, in which the individual secretes normal levels of insulin, but his or her response to high blood glucose levels is delayed.

REVIEW QUESTIONS

1. We have presented a number of important concepts and terms in this chapter. Define (or briefly explain) each of the following terms:
   a. Cellular metabolism
   b. Catabolism
   c. Anabolism
   d. Bioenergetics
   e. Gibb's free energy

f. Actual free-energy change (ΔG)
g. Standard free-energy change ($\Delta G^{\circ\prime}$)
h. Exergonic reaction
i. Coupled reaction
j. Adenosine triphosphate
k. Orthophosphate cleavage
l. Phosphate group transfer
m. Brush border enzyme
n. Facilitated diffusion
o. Lactose intolerance
p. Insulin
q. Glucose tolerance test
r. Hypoglycemia
s. Hyperglycemia
t. Juvenile-onset diabetes
u. Maturity-onset diabetes

2. Consider the following processes:
   a. A moving car (burning gasoline that turns the driving wheels).
   b. A refrigerator in operation (moving electrons that turn a heat pump)
   c. A toaster in operation (moving electrons that heat up a high-resistance wire)
   d. A person jogging (oxidizing glucose that contracts skeletal muscles)

   Reflect on the physical processes involved in these items or events. Suggest how a portion of the total energy is used to perform useful work. How is the unuseable portion lost or wasted?

3. It has been suggested that "living systems create and maintain a high degree of molecular order at the expense of their environment." In terms of cellular metabolism and energy utilization, explain what this statement means.

4. Consider the following reaction

$$A \underset{}{\overset{K_{eq}}{\rightleftharpoons}} B \quad (\text{where } K_{eq} = 10^{-4})$$

   a. Is this reaction spontaneous or nonspontaneous? Why?
   b. Calculate the standard free-energy change ($\Delta G^{\circ\prime}$) for this reaction using Equation 8–7, p. 140, given that the temperature is 25°C (298 K).

5. Consider the following reaction:

$$\text{Glucose} \xrightarrow[\text{ATP} \;\rightarrow\; \text{ADP}]{} \text{Glucose 6-phosphate}$$

$$(\Delta G^{\circ\prime} = -4.0 \text{ kcal/mol})$$

   a. Calculate the K_{eq} for this reaction using Equation 8–7.
   b. What do the K_{eq} value and the standard free-energy change value tell you about the spontaneity of the reaction?

6. Consider the following reaction:

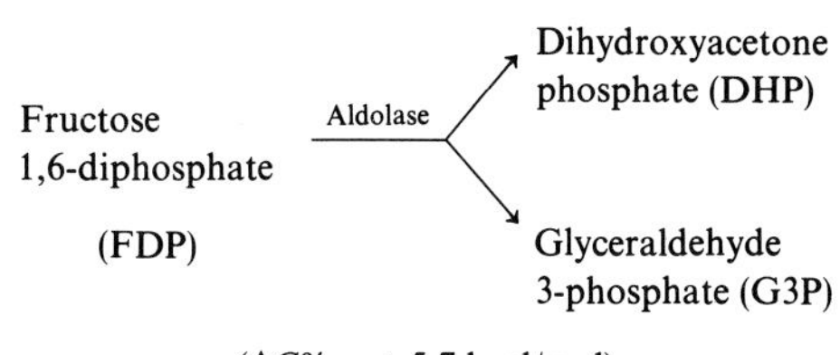

$$(\Delta G^{\circ\prime} = +5.7 \text{ kcal/mol})$$

   a. Calculate the actual free-energy change (ΔG) when $T = 25\,^{\circ}\text{C}$ (298 K); $[\text{FDP}] = 1.56 \times 10^{-4}$ M; $[\text{DHP}] = 1.4 \times 10^{-4}$ M; and $[\text{G3P}] = 1.9 \times 10^{-5}$ M. Use Equation 8–3.
   b. Compare your ΔG value with the $\Delta G^{\circ\prime}$ value for this reaction. What does this tell you about the spontaneity of the aldolase reaction under *intracellular conditions*?

7. A compound called *phosphocreatine* has a $\Delta G^{\circ\prime}$ value of -10.3 kcal/mol. Using appropriate equations, show how ATP mediates the transfer of a phosphate group from phosphocreatine to glucose to yield glucose 6-phosphate.

$$^{-}O-\overset{\overset{O}{\|}}{\underset{\underset{O^{-}}{|}}{P}}-NH-\underset{\underset{NH}{\|}}{C}-\overset{\overset{CH_3}{|}}{N}-CH_2-COO^{-}$$

Phosphocreatine

8. Divalent magnesium (Mg^{2+}) binds to ATP to form an ATP/Mg complex. Referring to the structure of ATP, draw a possible structure for this complex.

9. What type of glycosidic bonds are cleaved by the enzymes noted in Figure 8–6 of this chapter? (Refer back to Chapter 7, Carbohydrates.)

10. Indicate whether blood glucose levels would be elevated, decreased, or unchanged (as compared to a normal fasting level of 80 mg/100 ml) for the following hypothetical situations (be prepared to support your answer):

a. Excessive glucagon in the blood, no insulin present.
b. Lack of glucagon in the blood, no insulin present.
c. Pancreatic tumor producing constant and excessive amounts of both glucagon and insulin. The insulin concentration in the blood is greater than the glucagon concentration.

SUGGESTED READING

Bioenergetics and the Chemistry of ATP

Lehninger, A. L. *Bioenergetics*, 2nd ed. Menlo Park, CA: W. A. Benjamin, 1971.

Lehninger, A. L. *Biochemistry*, 2nd ed., Chs. 14 and 15. New York: Worth, 1975.

Segel, I. H. *Biochemical Calculations*, 2nd ed., Ch. 3. New York: John Wiley, 1976.

Introduction to Carbohydrate Metabolism

Bhagavan, N. V. *Biochemistry*, 2nd ed., pp. 176–85; 316–53. Philadelphia: J. B. Lippincott, 1978.

Montgomery, R., Dryer, R. L., Conway, T. W., and Spector, A. A. *Biochemistry: A Case-Oriented Approach*, 3rd ed., Ch. 7. St. Louis: C. V. Mosby., 1980.

Tietz, N., ed. *Fundamentals of Clinical Chemistry*, 2nd ed., pp. 240–63. Philadelphia: W. B. Saunders, 1976.

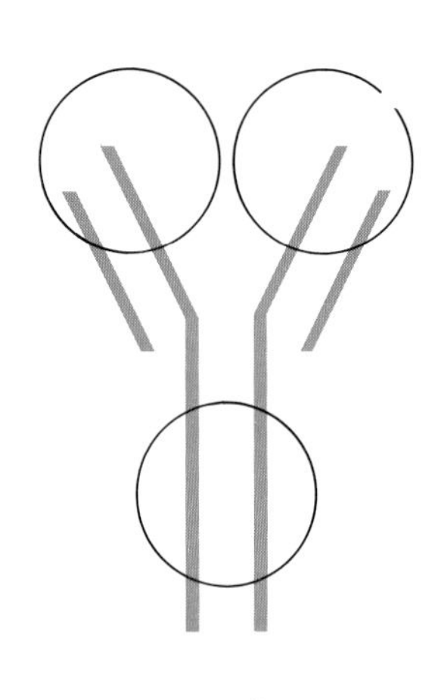

Chapter 9

Carbohydrate Metabolism: Glycolysis

9.1 INTRODUCTION

In the previous chapter, we discussed the initial degradation and absorption of carbohydrates. Let us now consider the metabolic pathways that degrade glucose and extract some of its chemical energy in the form of ATP. Glucose is a primary cellular-fuel molecule used by virtually every cell in the body. Glucose is degraded by a rather complex sequence of enzymatic steps commonly known as **glycolysis** (glycolysis means the *lysis*, or "rupture," of sugar). Some of the chemical energy contained in the glucose molecule is extracted in the form of ATP. Other simple sugars (galactose, mannose, and fructose) are converted into glycolytic intermediates and metabolized in that fashion.

glycolysis

In this chapter, we will discuss the glycolytic pathway in detail and describe how the glucose molecule is enzymatically degraded. The regulation of the pathway and the chemistry of important glycolytic enzymes will be presented, along with a brief discussion of alcoholic fermentation.

9.2 GLYCOLYSIS AND ALCOHOLIC FERMENTATION

Embden-Meyerhoff pathway

anaerobic conditions

Glycolysis (also known as the **Embden-Meyerhof pathway**) is the anaerobic, enzymatic disruption of the 6-carbon glucose molecule. It yields two 3-carbon fragments and two ATP molecules. Because this process does not require oxygen, it can be carried out under **anaerobic conditions** (without oxygen). Anaerobic glycolysis (glucose → 2 lactate) takes place in the cytoplasm of the cell and can only release 47.0 kcal/mol of free energy under standard conditions. Some of this free energy is conserved in the form of ATP.

$$\text{Glucose (6-carbon)} \xrightarrow[\searrow \text{ 2 ATP (net)}]{\text{Anaerobic glycolysis}} \text{2 lactic acid (3-carbon)}$$

In our bodies, the glycolytic pathway provides a means for obtaining *additional* ATP for muscular contraction during periods of temporary oxygen deficit, such as those occurring during vigorous exercise.

aerobic conditions

Eukaryotic cells generally utilize glucose under **aerobic conditions** (oxygen is present and is required). Under aerobic conditions, the cells form pyruvic acid instead of lactic acid (2 molecules of pyruvic acid per glucose). These pyruvic acid molecules are transported from the cytoplasm into the mitochondrion and are converted into *Acetyl-coenzyme A* (Acetyl-CoA). Acetyl-CoA then "fuels" the *t*ricarboxylic *a*cid *c*ycle (TCA, or Krebs cycle) and is *oxidized* (electrons are removed). Electrons are removed from the TCA cycle intermediates and passed on to the **electron transport system**, where ATP molecules are made via **oxidative phosphorylation**. This process is called **mitochondrial respiration** and requires oxygen as the final electron acceptor. When glucose is completely oxidized via aerobic glycolysis (glucose + $6\,O_2 \rightarrow 6\,CO_2 + 6\,H_2O$), 686 kcal/mol of free energy is released under standard conditions. Since significantly more free energy is released than under anaerobic glycolysis, a larger number of ATP molecules can be made—in fact, almost twenty times as many. (36 or 38 molecules[1] of ATP per glucose molecule are produced under aerobic conditions, as compared to 2 under anaerobic conditions.) Aerobic glycolysis and its relationship with the mitochondrial systems will be discussed in greater detail in Chapter 11, The Tricarboxylic Acid Cycle. Figure 9–1 summarizes the primary differences between aerobic and anaerobic glycolysis.

electron transport system
oxidative phosphorylation
mitochondrial respiration

Clearly, aerobic glycolysis is more efficient than anaerobic glycolysis with respect to the number of ATP molecules made. As stated previously, the anaerobic pathway is important in humans to insure continued production of ATP during periods of oxygen deficit, such as during physical exercise.

heterotrophic cells

It should be stressed that *all* **heterotrophic cells** (*cells feeding on others*) *extract energy via oxidation-reduction reactions.* If you recall, heterotrophic cells use various organic molecules (such as glucose) as their carbon and energy source. Oxidation-reduction reactions involve the transfer of electrons from an electron donor (reducing agent) to an electron acceptor (oxidizing agent). During cellular respiration, the electrons from a cellular fuel, such as glucose, are transferred from one intermediate to another and finally to molecular oxygen. Thus, *oxygen* serves as the utimate *electron acceptor*, or *oxidant*. Some of the energy contained in the fuel molecule is conserved in the form of ATP. During anaerobic glycolysis, oxidation-reduction reactions still take place even though oxygen is not used as an electron acceptor.

Certain organisms (yeasts) possess additional enzymes that convert the pyruvic acid into two molecules of ethanol and 2 CO_2 gas molecules. This process, called **alcoholic fermentation**, is shown in the following diagram:

alcoholic fermentation

$$\underset{\text{(6-carbon)}}{\text{Glucose}} \xrightarrow[\text{2 ATP (net)}]{\text{Alcoholic fermentation}} \underset{\text{(2-carbon)}}{2\ \text{Ethanol} + 2\ CO_2} \Big\}\ \text{wastes}$$

The wine, beer, and liquor industries have capitalized on this metabolic pathway to generate alcohol for human consumption.

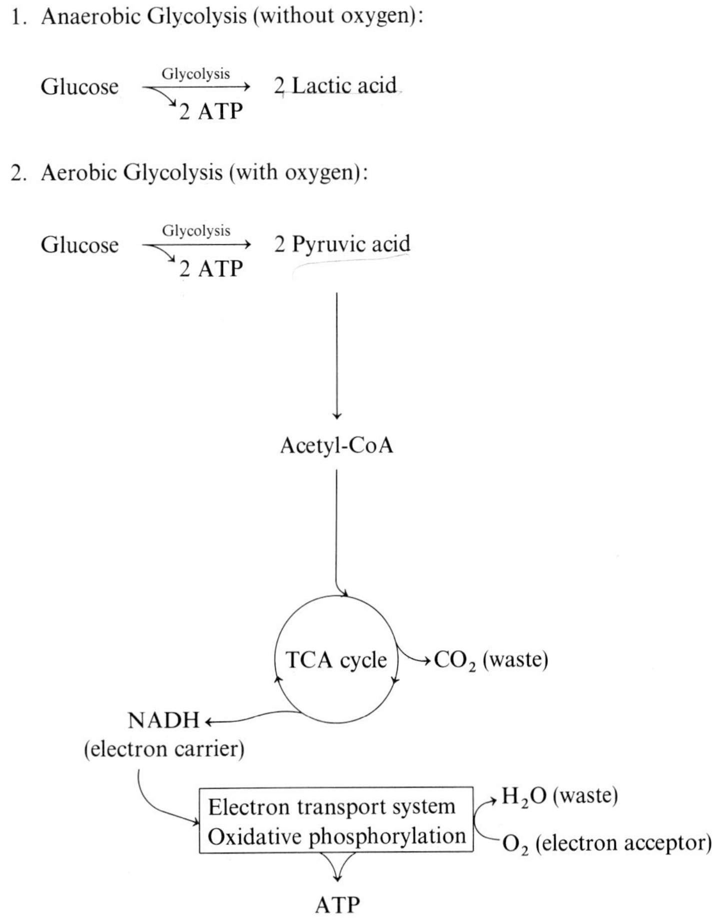

1. Anaerobic Glycolysis (without oxygen):

$$\text{Glucose} \xrightarrow[\text{2 ATP}]{\text{Glycolysis}} 2\ \text{Lactic acid}$$

2. Aerobic Glycolysis (with oxygen):

$$\text{Glucose} \xrightarrow[\text{2 ATP}]{\text{Glycolysis}} 2\ \text{Pyruvic acid}$$

FIGURE 9–1 A comparison between anaerobic and aerobic glycolysis. The electron carrier is *NADH* (*n*icotinamide *a*denine *d*inucleotide in reduced (*H*) form).

9.3 THE GLYCOLYTIC PATHWAY IN DETAIL

The glycolytic pathway is depicted in Figure 9–2. It consists of 11 separate enzymatic steps and can be divided into several important sections. First,

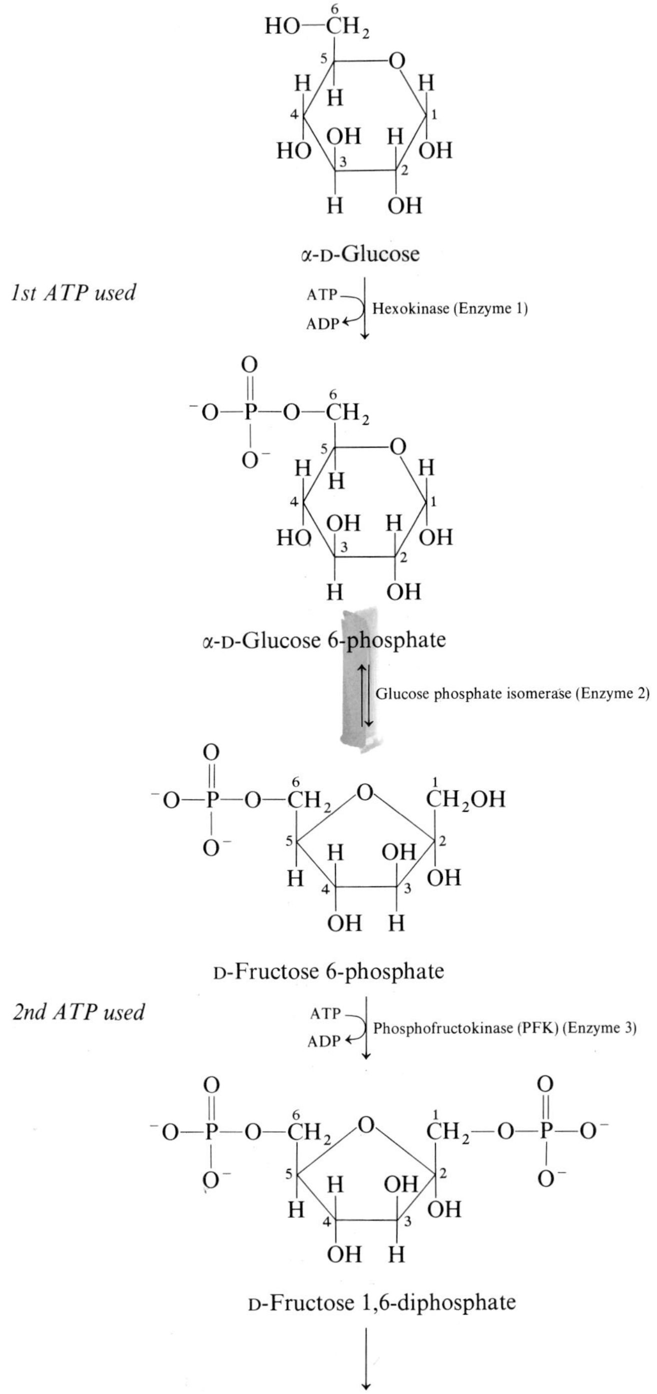

α-D-Glucose

1st ATP used

ATP → ADP — Hexokinase (Enzyme 1)

α-D-Glucose 6-phosphate

Glucose phosphate isomerase (Enzyme 2)

D-Fructose 6-phosphate

2nd ATP used

ATP → ADP — Phosphofructokinase (PFK) (Enzyme 3)

D-Fructose 1,6-diphosphate

FIGURE 9–2
The glycolytic pathway.

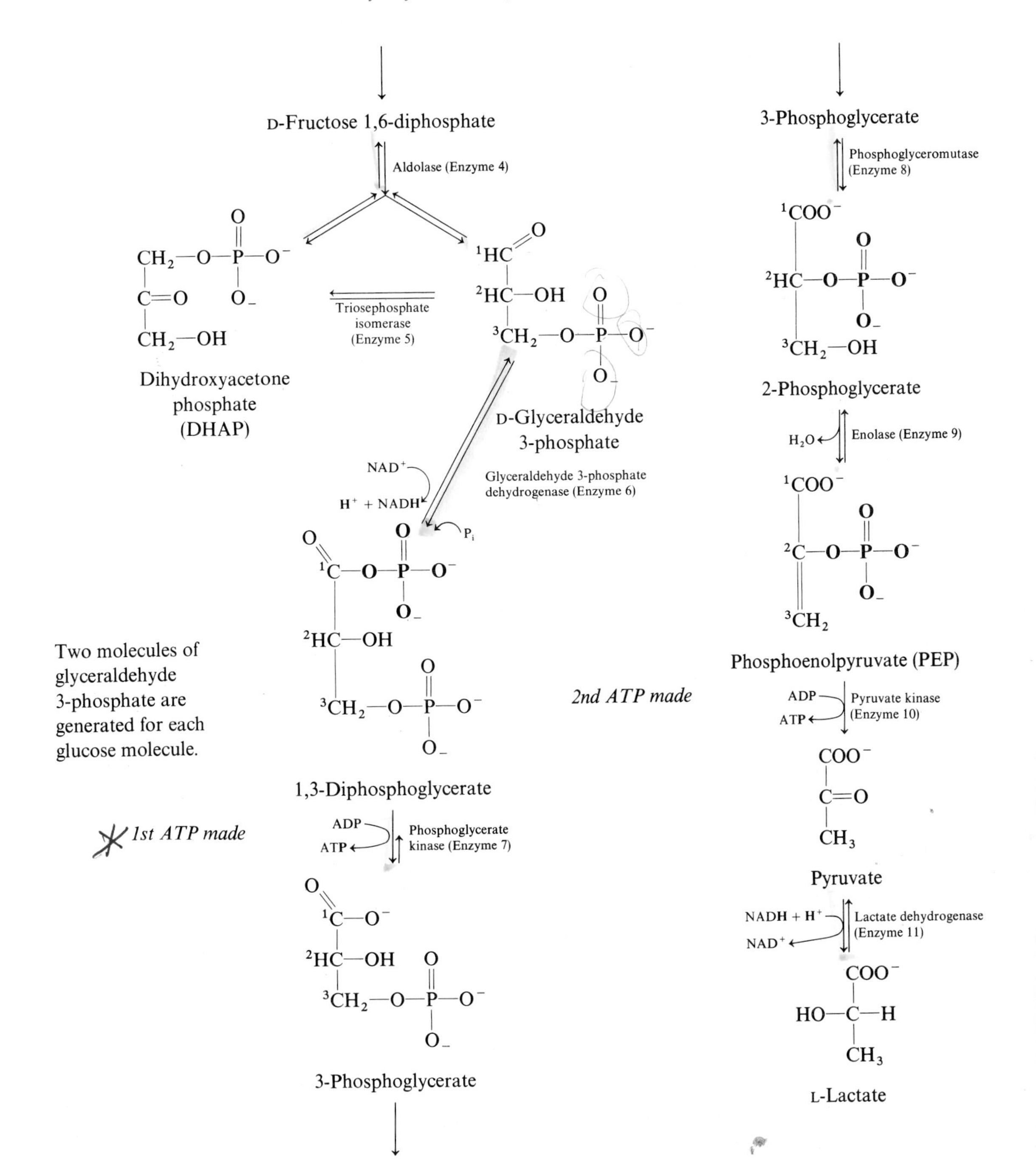

FIGURE 9–2 *Continued*

glucose (and other sugars) are *mobilized and transported* into the cell. Second, the 6-carbon sugar molecule is *phosphorylated* by ATP (phosphate groups are transferred from ATP to the molecule). Third, the 6-carbon sugar is *fragmented* into two 3-carbon intermediates (glyceraldehyde 3-phosphate). Fourth, an *ATP molecule is made* from ADP and a high-energy phosphate group donor (1,3-diphosphoglycerate). Fifth, a *second ATP molecule is made* using ADP and a high-energy phosphate group donor (phosphoenolpyruvate). Sixth, and last, *lactic acid is generated.*

When studying the glycolytic pathway, one should note the following points: (1) how the carbon skeleton of glucose is changed, (2) where and how many ATP molecules are used or made during the process, and (3) where and how many NADH molecules are used or made.

SOURCES OF GLUCOSE AND ITS TRANSPORT INTO THE CELL

There are two main sources of glucose: (1) dietary intake of glucose (or of glucose in the form of starches), and (2) glucose already stored in the form of glycogen. Dietary glucose is transported from the intestinal mucosal cells into the blood. This glucose is ultimately absorbed by the tissues, either directly or under the influence of insulin. Glycogen (found in the liver and in skeletal muscles) is the primary storage form of glucose in animals. Glycogen, you will recall from Chapter 7, is similar in structure to amylopectin: individual glucose molecules are bound together in a large, highly branched structure via $\alpha(1 \rightarrow 4)$ and $\alpha(1 \rightarrow 6)$ glycosidic bonds. Liver contains approximately 5 percent glycogen, by weight; skeletal muscle contains about 1 percent. In humans, this represents about 0.23 kilograms of carbohydrate

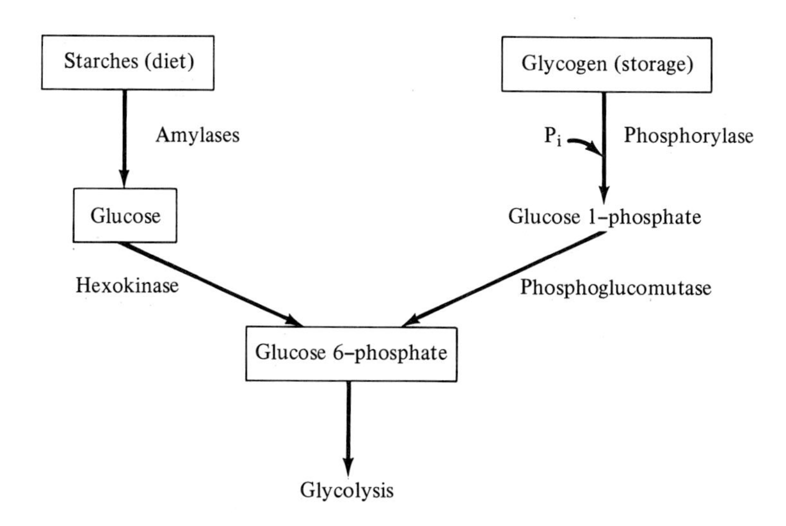

FIGURE 9–3
The entry of dietary and storage forms of glucose into glycolysis.

phosphorylase

stored in extremely accessible form. Liver and muscle glycogen is degraded into *glucose 1-phosphate* by an enzyme called **phosphorylase**. (The metabolism of glycogen will be discussed in Chapter 10.) Glucose 1-phosphate can be converted directly into glucose 6-phosphate via phosphoglucomutase. These interrelationships are summarized in Figure 9–3.

PHOSPHORYLATION AND ENZYMATIC MODIFICATION OF THE GLUCOSE MOLECULE

The Hexokinase Step. Once the glucose has been transported into the cytoplasm of the cell, it is immediately *phosphorylated* by the enzyme *hexokinase*. In this step, a phosphate group is transferred from an ATP molecule to the #6 carbon hydroxyl group of glucose, yielding ADP and a molecule of α-D-glucose 6-phosphate.

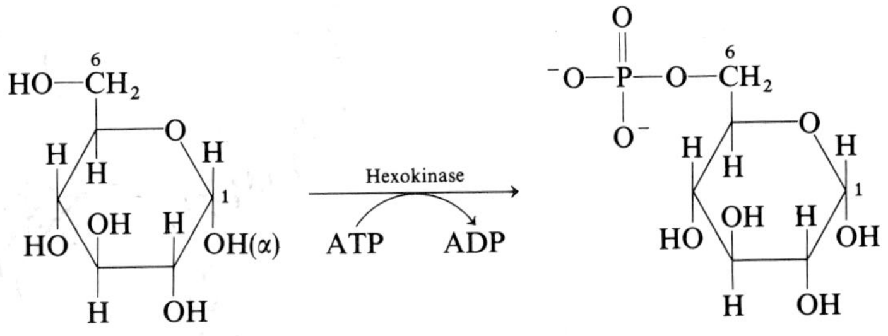

α-D-Glucose $\xrightarrow[\text{ATP} \quad \text{ADP}]{\text{Hexokinase}}$ α-D-Glucose 6-phosphate

α-D-Glucose

α-D-Glucose 6-phosphate
($\Delta G^{\circ\prime} = -4.0$ kcal/mol)

(*Note*: The phosphate group will be symbolized as Ⓟ in all remaining reactions.) This reaction is essentially irreversible at physiological concentrations of reactants and products. Hexokinase is rather nonspecific; it can phosphorylate not only glucose but other 6-carbon sugars as well (*hexo* means "six"). The term *kinase* identifies this enzyme as a "phosphotransferase." Note, too, that we showed hexokinase phosphorylating the α-anomer of glucose. This notation is for convenience only; hexokinase can phosphorylate both the α- and β-anomers of glucose. The anomeric specificity of the other enzymes of glycolysis will not be discussed.

At this point, you might be asking why an ATP molecule is *used*. After all, is not the synthesis of ATP the primary reason for the existence of this pathway? Yes, it is—as we shall see, the pathway will make more ATP than it uses. Glucose must be phosphorylated by ATP for two reasons. First, electrically neutral molecules, such as glucose, easily diffuse across cell membranes. Therefore, the cell would lose some of its cytoplasmic glucose by simple diffusion. Electrically charged molecules cannot easily diffuse across the cell membrane; thus, cellular glucose is trapped in the cell as negatively charged glucose 6-phosphate. In fact, the intracellular concentration of glucose itself is very low, because glucose is immediately phosphorylated by hexokinase as it enters the cell. Since the concentration of glucose within the cell is maintained at such a low level, extracellular glucose can easily flow into the cell *without* input of energy. Second, glucose is "primed" by phosphorylation. An input of chemical energy is needed to initiate the entire process. As we shall see, the formation of *high-energy phosphorylated intermediates* will generate *additional* ATP molecules.

The Isomerase Step

D-Glucose 6-phosphate is converted into D-*fructose 6-phosphate* by *glucose phosphate isomerase*. The 6-membered (pyranose) ring of glucose is converted into a 5-membered (furanose) ring by the enzyme.

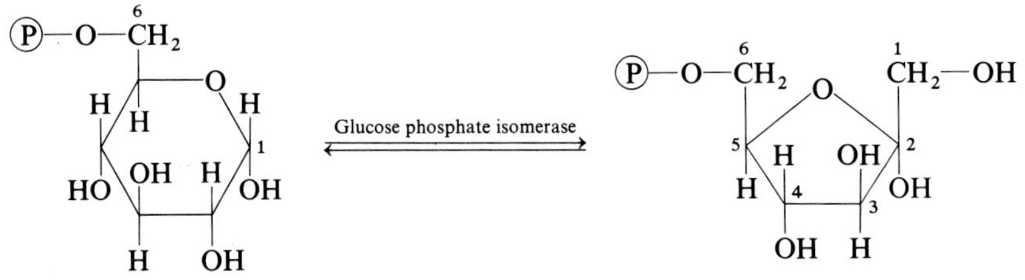

Glucose phosphate isomerase

α-D-Glucose 6-phosphate (pyranose ring) ⇌ D-Fructose 6-phosphate (furanose ring)

($\Delta G^{\circ\prime} = +0.4$ kcal/mol)

The Phosphofructokinase Step

phosphofructokinase (PFK)

Phosphofructokinase (PFK) catalyzes the addition of a phosphate group to the #1 carbon hydroxyl group of D-fructose 6-phosphate, generating D-*fructose 1,6-diphosphate*. A second ATP molecule is used at this point. This reaction is irreversible.

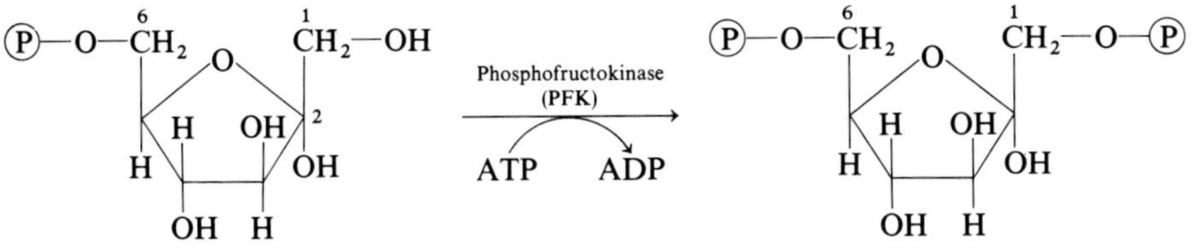

Phosphofructokinase (PFK)

D-Fructose 6-phosphate + ATP → D-Fructose 1,6-diphosphate + ADP

($\Delta G^{\circ\prime} = -3.4$ kcal/mol)

Phosphofructokinase is an allosteric enzyme controlled by the allosteric effectors ATP, ADP, and AMP. The importance of the PFK step in regulating glycolysis will be discussed in greater detail in Section 9.4.

The Aldolase and Triosephosphate Isomerase Steps

Aldolase *cleaves* the D-fructose 1,6-diphosphate molecule into two 3-carbon fragments (*dihydroxyacetone phosphate* and D-*glyceraldehyde 3-phosphate*).

Ⓟ—O—$^{6}CH_2$ … $^{1}CH_2$—O—Ⓟ

Cleavage point

D-Fructose 1,6-diphosphate

Aldolase
($\Delta G^{\circ\prime} = +5.7$ kcal/mol)

$^{1}CH_2$—O—Ⓟ
^{2}C=O
$^{3}CH_2$—OH

Dihydroxyacetone phosphate (DHAP)

Triose phosphate isomerase

^{1}HC=O (4)
^{2}HC—OH (5)
$^{3}CH_2$—O—Ⓟ (6)

D-Glyceraldehyde 3-phosphate

The dihydroxyacetone phosphate cannot be used in the next step of glycolysis; therefore, it is converted into an additional molecule of D-glyceraldehyde 3-phosphate by the enzyme *triosephosphate isomerase*. Thus, the 6-carbon sugar D-fructose 1,6-diphosphate is ultimately converted into two molecules of D-glyceraldehyde 3-phosphate by aldolase and triosephosphate isomerase, working together. It should be noted that both the aldolase and triosephosphate isomerase reactions are reversible; single-direction arrows were written in the reaction equation only to indicate the net formation of D-glyceraldehyde 3-phosphate from D-fructose 1,6-diphosphate. The standard free-energy change for the aldolase reaction is very positive ($\Delta G^{\circ\prime} = +5.7$ kcal/mol); therefore, the aldolase reaction is rather nonspontaneous under standard conditions. Since the *actual* cellular concentrations of the substrates and products are significantly less than 1.0 M, the actual free-energy change ($\Delta G'$) is negative and the reaction is rather spontaneous, resulting in over 50 percent conversion of D-fructose 1,6-diphosphate into products. Subsequent reactions in the glycolytic pathway favor the formation of D-glyceraldehyde 3-phosphate.

The Glyceraldehyde 3-Phosphate Dehydrogenase Step

*n*icotinamide *a*denine *d*inucleotide (NAD^+)

D-Glyceraldehyde 3-phosphate is *oxidized* (electrons removed) by *glyceraldehyde 3-phosphate dehydrogenase*. **Nicotinamide adenine dinucleotide (NAD^+)** acts as an oxidizing agent (that is, it accepts electrons from the substrate) and is reduced to NADH. The structure and chemistry of NAD^+ will be discussed in greater detail in Section 9.5.

D-Glyceraldehyde 3-phosphate: 1 HC=O; 2 HC—OH; 3 CH_2—O—Ⓟ

Glyceraldehyde 3-phosphate dehydrogenase; NAD^+ → NADH + H^+; P_i

1,3-Diphosphoglycerate: 1 O=C—O—Ⓟ; 2 HC—OH; 3 CH_2—O—Ⓟ

($\Delta G^{\circ\prime} = +1.5$ kcal/mol)

The glyceraldehyde 3-phosphate dehydrogenase reaction works by coupling an endergonic phosphorylation reaction ($\Delta G^{\circ\prime} = +11.8$ kcal/mol) to an exergonic oxidation reaction ($\Delta G^{\circ\prime} = -10.3$ kcal/mol). The aldehyde group of D-glyceraldehyde 3-phosphate is first oxidized to a carboxyl group. The energy of oxidation is then conserved by phosphorylating the product with inorganic phosphate (P_i), forming a new high-energy mixed anhydride bond. The product of the reaction couple is *1,3-diphosphoglycerate*, a high-energy phosphate group donor ($\Delta G^{\circ\prime} = -11.8$ kcal/mol) that will help generate an ATP molecule in the next step of the sequence. One should note that the previous phosphorylation reactions required an ATP molecule as phosphate group donor. In this reaction, inorganic phosphate is added directly to the substrate. The glyceraldehyde 3-phosphate dehydrogenase step is one of the most important steps in glycolysis because of this unique oxidation reaction, coupled as it is to phosphorylation by inorganic phosphate.

The Phosphoglycerate Kinase Step (The Formation of the First ATP)

In this reaction, *phosphoglycerate kinase* catalyzes the transfer of the phosphate group on the #1 carbon of 1,3-diphosphoglycerate to ADP, forming the first ATP.

1,3-Diphosphoglycerate: 1 O=C—O—Ⓟ; 2 HC—OH; 3 CH_2—O—Ⓟ

Phosphoglycerate kinase; ADP → ATP

3-Phosphoglycerate: O=C—O^-; HC—OH; CH_2—O—Ⓟ

($\Delta G^{\circ\prime} = -4.5$ kcal/mol)

If you recall, 7.3 kcal/mol of free energy are required under standard conditions to make ATP from ADP and P_i. 1,3-Diphosphoglycerate is a high-energy phosphate group donor ($\Delta G^{\circ\prime} = -11.8$ kcal/mol) and releases more than enough free energy to donate a phosphate to ADP and form ATP. The reaction is very exergonic ($\Delta G^{\circ\prime} = -4.5$ kcal/mol). This "pulls"

substrate-level phosphorylation
oxidative phosphorylation

glycolysis to the right, to favor ATP synthesis. This process is called **substrate-level phosphorylation**, because a high-energy chemical substrate phosphorylates ADP to form ATP. A second type of phosphorylation, called **oxidative phosphorylation**, occurs in mitochondria; this reaction requires the transport of electrons (see Chapter 12, Electron Transport and Oxidative Phosphorylation).

To summarize events to this point in the sequence: *one* glucose molecule has been split into *two* D-glyceraldehyde 3-phosphate molecules as a result of the aldolase and triosephosphate isomerase reactions, in the process using up two ATP molecules. Each D-glyceraldehyde 3-phosphate molecule yields one 1,3-diphosphoglycerate molecule and, eventually, one ATP molecule at the phosphoglycerate kinase step. The reaction has now broken even.

The Phosphoglyceromutase Step

Phosphoglyceromutase catalyzes the transfer of the phosphate group from the #3 carbon of *3-phosphoglycerate* to the #2 carbon, resulting in the formation of *2-phosphoglycerate*. This movement is simple, but important—2-phosphoglycerate is the precursor of a high-energy intermediate, *phosphoenolpyruvate*, which will be generated in the next step.

$$\begin{array}{l} 1\ COO^- \\ 2\ HC{-}OH \\ 3\ CH_2{-}O{-}Ⓟ \end{array} \xrightleftharpoons{\text{Phosphoglyceromutase}} \begin{array}{l} COO^- \\ 2\ HC{-}O{-}Ⓟ \\ CH_2OH \end{array}$$

3-Phosphoglycerate — 2-Phosphoglycerate

($\Delta G^{\circ\prime} = +1.06$ kcal/mol)

The Enolase Step

Enolase catalyzes the removal of a water molecule (*dehydration reaction*) from 2-phosphoglycerate to form phosphoenolpyruvate (PEP).

$$\begin{array}{l} COO^- \\ HC{-}O{-}Ⓟ \\ CH_2OH \end{array} \xrightleftharpoons[H_2O]{\text{Enolase}} \begin{array}{l} COO^- \\ C{-}O{-}Ⓟ \\ \| \\ CH_2 \end{array}$$

2-Phosphoglycerate — Phosphoenolpyruvate

($\Delta G^{\circ\prime} = +0.44$ kcal/mol)

Phosphoenolpyruvate has a rather large $\Delta G^{\circ\prime}$ of hydrolysis ($\Delta G^{\circ\prime} = -14.8$ kcal/mol). Thus, it is a high-energy phosphate group donor, capable of donating its phosphate group to ADP to form ATP. Phosphoenolpyruvate exhibits this large $\Delta G^{\circ\prime}$ because the molecule is constrained to exist in the *enol* (less stable) form instead of the *keto* (more stable) form. The conversion from the enol to the keto form is considered the primary driving force behind PEP hydrolysis.

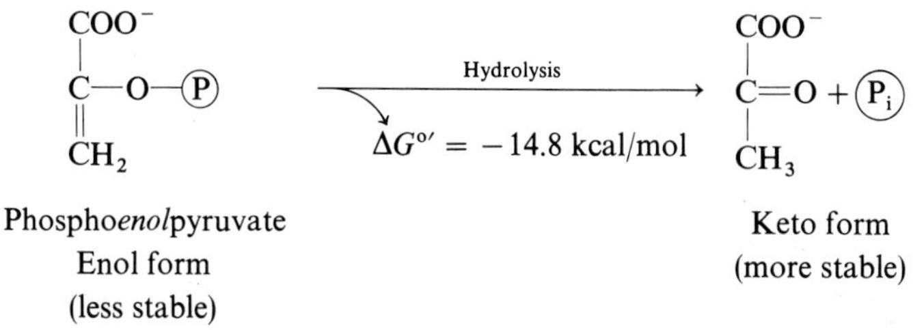

$$\underset{\substack{\text{Phospho}\textit{enol}\text{pyruvate}\\ \text{Enol form}\\ \text{(less stable)}}}{\begin{array}{l} COO^- \\ | \\ C{-}O{-}\text{(P)} \\ \| \\ CH_2 \end{array}} \xrightarrow[\Delta G^{o\prime} = -14.8 \text{ kcal/mol}]{\text{Hydrolysis}} \underset{\substack{\text{Keto form}\\ \text{(more stable)}}}{\begin{array}{l} COO^- \\ | \\ C{=}O \\ | \\ CH_3 \end{array}} + \text{(}P_i\text{)}$$

The Pyruvate Kinase Step (The Formation of the Second ATP)

Pyruvate kinase catalyzes the transfer of a phosphate group from phosphoenolpyruvate to ADP to form the second ATP.

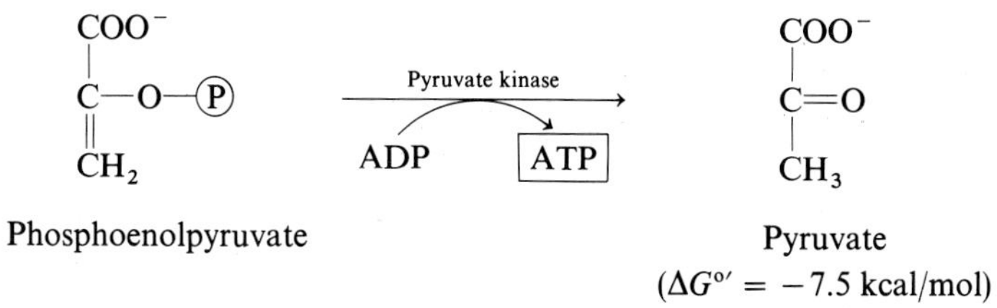

$$\underset{\text{Phosphoenolpyruvate}}{\begin{array}{l} COO^- \\ | \\ C{-}O{-}\text{(P)} \\ \| \\ CH_2 \end{array}} \xrightarrow[\text{ADP} \;\rightarrow\; \boxed{\text{ATP}}]{\text{Pyruvate kinase}} \underset{\substack{\text{Pyruvate}\\ (\Delta G^{o\prime} = -7.5 \text{ kcal/mol})}}{\begin{array}{l} COO^- \\ | \\ C{=}O \\ | \\ CH_3 \end{array}}$$

The reaction is essentially irreversible. A large negative standard free-energy change ($\Delta G^{o\prime} = -7.5$ kcal/mol) is observed for this reaction. Because *two* PEP molecules per original glucose molecule were made, *two* ATP molecules per glucose are, thus, generated at this point. We might summarize where and how many ATP molecules are either used or made for each glucose molecule.

2 ATP molecules used in "priming" steps of glycolysis:

- 1 ATP: Hexokinase step
- 1 ATP: Phosphofructokinase step

4 ATP molecules made in the latter steps of glycolysis:

- 2 ATP: Phosphoglycerate kinase step
- 2 ATP: Pyruvate kinase step

Therefore, 2 (net) ATP molecules are made for each glucose molecule.

The Lactate Dehydrogenase Step

Lactate dehydrogenase (*LDH*) catalyzes the reduction of pyruvate (that is, the addition of electrons and hydrogens to it) to form L-lactate.

$$\underset{\text{Pyruvate}}{\begin{array}{c}COO^-\\ |\\ C{=}O\\ |\\ CH_3\end{array}} \underset{NADH + H^+ \quad\; NAD^+}{\overset{\text{Lactate dehydrogenase}}{\rightleftharpoons}} \underset{\text{L-Lactate (waste)}}{\begin{array}{c}COO^-\\ |\\ HO{-}C{-}H\\ |\\ CH_3\end{array}}$$

*N*icotinamide *a*denine *d*inucleotide, reduced form (NADH), is the electron/hydrogen donor in this reaction. It is oxidized to NAD^+. Note that the molecules of NAD^+ used in the glyceraldehyde 3-phosphate dehydrogenase step are regenerated at the lactate dehydrogenase step and that, therefore, no net gain or loss of either NAD^+ or NADH occurs. The lactate produced at this step diffuses out of the cell as a waste product. This waste lactic acid can produce the characteristic muscle fatigue and soreness normally associated with very strenuous exercise. Some eukaryotic cells (liver cells) can reutilize lactic acid by resynthesizing glucose via **gluconeogenesis**. More will be said of gluconeogenesis in Chapter 10.

gluconeogenesis

ALCOHOLIC FERMENTATION

Certain organisms (yeasts) metabolize glucose via the alcoholic fermentation pathway. In this pathway, glucose is degraded into two molecules of *ethanol* and two molecules of CO_2 gas, in the process forming two ATP molecules. The pathway is identical to that described for glycolysis, with the following exceptions. First, these organisms do not contain the enzyme lactate dehydrogenase and, therefore, do not form lactic acid as the final waste product. Second, these organisms contain two additional enzymes, *pyruvate decarboxylase* and *alcohol dehydrogenase*, which together metabolize pyruvate.

Pyruvate decarboxylase catalyzes the *decarboxylation* of pyruvate (the removal of CO_2 from it) to form *acetaldehyde*.

$$\underset{\text{Pyruvate}}{\begin{array}{c}COO^-\\ |\\ C{=}O\\ |\\ CH_3\end{array}} \underset{CO_2\ \text{(waste)}}{\xrightarrow{\text{Pyruvate decarboxylase}}} \underset{\text{Acetaldehyde}}{\begin{array}{c}HC{=}O\\ |\\ CH_3\end{array}}$$

Alcohol dehydrogenase catalyzes the reduction of acetaldehyde to form *ethanol*. NADH is the electron/hydrogen donor. It should be noted that a dehydrogenase serves as the final enzyme in both the glycolytic and alcoholic fermentation pathways.

$$\underset{\text{Acetaldehyde}}{\begin{array}{c}HC{=}O\\ |\\ CH_3\end{array}} \underset{NADH + H \quad\; NAD^+}{\overset{\text{Alcohol dehydrogenase}}{\rightleftharpoons}} \underset{\text{Ethanol (waste)}}{\begin{array}{c}CH_2{-}OH\\ |\\ CH_3\end{array}}$$

ENTRY OF OTHER MONOSACCHARIDES INTO THE GLYCOLYTIC PATHWAY

Glucose is not the only monosaccharide that can be metabolized by the glycolytic pathway. Other monosaccharides are first phosphorylated by ATP and a kinase, then converted into a specific intermediate of the glycolytic pathway. The metabolism of a few representative monosaccharides is shown in Figure 9–4.

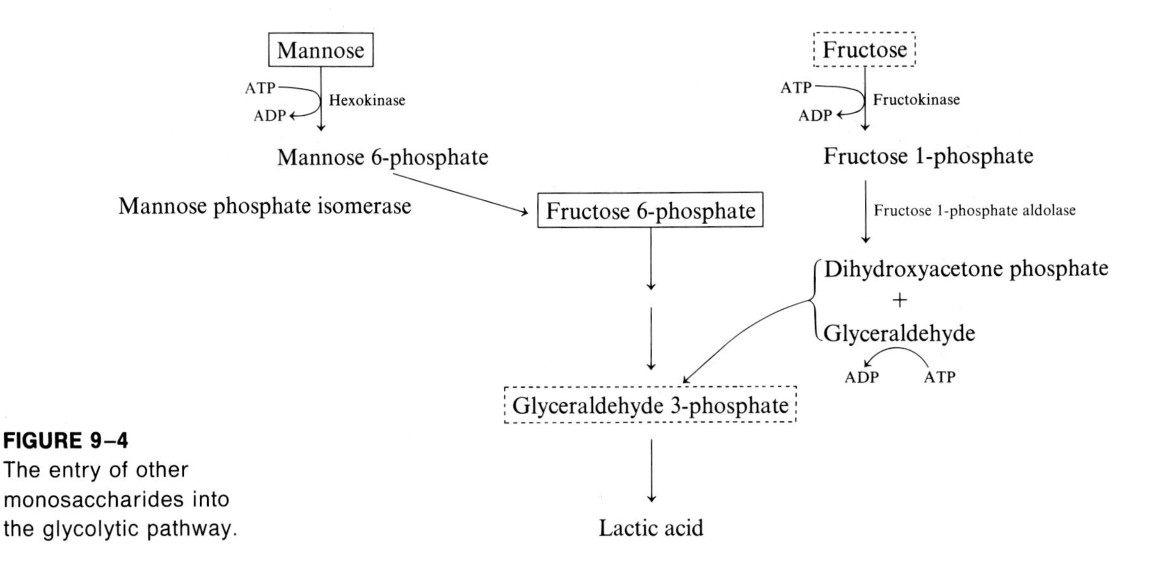

FIGURE 9–4
The entry of other monosaccharides into the glycolytic pathway.

9.4 THE REGULATION OF GLYCOLYSIS

The glycolytic pathway, like all other metabolic pathways, is under constant moment-by-moment control. Precise regulation of the concentrations of the pathway intermediates is not only efficient, but also allows for sudden changes in the output level of the pathway when the need arises. As stated in Chapter 6 (Enzymes), most metabolic pathways do not always run at a 100 percent rate. If the cell does not need a certain end product, the production of this product will slow down or stop so that energy and starting materials are not wasted.

The glycolytic pathway is controlled by certain allosteric enzymes along the pathway. These respond to the presence of either positive or negative allosteric effectors (see Chapter 6, Enzymes). The glycolytic pathway is regulated by three enzymes: (1) hexokinase (nonallosteric), (2) phosphofructokinase (PFK) (allosteric), and (3) pyruvate kinase (allosteric). In addition, *glycogen phosphorylase*, which converts stored glycogen into glucose, is allosterically modulated by AMP, ADP, and ATP. The regulation of the pathway is shown in Figure 9–5. We will only discuss the regulation of the phosphofructokinase step in detail.

Referring to Figure 9–5, you will note that ATP is a negative allosteric effector for phosphofructokinase and pyruvate kinase. When the glycolytic pathway is operating at maximum, large amounts of ATP will be made. The pathway will continue to operate at this rate as long as the ATP continues to be consumed by processes within the cell. However, if the ATP is not used up, its concentration will increase, and more ATP will diffuse to the allosteric binding sites on the enzymes sensitive to ATP, shutting these enzymes down. When enough phosphofructokinase molecules have shut down, the entire glycolytic pathway will slow or stop, along with ATP

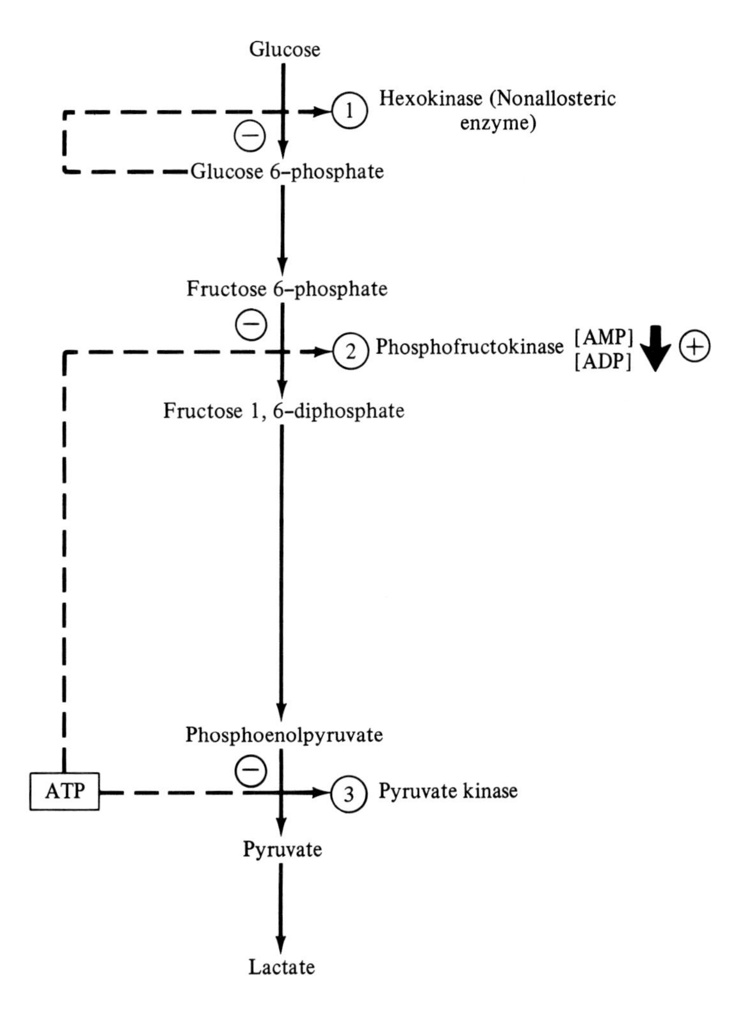

FIGURE 9–5
The regulation of the glycolytic pathway. Phosphofructokinase and pyruvate kinase are allosteric enzymes that respond to various allosteric effectors. ATP is a negative (−) allosteric effector; AMP and ADP are positive (+) allosteric effectors. Glucose 6-phosphate is a noncompetitive inhibitor of hexokinase.

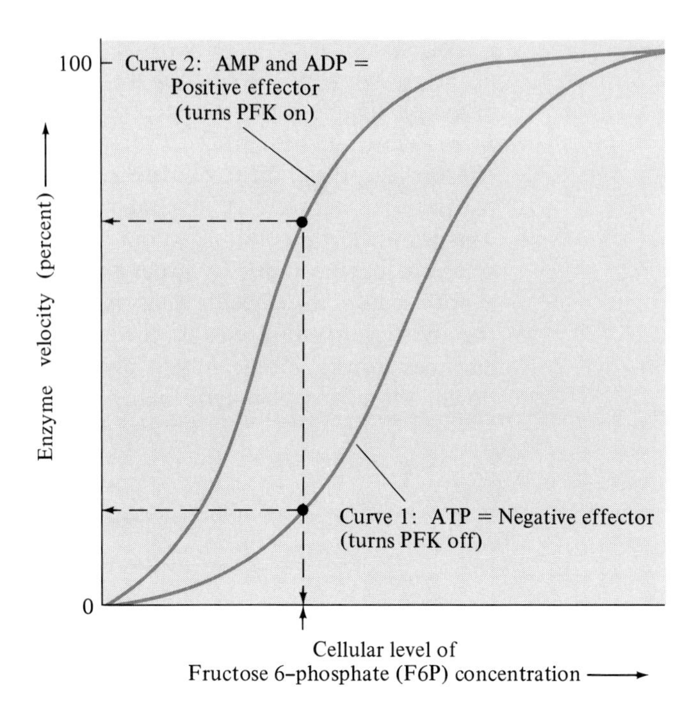

FIGURE 9–6
The allosteric regulation of phosphofructokinase (PFK). ATP is a negative allosteric effector and turns PFK off; ADP and AMP are positive allosteric effectors and turn PFK on. (Adapted from Page, D. S., *Principles of Biological Chemistry* (Boston: Willard Grant, 1976), p. 195.)

production. This process is called *feedback inhibition*. Conversely, when ATP concentrations are low and the concentrations of AMP and ADP are high, AMP and ADP act as *positive allosteric effectors* and turn on phosphofructokinase. The effects of AMP, ADP, and ATP on the activity of phosphofructokinase are summarized in Figure 9–6.

9.5 THE MECHANISM OF LACTATE DEHYDROGENASE

Two very important dehydrogenases were presented in our discussion of the glycolytic pathway: glyceraldehyde 3-phosphate dehydrogenase (GPDH) and lactate dehydrogenase (LDH). We will only discuss LDH in this section. Both of these enzymes (and a number of the other dehydrogenases) require *n*icotinamide *a*denine *d*inucleotide, NAD^+, as an enzyme cofactor. The cofactor acts as an electron/hydrogen acceptor, while the reduced form of the cofactor (NADH) acts as an electron/hydrogen donor. Nicotinamide adenine dinucleotide is derived from the vitamin **niacin** (nicotinic acid), as shown in the following diagram:

niacin

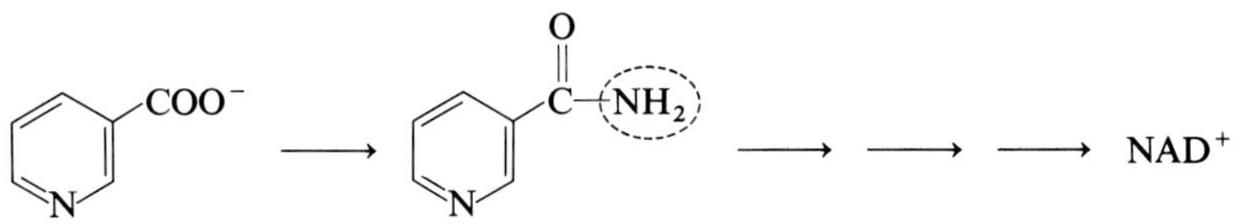

Nicotinic acid (Niacin) → Nicotinamide → → → NAD^+

(Many steps)

Location of the additional phosphate group of $NADP^+$

Nicotinamide ring (The "business end")

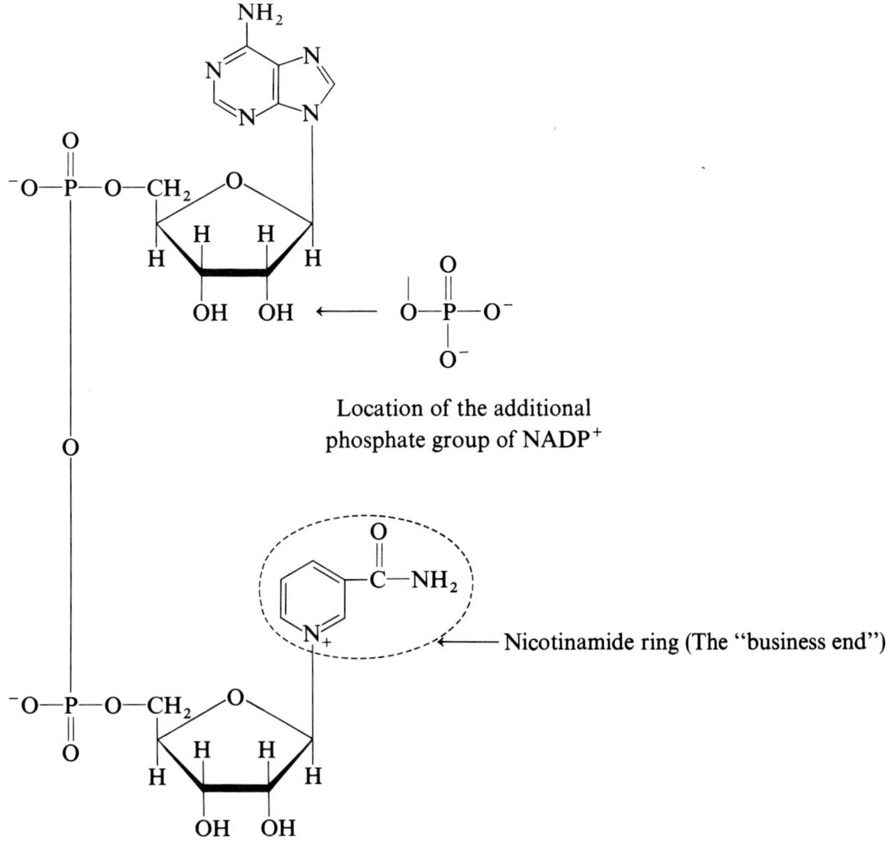

Nicotinamide Adenine Dinucleotide (NAD^+)

We will refer to the nicotinamide ring as the "business end" of the molecule, because it is here that electrons are either added or removed by the dehydrogenase. The remaining part of the NAD^+ molecule is important in binding the cofactor into the active site of the dehydrogenase. Basically, a dehydrogenase uses the NAD^+ cofactor as an electron acceptor in a reaction by which two electrons are removed from the substrate and transferred to the nicotinamide ring of NAD^+, reducing it to NADH. This two-electron

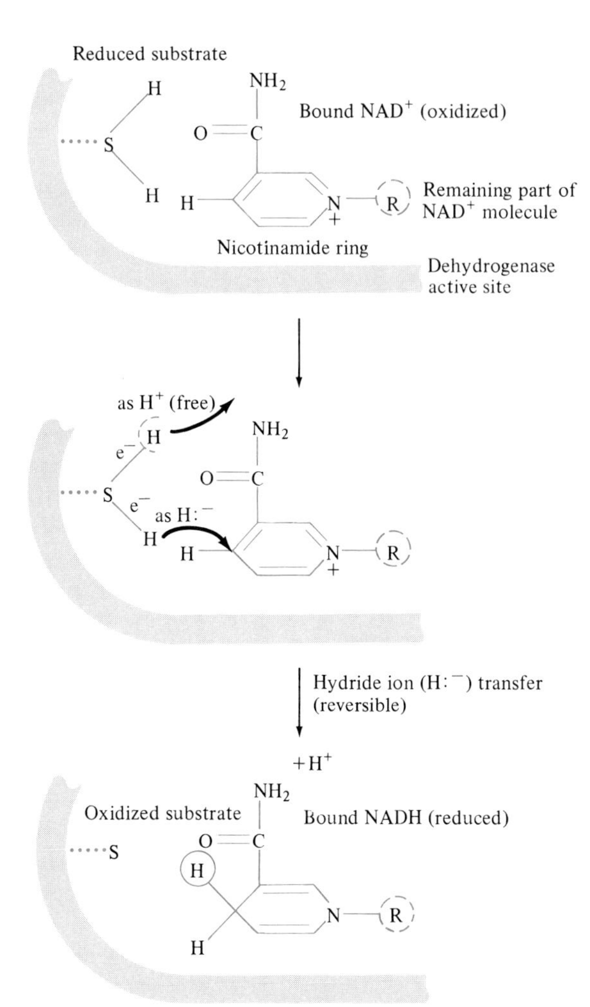

FIGURE 9–7
The reduction of NAD^+ by a dehydrogenase. Two electrons are removed from the substrate (via the hydride ion, $H{:}^-$) and added to the nicotinamide ring of NAD^+, thus oxidizing the substrate and reducing the NAD^+ to NADH. This process can be summarized as follows:

$$\underset{\text{Reduced}}{SH_2} + NAD^+ \underset{\text{Dehydrogenase}}{\rightleftharpoons} \underset{\text{Oxidized}}{S} + NADH + H^+$$

transfer occurs via an intermediate called the *hydride ion* ($H{:}^-$, a proton with two electrons) as shown in Figure 9-7. Dehydrogenases can also catalyze a reverse reaction in which NADH donates two electrons (via $H{:}^-$) to a substrate, thus reducing the substrate.

Lactate dehydrogenase uses NADH to reduce pyruvic acid to lactic acid in the final step of glycolysis. As was stated in Chapter 6 (Enzymes), LDH exists in five distinct forms, called **isoenzymes**. Each isoenzyme catalyzes the same reaction, but at a different rate. It should be noted that LDH is actually a tetramer (that is, it is composed of 4 subunits). There are two different kinds of subunits, a "heart" or "H" subunit and a "muscle" or "M" subunit. There are five possible combinations of these two types of subunits, as shown in Figure 9–8.

isoenzymes

Each isoenzyme differs from the others in electrical charge and can be separated from the others by **electrophoresis**. The LDH_1 and LDH_2 isoenzymes tend to migrate very rapidly to the positive anode; thus, they are called "fast moving" isoenzymes. The LDH_4 and LDH_5 isoenzymes do not

electrophoresis

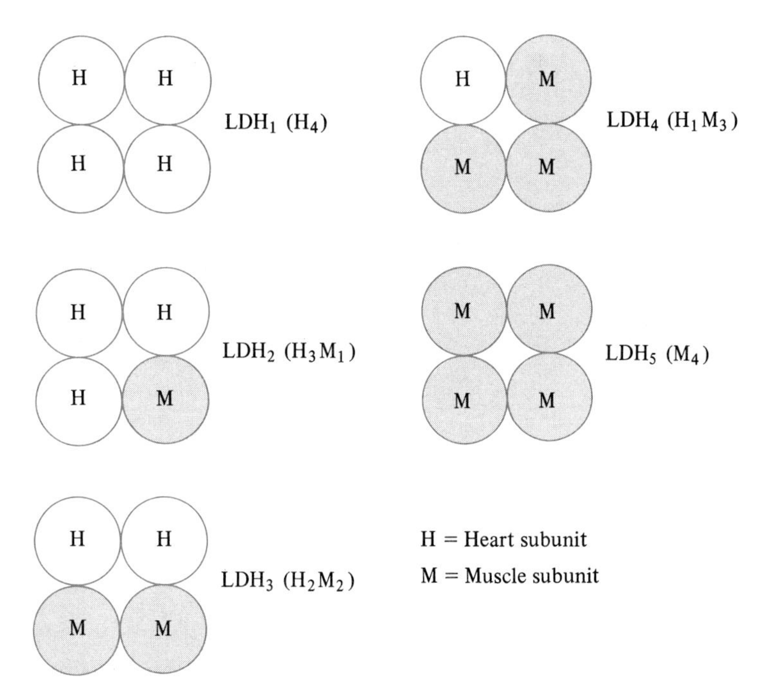

FIGURE 9–8
Lactate dehydrogenase isoenzymes.

migrate very readily and tend to stay near the origin (LDH_5 actually migrates slightly to the negative cathode). These isoenzymes are called "slow moving" isoenzymes.

Certain organs have different concentrations of the five LDH isoenzymes, as shown in Table 9–1. Heart tissue contains large amounts of LDH_1 and LDH_2 (which are primarily composed of the heart subunit). Liver and skeletal muscle contain large amounts of LDH_4 and LDH_5 (which are primarily composed of the muscle subunit). When disease affects a specific organ, some of the cells of the diseased organ die and rupture (see Chapter 6, Enzymes), releasing LDH isoenzymes into the blood. For example, when a **myocardial infarction** (heart attack) occurs, the cells of the affected heart muscle begin to die of oxygen starvation and release large amounts of LDH_1 and LDH_2 into the blood. The increase in concentration can be readily detected by analyzing a blood sample by electro-

myocardial infarction

TABLE 9–1 Distribution of LDH Isoenzymes in Various Tissues (percent)

Isoenzyme	Heart	Skeletal Muscle	Liver
LDH_1 (H_4)	60	4	2
LDH_2 (H_3M)	33	7	6
LDH_3 (H_2M_2)	7	17	15
LDH_4 (HM_3)	Trace	16	13
LDH_5 (M_4)	Trace	56	64

Note: H = heart subunit; M = muscle subunit.

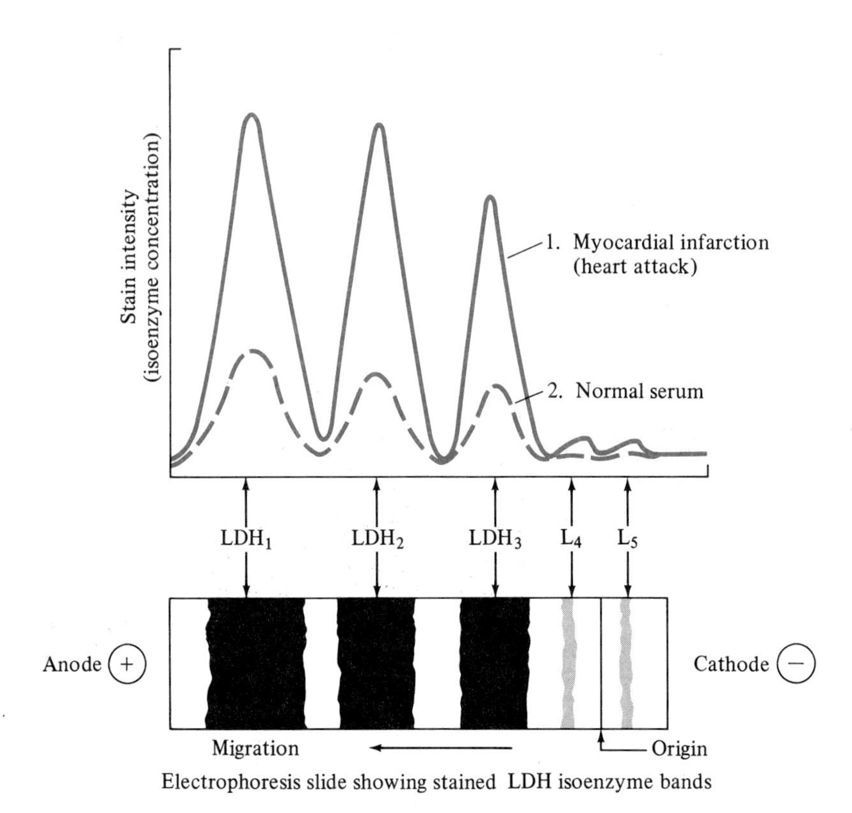

FIGURE 9–9
LDH isoenzyme electrophoresis pattern consistent with myocardial infarction.

phoresis. An electrophoresis pattern consistent with myocardial infarction is shown in Figure 9–9. Increased *total* serum LDH (the total activities of LDH_1–LDH_5) will be detected and will reach a maximum level 24–48 hours after the attack.

Other disease states, in other organs, produce different LDH isoenzyme patterns. For example, liver hepatitis will produce large amounts of LDH isoenzyme 5. (Why?)

The five LDH isoenzymes exhibit different kinetic parameters for the same chemical reaction—that is, they have different K_M and V_{max} values for their substrates.) Thus, they control the rate of conversion of pyruvate to lactate during glycolysis. For example, the LDH_5 isoenzyme (M_4) has a low K_M for pyruvate and a high V_{max} value. This means that LDH_5 catalyzes the conversion of pyruvate to lactate at a very high rate. Therefore, tissues that contain large amounts of LDH_4 and LDH_5 will tend to utilize glucose anaerobically via anaerobic glycolysis (glucose to lactate with the net formation of 2 ATP molecules/glucose). On the other hand, the LDH_1 (H_4) isoenzyme exhibits a very high K_M for pyruvate. This means that this isoenzyme catalyzes the conversion of pyruvate to lactate at a low rate. Since pyruvate is not being converted into lactate, its concentration increases and fuels the TCA cycle. Aerobic glycolysis will be favored, with the complete oxidation of glucose into CO_2 and H_2O and the formation of either 36 or 38 ATP molecules for each glucose molecule. Tissues that contain large amounts of LDH_1 and LDH_2 ("heart" subunit isoenzymes) will exhibit aerobic glycolysis; tissues with large amounts of LDH_4 and LDH_5 ("muscle" subunit isoenzymes) will carry out anaerobic glycolysis.

SUMMARY

Some of the chemical energy contained in the glucose molecule is extracted in the form of adenosine triphosphate (ATP) via a process called glycolysis.

Glycolysis (the Embden-Meyerhof pathway) can be defined as the anaerobic, enzymatic splitting of the 6-carbon glucose molecule into two 3-carbon fragments, in the process forming two ATP molecules from ADP. When glycolysis is operating under aerobic conditions, a total of 36 molecules of ATP are produced per original glucose molecule. Alcoholic fermentation is a modification of the anaerobic glycolytic pathway. In this modification, glucose is disrupted into two molecules of ethanol (alcohol) and two molecules of CO_2, in the process forming two ATP molecules from ADP.

The glycolytic pathway was depicted in Figure 9–2. First, glucose is mobilized and transported into the cell. Second, the glucose molecule is phosphorylated. Third, the 6-carbon sugar is fragmented into two 3-carbon intermediates (glyceraldehyde 3-phosphate). Fourth, an ATP molecule is made from ADP and a high-energy phosphate group donor (1,3-diphosphoglycerate). Fifth, a second ATP molecule is made from ADP and a high-energy phosphate group donor (phosphoenolpyruvate). Sixth, lactic acid is generated as the final product of the pathway.

The glycolytic pathway is regulated by three enzymes; (1) hexokinase, (2) phosphofructokinase (PFK), and (3) pyruvate kinase. ATP acts on PFK and pyruvate kinase as a negative allosteric effector, thus turning them off. Low concentrations of ATP and high concentrations of AMP and ADP turn on the PFK enzyme, thus stimulating the pathway.

Two dehydrogenases are present in the glycolytic pathway: glyceraldehyde 3-phosphate dehydrogenase (GPDH) and lactate dehydrogenase (LDH). Both of these enzymes require NAD^+ as a cofactor. The NAD^+ (oxidized form) acts as an electron acceptor and accepts two electrons from the substrate (transferred via the hydride ion, $H:^-$). In this way, the reduced form of the coenzyme, NADH, is produced. Conversely, NADH can act as an electron donor and, thus, reduce a dehydrogenase substrate.

Lactate dehydrogenase uses NADH to reduce pyruvate into lactate (this is the final step in anaerobic glycolysis). The enzyme exists in five forms, known as *isoenzymes* LDH_1–LDH_5. Each isoenzyme catalyzes the same chemical reaction, but at a different rate. Analysis of total serum LDH levels and of the individual LDH isoenzyme levels is used to diagnose various disease states. A heart attack (myocardial infarction) produces increased serum levels of LDH_1 and LDH_2, indicating damage to the heart muscle cells.

REVIEW QUESTIONS

1. We have presented a number of important concepts and terms in this chapter. Define (or briefly explain) each of the following terms:

a. Glycolysis
b. Anaerobic
c. Aerobic
d. Alcoholic fermentation
e. Reducing agent
f. Phosphorylation
g. Substrate-level phosphorylation
h. Phosphoenolpyruvate
i. NAD^+ and NADH
j. Dehydrogenase
k. Isoenzyme
l. Hydride ion ($H:^-$)

2. We have presented a number of enzymatic reactions in this chapter. Write out the complete chemical reaction catalyzed by each of the following enzymes. Draw the complete structures of all substrates and products; include important cofactors.
   a. Hexokinase
   b. Phosphofructokinase (PFK)
   c. Glyceraldehyde 3-phosphate dehydrogenase (GPDH)
   d. Pyruvate kinase
   e. Lactate dehydrogenase (LDH)

*3. A sample of glucose has been labeled with ^{14}C in the #1 position.
   a. Where would the label appear if this glucose were metabolized into lactate?
   b. Where would the label appear if this glucose were metabolized into ethanol and CO_2 via alcoholic fermentation?

*4. The metabolism of sucrose in a certain bacterium first involves the action of the enzyme *sucrose phosphorylase* (and P_i) to yield glucose 1-phosphate and fructose. Assuming that both the glucose 1-phosphate and fructose molecules are further metabolized into lactate, suggest:
   a. Where would ATP molecules be used, and how many?
   b. Where would ATP molecules be made, and how many?
   c. Where would NADH molecules be made, and how many (net)?

   Be sure to indicate at which specific step in glycolysis these compounds are being used or made. Also, note that no ATP is used to generate the glucose 1-phosphate during the initial hydrolysis of sucrose.

*5. A cardiologist is studying the effect of alcohol on triglyceride (lipid) accumulation in rat heart tissue. The triglycerides are degraded and the resulting glycerol reacted with ATP to form glycerol phosphate, according to the following reaction:

$$\text{Glycerol} \xrightarrow[\text{ATP} \quad \boxed{\text{ADP}}]{\text{Glycerol kinase}} \text{Glycerol phosphate}$$

   The cardiologist decided to devise a *coupled enzyme assay* using the ADP formed in the above reaction and other purified enzymes, substrates, and coenzymes of the glycolytic pathway. Suggest a coupled enzyme assay that will monitor the original glycerol concentration. (*Hint*: What glycolytic enzymes use ADP, and how can these enzymes/cofactors be utilized to devise a workable assay system? Note that the cofactor NADH has a spectrophotometer absorbance peak at 340 nm.)[2]

6. Cancer cells often contain large amounts of lactate dehydrogenase; furthermore, they utilize 5 to 10 times as much glucose as normal cells and convert most of it into lactate. From what you know about glycolysis and the LDH isoenzymes, which LDH isoenzymes would predominate in cancer cells? Draw an electrophoresis pattern for a cancer cell homogenate. Be able to support your conclusions.

SUGGESTED READING

Bhagavan, N. V. *Biochemistry*, 2nd ed., Ch. 4. Philadelphia: J. B. Lippincott, 1978.

Lehninger, A. L. *Biochemistry*, 2nd ed., Ch. 16. New York: Worth 1975.

Montgomery, R., Dryer, R. L., Conway, T. W., and Spector, A. A. *Biochemistry: A Case-Oriented Approach*, 3rd ed., Ch. 7, pp. 281–335. St. Louis: C. V. Mosby, 1980.

Newsholme, E. A., and Start, C. *Regulation in Metabolism*, Chs. 3 and 6. New York: John Wiley, 1973.

NOTES

1. Aerobic glycolysis generates 36 ATP molecules per glucose molecule when the *glycerol phosphate shuttle* system is operating and 38 ATP molecules per glucose molecule when the *malate shuttle* is operating. These shuttles will be discussed in Chapter 11, The Tricarboxylic Acid Cycle.
2. Segel, I. I. *Biochemical Calculations*, 2nd ed. (New York: John Wiley, 1976), p. 292.

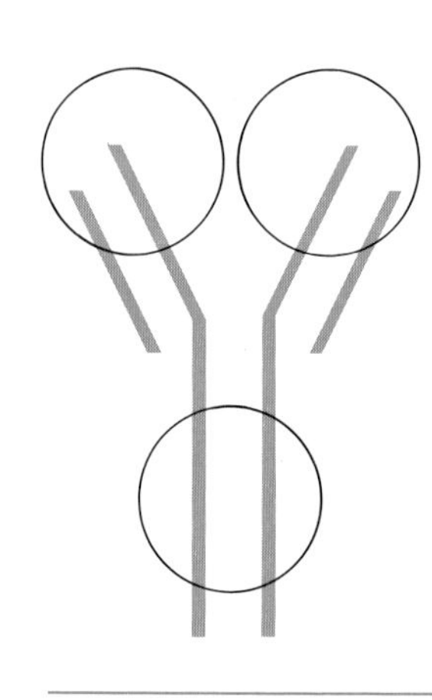

Chapter 10

Additional Aspects of Carbohydrate Metabolism

10.1 INTRODUCTION

The glycolytic pathway was discussed in detail in the previous chapter. In this chapter, other important pathways for carbohydrate metabolism will be presented, including (1) the process by which glucose is resynthesized from lactate (gluconeogenesis); (2) an alternative pathway for glucose metabolism (the pentose phosphate pathway); and (3) the biosynthesis and degradation of glycogen.

10.2 THE CORI CYCLE AND GLUCONEOGENESIS

gluconeogenesis

Cori cycle

When muscles are actively exercising, they produce large amounts of lactic acid. This lactate diffuses from the skeletal muscle cells into the bloodstream. During rest, the excess lactic acid is absorbed by the liver cells and used to synthesize glucose via a process called **gluconeogenesis**. The term gluconeogenesis literally means "the synthesis of new glucose." The relationship between glycolysis and gluconeogenesis was first established by C. Cori and G. Cori in the 1930s; therefore, this relationship is called the **Cori cycle** (Figure 10–1). We should stress that gluconeogenesis is *not* a reversal of the glycolytic pathway. Three key glycolytic steps are *irreversible*: (1) the hexokinase step, (2) the phosphofructokinase step, and (3) the pyruvate kinase step. These irreversible steps must be bypassed by a separate set of enzymes if glucose is to be resynthesized from the lactic acid. These bypass steps are shown in Figure 10–2 (page 182).

Once lactic acid has diffused from the bloodstream into the cytoplasm

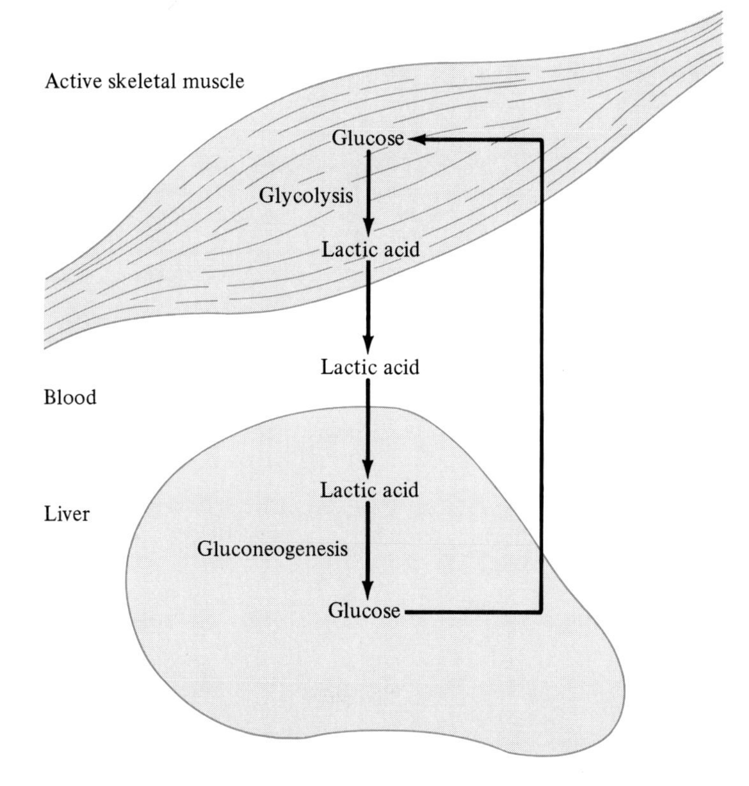

FIGURE 10–1
The Cori cycle giving the relationship between glycolysis and gluconeogenesis.

of a liver cell, it is converted to pyruvic acid by a reversal of the lactate dehydrogenase step. The pyruvic acid is then transported into a mitochondrion. In the mitochondrion, the pyruvate is converted into *oxaloacetate* via the enzyme *pyruvate carboxylase* according to the following reaction:

$$\underset{\substack{\text{Pyruvate}\\\text{(in mitochondrion)}}}{\begin{array}{c}COO^-\\|\\C{=}O\\|\\CH_3\end{array}} + \underset{\text{Bicarbonate}}{HCO_3^-} \xrightarrow[\text{ATP} \;\to\; \text{ADP} + \text{P}_i]{\text{Pyruvate carboxylase}} \underset{\text{Oxaloacetate}}{\begin{array}{c}COO^-\\|\\C{=}O\\|\\CH_2\\|\\COO^- \;(\text{From } HCO_3^-)\end{array}}$$

biotin
carboxylation

Pyruvate carboxylase requires the vitamin cofactor **biotin** for its activity (see Chapter 20, Vitamins and Nutrition). It catalyzes a **carboxylation** reaction in which CO_2 (supplied by bicarbonate, HCO_3^-) is added to the methyl group of pyruvate. The hydrolysis of an ATP molecule is required to drive the reaction. The oxaloacetate is then converted (by reduction) into *malate*, which is transported out of the mitochondrion into the cytoplasm. In the cytoplasm, the malate is reconverted (oxidized) to oxaloacetate (see Chapter 11, The Tricarboxylic Acid Cycle).

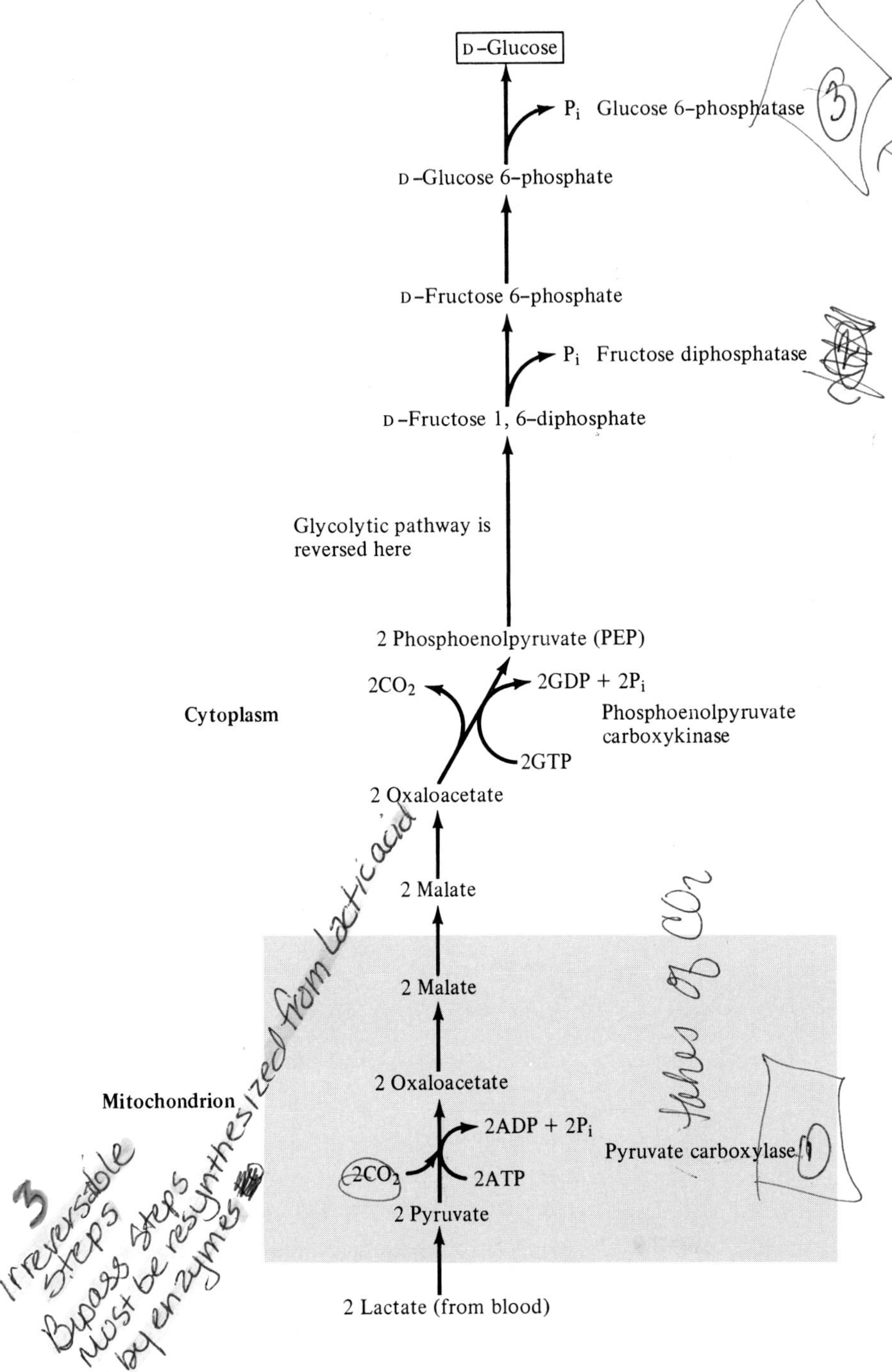

FIGURE 10–2 The gluconeogenic pathway leading to the resynthesis of glucose. The following irreversible steps in glycolysis are bypassed: (1) Pyruvate kinase is bypassed with pyruvate carboxylase and other enzymes. (2) Phosphofructokinase is bypassed with fructose diphosphatase. (3) Hexokinase is bypassed with glucose 6-phosphatase.

Oxaloacetate is converted into phosphoenolpyruvate (PEP) by *phosphoenolpyruvate carboxykinase*. The enzyme uses *guanosine triphosphate* (GTP) as a high-energy phosphate group donor to phosphorylate oxaloacetate. Phosphorylation is accompanied by a *decarboxylation* reaction by which one of the carboxyl groups of oxaloacetate is lost as CO_2.

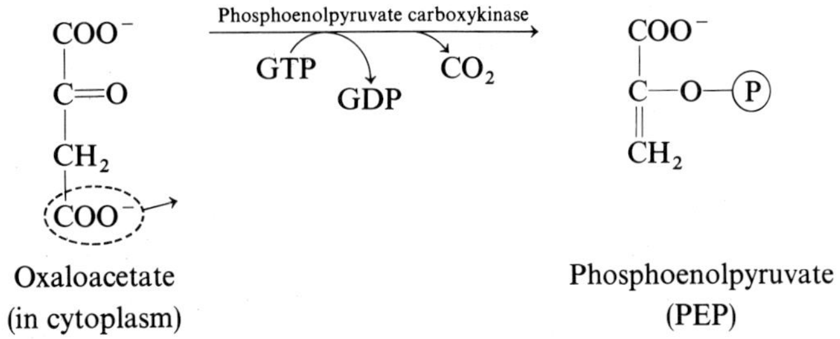

Phosphoenolpyruvate carboxykinase

GTP → GDP; CO_2

COO^- — C=O — CH_2 — COO^-

Oxaloacetate
(in cytoplasm)

COO^- — C—O—(P) — ‖ CH_2

Phosphoenolpyruvate
(PEP)

This complicated series of cytoplasmic and mitochondrial steps is needed to regenerate the high-energy glycolytic intermediate *phosphoenolpyruvate*. Once phosphoenolpyruvate has been formed, it can be used to generate the other glycolytic intermediates. A *reversal* of the enzyme steps between phosphoenolpyruvate and D-fructose 1,6-diphosphate generates D-fructose 1,6-diphosphate. Since D-fructose 6-phosphate cannot be formed from D-fructose 1,6-diphosphate by reversing the phosphofructokinase step, D-fructose 6-phosphate is generated from D-fructose 1,6-diphosphate using *fructose diphosphatase*.

Fructose diphosphatase

(P)—O—$\overset{6}{C}H_2$ O $\overset{1}{C}H_2$—O—(P); H, OH, H, OH, OH, H

→ (P_i)

(P)—O—$\overset{6}{C}H_2$ O $\overset{1}{C}H_2$—OH; H, OH, H, OH, OH, H

D-Fructose 1,6-diphosphate

D-Fructose 6-phosphate

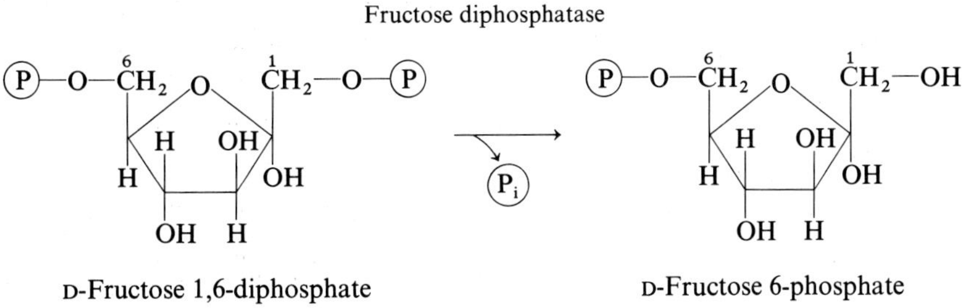

Finally, D-glucose 6-phosphate is formed from D-fructose 6-phosphate by reversing the glucose-phosphate isomerase step. D-Glucose is generated from D-glucose 6-phosphate by D-*glucose 6-phosphatase*, thus bypassing the third irreversible step in glycolysis.

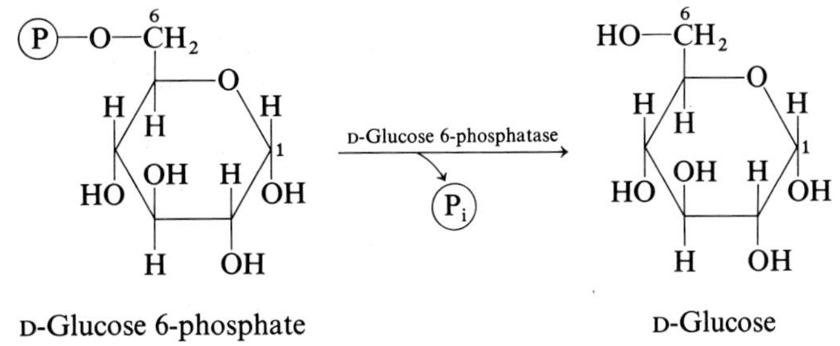

(P)—O—$\overset{6}{C}H_2$; O; H, H, H, 1; HO, OH, H, OH; H, OH

D-Glucose 6-phosphatase → (P_i)

HO—$\overset{6}{C}H_2$; O; H, H, H, 1; HO, OH, H, OH; H, OH

D-Glucose 6-phosphate

D-Glucose

The newly formed glucose is either excreted out of the liver cells to restore low blood-glucose levels or stored as glycogen. It should be noted that a total of six high-energy phosphate bonds are required to make one glucose molecule from two molecules of lactic acid.

10.3 THE PENTOSE PHOSPHATE PATHWAY

The pentose phosphate pathway is an alternative route for glucose oxidation in certain tissues. This pathway is also called the *pentose shunt*, the *hexose monophosphate pathway*, and the *phosphogluconate oxidation pathway*. Such organs as the liver, mammary glands, testes, and adrenal cortex contain the necessary enzymes for this pathway. Between 30 and 60 percent of all the glucose normally metabolized by these organs is oxidized by the pentose phosphate pathway.

We can summarize the important functions of the pentose phosphate pathway. First, the pathway is responsible for producing the *extramitochondrial reducing power* (*cytoplasmic NADPH*) necessary for various anabolic (biosynthetic) reactions. An additional phosphate group is bound to one of the ribose residues of NADH, forming **NADPH** (see Chapter 9). The NADPH is the cofactor for NADPH-specific dehydrogenases and acts as an electron/hydrogen donor in various reductive biosynthetic reactions. For example, NADPH is required in the biosynthesis of fatty acids and steroids. Since the liver, mammary glands, testes, and adrenal cortex are active producers of such compounds, these organs obviously contain the enzymes that make NADPH.

NADPH

Second, the pentose phosphate pathway converts hexoses (6-carbon sugars) into pentoses (5-carbon sugars) and provides a means by which 3-, 4-, 5-, 6-, and 7-carbon sugars can be interconverted for entrance into the glycolytic pathway.

Third, the pathway is responsible for the biosynthesis of *ribose 5-phosphate* (a pentose). This sugar is used in constructing *ribonucleic acid* (*RNA*). After it is converted to the *deoxyribose form*, it is also used in constructing *deoxyribonucleic acid* (*DNA*) (see Chapter 7).

Fourth, the pathway is an alternative route for glucose oxidation in certain tissues. In essence, each carbon atom of glucose 6-phosphate can be oxidized to CO_2, in the process forming *two NADPH molecules*. As we have already stated, these NADPH molecules can be used to reduce intermediates in various biosynthetic reactions. Alternatively, the NADPH can be used to generate NADH molecules via the *transhydrogenase reaction*:

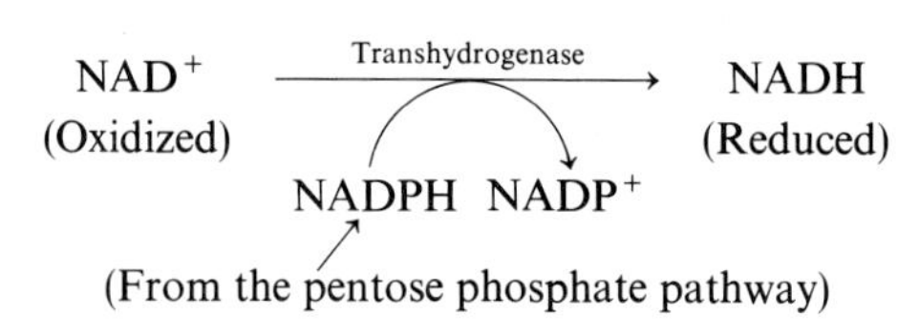

As we shall see in Chapter 11 (The Tricarboxylic Acid Cycle), a total of three ATP molecules can be made for *each* NADH molecule. If 12 NADPH molecules can be generated from the 6-carbon glucose (2 NADPH/carbon × 6 carbon atoms), then 12 NADH molecules would also be produced via the transhydrogenase step. Thus, a total of 36 ATP molecules could be made if one molecule of glucose were completely oxidized to CO_2 by the pentose phosphate pathway (3 ATP/NADH × 12 NADH). The pentose phosphate pathway, therefore, is equivalent to aerobic glycolysis with respect to the number of ATP molecules that can be generated for each glucose molecule.

10.4 GLYCOGEN METABOLISM

Approximately 0.23 kilograms of glycogen is stored in the liver cells and skeletal muscle cells of the average adult. This glycogen is actually stored in large, amorphous granules within the cytoplasm of these cells. In the granules are also found the enzymes responsible for both the degradation and the biosynthesis of glycogen.

Cascade *

A complex series of enzymatic steps degrade glycogen into *glucose 1-phosphate*, which is converted into glucose 6-phosphate (for glycolysis). The complete glycogen degradation pathway is shown in Figure 10–3 (page 186). This pathway is called a "cascade" sequence, because one enzyme of the pathway activates another enzyme, which activates another enzyme, and so on, until glycogen is completely degraded into glucose 1-phosphate. Other enzymatic pathways in our bodies also operate via "cascade" activation mechanisms. For example, the clotting of blood requires 11 separate steps before the fibrin clot actually forms.

Animals (and humans) produce *epinephrine* (*adrenaline*) in response to frightful or stressful situations. The production of epinephrine by the adrenal medulla triggers the degradation of glycogen and yields glucose 1-phosphate for immediate energy needs. This is the classic "fight-or-flight" response, in which reserve liver and muscle glycogen is mobilized to yield glucose. Specifically, epinephrine molecules travel from the adrenal medulla to liver and skeletal muscle cells via the blood stream. There, they bind to specific *receptor proteins* on the outer surfaces of cells, activating a membrane-bound enzyme called *adenylate cyclase*. Adenylate cyclase synthesizes a substance called **cyclic AMP (cAMP)** from ATP. Epinephrine is called the "first chemical messenger" because it transmits its chemical message from the *outside of the cell*. Cyclic AMP, the "second messenger," transmits its chemical message from the *inside of the cell*. The chemical events that have been described so far are shown in Figure 10–4 (page 187), along with the structures of epinephrine and cAMP.

cyclic AMP (cAMP)

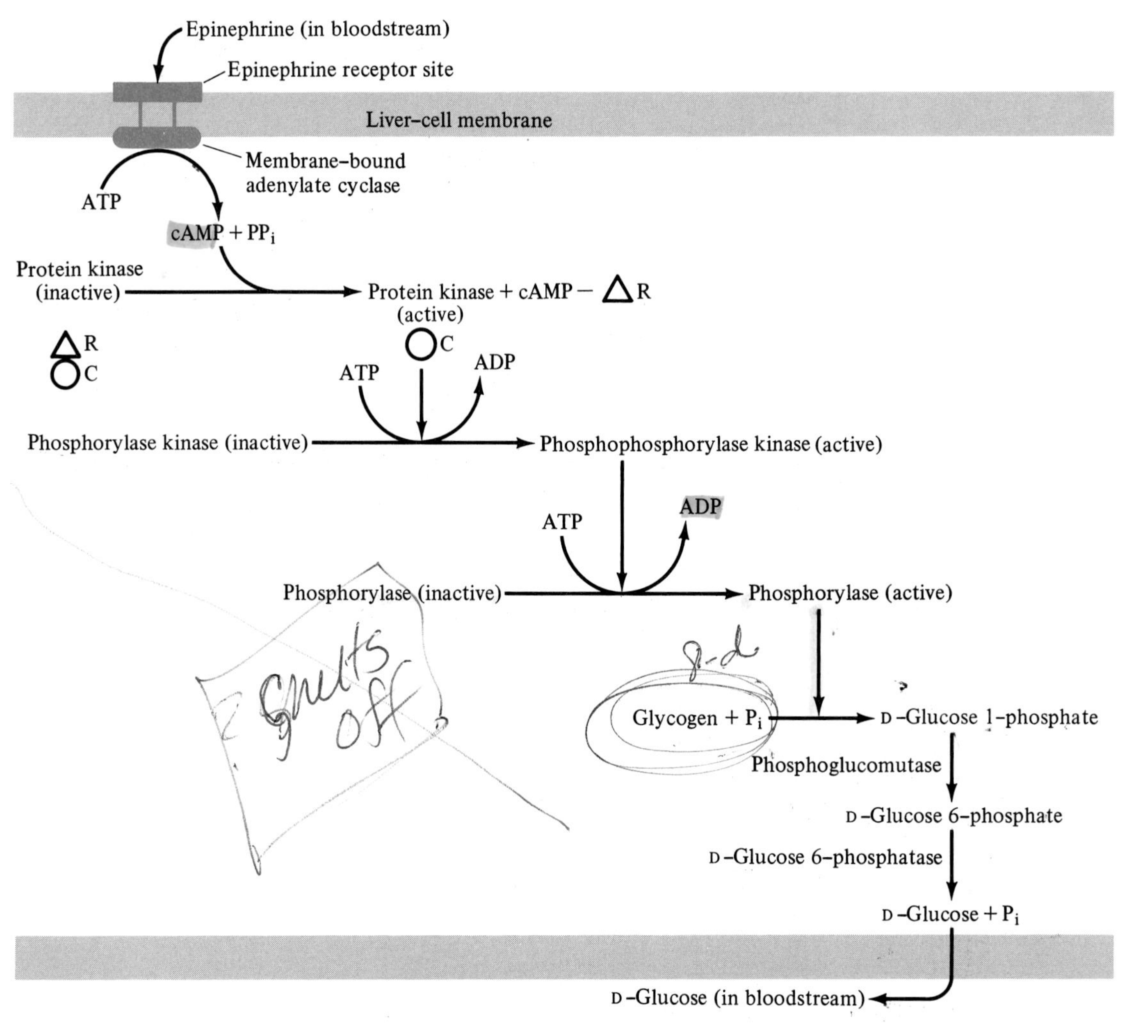

FIGURE 10–3 Schematic representation of the "cascade" sequence of glycogen degradation. (Adapted from Armstrong, F.B., and Bennett, T.P., *Biochemistry* (New York: Oxford University Press, 1979), p. 380.)

Once cAMP has been produced by the membrane-bound adenylate cyclase, it activates *protein kinase*. Protein kinase is a two-subunit enzyme, the dimer form of which is inactive. One subunit is called the *catalytic* (*C*) *subunit*, because this subunit is enzymatically active. The other subunit is called the *regulatory* (*R*) *subunit*, because it binds a molecule of cAMP. When the cAMP molecule binds to the regulatory subunit, the dimer *dissociates* into separate C and R subunits. The C subunit, now activated, catalyzes the activation of the next enzyme in the sequence.

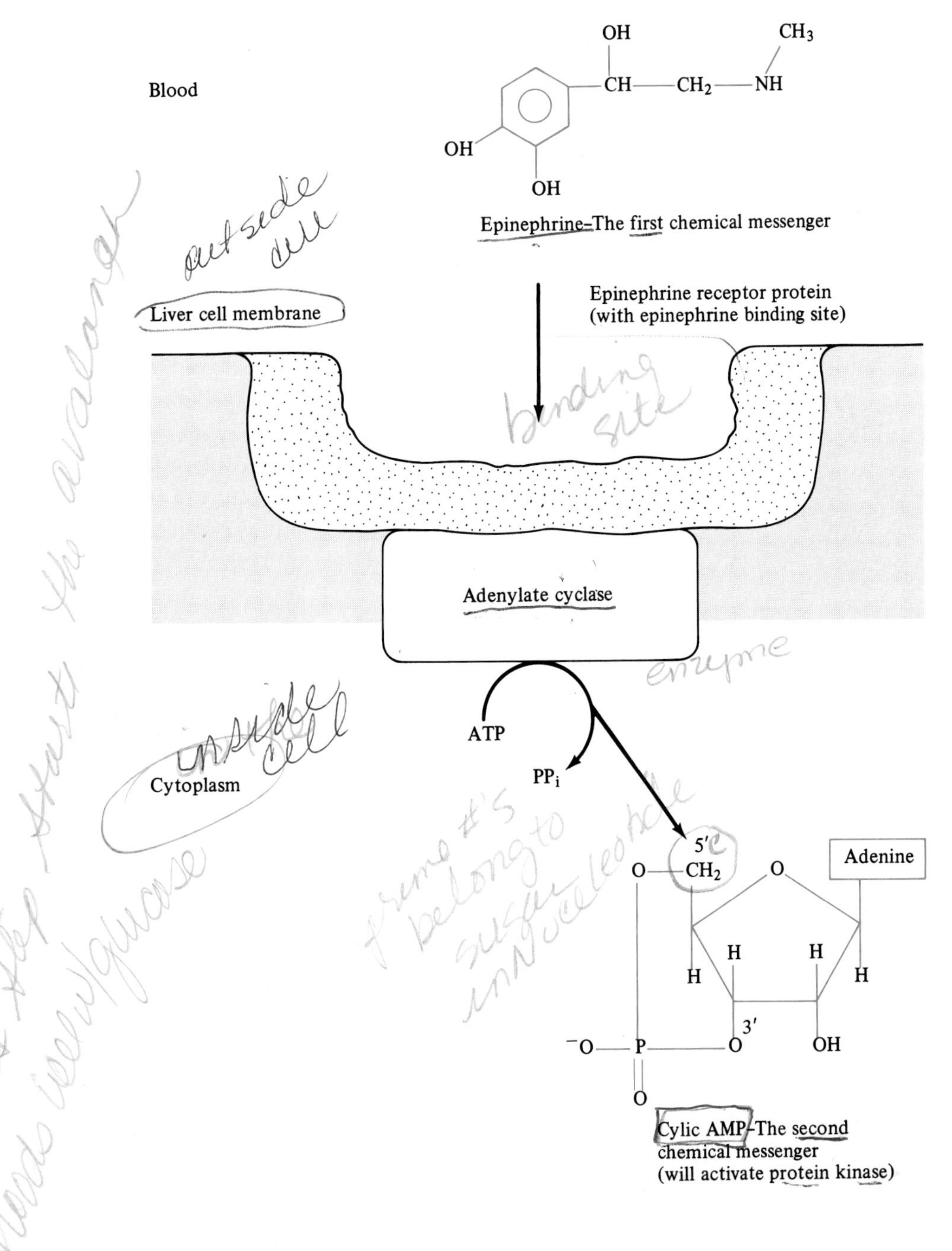

FIGURE 10–4 The activation of adenylate cyclase and the production of cyclic AMP. After epinephrine binds to the epinephrine receptor protein, adenylate cyclase is activated by a conformational change. Adenylate cyclase catalyzes the formation of cyclic AMP (cAMP) from ATP. Cyclic AMP, in turn, activates protein kinase.

Once protein kinase has been activated, it catalyzes the activation of *phosphorylase kinase* according to the following reaction:

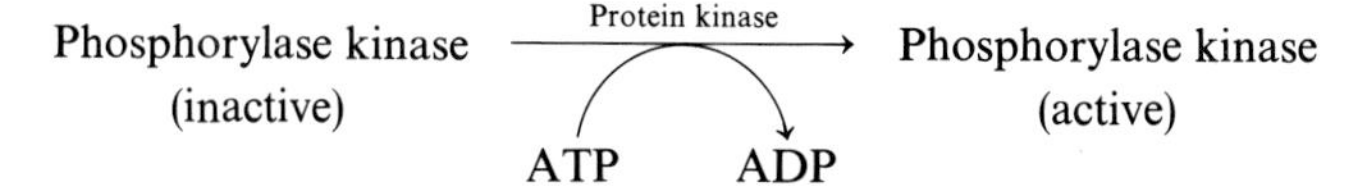

Protein kinase activates phosphorylase kinase by catalyzing the phosphorylation of specific serine residues on the phosphorylase kinase enzyme. This requires ATP as a high-energy phosphate group donor. Once phosphorylation has occurred, a conformational change in the molecule activates the enzyme. This form of enzyme activation was discussed in Chapter 6 (Enzymes).

Once phosphorylase kinase has been activated, it catalyzes the activation of *phosphorylase* (the enzyme that degrades glycogen). Again, specific serine residues are phosphorylated by ATP to activate the enzyme. In this case, inactive dimers aggregate into the active tetramer (see Figure 10–5). Recent evidence suggests that in muscle cells, the enzyme exists only as a dimer and that phosphorylation of this dimer activates it.

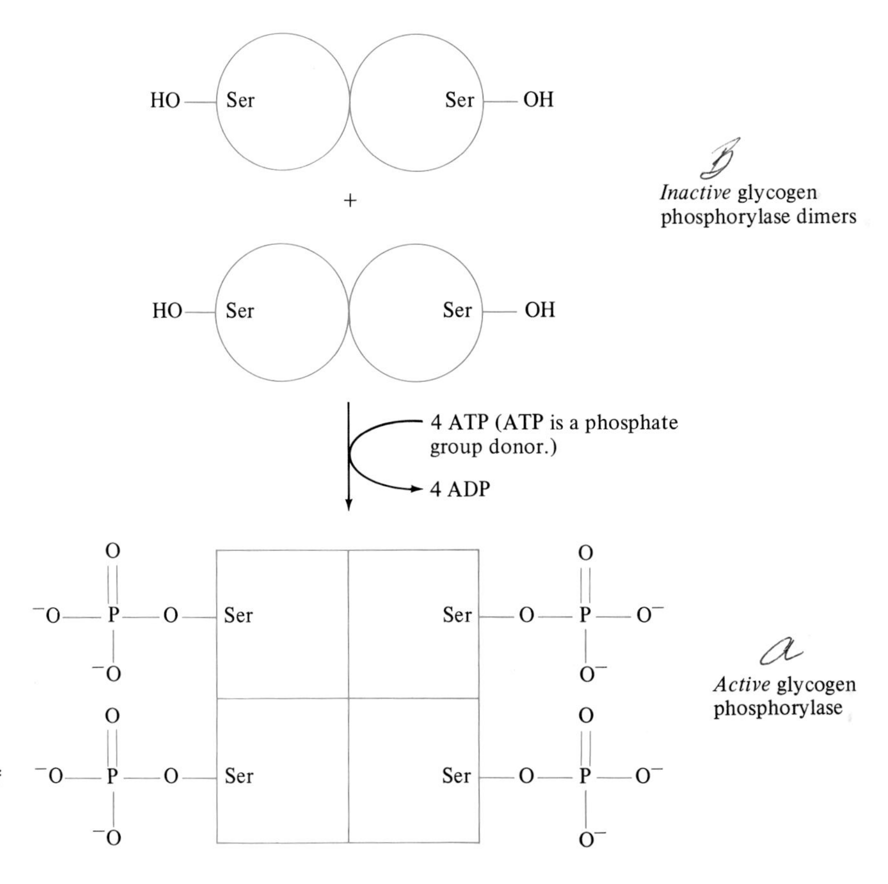

FIGURE 10–5
Covalent modification of glycogen phosphorylase.

Once active phosphorylase has been formed, it catalyzes the degradation (phosphorylysis) of the glycogen molecule to yield glucose 1-phosphate. It should be noted that phosphorylase can catalyze the phosphorolysis of the $\alpha(1 \rightarrow 4)$ glycosidic bonds, but *not* the $\alpha(1 \rightarrow 6)$ bonds. Associated with the phosphorylase enzyme, therefore, is a second enzyme, called the **debranching enzyme**. This debranching enzyme catalyzes the hydrolysis of the $\alpha(1 \rightarrow 6)$ glycosidic bonds and the resynthesis of new $\alpha(1 \rightarrow 4)$ bonds. This eliminates the $\alpha(1 \rightarrow 6)$ branch point. The phosphorylase can then continue and degrade the rest of the chain. Thus, glycogen degradation results from the combination of two specific enzymes, the phosphorylase enzyme and the debranching enzyme, which work together to yield glucose 1-phosphate.

debranching enzyme

Finally, the glucose 1-phosphate can be converted into glucose 6-phosphate, which can directly enter glycolysis.

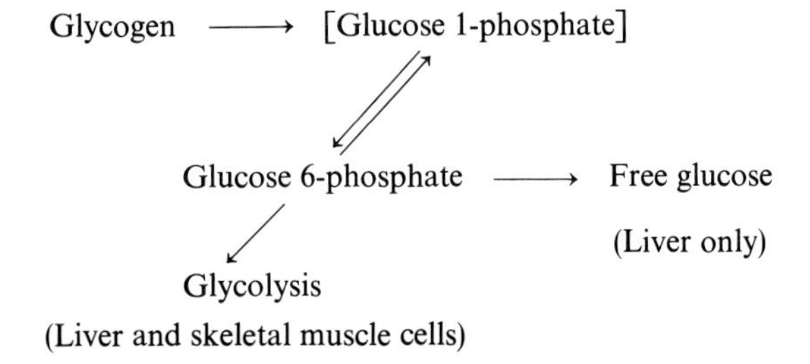

Skeletal muscle cells do not contain glucose 6-phosphatase; therefore, they cannot convert glucose 6-phosphate to free glucose.

Up to this point, we have only discussed the degradation of glycogen by the complex phosphorylase pathway. Glycogen *synthesis* can also occur, but not at the same time as glycogen degradation. Obviously, if the body needs free glucose to fulfill immediate energy needs, it will be involved in degrading glycogen and not in synthesizing it. Likewise, when the body is resting and its energy demands are lower, glycogen degradation is shut off and glycogen synthesis can take place. One can think of this complex pathway as containing an on/off switch. The *glycogen synthetase* molecule is in an *inactive* state when the phosphorylase system is *active*, and vice versa. Interestingly, the glycogen synthetase molecule is *deactivated* by phosphorylation—in contrast to the phosphorylase molecule, which is activated by phosphorylation. When glycogen is to be synthesized, the glycogen synthetase molecule is first dephosphorylated by a phosphatase, thus activating the molecule.

Glycogen synthesis takes place when an active glycogen synthetase molecule catalyzes the formation of glycogen from a substance called *uridine diphosphate glucose* (*UDP-glucose*) according to the following reaction:

$$[\text{Glycogen}]_n \xrightarrow{\text{Active glycogen synthetase}} [\text{Glycogen}]_{n+1}$$

UDP-Glucose → UDP

where n is the number of glucose residues in the glycogen molecule.

Thus, the glycogen molecule is lengthened at each step by one glucose residue, in the process forming a new $\alpha(1 \rightarrow 4)$ glycosidic bond. The $\alpha(1 \rightarrow 6)$ branch points are inserted by a "branching enzyme."

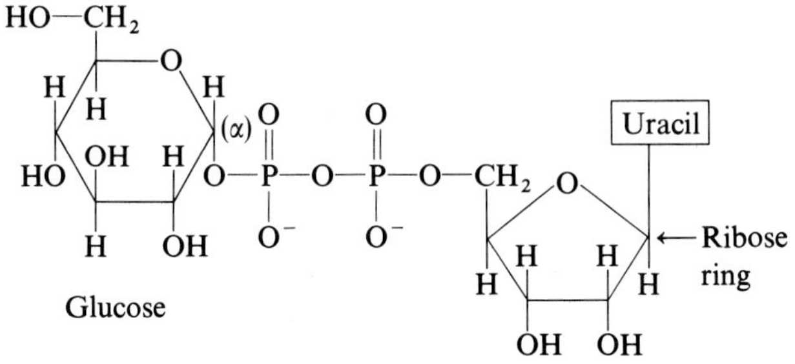

HO—CH$_2$ (α) O O Uracil; HO, OH, H, O—P—O—P—O—CH$_2$, O$^-$ O$^-$, ← Ribose ring, OH OH

Glucose

Uridine Diphosphate-Glucose
(UDP-glucose)

GLYCOGEN STORAGE DISEASES

glycogen storage diseases

The **glycogen storage diseases** result from abnormalities in the storage and metabolism of glycogen. *Von Gierke's disease* (a common glycogen storage disease) is characterized by an enlarged liver, generalized weakness relieved by eating (indicating a moderate-to-severe form of *hypoglycemia*), and mental retardation. Results from blood chemistry tests reveal a low blood-glucose level, high triglyceride and fatty acid levels, a high cholesterol level, a high lactate level, and a high uric-acid level. Thus, a rather large variety of biochemical abnormalities are exhibited and must be explained.

The liver is enlarged by an excessive amount of liver glycogen (hence the term glycogen *storage* disease). A low blood-glucose level indicates that glycogen is not being metabolized normally; therefore, little glucose is being released into the blood from the liver glycogen stores. Abnormal glycogen storage results from a deficiency of *glucose 6-phosphatase*.

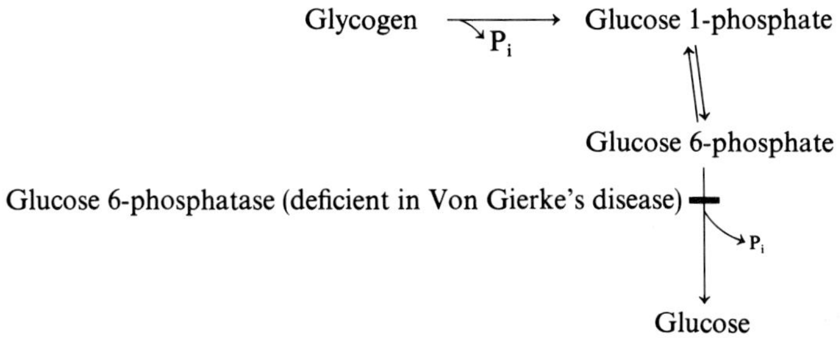

Glycogen $\xrightarrow{P_i}$ Glucose 1-phosphate

$\updownarrow$

Glucose 6-phosphate

Glucose 6-phosphatase (deficient in Von Gierke's disease) ⟶ P_i

$\downarrow$

Glucose

As a result of this enzymatic block, the glucose 6-phosphate level rises, and large amounts of glycogen are synthesized and stored.

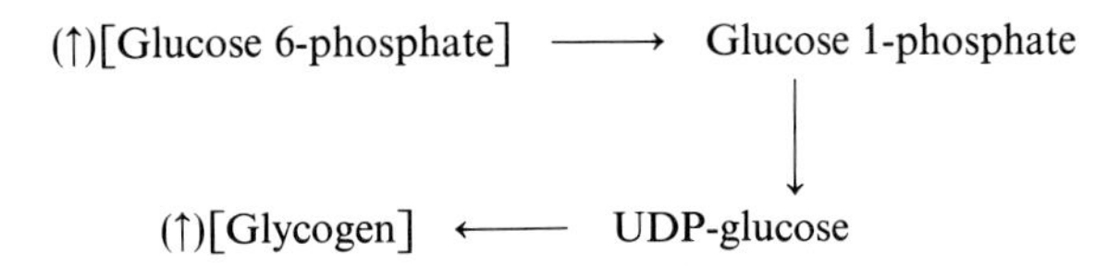

In addition, the high glucose 6-phosphate level floods the liver glycolytic pathway with large amounts of glucose 6-phosphate, which is broken down into lactate and causes a high blood-lactate level. Decreased blood glucose results, leading to severe hypoglycemia. Because of the low blood-glucose level, the adrenal cortex is stimulated to secrete adrenalin, which stimulates the adipose (fatty tissue) to mobilize triglycerides. Thus, high triglyceride and free fatty acid levels are found in the blood. Because of the high glucose 6-phosphate level, the pentose phosphate pathway is also flooded with this substance, causing the production of excessive amounts of *ribose 5-phosphate* (see Section 10.3). Ribose 5-phosphate stimulates the synthesis of *purines* (see Chapter 16); increased degradation of purines leads to the increased production of **uric acid**. If large amounts of uric acid are present in the blood and in the joints, it crystallizes and causes a painful disease commonly known as **gout**. The glycogen storage disease is obviously very serious. It is remarkable to note how so many diverse metabolic pathways are affected by the absence of *one* enzyme in just *one* pathway.

uric acid

gout

SUMMARY

Lactic acid diffuses from actively exercising skeletal muscles into the blood stream and hence to the liver. If glucose is to be resynthesized from lactate via gluconeogenesis, three irreversible steps in glycolysis must be bypassed. The pyruvate kinase step is bypassed via mitochondrial pyruvate carboxylase and by oxaloacetate/malate interconversion. The phosphofructokinase step is bypassed by fructose 1,6-diphosphatase. The hexokinase step is bypassed by glucose 6-phosphatase.

Glucose can be oxidized into CO_2 and water via an alternative pathway, known as the pentose phosphate pathway. The pentose phosphate pathway produces NADPH molecules, which are used to reduce intermediates in various biosynthetic reactions. In addition, the pathway converts hexoses (6-carbon sugars) into pentoses (5-carbon sugars) and is responsible for synthesizing ribose 5-phosphate, which is used in forming the nucleic acids RNA and DNA. The pentose phosphate pathway is approximately equivalent to aerobic glycolysis, because 36 ATP molecules can be produced for each completely oxidized glucose molecule.

Glycogen is the primary storage form of glucose in the liver and in the skeletal muscles. Glycogen is degraded via a rather complex "cascade" pathway summarized as follows: The adrenal medulla secretes epinephrine (adrenaline), which travels via the blood stream to the liver or skeletal muscle cells. There, it binds to epinephrine receptor sites and activates adenylate cyclase, which produces cyclic AMP. Cyclic AMP activates protein kinase; protein kinase activates phosphorylase kinase; and phosphorylase kinase activates phosphorylase. Phosphorylase, in combination with a debranching enzyme, catalyzes the complete degradation of glycogen into glucose 1-phosphate. Glucose 1-phosphate can then be converted into glucose 6-phosphate.

Glycogen synthesis does not take place while glycogen degradation is occurring. Glycogen synthesis occurs when glycogen synthetase is activated by dephosphorylation. Von Gierke's disease, a very serious glycogen storage disease, is a result of a deficiency in glucose 6-phosphatase. Large quantities of glucose 6-phosphate are produced; this causes excessive amounts of glycogen to be synthesized and deposited in the liver. Hypoglycemia, increased mobilization of lipids and crystallization of uric acid in joints also accompany this serious disease state.

REVIEW QUESTIONS

1. We have presented a number of important concepts and terms in this chapter. Define (or briefly explain) each of the following terms:
   a. Cori cycle
   b. Gluconeogenesis
   c. Carboxylation
   d. Pentose phosphate pathway
   e. NADPH
   f. Ribose 5-phosphate
   g. Transhydrogenase
   h. "Cascade" sequence
   i. Adenylate cyclase
   j. Cyclic AMP
   k. Protein kinase
   l. Phosphorylation
   m. Debranching enzyme
   n. UDP-glucose
   o. Glycogen synthetase
   p. Glycogen storage disease

2. We have presented a number of enzymatic reactions in this chapter. Write out the complete chemical reaction catalyzed by each of the following enzymes. Draw the complete structures of all substrates and products; include important cofactors.
   a. Pyruvate carboxylase
   b. PEP carboxykinase
   c. Glucose 6-phosphatase
   d. Adenylate cyclase
   e. Glycogen phosphorylase (active)
   f. Glycogen synthetase (active)

*3. Gluconeogenesis.
   a. Suppose the HCO_3^- used in the pyruvate carboxylase reaction were radioactively labeled with ^{14}C (noted as $H^{14}CO_3^-$). Where would the label appear in a newly synthesized glucose molecule?
   b. Suppose pyruvate were labeled with ^{14}C at the #3 methyl position. Where would the label appear in a newly synthesized glucose molecule?

4. How many ATP (and GTP) molecules are required to resynthesize one glucose molecule from lactate?

5. Draw a complete structure for the NADPH molecule.

6. You are strenuously jogging. Which of the following best describes the chemical environment in your skeletal muscle cells during your jog? Be able to support your answer.
   a. Low cyclic AMP level and low active protein kinase level.
   b. Low lactate level.
   c. High inactive phosphorylase level.
   d. High active phosphorylase kinase level and high inactive phosphorylase level.
   e. High active protein kinase level and high active phosphorylase level.

*7. Suggest how UDP-glucose might be synthesized in the cell.

SUGGESTED READING

Bhagavan, N. V. *Biochemistry*, 2nd ed., Ch. 4. Philadelphia: J. B. Lippincott, 1978.

Lehninger, A. L. *Biochemistry*, 2nd ed., Ch. 23. New York: Worth, 1975.

Newsholme, E. A., and Start, C. *Regulation in Metabolism*, Chs. 4 and 6. New York: John Wiley, 1973.

Stryer, L. *Biochemistry*, 2nd ed., Chs. 15 and 16. San Francisco: W. H. Freeman, 1981.

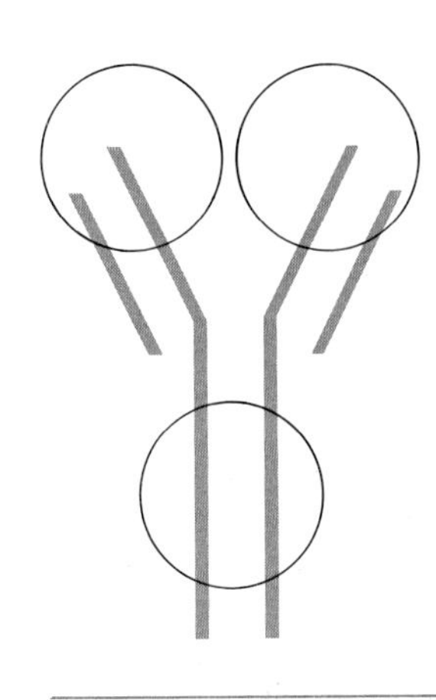

Chapter 11

The Tricarboxylic Acid Cycle

11.1 INTRODUCTION

During aerobic glycolysis, glucose is first partially oxidized into pyruvate. This pyruvate is then transported into the mitochondrion, where it is completely **oxidized** into carbon dioxide and water. Once inside the mitochondrion, pyruvate is converted into a substance called *acetyl-coenzyme A* (*acetyl-CoA*). This acetyl-CoA enters the *tricarboxylic acid cycle*, and some of the electrons originally contained in the pyruvic acid molecule are extracted in the form of NADH and $FADH_2$. These electrons are used to drive ATP synthesis via the processes of *electron transport* and *oxidative phosphorylation.*

oxidized

In this chapter, we will discuss the structure and functions of mitochondria (the "powerhouses" of the cell). Specifically, we will describe the transport of pyruvate into the mitochondrion and its conversion there into acetyl-CoA. We will then describe the tricarboxylic acid (TCA) cycle, including the enzymatic steps comprising it and the chemical regulators governing it. We will then summarize the number of molecules of ATP that can be generated from one molecule of glucose when it is completely oxidized by aerobic glycolysis.

11.2 THE STRUCTURE OF THE MITOCHONDRION

With the invention of the electron microscope, scientists have been able to photograph the internal structure of the cell with ever-increasing clarity and detail. The various cellular organelles in the typical eukaryotic cell exhibit

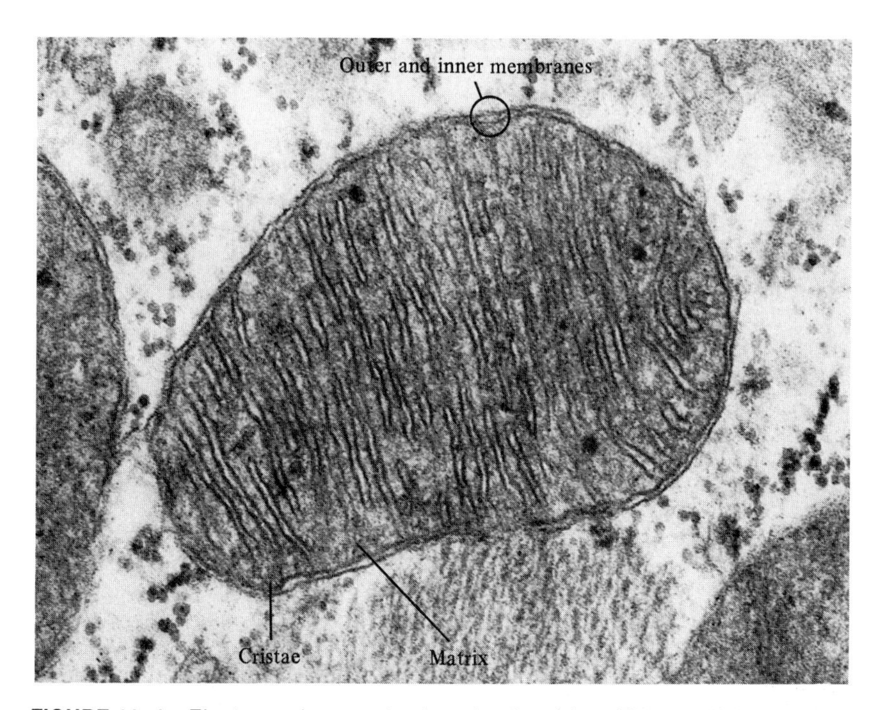

FIGURE 11–1 Electron micrograph of a mitochondrion. The parallel arrays of membranes, the *cristae*, contain the electron transport enzymes. The internal portion of the mitochondrion, the *matrix*, contains the TCA cycle enzymes. (Mitochondrion from rat cardiac tissue, magnification 109 200 ×. Photograph courtesy of Mrs. Nancy Crise-Benson, Department of Biological Sciences, California State University, Hayward.)

an internal substructure of their own. Such is the case with the mitochondrion. An electron micrograph of a typical mitochondrion is shown in Figure 11–1. An artist's conception of the three-dimensional structure of the mitochondrion is shown in Figure 11–2. The term **mitochondrion** comes from the Greek words *mitos*, "thread" and *chondrion*, "granule." Upon examining Figure 11–1, you should see why the word "mitochondrion" aptly describes this organelle. There are many mitochondria in a typical cell. Liver cells may contain as many as 800 individual mitochondria randomly distributed through the cytoplasm; other cells may contain fewer. Mitochondria may be randomly scattered throughout the cytoplasm, or they may be organized in regular rows or clusters. For example, the mitochondria in insect flight-muscle cells are arranged in regular rows close to the actin/myosin muscle filaments, whose contraction is powered by ATP produced by mitochondria. The tails of sperm cells contain mitochondria stacked in neat piles along the length of the tail.

mitochondrion

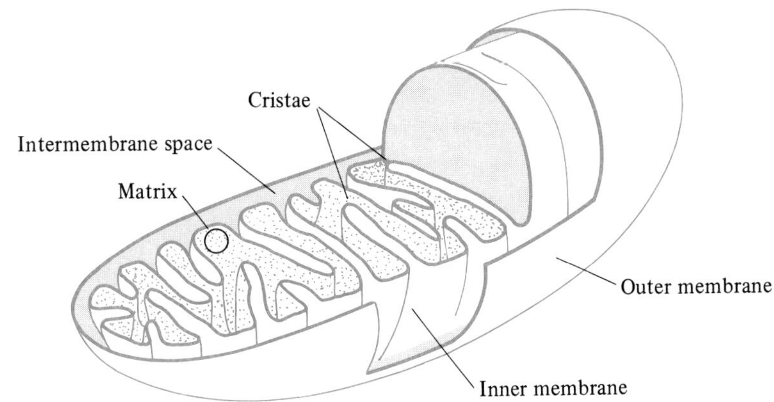

FIGURE 11–2 An artist's diagram of a mitochondrion. (Adapted from S. L. Wolfe, *Biology of the Cell* (Belmont, CA: Wadsworth, 1972).)

A typical mitochondrion has a rather complex internal structure (see Figures 11–1 and 11–2). Most mitochondria are about the same size as a bacterium (approximately 1 × 2 microns). The mitochondrion contains two lipid/protein membranes, the *outer* and *inner membranes*. The outer membrane is permeable to most small-molecular-weight substances; the inner membrane is *not* permeable to such substances as sugars, ions, NAD^+, NADH, CoA, and so on. Specific transport mechanisms regulate the movement of various substances from the space between the two membranes into the internal space of the mitochondrion. The inner mitochondrial membrane is highly folded and convoluted. These folds are called *cristae* ("crest") and their inner surfaces contain the enzymes of the electron transport system and also those responsible for ATP synthesis. The internal portion of the mitochondrion is called the *matrix* (which means "womb," a place where something originates). The matrix is a solution of proteins, enzymes, and small-molecular-weight substances. It contains most of the enzymes that catalyze the tricarboxylic acid cycle. Mitochondria can undergo dramatic changes in size and shape during various phases of mitochondrial respiration. Specifically, the matrix oscillates from a rather open to a more condensed structure, and back again, during ATP synthesis.

11.3 THE FORMATION OF ACETYL-COENZYME A

Because the conversion of pyruvate into lactate is not favored during aerobic glycolysis, the pyruvate concentration builds up in the cytoplasm. This pyruvate is transported into the mitochondrion. In turn, a specific pyruvate transport system in the inner mitochondrial membrane transports pyruvate from the cytoplasm into the matrix of the mitochondrion.

Once inside the mitochondrial matrix, pyruvate is converted into acetyl-CoA by the *pyruvate dehydrogenase complex* according to the following reaction:

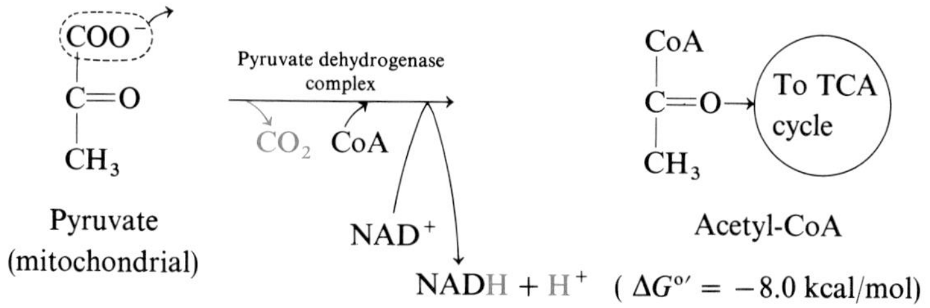

$$\text{Pyruvate (mitochondrial)}\ \ COO^- - C{=}O - CH_3 \xrightarrow[\text{CoA, } NAD^+ \to NADH + H^+ \ (-CO_2)]{\text{Pyruvate dehydrogenase complex}} CoA - C{=}O - CH_3\ \text{(Acetyl-CoA)} \to \text{To TCA cycle}$$

$(\Delta G^{\circ\prime} = -8.0 \text{ kcal/mol})$

Let us consider the pyruvate dehydrogenase reaction in more detail. The pyruvate dehydrogenase complex is a large, multienzyme complex with a molecular weight of several million daltons. (The molecular weight of heart and kidney mitochondrial pyruvate dehydrogenase is 7 million daltons.) An idealized diagram of the complex is shown in Figure 11–3. The complex, which forms a 20-sided polyhedron, contains a core of 60 identical polypeptide chains (each called *dihydrolipoyl transacetylase*). Surrounding the core are 20 molecules of *pyruvate dehydrogenase*, 5–6 molecules of *dihydrolipoyl dehydrogenase*, and a few regulatory enzyme molecules.

coenzyme A
acetyl-CoA
oxidative decarboxylation

A number of important enzymatic reactions take place while pyruvate is being converted into acetyl-CoA. Basically, the carboxyl group of pyruvate is removed as CO_2 and the two remaining carbon atoms of the original pyruvate molecule are transferred to **coenzyme A** forming **acetyl-CoA**. This entire process is called **oxidative decarboxylation**, because CO_2 is removed via oxidation-reduction reactions.

Five different enzyme cofactors are bound to the pyruvate dehydrogenase complex. Each participates in a biochemical reaction necessary for converting pyruvate into acetyl-CoA. A number of these cofactors are actually derivatives of vitamins (see Chapter 20).

thiamine pyrophosphate (TPP)
lipoic acid
swinging arm

The first step of the sequence is catalyzed by pyruvate dehydrogenase (enzyme E_1). A vitamin coenzyme called **thiamine pyrophosphate (TPP)** is bound into the active site of the pyruvate dehydrogenase molecule and catalyzes the decarboxylation of pyruvate. The remaining 2-carbon fragment is then oxidized by the 5-membered sulfur ring of **lipoic acid** (the **swinging arm**). This swinging arm transfers the 2-carbon fragment (the acetyl group) to *coenzyme A*, forming *acetyl-CoA*. This entire reaction is

FIGURE 11–3 The pyruvate dehydrogenase complex. The mechanism of the pyruvate dehydrogenase complex is shown. Pyruvate is decarboxylated by pyruvate dehydrogenase (E_1) and thiamine pyrophosphate (TPP). The lipoyllysyl side chain (the "swinging arm") picks up the remaining two-carbon fragment and transfers it to coenzyme A to form acetyl-CoA. The swinging arm is reoxidized by dihydrolipoyl dehydrogenase (E_3). (Adapted from A. L. Lehninger, *Biochemistry*, 2nd ed. (New York: Worth, 1975), p. 452.)

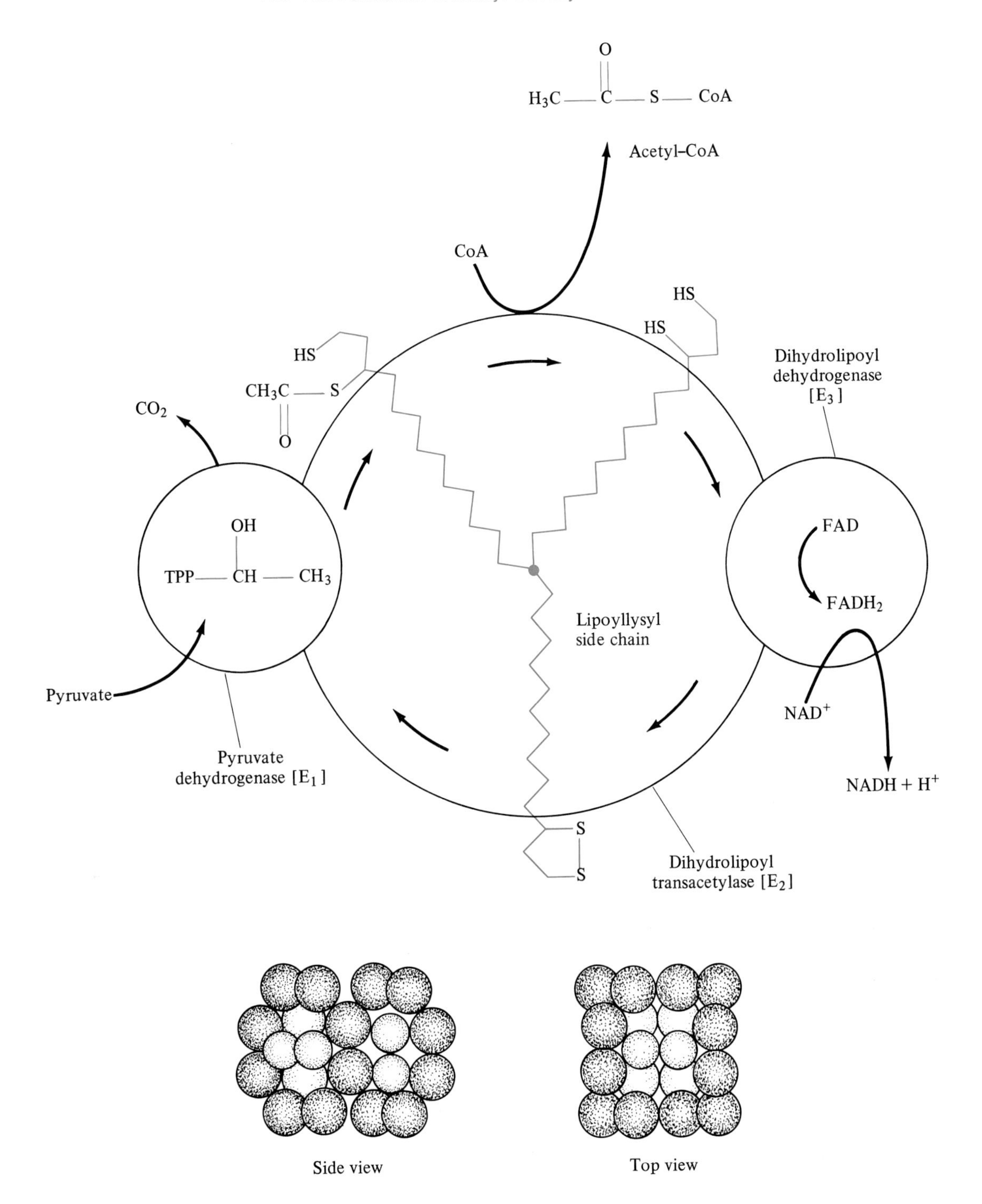
O
$H_3C-C-S-CoA$
Acetyl–CoA
CoA
HS
HS
Dihydrolipoyl
dehydrogenase
[E_3]
HS
CH_3C-S
O
CO_2
OH
TPP— CH — CH_3
FAD
$FADH_2$
Lipoyllysyl
side chain
Pyruvate
NAD^+
Pyruvate
dehydrogenase [E_1]
$NADH + H^+$
S
S
Dihydrolipoyl
transacetylase [E_2]
Side view
Top view

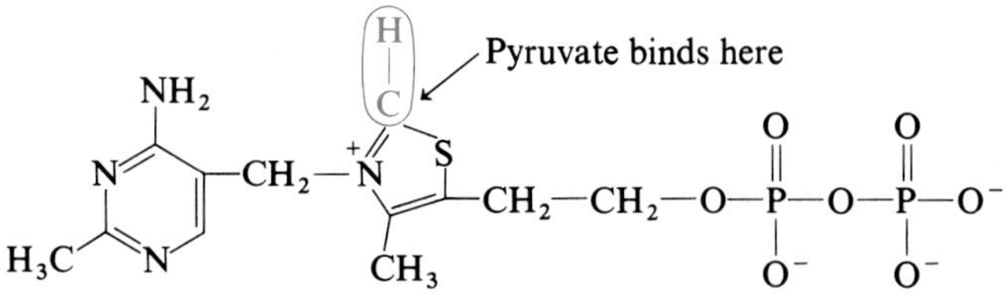

Pyruvate binds here

Thiamine pyrophosphate (TPP)

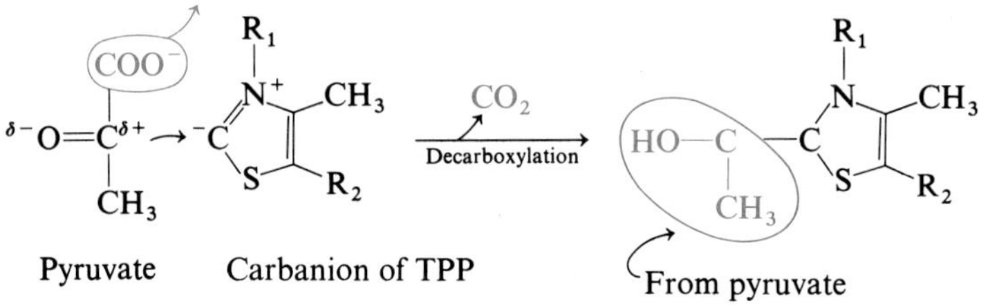

Decarboxylation

Pyruvate   Carbanion of TPP   From pyruvate

catalyzed by the core enzyme complex *dihydrolipoyl transacetylase* (enzyme E_2—see Figure 11–3). Once acetyl-CoA has been formed, it will leave the

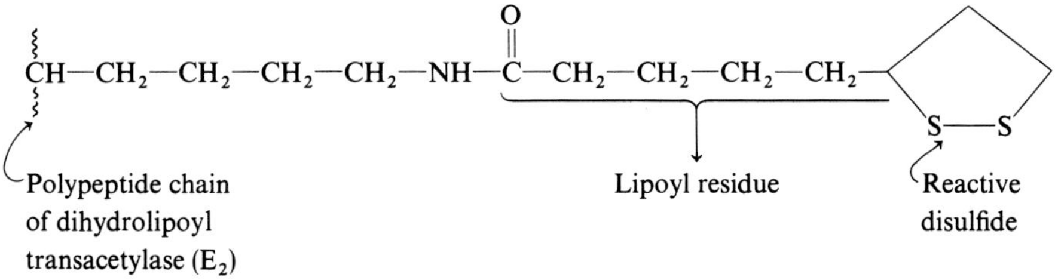

Polypeptide chain of dihydrolipoyl transacetylase (E_2)   Lipoyl residue   Reactive disulfide

The swinging arm

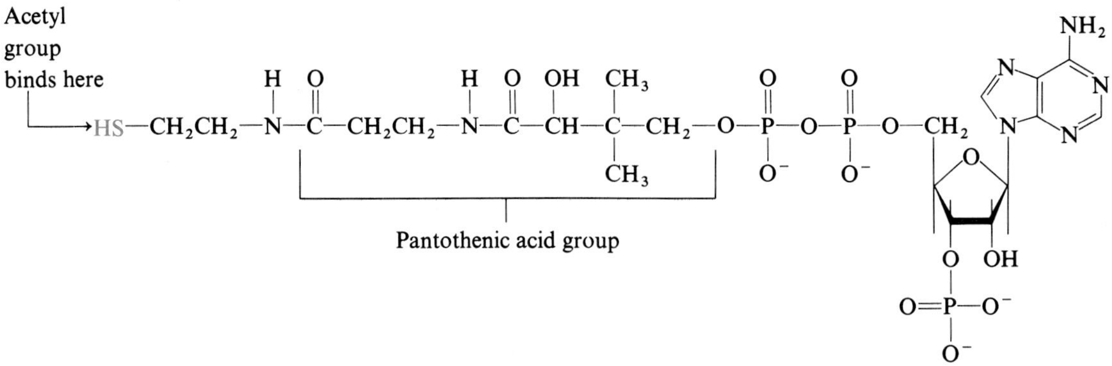

Acetyl group binds here

Pantothenic acid group

Coenzyme A (CoA-SH)

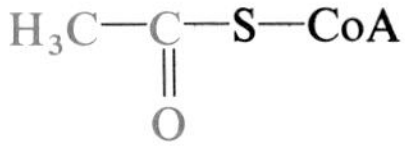

$H_3C-C(=O)-S-CoA$ *Acetyl*-CoA

dihydrolipoyl dehydrogenase
flavin adenine dinucleotide (FAD)

active site of enzyme E_2 and enter the TCA cycle. The lipoyl swinging arm is now in a reduced state and, therefore, must be reoxidized into the closed disulfide ring. This oxidation is catalyzed by the third enzyme of the complex, **dihydrolipoyl dehydrogenase** (E_3). This enzyme has a bound cofactor called **flavin adenine dinucleotide (FAD)** that accepts the two hydrogens and two electrons from the reduced form of the lipoyl swinging arm. As a result, the 5-membered ring closes and the lipoyl swinging arm is ready for a second cycle, in which it will accept a second acetyl group from pyruvate dehydrogenase (E_1). The structure of flavin adenine dinucleotide (FAD) is shown:

Finally, the two electrons contained on $FADH_2$ are transferred to NAD^+ to form NADH according to the following reaction:

$$\underset{\text{(Reduced)}}{FADH_2} \xrightarrow{NAD^+ \;\to\; NADH + H^+} \underset{\text{(Oxidized)}}{FAD}$$

11.4 THE TRICARBOXYLIC ACID CYCLE

The *tri*carboxylic *a*cid (*TCA*) cycle is so named because citric acid, one of the reaction intermediates, contains three carboxylic acid groups.

$$\begin{array}{c} COO^- \\ | \\ CH_2 \\ | \\ HO-C-COO^- \\ | \\ CH_2 \\ | \\ COO^- \end{array}$$

Citric acid (with three carboxylic acid groups)

The pathway is also called the *citric acid cycle* and the *Krebs cycle* (named after the biochemist who first elucidated the enzymatic sequence of the pathway). The primary function of the TCA cycle is (1) to completely oxidize acetyl-CoA to CO_2 and H_2O and (2) to recover some of the free energy of the acetyl-CoA in biologically useful form (NADH or $FADH_2$). Carbohydrates are not the only source of acetyl-CoA. Figure 11–4 summarizes other reactions that can provide the mitochondrion with acetyl-CoA for complete oxidation via the TCA cycle.

The TCA cycle is an unusual metabolic pathway because the pathway is *cyclical*—that is, the *oxaloacetate* used in the initial enzymatic step is regenerated by the last enzymatic step. Because the pathway is cyclic in nature, two levels of catalysis are evident. First, the enzymes of the pathway are catalytic. Second, because oxaloacetate is regenerated by the pathway, the pathway *as a whole is catalytic*. The complete pathway is summarized

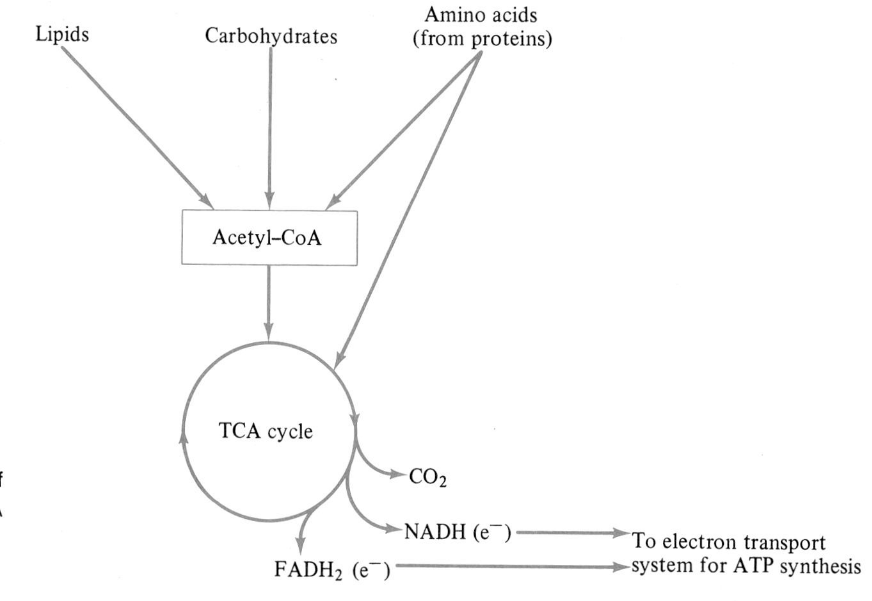

FIGURE 11–4
Lipids, carbohydrates, and amino acids are degraded into acetyl-CoA. This figure illustrates the central role of acetyl-CoA and the TCA cycle in cellular metabolism.

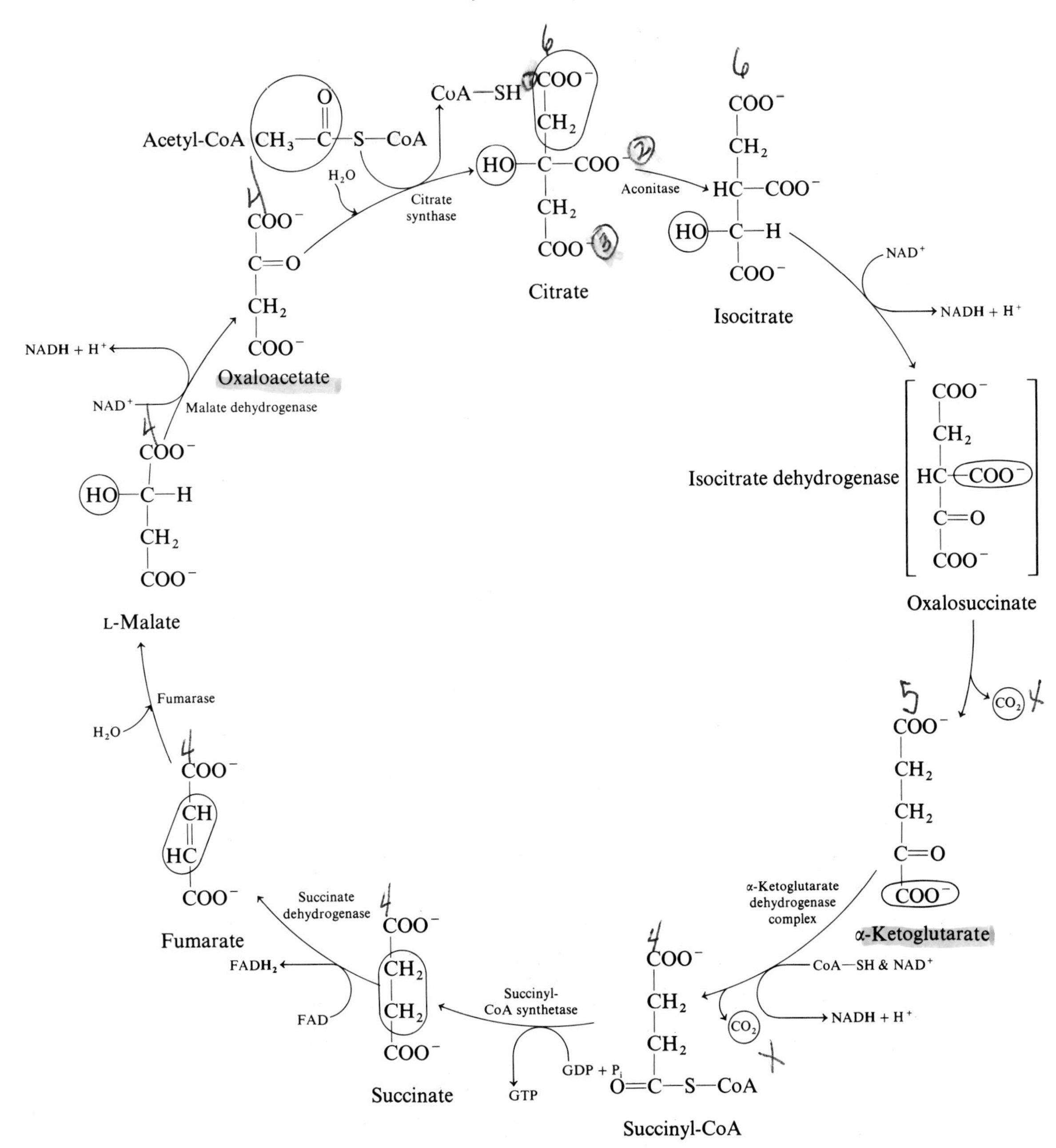

FIGURE 11–5 The tricarboxylic acid (TCA) cycle. The individual enzymatic steps are discussed in the text.

in Figure 11–5. Study the pathway carefully and note (1) how the carbon skeleton of the TCA cycle intermediates is changed, (2) where CO_2 is released, and (3) where NADH (and $FADH_2$) are generated. The NADH (and $FADH_2$) molecules transfer the electrons from the TCA cycle intermediates to the electron transport system for ATP synthesis. Three ATP molecules can be formed for every molecule of NADH (which actually represents two electrons); only two ATP molecules can be formed from each $FADH_2$ molecule. The difference in yield is explained in Chapter 12.

The individual enzymatic steps of the TCA cycle are summarized in the following pages.

STEP 1: CITRATE SYNTHASE

This enzyme (also known as the "condensing enzyme") catalyzes the condensation of *acetyl-CoA with oxaloacetate to yield citrate and CoA-SH.* The reaction can be classified as an *aldol condensation* in which the methyl group of acetyl-CoA attacks the carbonyl group of oxaloacetate as follows:

COO⁻ Citrate synthase
$^{\delta-}O=C^{\delta+}$
CH_2
COO⁻
Oxaloacetate

O^- CoA-SH
$:\bar{C}H_2=C-$(S—CoA)
Enolate form
H^+
$H_3C-\overset{O}{\overset{\|}{C}}-S-CoA$
Acetyl-CoA

$\xrightarrow{H_2O}$

COO⁻
HO—C—[CH_2—COO⁻]
CH_2 From acetyl-CoA
COO⁻
Citrate

($\Delta G^{\circ\prime} = -7.7$ kcal/mol)

The hydrolysis of a high-energy *thioester* bond, releasing CoA, favors citrate formation. As we shall see in Section 11.5, citrate synthase is an allosteric enzyme. It has been called the rate limiting or "pacemaker" enzyme of the TCA cycle.

STEP 2: THE ACONITASE REACTION

Aconitase catalyzes the stereospecific removal of a water molecule from citrate to yield *cis-aconitate*. The *cis*-aconitate is stereospecifically rehydrated by a water molecule to yield *isocitrate*. In essence, aconitase catalyzes the removal of an OH group from the #3 carbon atom and reinserts an OH group on the #4 carbon atom. The reaction is reversible.

1 COO⁻
2 CH_2
3 (HO)—C—COO⁻
4 CH_2
5 COO⁻
Citrate

$\xrightarrow{\text{Aconitase}}$ H_2O

[COO⁻
CH_2
C—COO⁻
‖
4 CH
COO⁻]
cis-Aconitate

$\xrightarrow{\text{Aconitase}}$ H_2O

1 COO⁻
2 CH_2
3 HC—COO⁻
4 (HO)—CH
COO⁻
Isocitrate

Aconitase has been referred to as the "ferrous wheel" enzyme, because an iron atom (ferrous, Fe^{2+}) rotates within the active site of the enzyme. First, it removes an OH group from citrate and releases it as H_2O; then it removes an OH group from a nearby H_2O molecule. Rotating, it brings the second OH into the vicinity of the #4 carbon atom of *cis*-aconitate. You can think of the water molecule as the "seat" of a Ferris wheel—hence the name. This interesting reaction is diagrammed in Figure 11–6.

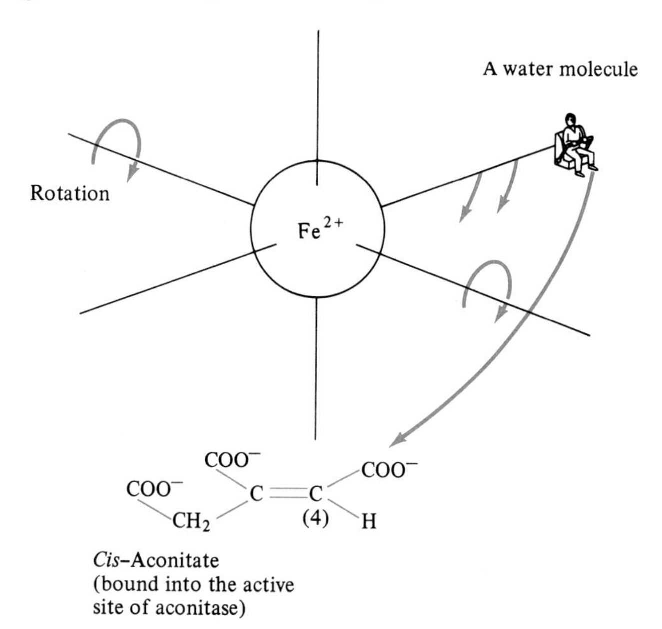

FIGURE 11–6
The "ferrous wheel" mechanism. An OH group from a water molecule is added to *cis*-aconitate to form isocitrate. This is done by the rotation of an iron (ferrous, Fe^{2+}) atom within the active site of aconitase, bringing the water molecule close to the C—C bond of *cis*-aconitate.

STEP 3: ISOCITRATE DEHYDROGENASE

This enzyme catalyzes an *oxidative decarboxylation* reaction in which a CO_2 molecule is removed from *isocitrate* to yield *α-ketoglutarate*. Two hydrogens (with two electrons) are also removed, generating one molecule of NADH. The oxalosuccinate intermediate is not shown.

COO^- — CH_2 — HC—(COO^-) — (HO—CH) — COO^- (Isocitrate) $\xrightarrow{\text{Isocitrate dehydrogenase}}$ COO^- — CH_2 — CH_2 — O=C — COO^- (α-Ketoglutarate)

NAD^+ → $NADH + H^+$; CO_2 released

Isocitrate

α-Ketoglutarate

($\Delta G^{\circ\prime} = -5.0$ kcal/mol)

STEP 4: α-KETOGLUTARATE DEHYDROGENASE

This enzyme catalyzes the oxidative decarboxylation of *α-ketoglutarate* to yield *succinyl-CoA*. This reaction is irreversible. A second CO_2 molecule is released at this step, and a second molecule of NADH is generated. The α-ketoglutarate dehydrogenase complex is very similar in structure to the pyruvate dehydrogenase complex described previously. It catalyzes a similar reaction, and uses the same cofactors (TPP, lipoyl "swinging arm," CoA, FAD, and NAD^+).

$$\underset{\alpha\text{-Ketoglutarate}}{^{-}OOC{-}CH_2{-}CH_2{-}C(=O){-}COO^-} \xrightarrow[CO_2 \quad CoA \quad NAD^+ \to NADH + H^+]{\alpha\text{-Ketoglutarate dehydrogenase complex (``Swinging arm'' mechanism)}} \underset{\text{Succinyl-CoA}}{^{-}OOC{-}CH_2{-}CH_2{-}C(=O){-}S{-}CoA}$$

($\Delta G^{o\prime} = -8.0$ kcal/mol)

Some of the energy of oxidation is conserved in the high-energy thiol compound, succinyl-CoA.

STEP 5: SUCCINYL-CoA SYNTHETASE

This enzyme catalyzes the formation of *succinate* from *succinyl-CoA*. Since succinyl-CoA has a high-energy thioester bond, some of the free energy released when this bond is hydrolyzed is conserved by substrate-level phosphorylation. In this reaction, *guanosine diphosphate* (*GDP*) is phosphorylated (by P_i) to yield **guanosine triphosphate (GTP)**.

guanosine triphosphate (GTP)

$$\underset{\text{Succinyl-CoA}}{^{-}OOC{-}CH_2{-}CH_2{-}C(=O){-}S{-}CoA} \xrightarrow[CoA \quad GDP + P_i \to \boxed{GTP}]{\text{Succinyl-CoA Synthetase}} \underset{\text{Succinate}}{^{-}OOC{-}CH_2{-}CH_2{-}COO^-}$$

($\Delta G^{o\prime} = -0.8$ kcal/mol)

The GTP formed is energetically equivalent to an ATP molecule, since a high-energy phosphate bond can be transferred via the following reaction:

$$GDP \xrightarrow[ADP \to \boxed{ATP}]{\text{Nucleoside diphosphate kinase}} GTP$$

The reaction catalyzed by succinyl-CoA synthetase is reversible. Therefore, succinyl-CoA can be made from succinate and CoA at the expense of a high-energy phosphate bond (GTP $\rightarrow$ GDP + P_i). Hence, the name *succinyl-CoA synthetase* is used.

STEP 6: SUCCINATE DEHYDROGENASE

This enzyme catalyzes the oxidation of succinate to yield *fumarate*. Specifically, two protons and two electrons are removed from succinate and transferred to the electron acceptor FAD, yielding $FADH_2$. FAD is tightly bound to the enzyme. The electrons on $FADH_2$ enter the electron transport system and can generate two ATP molecules.

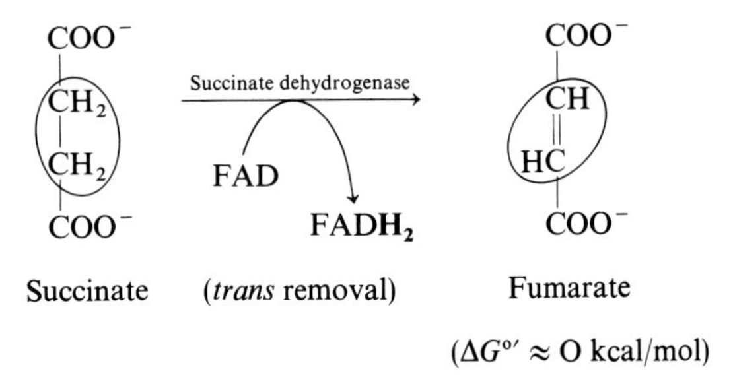

The reaction is reversible. It should be noted that *malonate* has a similar structure to succinate and, therefore, can bind to the active site of the enzyme. This prevents the binding of succinate (competitive inhibition), blocking succinate to fumarate conversion and poisoning the TCA cycle.

COO^- — CH_2 — CH_2 — COO^-	versus	COO^- — CH_2 — COO^-
Succinate (Substrate of succinate dehydrogenase)		Malonate (Competitive inhibitor of succinate dehydrogenase)

STEP 7: FUMARASE

This enzyme catalyzes the stereospecific hydration (addition of a water molecule) of fumarate to yield L-*malate*. The reaction is reversible.

$$\begin{array}{c} COO^- \\ | \\ CH \\ \| \\ HC \\ | \\ COO^- \end{array} \xrightarrow[H_2O]{\text{Fumarase}} \begin{array}{c} COO^- \\ | \\ HO-CH \\ | \\ CH_2 \\ | \\ COO^- \end{array}$$

Fumarate L-Malate

($\Delta G^{\circ\prime} \approx 0$ kcal/mol)

STEP 8: MALATE DEHYDROGENASE

This enzyme catalyzes the oxidation of L-malate, removing two hydrogens to regenerate *oxaloacetate*. One molecule of NADH is generated at this step.

$$\begin{array}{c} COO^- \\ | \\ HO-CH \\ | \\ CH_2 \\ | \\ COO^- \end{array} \xrightarrow[NAD^+ \;\to\; NADH + H^+]{\text{Malate dehydrogenase}} \begin{array}{c} COO^- \\ | \\ O=C \\ | \\ CH_2 \\ | \\ COO^- \end{array}$$

L-Malate Oxaloacetate

($\Delta G^{\circ\prime} = +7.1$ kcal/mol)

Since oxaloacetate is rapidly used by citrate synthase, the malate dehydrogenase reaction is pulled to the right, favoring oxaloacetate formation.

At this point, we should explain exactly how acetyl-CoA is completely oxidized to CO_2 and water. To do this, we must refer back to Figure 11–5 and also to Figure 11–7. Let us assume that the carbonyl carbon atom of acetyl-CoA is *radioactively labeled* with carbon-14 (^{14}C). This label is noted with an asterisk.

$$H_3C-\overset{O}{\overset{\|}{\underset{*}{C}}}-S-CoA$$

Labeled acetyl-CoA

The labeled acetyl-CoA is added to a mitochondrial suspension, and the TCA cycle is allowed to operate. The two carbon atoms of acetyl-CoA will enter the TCA cycle and ultimately be lost as CO_2 (via decarboxylation reactions). Referring to Figure 11–7, one can follow the fate of the carbon atoms of acetyl-CoA. As one proceeds clockwise around the TCA cycle, the carbon atoms originally contributed by acetyl-CoA end up as carbon atoms #1 and #2 of citrate. This labeled citrate is then converted into oxaloacetate (OAA) by the enzymes of the TCA cycle. Because succinate is a *symmetrical* molecule, 50 percent of the L-malate will be labeled in one manner, while the remaining 50 percent will be labeled differently (see Figure

11–7). The labeled oxaloacetate can now accept a second molecule of acetyl-CoA. A second cycle (or turn) of the TCA cycle is thus begun. The ^{14}C atom that was originally the carbonyl carbon of the first acetyl-CoA molecule is lost as CO_2 at the isocitrate dehydrogenase step and at the α-ketoglutarate dehydrogenase step during this second turn of the cycle. The methyl carbon atom of the first acetyl-CoA molecule is lost as CO_2 after other turns of the cycle.

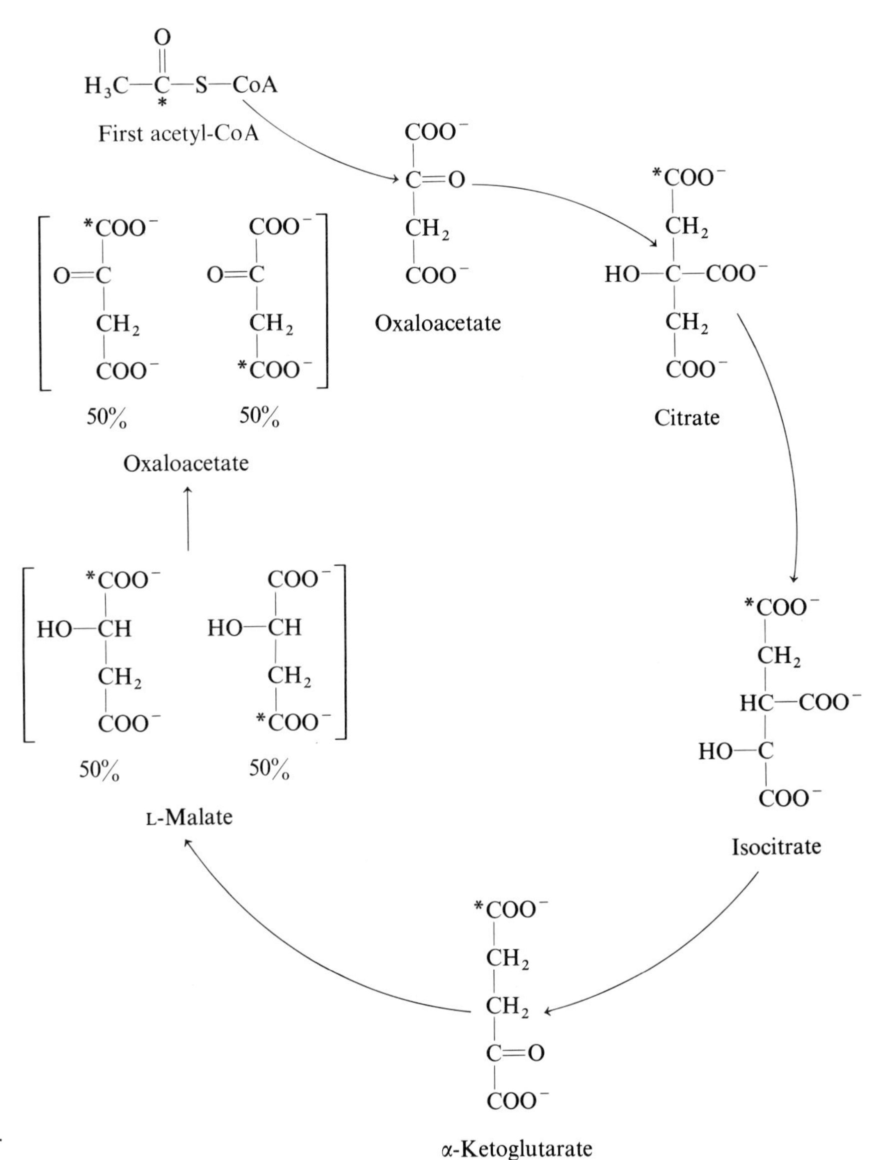

FIGURE 11–7
The movement of a radioactively labeled carbon atom through two turns of the TCA cycle.

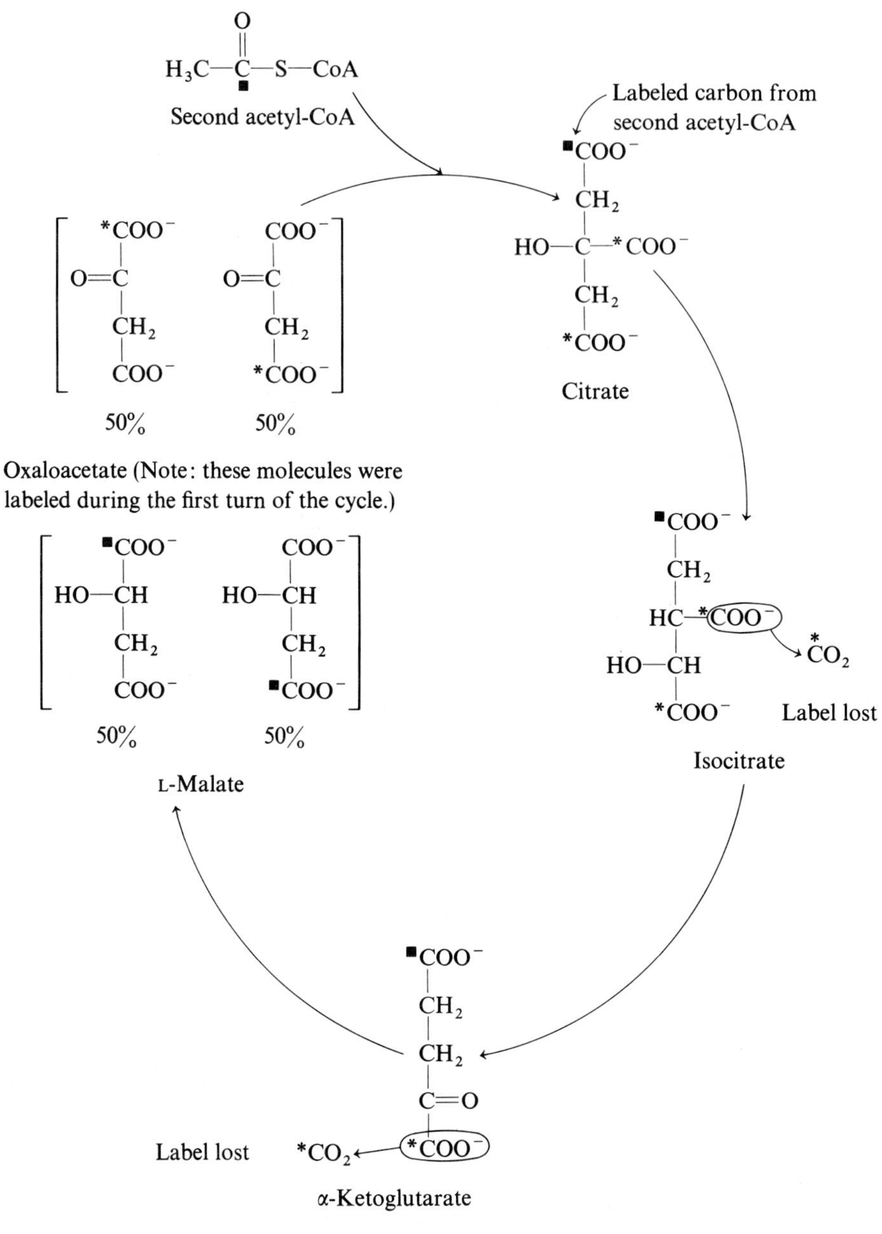

FIGURE 11–7
(continued)

Some of the electrons originally contained in the acetyl-CoA molecule are removed via the dehydrogenase steps in the form of either NADH or $FADH_2$. These electrons are transported to the enzymes of the electron transport system and shuttled through the system. During this process, ATP

oxygen

is synthesized. **Oxygen** *is the final electron acceptor*; it will accept 2 electrons and 2 protons (H^+), forming water. Thus, carbon dioxide and water are the final waste products of aerobic glycolysis.

11.5 THE REGULATION OF THE TCA CYCLE AND THE AMPHIBOLIC NATURE OF THE PATHWAY

The TCA cycle is under critical metabolic regulation, as carbohydrate, lipid, and amino acid metabolic pathways all feed into it. A number of TCA cycle enzymes respond to various inhibitors and to allosteric effector molecules. Some of these enzymes are shown in Figure 11–8.

Pyruvate dehydrogenase, although not really a part of the actual TCA cycle, is controlled by a number of inhibitors. For example, NADH and acetyl-CoA (both products of the enzyme) inhibit it. In addition, the complex is subject to *covalent modification* modulated by ATP, NADH, and acetyl-CoA (see Chapter 6, Enzymes).

Citrate synthase and *isocitrate dehydrogenase* have been called the "pacemaker" enzymes of the pathway. High levels of NADH and ATP shut these enzymes down, slowing down the production of NADH and, therefore, of ATP. ATP acts as a negative allosteric effector on citrate synthase, whereas both ATP and NADH are negative allosteric effectors for isocitrate dehydrogenase. ADP acts as a positive allosteric effector for isocitrate dehydrogenase.

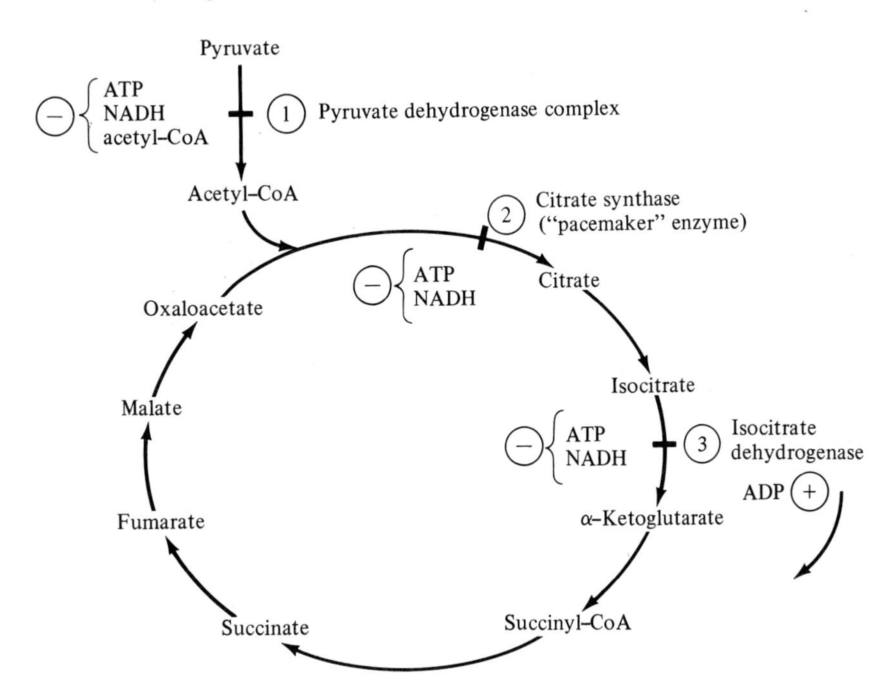

FIGURE 11–8
The regulation of the TCA cycle.

THE AMPHIBOLIC NATURE OF THE TCA CYCLE

amphibolic

anabolic

The TCA cycle is an example of an **amphibolic** *metabolic pathway* (*amphi* means "both"). It is so classified because it can operate as either (1) a *catabolic pathway* by degrading acetyl-CoA to yield NADH and $FADH_2$ for energy, or (2) as an **anabolic** (biosynthetic) *pathway* by supplying intermediates for biosynthesis. If the cell needs energy, the TCA cycle will operate catabolically and supply NADH to the electron transport/oxidative phosphorylation system for producing ATP. Lipids, carbohydrates, and amino acids will thus be degraded to acetyl-CoA to fuel the TCA cycle. (Some amino acids are degraded directly into various TCA cycle intermediates—see Chapter 15.) On the other hand, if the cell has an adequate supply of ATP and is con-

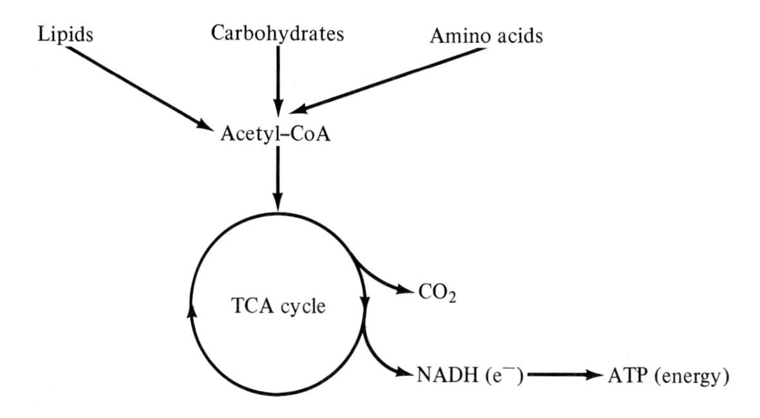

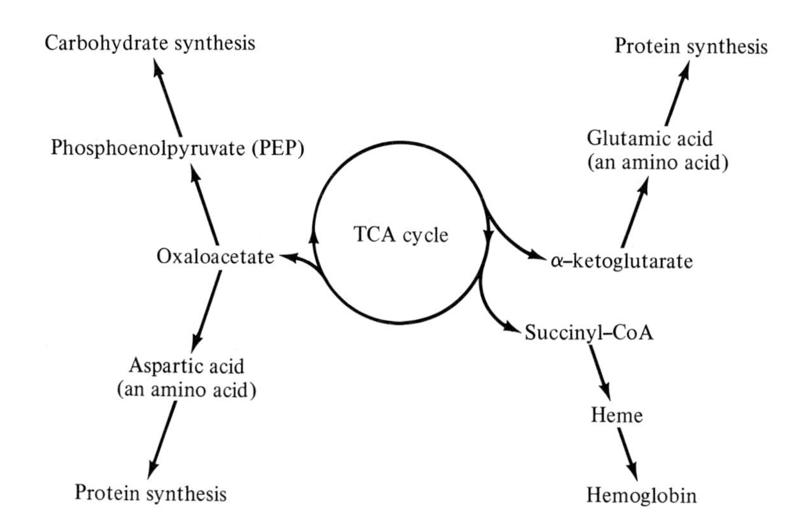

FIGURE 11–9
The amphibolic nature of the TCA cycle. The TCA cycle can operate both catabolically and anabolically, depending on the particular needs of the cell.

ducting biosynthetic activities, the TCA cycle will operate anabolically and supply its intermediates for the biosynthesis of other substances needed by the cell or by other cells. For example, oxaloacetate is the direct precursor of the amino acid *aspartic acid*, α-ketoglutarate is the precursor of *glutamic acid*, and succinyl-CoA is required for the biosynthesis of *heme*, an essential part of hemoglobin. The amphibolic nature of the TCA cycle is diagrammed in Figure 11–9.

anaplerotic

biotin

If the TCA cycle is supplying its intermediates for other biosynthetic steps, it will obviously run out of intermediates. Thus, **anaplerotic** ("filling-up") *reactions* are present that resupply the TCA cycle with intermediates. The most common anaplerotic reaction is catalyzed by *pyruvate carboxylase*. In this reaction, pyruvate is carboxylated using bicarbonate (HCO_3^-) and ATP to form oxaloacetate. The enzyme pyruvate carboxylase requires **biotin**. The reaction and structure of biotin is shown in the following reaction diagram:

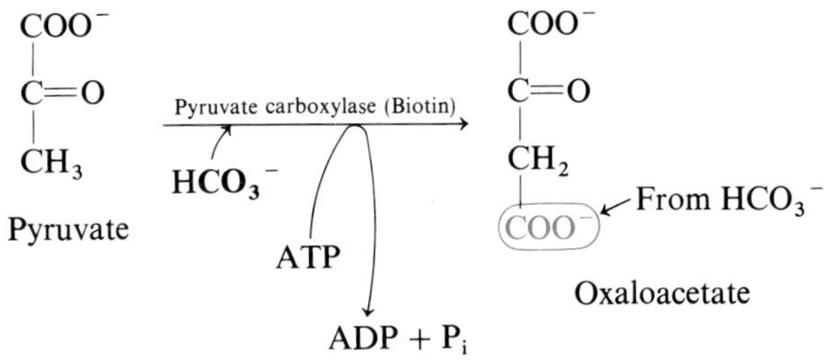

COO⁻ | C=O | CH_3 (Pyruvate) —Pyruvate carboxylase (Biotin); HCO_3^-; ATP → ADP + P_i→ COO⁻ | C=O | CH_2 | COO⁻ (Oxaloacetate) ← From HCO_3^-

The structure of biotin: ^-OOC–(chain)–ring with HN, C=O, NH ← CO_2 binds here, S

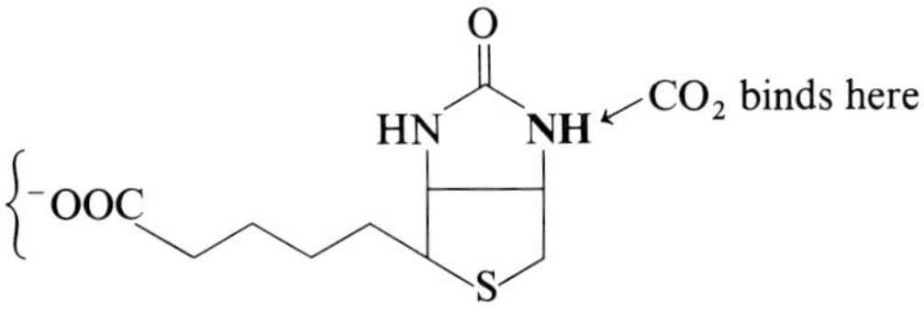

11.6 THE ATP BALANCE SHEET

At this point, we should summarize exactly how many ATP molecules are generated during aerobic glycolysis. You will recall that during anaerobic glycolysis, only 2 ATP molecules are made for each glucose molecule. Aerobic glycolysis is significantly more efficient than anaerobic glycolysis, because either 36 or 38 ATP molecules can be generated for each glucose molecule, depending on the type of cytoplasmic "shuttle" system that is operating in the particular cell. Table 11–1 summarizes the number of ATP molecules generated for each glucose molecule via the process of aerobic glycolysis. The TCA cycle portion is summarized in Table 11–2.

Referring to Table 11–1, we can outline where ATP molecules are made and how many. Two ATP molecules are required to phosphorylate the glucose molecule (by *hexokinase* and *phosphofructokinase*). Since aerobic

TABLE 11–1 ATP Molecules Generated during Aerobic Glycolysis (Glucose → 2 Pyruvate → 6 CO_2 + 6 H_2O).

Reaction	Intermediate	Number of ATP Molecules
1. Hexokinase	——	−1[a]
2. Phosphofructokinase	——	−1[a]
3. Glyceraldehyde 3-phosphate dehydrogenase	2 NADH	4[b]
4. Phosphoglycerate kinase	——	2
5. Pyruvate kinase	——	2
6. TCA cycle[c]	2 acetyl-CoA	30
TOTAL		36

[a] Priming step.
[b] This assumes that the cytoplasmic glycerol shuttle is in operation. If the malate shuttle is in operation, 6 ATPs will be made for each pair of cytoplasmic NADH, yielding a total of 38 ATPs instead of 36.
[c] See Table 11–2 for a summary of the mitochondrial TCA cycle for one acetyl-CoA molecule.

glycerol shuttle
malate shuttle

glycolysis is now being considered, *two* molecules of cytoplasmic NADH are generated at the *glyceraldehyde 3-phosphate dehydrogenase* step for each original glucose molecule. The electrons in this cytoplasmic NADH are shuttled into the matrix of the mitochondrion via either (1) the **glycerol shuttle** (found in muscle cells) or (2) the **malate shuttle** (found in liver cells). If the glycerol shuttle is in operation, only 2 molecules of ATP can be generated for each cytoplasmic NADH; if the malate shuttle is in operation, 3 molecules can be generated. Since 2 molecules of cytoplasmic NADH are formed for each original glucose molecule, then either 4 or 6 ATP molecules can be generated at this step, depending on the type of shuttle system in operation. ATP is formed at two steps within glycolysis—at the *phosphoglycerate kinase* step (2 ATP molecules) and at the *pyruvate kinase* step (2 ATP molecules). This ATP is formed by substrate-level phosphorylation. At this point, two molecules of pyruvate remain for each original glucose

TABLE 11–2 TCA Cycle Summary for One Acetyl-CoA Molecule

Reaction	Intermediate	Number of ATP Molecules
1. Pyruvate dehydrogenase	1 NADH	3
2. Isocitrate dehydrogenase	1 NADH	3
3. α-Ketoglutarate dehydrogenase	1 NADH	3
4. Succinyl-CoA synthetase	——	1[a]
5. Succinate dehydrogenase	1 $FADH_2$	2
6. Malate dehydrogenase	1 NADH	3
TOTAL		15

[a] 1 GTP = 1 ATP.

molecule. These two pyruvate molecules are transported from the cytoplasm into the mitochondrial matrix, where they are converted into two acetyl-CoA molecules by the *pyruvate dehydrogenase complex*. Because the pyruvate dehydrogenase complex and the dehydrogenases of the TCA cycle produce a number of NADH (and $FADH_2$) molecules, a large number of ATP molecules can be made (see Table 11–2). Specifically, 30 ATP molecules are generated from the NADH (and $FADH_2$) produced by the TCA cycle (including the NADH generated by the pyruvate dehydrogenase complex). The overall result is the formation of either 36 or 38 ATP molecules for each original glucose molecule oxidized.

SUMMARY

During aerobic glycolysis, glucose is partially oxidized into pyruvate, which is then transported into the mitochondrion for more complete oxidation into carbon dioxide and water. Some of the electrons originally contained in the pyruvate molecule are extracted in the form of NADH and $FADH_2$. These electrons drive ATP synthesis via the processes of electron transport and oxidative phosphorylation.

The mitochrondrion is a cigar shaped organelle composed of two membranes—an outer, relatively permeable membrane and an inner, impermeable membrane. The inner membrane is highly convoluted; its folds forming structures called cristae. The proteins of the electron transport system are located on the inner surface of this membrane. The internal space of the mitochondrion is a gel-like solution of proteins called the matrix, containing the TCA cycle enzymes.

Pyruvate is transported into the mitochondrial matrix and there converted into acetyl-CoA by the pyruvate dehydrogenase complex. The pyruvate dehydrogenase complex is a large multienzyme complex that catalyzes the oxidative decarboxylation of pyruvate and the formation of acetyl-CoA via a "swinging arm" mechanism. The oxidation of the swinging arm requires FAD and NAD^+.

The tricarboxylic acid (TCA) cycle oxidizes acetyl-CoA into CO_2 and water, in the process recovering some of the free energy inherent within acetyl-CoA as NADH or $FADH_2$. Carbohydrates, lipids, and amino acids are all degraded into acetyl-CoA; thus, a variety of cellular fuels can enter this very important central pathway. The pathway is cyclical. Simply stated, acetyl-CoA condenses with oxaloacetate to form citrate. This citrate is converted into isocitrate via the aconitase reaction. Isocitrate is converted into α-ketoglutarate, in the process forming 1 NADH molecule. α-Ketoglutarate is in turn converted into succinyl-CoA, again forming 1 NADH molecule. Succinyl-CoA is converted into succinate, forming 1 molecule of GTP by substrate-level phosphorylation. One molecule of $FADH_2$ is generated when succinate is converted into fumarate. Fumarate is converted into malate, which is then converted back into oxaloacetate with the formation of 1 molecule of NADH. A second molecule of acetyl-CoA can then enter the cycle, and the process is repeated. The two carbons of acetyl-CoA are ultimately lost as CO_2 during repeated cycles of the pathway.

The TCA cycle is under precise metabolic control. Citrate synthase and isocitrate dehydrogenase are called the "pacemaker" enzymes of the pathway, because high levels of NADH and ATP will shut down these enzymes, slowing down the pathway. Other TCA cycle enzymes are under allosteric control. The pyruvate dehydrogenase complex is also regulated and is subject to covalent modification modulated by ATP, NADH, and acetyl-CoA.

The TCA cycle is an amphibolic pathway—that is, it can operate anabolically or catabolically, depending on the specific needs of the cell. When the TCA cycle is operating anabolically, anaplerotic reactions, such as that catalyzed by pyruvate carboxylase, help fill up the pathway with new TCA cycle intermediates, such as oxaloacetate.

Aerobic glycolysis is significantly more efficient than anaerobic glycolysis—either 36 or 38 ATP molecules can be generated per glucose molecule, depending on the type of cytoplasmic shuttle system operating within the cell.

REVIEW QUESTIONS

1. We have presented a number of important concepts and terms in this chapter. Define (or briefly explain) each of the following terms:
   a. Mitochondrion
   b. Cristae
   c. Matrix
   d. Pyruvate dehydrogenase complex
   e. Acetyl-CoA
   f. Tricarboxylic acid cycle
   g. "Ferrous wheel" enzyme
   h. Oxidative decarboxylation
   i. Amphibolic pathway
   j. Anaplerotic reactions
   k. Glycerol and malate shuttles
2. In this chapter, we have presented a number of enzyme cofactors derived from vitamins. Briefly describe the biochemical function of each of the following coenzymes (see Chapter 20):
   a. $NAD^+/NADH$
   b. Thiamin pyrophosphate (TPP)
   c. $FAD/FADH_2$
   d. Coenzyme A
   e. Biotin
3. We have presented a number of enzyme reactions in this chapter. Explain what each of these enzymes does. Give an appropriate reaction, with structures.
   a. Pyruvate dehydrogenase complex
   b. Citrate synthase ("condensing enzyme")
   c. Aconitase
   d. Dehydrogenases (in general)
   e. Pyruvate carboxylase
4. From what you know about the pyruvate dehydrogenase complex, suggest how the α-ketoglutarate dehydrogenase complex may operate. Show representative structures and reactions.

*5. A suspension of rat liver mitochondria was incubated in a buffer containing ^{14}C-pyruvate. The suspension was aerated with oxygen, and $^{14}CO_2$ gas was detected after a short period of time. The ^{14}C label on the pyruvate is noted with an asterisk.

$$H_3\overset{*}{C}-\overset{O}{\overset{\|}{C}}-COO^- \quad \text{Pyruvate}$$

Diagram the complete TCA cycle (with structures) and show the position of the ^{14}C label through two turns of the cycle.

6. Malonate competitively inhibits succinate dehydrogenase.
   a. What happens to the TCA cycle when malonate is added to a mitochondrial suspension?
   b. What happens to the succinate concentration when malonate is present?
   c. If fumarate is added to a mitochondrial suspension (also containing malonate), will citrate be formed? Will succinate also accumulate?
   d. Suggest how the malonate inhibition can be relieved.
7. Avidin, a protein in egg whites, readily binds biotin. Thus, chronic ingestion of raw egg whites can cause a biotin deficiency. Which important enzymatic reaction would be inhibited if such a deficiency occurred? What problems in the operation of the TCA cycle might result?

SUGGESTED READING

Lehninger, A. L. *The Mitochondrion: Molecular Basis of Structure and Function.* New York: W. A. Benjamin, 1964.

Lehninger, A. L. *Biochemistry*, 2nd ed., Chs. 17 and 19. New York: Worth, 1975.

Montgomery, R., Dryer, R. L., Conway, T. W., and Spector, A. A. *Biochemistry: A Case-Oriented Approach*, 3rd ed., Ch. 6. St. Louis: C. V. Mosby, 1980.

Stryer, L. *Biochemistry*, 2nd ed., Ch. 13. San Francisco: W. H. Freeman, 1981.

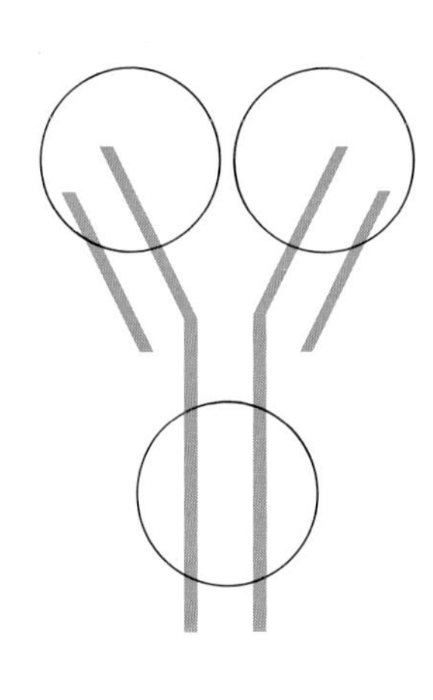

Chapter 12
Electron Transport and Oxidative Phosphorylation

12.1 INTRODUCTION

electron transport system

oxidative phosphorylation

Acetyl-CoA is completely oxidized by means of the TCA cycle. During this process, electrons are removed from TCA cycle intermediates and transferred to either NAD^+ or FAD, generating NADH and $FADH_2$. These electron carriers diffuse to the mitochondrial **electron transport system**. This system is a series of enzymes bound to the inner surface of the inner mitochondrial membrane, close to the matrix. As the name implies, electrons from NADH and $FADH_2$ are transported through this system. *Oxygen serves as the final electron acceptor* and, in the process, is reduced to water. During electron transport, a large amount of *free energy is released, some of which is conserved in the form of ATP*. ATP is synthesized from ADP and P_i via the process known as **oxidative phosphorylation**. In essence, the energy released during electron transport (and, thus, during the subsequent oxidation of electron carriers) promotes the phosphorylation of ADP, yielding ATP.

In this chapter, we will discuss the individual components of the electron transport system and describe how electrons are transported through it. To conclude the chapter, we will discuss how ATP is synthesized.

12.2 OXIDATION-REDUCTION REACTIONS

oxidation-reduction reactions
reducing agent
oxidizing agent

Oxidation-reduction reactions involve the transfer of electrons. A **reducing agent** can be defined as an electron *donor* and an **oxidizing agent** as an electron *acceptor*. When a reducing agent donates its electrons, the process is called reduction and the reducing agent is said to be oxidized (that is, it has lost its electrons). As the oxidizing agent accepts electrons, it becomes reduced.

The tendency of a reducing agent to donate electrons or of an oxidizing agent to gain electrons can be measured in the laboratory and quantified in terms of **standard oxidation-reduction potentials** (E_o'). A standard oxidation-reduction potential can be defined as the relative tendency of a reducing agent to donate electrons or of an oxidizing agent to accept electrons. For example, the ability of hydrogen (H^+) to accept electrons is given by the following equation:

standard oxidation-reduction potentials

$$2H^+ + 2e^- \rightleftharpoons H_2(g)$$

(Oxidizing agent) (Reduced) ($E_o' = -0.42$ volts)

It should be noted that the prime (′) designation of E_o' identifies an oxidation-reduction reaction that takes place at pH 7.0.

Substances that exhibit *greater* negative standard oxidation-reduction potentials exhibit a greater tendency to donate electrons and are, therefore, defined as reducing agents. Substances that exhibit *more positive* standard oxidation-reduction potentials exhibit a greater tendency to accept electrons and, therefore, are defined as oxidizing agents. For example, hydrogen ($E_o' = -0.42$ volts) is a very good reducing agent. NADH ($E_o' = -0.32$ volts) is also a very good reducing agent and will easily donate its electrons to appropriate electron acceptors. Oxygen ($E_o' = +0.82$ volts) is a very strong oxidizing agent and will readily accept electrons from appropriate electron donors.

Thus, NADH will donate its electrons to appropriate electron carriers of the electron transport system. These electron carriers have more positive standard E_o' values as compared to the value for NADH; they therefore accept electrons from NADH. The components of the electron transport system have increasingly positive E_o' values, with oxygen exhibiting the largest positive E_o' value (+0.82 volts). Thus, oxygen is the final electron acceptor and accepts electrons from the electron transport system. Oxygen is reduced to water by accepting 2 electrons and 2 protons. This is the primary reason why we must breathe oxygen—to supply respiring cells with oxygen to serve as the final electron acceptor during cellular respiration. It must be emphasized that the electron transport system represents a *gradient* of increasing oxidation potential; NADH donates electrons to the electron transport system and the electron transport system donates these electrons to oxygen (which has the largest positive E_o' value). This process is shown in the following diagram:

$$\text{NADH} \xrightarrow{(e^-)} \text{Electron carriers} \xrightarrow{(e^-)} O_2$$

(Good reducing agent) (Electron transport system) (Good oxidizing agent)

As electrons are transported from one enzyme to the next through the electron transport system, a certain amount of free energy is released. We can calculate the actual number of calories released by means of the following equation, which relates the standard free-energy change to the number of electrons transferred:

$$\Delta G^{o\prime} = -nF\Delta E_o' \tag{12–1}$$

where $\Delta G^{o\prime}$ = Standard free-energy change; n = number of electrons transported; F = Faraday's constant = 23 062 cal/volt;

$$\begin{aligned} \Delta E_o' &= [E_o' \text{ (electron acceptor)} - E_o' \text{ (electron donor)}] \qquad (12\text{–}2) \\ &= [E_o' (O_2) \qquad\qquad\qquad - E_o' (NADH)] \\ &= [(+0.82 \text{ volts}) \qquad\qquad - (-0.32 \text{ volts})] \\ &= +1.14 \text{ volts} \end{aligned}$$

Thus;

$$\begin{aligned} \Delta G^{o\prime} &= -(2)(23\,062 \text{ cal/volt})(+1.14 \text{ volts}) \qquad (12\text{–}3) \\ &= -52\,700 \text{ cal/mol} \\ &= -52.7 \text{ kcal/mol} \end{aligned}$$

Some of this free energy can be conserved and used to synthesize ATP from ADP and P_i. The mechanism of ATP synthesis will be discussed in Section 12.4.

12.3 THE ELECTRON TRANSPORT SYSTEM

The complete electron transport system is diagrammed in Figure 12–1. The system consists of a number of enzymes embedded in the inner mitochondrial membrane of the mitochondrial cristae. The separate components are not

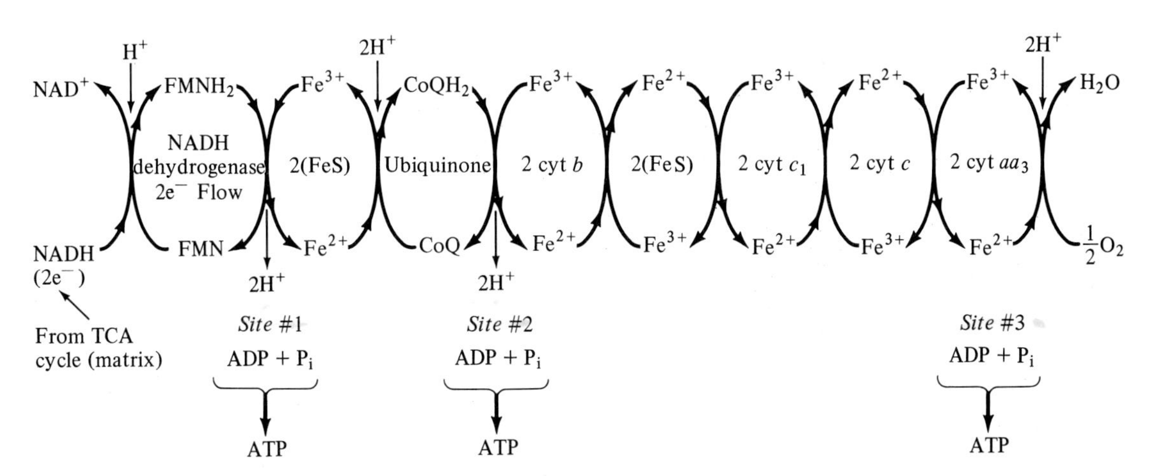

FIGURE 12–1 The sequence of the electron transport system. The electrons from NADH are shuttled through the electron transport system. Oxygen, the final acceptor, is reduced to water (a waste product). As each oxidized component of the system accepts electrons, that component is reduced. Reoxidation occurs when the electrons are donated to the next component of the system. Protons are pumped across the inner mitochondrial membrane at three distinct sites (sites 1, 2, and 3). ATP is synthesized from ADP and P_i at these sites as a *result of electron transport.* The FeS represents the iron-sulfur flavoprotein.

organized in nice rows as Figure 12–1 might suggest. Rather, the components reside in the membrane in a more random pattern. The individual components of the electron transport system are described in the next few pages.

NADH DEHYDROGENASE

flavin mononucleotide

This is the first component of the electron transport system. NADH dehydrogenase is a large, membrane-bound enzyme that accepts two electrons (and two protons) from NADH and transfers them to **flavin mononucleotide** (*FMN*) to yield $FMNH_2$, according to Figure 12–2. A flavoprotein containing *iron-sulfur* "clusters" (abbreviated FeS) accepts the electrons from $FMNH_2$. The iron atoms in these clusters are in the ferric (Fe^{3+}) state. When an electron is transferred from $FMNH_2$, *one* of the (Fe^{3+}) iron atoms of the FeS cluster is reduced to the ferrous (Fe^{2+}) state. The structure of *flavin mononucleotide* and the chemistry of the FeS reduction step is shown in the following diagrams:

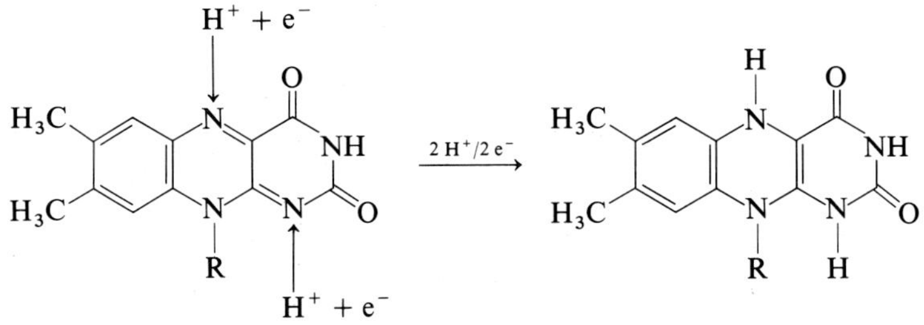

$H^+ + e^-$ $\xrightarrow{2\,H^+/2\,e^-}$

FMN (Ox) FMNH$_2$ (Red)

Fe^{3+} Fe^{3+} $\xrightarrow{(1\,e^-\text{ steps})}$ Fe^{3+} Fe^{2+}

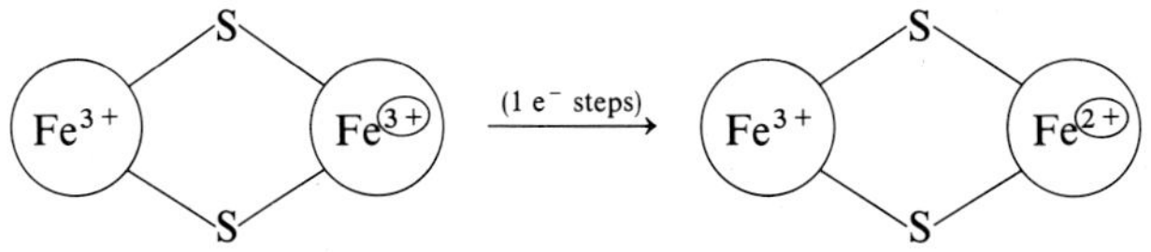

FeS (Ox) FeS (Red)

The FeS cluster is part of the flavoprotein.

Note: Iron-sulfur clusters containing 4 iron and 4 sulfur atoms are found in most mammalian systems.

COENZYME Q (UBIQUINONE)

coenzyme Q
ubiquinone

This substance is a lipid-like cofactor that accepts electrons from the FeS flavoprotein and transfers them to the *cytochromes*. **Coenzyme Q** is called **ubiquinone** because it is ubiquitous in various biological systems (*ubiquitous*

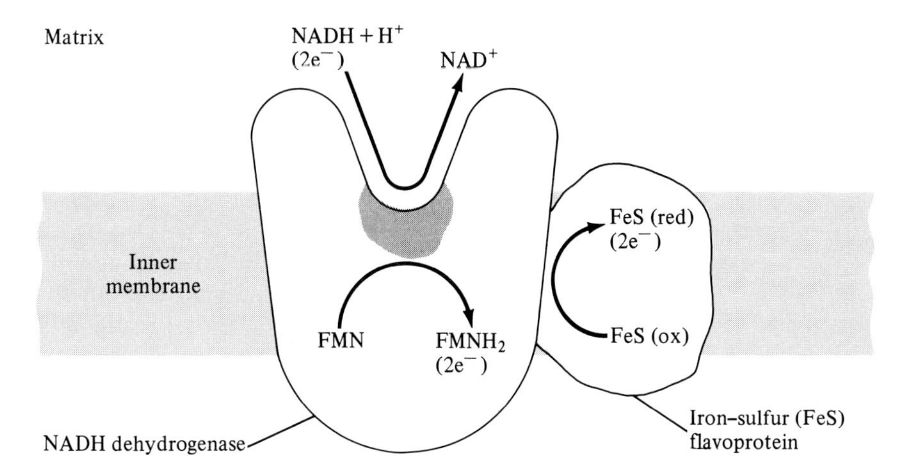

FIGURE 12–2 A hypothetical representation of the organization of NADH dehydrogenase and an iron-sulfur flavoprotein within the inner mitochondrial membrane.

means "present everywhere"). Because Coenzyme Q has a hydrophobic *isoprenoid* tail (see Chapter 14, Lipids and Membranes), it is readily soluble in the hydrophobic bilayer of the inner mitochondrial membrane. Coenzyme Q is, therefore, highly mobile within the membrane and serves to shuttle electrons from the flavoproteins to the cytochromes.

$$\text{Coenzyme Q (CoQ) (Ox): } H_3CO,\ H_3CO,\ CH_3\text{-substituted quinone with isoprenoid tail (R) } \left[CH_2{-}CH{=}\underset{|}{\overset{CH_3}{C}}{-}CH_2\right]_n{-}H \;\; (\text{each } O \leftarrow H^+ + e^-) \xrightarrow[\text{Reduction}]{2\,H^+/2\,e^-} \text{CoQ-}H_2\text{ (Red): } H_3CO,\ H_3CO,\ CH_3,\ R,\ OH,\ OH$$

CoQ_6 ($n = 6$)
CoQ_8 ($n = 8$)
CoQ_{10} ($n = 10$)
} The n value depends upon the species.

THE CYTOCHROMES (CYTOCHROME b, c_1, c, AND a/a_3)

cytochromes

The **cytochromes** are membrane-bound enzymes that participate in *one-electron* transfer steps during electron transport.

$$CoQH_2 \xrightarrow{2\,e^-} \text{Cyto } b \longrightarrow [\text{FeS flavoprotein}] \longrightarrow \text{Cyto } c_1 \longrightarrow \text{Cyto } c \longrightarrow$$

$$\longrightarrow \text{Cyto } a \longrightarrow \text{Cyto } a_3 \quad \tfrac{1}{2}O_2 = \text{(Final electron acceptor)} \rightarrow H_2O \text{ (Waste)}$$

Each of these enzymes contains a *heme* group identical to that found in hemoglobin. A heme group, you will recall, consists of an iron atom (Fe^{3+}) located in the center of a porphyrin ring system (see Chapter 4, Functional Proteins: Hemoglobin and Collagen). Cytochromes *a* and a_3 contain a slightly different heme group (*Heme A*). The heme groups of cytochromes c_1 and *c* are covalently bound to the polypeptide chain of the cytochrome via a *thioether linkage* formed from a porphyrin side chain and a cysteine residue (see Figure 12–3). The heme groups of cytochromes *b* and *a* are not covalently bound to the protein. The 5th and 6th binding positions of cytochrome *c* are coordinated with a histidine and a methionine residue. This contrasts with the heme group of hemoglobin, in which only one of these two binding positions is occupied by a histidine residue (in the oxygenated state, the other position is usually occupied by oxygen). Cytochrome-mediated electron transfer is illustrated by the following diagram: Iron in the Fe^{3+} valence state accepts an electron and, thus, is converted to the Fe^{2+} valence state.

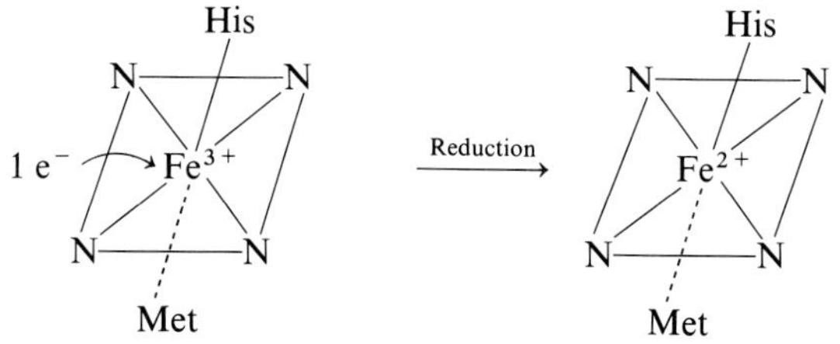

Heme group of cytochrome *c*

Cytochrome oxidase (also noted as the cytochrome a/a_3 complex) is the last cytochrome of the electron transport system. It accepts electrons from cytochrome *c* and transfers them to oxygen forming a water molecule (2 protons are also required). The cytochrome oxidase complex is a large multisubunit complex that contains not only heme A, but divalent copper

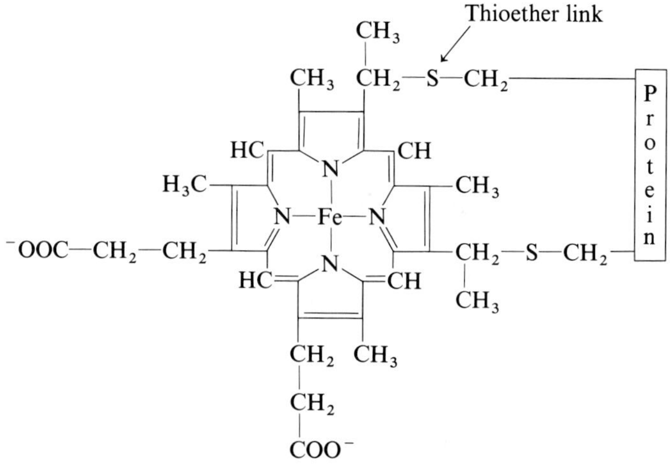

FIGURE 12–3
The structure of the heme group of cytochrome *c*. The heme group is covalently linked to the polypeptide chain by thioether bonds.

(Cu^{2+}). The heme A accepts electrons from cytochrome *c* and then transfers these electrons to Cu^{2+} according to the following reaction:

$$[\text{2 Heme A} \xrightarrow{1\,e^- \text{ steps}} 2\,Cu^{2+} \rightarrow 2\,Cu^{+}] \begin{cases} \nearrow H_2O \\ \frac{1}{2}O_2 + 2\,H^+ \end{cases}$$

$$(Fe^{3+} \rightleftarrows Fe^{2+})$$

Cytochrome oxidase
(Cytochrome a/a_3 complex)

The oxygen binds directly to the iron of the heme group. Cytochrome oxidase is inhibited by *cyanide* (*CN^-*) and *carbon monoxide* (*CO*). These very toxic substances bind to the iron of the cytochrome oxidase heme group, thus keeping oxygen from binding.

One should note that electron transport initially involved the transfer of two electrons (and two H^+). Two electrons were transferred from NADH to the NADH dehydrogenase enzyme and from the enzyme to the FeS flavoprotein and Coenzyme Q. After the Coenzyme Q step, the cytochromes can only transfer electrons one at a time ($Fe^{3+} \rightleftarrows Fe^{2+}$). It has been suggested that *pairs* of cytochromes transfer the two electrons, one electron per cytochrome.

12.4 OXIDATIVE PHOSPHORYLATION: THE SYNTHESIS OF ADENOSINE TRIPHOSPHATE

SITES OF ATP SYNTHESIS

As electrons are transported through the electron transport system, a significant amount of free energy is released (52.7 kcal/mol; see Equation 12–3). The electron transport system conserves some of this free energy by promoting the synthesis of ATP from ADP and P_i. Remember, 7.3 kcal/mol of free energy is required to synthesize 1 mole of ATP. Experimental evidence supports the contention that there are three distinct sites for ATP synthesis (sites 1, 2, and 3, as shown on Figure 12–1). As two electrons are transported through the electron transport system from 1 NADH molecule, a total of 3 ATP molecules are made, one at each site.

The standard oxidation-reduction potentials for the components of the electron transport system are summarized in Table 12–1. Upon examining Table 12–1, you will observe that *significant differences* in standard oxidation-reduction potential exist between certain components of the system. If these differences are great enough, they should represent *enough* of a change in free energy to drive ATP synthesis. This is indeed the case (see Figure 12–4). The three sites responsible for ATP synthesis are associated with specific regions of the electron transport system. *Site 1* is associated with the NADH dehydrogenase/flavoprotein complex, *site 2* with cytochrome *b* and c_1, and *site 3* with the cytochrome oxidase complex.

TABLE 12–1 Standard Oxidation-reduction Potentials for the Components of the Electron Transport System, at pH 7.0.

Component	E'_o (volts)	
1. NADH	−0.32	Site 1: 1 ATP
2. NADH dehydrogenase ($FMNH_2$, FeS)	−0.11	
3. Coenzyme Q	+0.10	
4. Cytochrome *b*	+0.06	Site 2: 1 ATP
5. Cytochrome c_1	+0.22	
6. Cytochrome *c*	+0.25	
7. Cytochrome a/a_3	+0.28	Site 3: 1 ATP
8. Oxygen	+0.82	

Note: $\Delta G^{\circ\prime} = -nF\,\Delta E'_o$ (Equation 12–1)

and

$\Delta E'_o = E'_o$ (electron acceptor) $-$ E'_o (electron donor) (Equation 12–2)

Finally, we should note that the electron donor $FADH_2$ can only produce 2 ATP molecules for each pair of electrons transported through the electron transport system, even though NADH can produce 3. Why is this so? If you recall our discussion on the TCA cycle, $FADH_2$ is produced at the succinate dehydrogenase step. Research has indicated that succinate dehydrogenase is physically associated with the *second* site of the electron transport system. Thus, the electrons are transferred to the electron transport system (via a flavoprotein) at the cytochrome b/c_1 region, *bypassing*

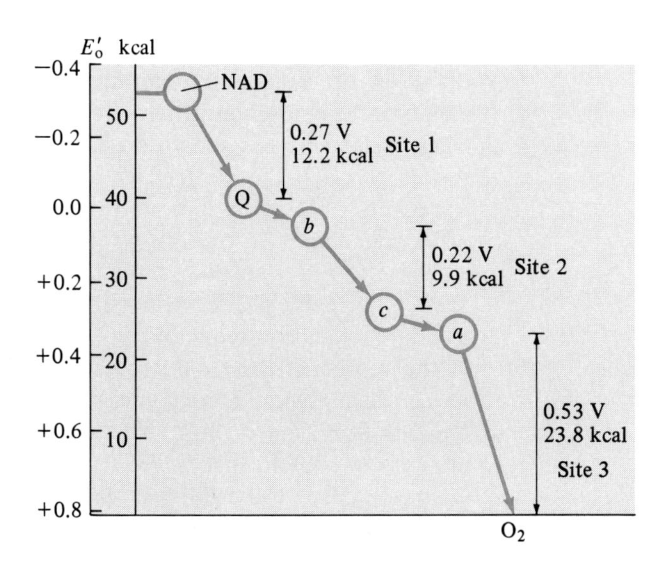

FIGURE 12–4 A representation of the decline in free energy as electrons are transported through the electron transport system. (Adapted from A. L. Lehninger, *Biochemistry*, 2nd ed. (New York: Worth, 1975), p. 516.)

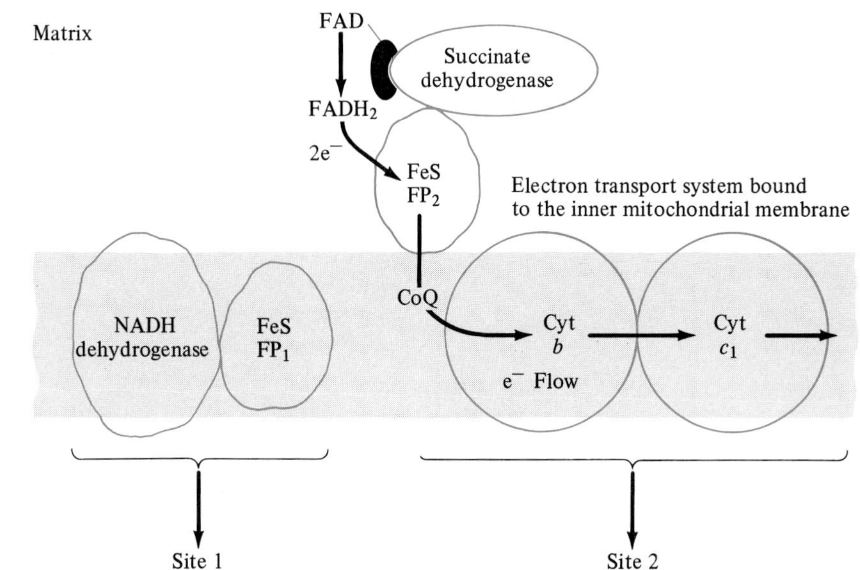

FIGURE 12–5
The entry of $FADH_2$ into the electron transport system. The electrons derived from $FADH_2$ enter the electron transport system at site 2 via an FeS flavoprotein (FP_2) and coenzyme Q (CoQ). Site 1 is thus bypassed; therefore, only two ATP molecules can be made for each two electrons transported.

the NADH dehydrogenase complex. Therefore, the first ATP synthesis site is also bypassed. Only two ATP molecules are made for each two electrons entering the electron transport system after site 1 (see Figure 12–5).

OXIDATIVE PHOSPHORYLATION

substrate-level phosphorylation
oxidative phosphorylation

Adenosine triphosphate can be synthesized from ADP and P_i by either of two types of phosphorylation, (1) **substrate-level phosphorylation** or (2) **oxidative phosphorylation**. Substrate-level phosphorylation was discussed in Chapter 9 (Carbohydrate Metabolism). In substrate-level phosphorylation, high-energy phosphorylated substances such as phosphoenolpyruvate donate a phosphate group to ADP to form the high-energy bond of ATP. In oxidative phosphorylation, on the other hand, electrons are shuttled through the electron transport system to drive the synthesis of ATP from ADP and P_i. As we have already stated, this process probably occurs at three distinct sites associated with specific components of the electron transport system.

THE CHEMIOSMOTIC COUPLING HYPOTHESIS

The chemiosmotic coupling hypothesis was first made by Dr. Peter Mitchell in 1961. It is simply stated as follows: The transport of electrons through the electron transport system causes the transport of protons across the inner mitochondrial membrane. Protons are "pumped" from the matrix across the membrane and into the space between the inner and outer mitochondrial membranes. This *increases* the hydrogen ion concentration on

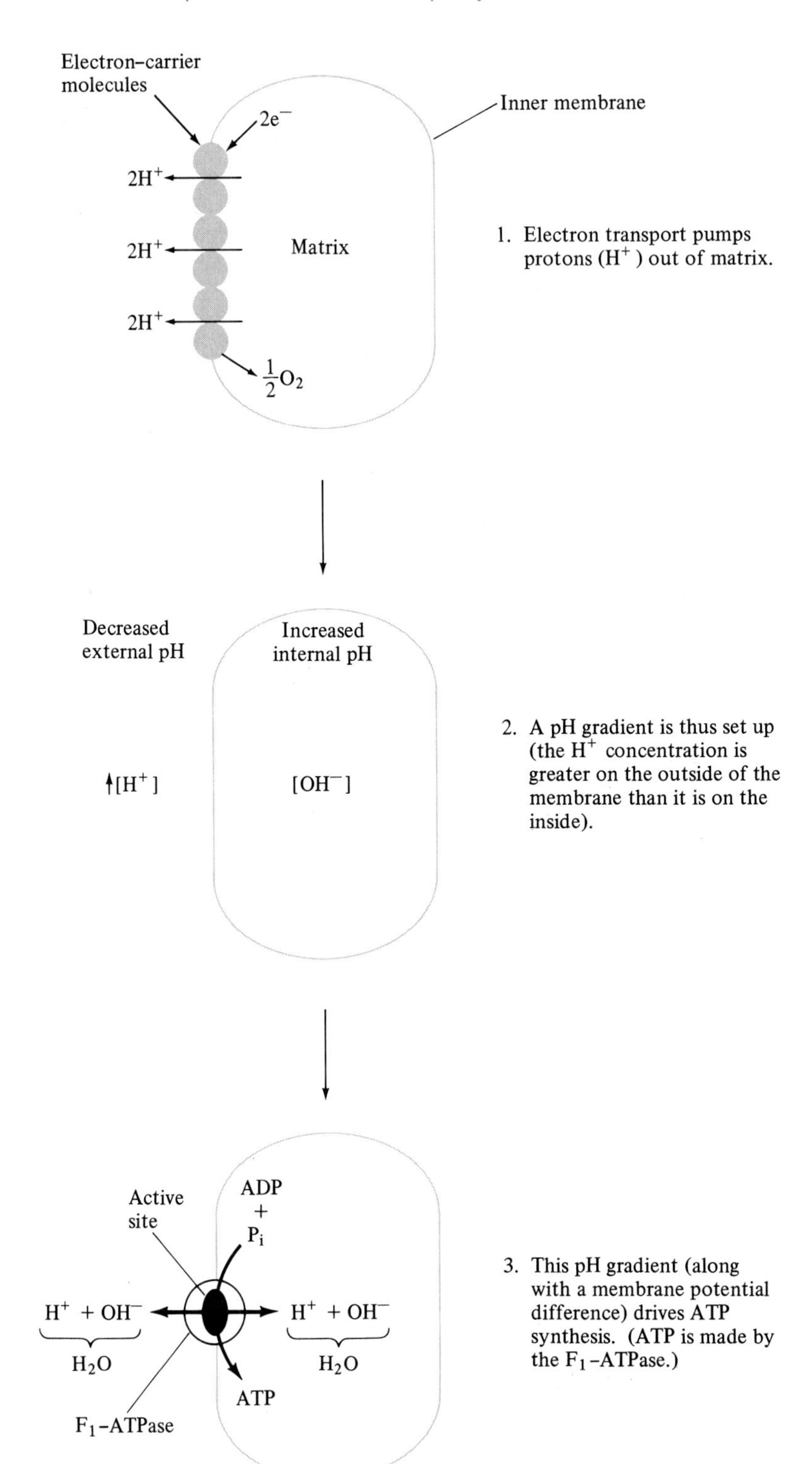

FIGURE 12–6
The chemiosmotic coupling hypothesis. (Adapted from A. L. Lehninger, *Biochemistry*, 2nd ed. (New York: Worth, 1975), p. 526.)

the outer side of the inner membrane and *decreases* it on the inner (matrix) sidc. A pH gradient (difference in pH) results across the inner mitochondrial membrane, along with an electrical potential difference. This pH gradient and electrical potential difference drives the synthesis of ATP. These chemical processes are summarized in Figure 12–6. The synthesis of ATP from ADP and P_i is catalyzed by an enzyme called **F_1-ATPase** associated with the electron transport system. F_1-ATPase catalyzes the hydrolysis of ATP into ADP and P_i as follows:

F_1-ATPase

$$\text{ATP} \xrightarrow[\text{(Hydrolysis)}]{\text{F}_1\text{-ATPase}} \text{ADP} + \text{P}_i$$

It is thought that the pH gradient and membrane potential produced by electron transport may cause the reaction to proceed in reverse:

$$\text{ATP} \xleftarrow{\text{F}_1\text{-ATPase}} \text{ADP} + \text{P}_i$$

H^+ flow through F_1-ATPase molecule

The protons pumped across the membrane may *flow back through* a channel in the F_1-ATPase enzyme and drive ATP synthesis.

THE TRANSPORT OF ATP OUT OF THE MITOCHONDRION

The ATP made during oxidative phosphorylation is still contained in the matrix of the mitochondrion. An ATP transport protein found in the inner mitochondrial membrane transports the ATP from the matrix across the inner mitochondrial membrane into the space between the inner and outer membranes. An *ADP* molecule must be transported into the matrix for each *ATP* that is pumped out. The outer membrane is very permeable to ATP, so the ATP can easily diffuse out of the mitochrondrion into the cytoplasm. This transport mechanism is shown in Figure 12–7.

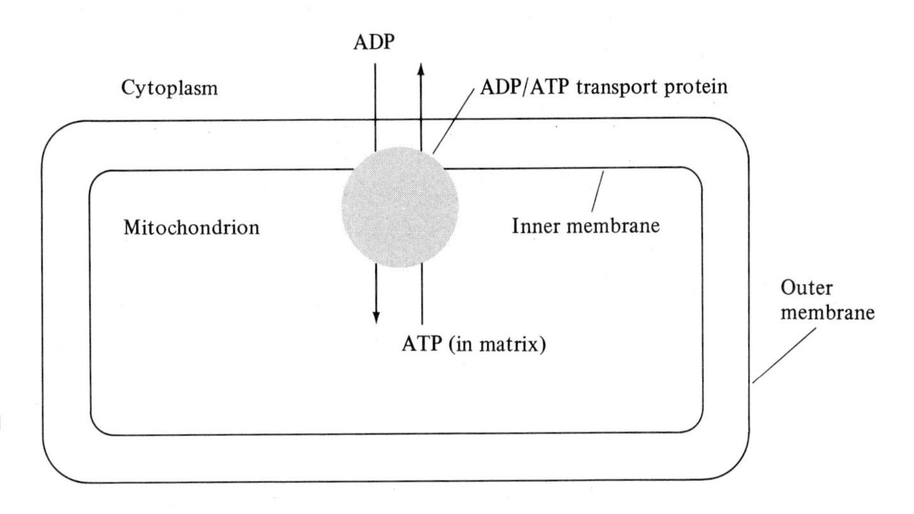

FIGURE 12–7
The transport of ATP from the mitochondrial matrix into the cytoplasm.

SUMMARY

Oxidation-reduction reactions involve the transfer of electrons. A reducing agent is an electron donor; an oxidizing agent is an electron acceptor. The tendency of a reducing agent to donate electrons or of an oxidizing agent to accept electrons is quantified in terms of standard oxidation-reduction potentials (E_0'). NADH is a good reducing agent ($E_0' = -0.32$ V); oxygen is an excellent oxidizing agent ($E_0' = +0.82$ V). NADH donates electrons to the electron transport system, and the electron transport system donates these electrons to oxygen. Oxygen, the final electron acceptor, is reduced to water.

The electrons contained in NADH (and also in $FADH_2$) are transferred through an electron transport system. This system is a series of membrane-bound enzymes arranged on the inner surface of the inner mitochondrial membrane. As electrons are transported through the system, some free energy is conserved and used in synthesizing ATP. The sequence of the electron transport system includes: NADH dehydrogenase, FeS flavorprotein, Coenzyme Q, cytochrome b, c_1, c, and cytochrome oxidase (a/a_3). Three sites for ATP synthesis are associated with the electron transport system. Site 1 is associated with the NADH dehydrogenase complex, site 2 with the cytochrome b and c_1 complex, and site 3 with the cytochrome oxidase complex. The electrons of $FADH_2$ enter the electron transport system after the first site; thus, only two ATP molecules can be made for every two electrons transported from $FADH_2$.

Adenosine triphosphate (ATP) can be synthesized by either of two types of phosphorylation, (1) substrate-level phosphorylation or (2) oxidative phosphorylation. In oxidative phosphorylation, electrons are transported through the electron transport system (via the oxidation of electron donating substances such as NADH) to drive the synthesis of ATP from ADP and P_i. The chemiosmotic coupling hypothesis states that electron transport sets up a pH gradient across the inner mitochrondrial membrane, and that this gradient (along with an electrochemical potential) drives F_1-ATPase, which joins an ADP and P_i molecule to form ATP. Once ATP has been made, the ATP is pumped out of the mitochondrion via an ATP transport protein.

REVIEW QUESTIONS

1. We have presented a number of important concepts and terms in this chapter. Define (or briefly explain) each of the following terms:

a. Reducing agent
b. Oxidizing agent
c. Oxidation-reduction potential (E_0')
d. Electron transport system
e. NADH
f. NADH dehydrogenase
g. Iron-sulfur (FeS) flavoprotein
h. Coenzyme Q (CoQ)
i. Cytochromes
j. Cytochrome oxidase
k. ATP synthesis site
l. Oxidative phosphorylation
m. Chemiosmotic coupling hypothesis
n. F_1-ATPase
o. ATP/ADP transport pump

2. From memory, reproduce the sequence of the electron transport system. In addition, show the three sites where ATP synthesis takes place.

*__3.__ Suppose life was found on Mars and that an analysis of Martian mitochondria demonstrated that they contain an electron transport system very similar to our own. This electron transport system can only make 2 ATP molecules for each 2 electron transported. The E_0' (volts) of each component of the Martian electron transport system is shown in the following table (in no particular order):

Component	E_0' (volts)
Oxygen (O_2)	+0.82
Martian Q	−0.05
DAN (reduced form)	−0.55
Delta Xi protein	−0.10
Protein X	+0.75
Protein Y	+0.65

Suggest a possible sequence for the Martian

electron transport system, and show where the two ATP synthesis sites are located in the sequence. Be able to support your answer.

4. Explain why the following substances are so toxic to living systems:
a. Cyanide (CN^-) **b.** Carbon monoxide (CO)

*** 5.** Snake venom contains enzymes (phospholipases) that digest membrane lipids, thus destroying cell membranes. It has been suggested that snake venom interrupts oxidative phosphorylation. Suggest how snake venom might do this.

SUGGESTED READING

Lehninger, A. L. *Biochemistry*, 2nd ed., Ch. 18 and 19. New York: Worth, 1975.
Stryer, L. *Biochemistry*, 2nd ed., Ch. 14. San Francisco: W. H. Freeman, 1981.

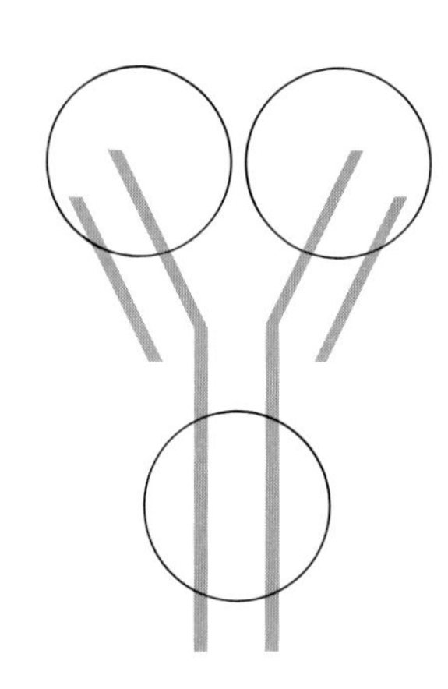

Chapter 13
Lipids and Membranes

13.1 INTRODUCTION

Lipids are a general class of biologically important organic molecules that are insoluble in water (that is, they are hydrophobic) but are soluble in a variety of organic solvents such as benzene and ether. Lipids are very important in many cellular functions. For example, they are structural components for cellular membranes and a form of cellular metabolic fuel. Some lipids serve as *hormones* (steroid hormones and prostaglandins); others are *vitamins* (fat-soluble vitamins A, D, E, and K) and are required for specific physiological processes. Other forms of lipids serve as cell-surface components or provide a protective coating.

In this chapter, we shall first discuss the various classes of lipids and describe their biological functions. We will then discuss membrane structure and function. Disease states related to abnormal lipid chemistry will be presented where appropriate.

13.2 THE CLASSES OF LIPIDS

saponifiable
nonsaponifiable

There are two main classes of lipids: (1) the **saponifiable** lipids and (2) the **nonsaponifiable** lipids. *Saponification* refers to the process by which lipid esters can be hydrolyzed into fatty acids (soaps) and an alcohol. Neutral lipids (triglycerides), phospholipids, sphingolipids, and waxes all belong to the first class of lipids. Nonsaponifiable lipids cannot be hydrolyzed to fatty acids and an alcohol. Terpenes, steroids, and prostaglandins belong to the second class of lipids. Table 13–1 summarizes the different types of lipids.

TABLE 13–1 The Major Types of Lipids.

I. Saponifiable Lipids
   A. Neutral Lipids (lipids without electrical charge).
      Mono-, di-, and triglycerides—serve as a storage and transport form of cellular fuel.
   B. Phospholipids (derivatives of phosphatidic acid—usually electrically charged). Usually associated with membranes or with lipoprotein complexes.
   C. Sphingolipids (derivatives of sphingosine). Serve as structural components of nerve cell membranes (nerve cell insulation).
   D. Waxes (long-chain lipid/alcohol esters). Serve as protective coatings on feathers, fur, skin, leaves, and fruits.

II. Nonsaponifiable Lipids
   A. Terpenes (constructed from 5-carbon isoprene units). Major components of plant and tree fragrances and pigments. The fat-soluble vitamins (vitamin A, E, and K) are terpene derivatives.
   B. Steroids (derivatives of a 4-ring hydrocarbon system).
      1. Cholesterol—component of membranes.
      2. Steroid hormones—chemical messengers produced by the endocrine glands.
      3. Bile salts—emulsifiers involved in the absorption of dietary lipids.
   C. Prostaglandins (20-carbon unsaturated fatty acids with a 5-membered ring). Serve as chemical messengers—regulate a number of important physiological functions.

13.3 FATTY ACIDS

Before we can actually discuss the chemical structures and properties of the various classes of lipids, we must first describe the chemistry of fatty acids. *Fatty acids* are rarely found free in nature; they usually occur in combination with an alcohol as esters. In fact, fatty acids are the main structural component of the saponifiable lipids.

Fatty acids are long-chain hydrocarbons with a carboxylic acid functional group on one end of the chain, as shown below:

$$\mathrm{H_3\overset{10}{C}{-}CH_2{-}CH_2{-}CH_2{-}CH_2{-}CH_2{-}CH_2{-}CH_2{-}CH_2{-}\overset{\overset{\displaystyle O}{\|}}{C^{1}}{-}OH}$$

Hydrophobic tail
("water-hating")

Hydrophilic head
(The carboxylic acid functional group, "water-loving")

Simplified diagram

Because fatty acids have both hydrophobic and hydrophilic regions, they are referred to as *amphipathic* molecules (*amphi-* means "both"). The carboxylic acid group can behave as an acid by donating its proton to the solution—hence the term "fatty acid."

$$H_3C\sim\sim\sim\overset{\overset{O}{\|}}{C}-OH \xrightarrow{\text{Dissociation}} H_3C\sim\sim\sim\overset{\overset{O}{\|}}{C}-O^- + H^+$$

micelles

Since fatty acids are amphipathic and can ionize in aqueous solution, they associate with each other in a rather unique manner. Specifically, the fatty acids form spherical clusters, called **micelles**, in which the hydrophobic chains are directed toward the interior of the structure, and the hydrophilic (polar) carboxylate groups are on the outside, in contact with the aqueous environment. Micelles can become rather large (on a molecular level)—many hundreds or thousands of fatty acid molecules can be associated with each other, held together by *weak noncovalent* attractive forces (called London dispersion forces or van der Waals forces). The structure of a micelle is shown in Figure 13–1. Micelle formation and micelle structure are involved in a number of very important biological and physiological functions. As we shall see in the next chapter, *micelles aid in the transport of insoluble lipids in the blood.*

At this point, a few generalizations about fatty acids are in order.

1. Fatty acids are usually straight-chain monocarboxylic acids.
2. Fatty acids are usually composed of *even* numbers of carbon atoms (this is because they are constructed from two-carbon units—(see Chapter 14, Lipid Metabolism).

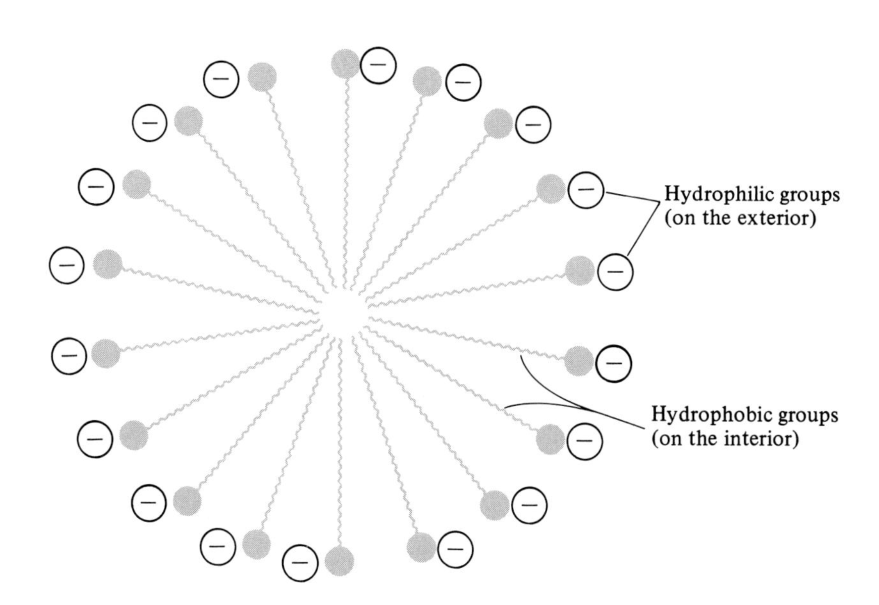

FIGURE 13–1
An idealized structure of a micelle. A micelle is approximately spherical in shape.

saturated
unsaturated

3. Fatty acids can be **saturated** (having no double bonds) or **unsaturated** (having one or more double bonds).
4. Unsaturated fatty acids usually predominate in the saponifiable lipids.
5. The hydrocarbon chain of an unsaturated fatty acid has a bend, or kink, due to a *cis* double bond in the chain.
6. Fatty acids with less than ten carbon atoms are usually liquids, at room temperature; longer fatty acids are usually solids. Longer-chain fatty acids (16-, 18-, and 20-carbons long) usually predominate in nature.
7. Unsaturated fatty acids are usually liquids at room temperature, even if they have relatively long hydrocarbon chains.

Table 13–2 lists some important fatty acids, along with their chemical names, formulas, and melting points.

Generally, unsaturated fatty acids contain double bonds in the *cis* configuration.

(H) (H)
C=C

Cis

(H)
C=C
(H)

Trans

TABLE 13–2 Some Important and Naturally Occurring Fatty Acids.

Symbol	Structure	Name	M.P. (°C)
	A. Saturated fatty acids		
14:0	$H_3C{-}(CH_2)_{12}{-}COOH$	Myristic acid	53.9
16:0	$H_3C{-}(CH_2)_{14}{-}COOH$	Palmitic acid	63.1
18:0	$H_3C{-}(CH_2)_{16}{-}COOH$	Stearic acid	69.6
	B. Unsaturated fatty acids		
16:1	$H_3C{-}(CH_2)_5{-}CH{=}CH{-}(CH_2)_7{-}COOH$	Palmitoleic acid	−0.5
18:1	$H_3C{-}(CH_2)_7{-}CH{=}CH{-}(CH_2)_7{-}COOH$	Oleic acid	13.4
18:2	$H_3C{-}(CH_2)_4{-}CH{=}CH{-}CH_2{-}CH{=}CH{-}(CH_2)_7{-}COOH$	Linoleic acid	−5.0

Note: The numerical symbol is a shorthand notation used to describe a particular fatty acid. The number on the left indicates the number of carbon atoms in the fatty acid; the number on the right indicates the number of double bonds in the molecule. For example, palmitoleic acid is a 16-carbon fatty acid with 1 double bond between the number 9 and 10 carbon atoms. We would refer to this fatty acid as a (16:1) fatty acid. It should also be noted that many short-chain fatty acids less than 12 carbons in length are also present in nature. Very short fatty acids include propionic acid (3C) and butyric acid (4C).

The *cis*-configuration in the fatty acid chain causes a characteristic bend, or kink.

Cis (common)

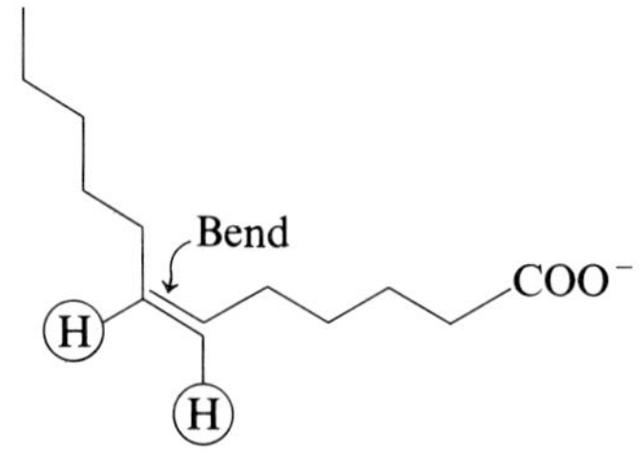

Bend

COO^-

(H)

(H)

Trans (uncommon)

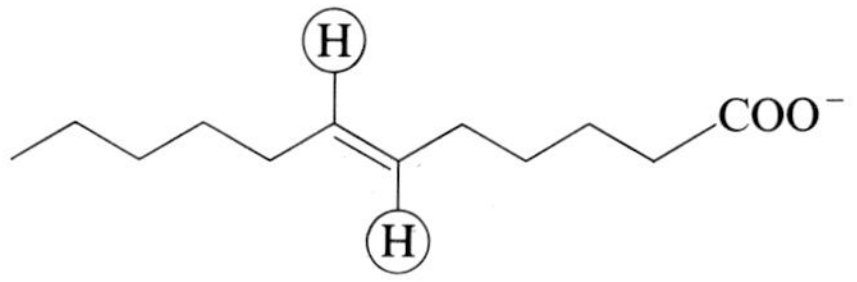

(H)

COO^-

(H)

This bend in the fatty acid chain keeps the hydrocarbon chains of different fatty acids apart, thus maintaining a higher degree of molecular motion. Because of this feature, long-chain unsaturated fatty acids are usually liquid at room temperature. (Refer to Table 13–2 and notice the low M.P. values for the unsaturated fatty acids.) One should note that the fluidity of biological membranes (an important feature to be discussed in a later section) occurs because the fatty acid chains of membrane lipids are unsaturated.

The double bonds in unsaturated fatty acids make them susceptible to oxidation by air or ozone and to degradation by bacteria. A postulated reaction with ozone is shown in the following equation:

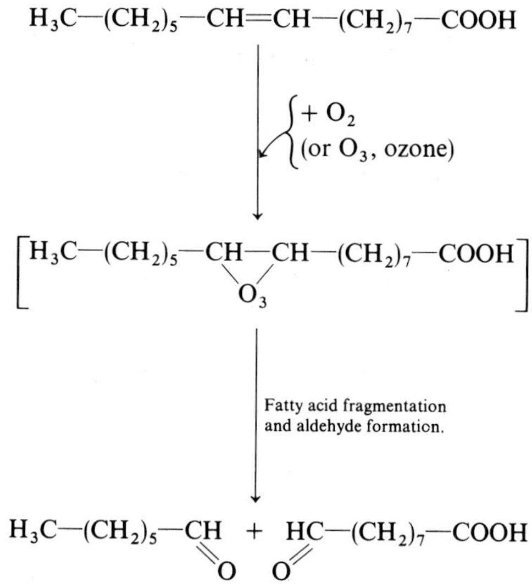

$$H_3C{-}(CH_2)_5{-}CH{=}CH{-}(CH_2)_7{-}COOH$$

$$\downarrow \left\{ \begin{array}{l} + O_2 \\ \text{(or } O_3\text{, ozone)} \end{array} \right.$$

$$\left[H_3C{-}(CH_2)_5{-}\underset{\diagdown O_3 \diagup}{CH{-}CH}{-}(CH_2)_7{-}COOH \right]$$

$$\downarrow \text{Fatty acid fragmentation and aldehyde formation.}$$

$$H_3C{-}(CH_2)_5{-}\underset{\| \atop O}{CH} + \underset{\| \atop O}{HC}{-}(CH_2)_7{-}COOH$$

Thus, fatty acid chains can cleave when exposed to ozone, producing *aldehydes*. This undesirable reaction could impair or destroy the biological function of a specific lipid containing these unsaturated fatty acids. If sufficient damage results, cellular or physiological function may be damaged. Ozone (present in polluted atmospheres) can destroy membrane lipids in the alveoli of the lungs; it is, therefore, of extreme concern to man.

13.4 NEUTRAL LIPIDS (TRIGLYCERIDES)

Neutral lipids are composed of a *glycerol backbone* with one, two, or three fatty acids esterified to the hydroxyl groups of the glycerol molecule.

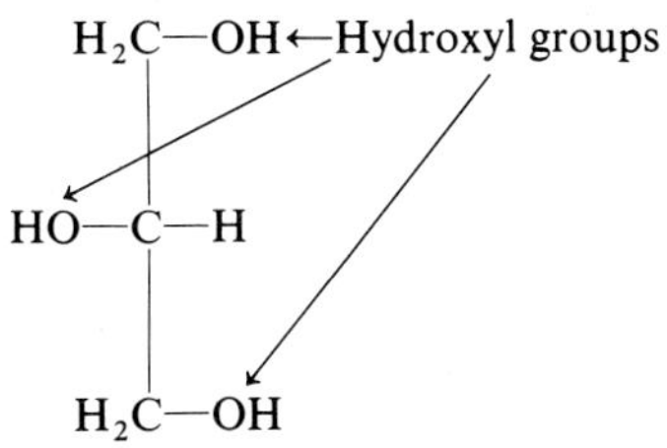

$H_2C—OH$ ← Hydroxyl groups
$HO—C—H$
$H_2C—OH$

Glycerol

$H_2—C—O—C(=O)—R_1$ ← Ester link
$HO—C—H$
$H_2C—OH$

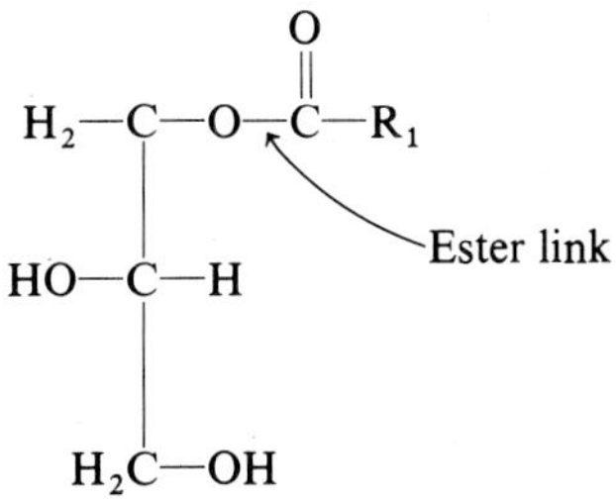

*Mono*glyceride (the R_1 group designates the hydrocarbon chain of the fatty acid)

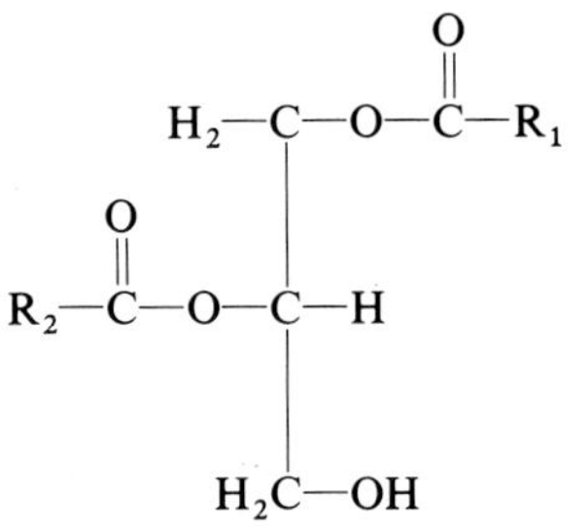

$H_2—C—O—C(=O)—R_1$
$R_2—C(=O)—O—C—H$
$H_2C—OH$

*Di*glyceride (two fatty acids)

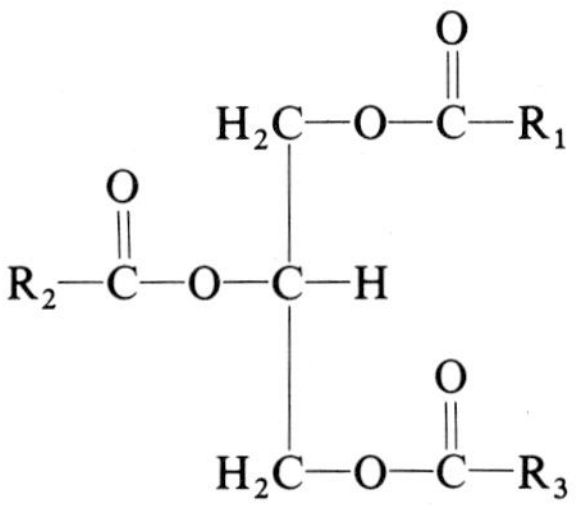

$H_2C—O—C(=O)—R_1$
$R_2—C(=O)—O—C—H$
$H_2C—O—C(=O)—R_3$

*Tri*glyceride (three fatty acids)

These substances do not possess an electrical charge; hence, they are called *neutral lipids*. Another common name for a triglyceride is *triacylglycerol*, where the *acyl* refers to the hydrocarbon chains of the fatty acids. Notice that the carboxylic acid group participates in an *ester linkage* with an alcohol group of the glycerol molecule. A typical triglyceride is *tripalmitin* (see p. 234). In tripalmitin, the three fatty acids are all *palmitic acid*.

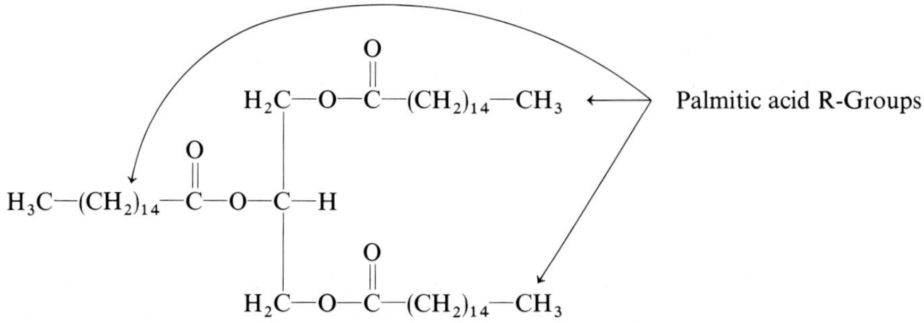

$$
\begin{array}{c}
\quad\quad\quad\quad\quad\quad H_2C{-}O{-}\overset{\overset{\displaystyle O}{\|}}{C}{-}(CH_2)_{14}{-}CH_3 \\
\quad\quad\quad\quad | \\
H_3C{-}(CH_2)_{14}{-}\overset{\overset{\displaystyle O}{\|}}{C}{-}O{-}C{-}H \\
\quad\quad\quad\quad | \\
\quad\quad\quad\quad\quad\quad H_2C{-}O{-}\overset{\overset{\displaystyle O}{\|}}{C}{-}(CH_2)_{14}{-}CH_3
\end{array}
$$

Palmitic acid R-Groups

Tripalmitin (a triglyceride)

Different fatty acids can be esterified to the same glycerol backbone to form different triglycerides. For example, tripalmitin contains three palmitic acid chains. However, other fatty acids such as *stearic acid* or *oleic acid* could be substituted in any of the three positions. Similarly, the three fatty acids can be either saturated or unsaturated. (The unsaturated fatty acids generally predominate.) Saturated and unsaturated substituents are demonstrated in the following diagram:

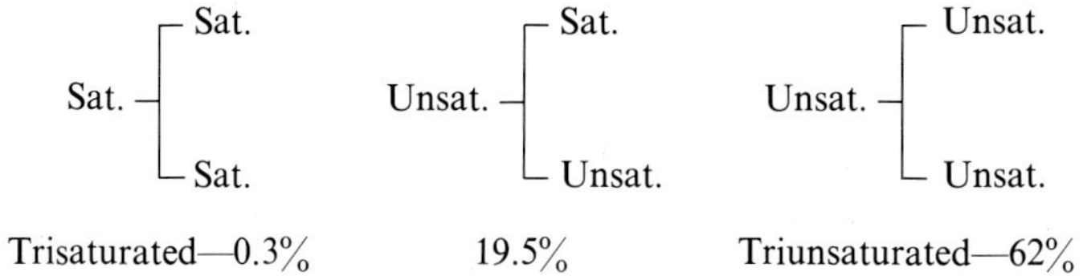

Trisaturated—0.3%	19.5%	Triunsaturated—62%
Sat. — [Sat., Sat.]	Unsat. — [Sat., Unsat.]	Unsat. — [Unsat., Unsat.]

adipose tissue Triglycerides are stored in the body in specialized tissue called **adipose tissue** (see Chapter 14, Lipid Metabolism). Because triglycerides contain a large number of carbon atoms that are highly **reduced** (bound with hydrogen atoms), they serve as an excellent energy-rich cellular fuel. In addition, these substances, because of their hydrophobicity, do not associate with water; thus, they are in anhydrous form (*anhydrous* means "without water"). Thus, large amounts of lipids can be stored in tissues without water contributing to the bulk and weight of the stored fuel. This is in contrast to carbohydrates, which are relatively hydrated (bound with water).

reduced

As mentioned previously, the length of the fatty acid chain and the degree of unsaturation ultimately determine whether the triglyceride will be a solid or a liquid at room temperature. Animal fats tend to be solids at room temperature. They are composed primarily of long-chain saturated fatty acids. On the other hand, vegetable oils such as olive oil and corn oil, are liquids at room temperature. These oils consist of triglycerides containing unsaturated fatty acids. The term "*polyunsaturated*" used on the labels of many commercial vegetable oils refers to the fact that these oils contain a large percentage of unsaturated fatty acids (that is, they have many double bonds). Because of these double bonds, polyunsaturated oils are susceptible to oxidation by air or to degradation by bacteria. Cleavage of ester links and cleavage of double bonds within the fatty acid chain produce a variety of volatile degradation products, all possessing rather disagreeable odors or flavors. This process is called **rancidification** and produces rancid foods.

rancidification

Saponification is a Latin term meaning "soap making." It refers to a process that degrades triglycerides into salts of fatty acids (*soaps*) and glycerol. The saponification process and reaction is shown in the following equation:

$$\begin{array}{c} H_2CO{-}\overset{\overset{O}{\|}}{C}{-}R_1 \\ | \\ R_2{-}\overset{\overset{O}{\|}}{C}{-}O{-}C{-}H \\ | \\ H_2CO{-}\overset{\overset{O}{\|}}{C}{-}R_3 \end{array} \xrightarrow[3\,H_2O]{\text{1. NaOH (lye)} \quad \text{2. Heat}} \begin{array}{c} H_2C{-}OH \\ | \\ HO{-}C{-}H \\ | \\ H_2C{-}OH \end{array}$$

Triglycerides (found in animal fat)

Glycerol + $3\,R{-}\overset{\overset{O}{\|}}{C}{-}O^-Na^+$

Fatty acids (soaps)

Lipases catalyze a similar reaction in the body, according to the following equation:

$$\begin{array}{c} H_2C{-}O\overset{\overset{O}{\|}}{C}{-}R_1 \\ | \\ R_2{-}\overset{\overset{O}{\|}}{C}{-}O{-}C{-}H \\ | \\ H_2C{-}O\overset{\overset{O}{\|}}{C}{-}R_3 \end{array} \xrightarrow{\text{Lipase}} \begin{array}{c} H_2C{-}OH \\ | \\ HO{-}C{-}H \\ | \\ H_2C{-}OH \end{array} + 3\,R{-}\overset{\overset{O}{\|}}{C}{-}O^-$$

Triglyceride — Glycerol — 3 Fatty acids

Obviously, the lipase does not need sodium hydroxide (NaOH) or heat to catalyze this reaction. In Chapter 14, we will elaborate on this important reaction.

13.5 PHOSPHOLIPIDS AND SPHINGOLIPIDS

phospholipid

A **phospholipid** is a molecule composed of a glycerol backbone with two fatty acids esterified to the #1 and #2 carbon hydroxyl groups. The #3 carbon hydroxyl group is involved in a *phosphoester linkage* with a phosphate group—hence the name. Notice that the following structure has a phosphate group with two negative charges:

```
                              O
                              ‖
                        ¹H₂C—O—C—R₁   ←——  Usually sat.
                 O        |
                 ‖        |
Usually unsat. ——→ R₂—C—O—²C—H
                          |        O
                          |        ‖
                        ³H₂C—O—P—O⁻   ←——  Phosphate group
                                   |
                                   O⁻
```

Phosphatidic acid (parent compound to a number of other phospholipids)

Other alcohol groups can bond to one of the oxygen atoms of the phosphate group; thus, a variety of different *phosphoglycerides* can be formed.

```
                 O
                 ‖
          H₂C—O—C—R₁          (X) = Choline;
  O        |                        Ethanolamine;
  ‖        |                        Serine (an amino acid);
R₂—C—O—C—H                          etc.
           |     O
           |     ‖
          H₂C—O—P—O—(X)
                 |
                 O⁻
```

For example, *phosphatidyl choline* (lecithin) is formed when choline is bound to the phosphate group.

```
                   O
                   ‖
            CH₂—O—C—R₁
  O          |
  ‖          |
R₂—C—O—C—H
             |                             Choline group
             |     O               CH₃ ↙
             |     ‖                |
            H₂C—O—P—O—(CH₂)₂—⁺N—CH₃
                   |                |
                   O⁻              CH₃
```

Phosphatidyl Choline (Lecithin)

Notice in this molecule that the phosphate oxygen has a negative charge and the nitrogen atom of the choline group has a positive charge. Lecithins are the most abundant phosphoglycerides in animal tissue, and serve a very important role as a structural component of most cell membranes (see Section 13.11). Lecithin is also found in the alveoli of lungs and is known as *lung surfactant* (see Section 13.6 regarding respiratory distress syndrome). Phosphatidyl ethanolamine, phosphatidyl serine, and phosphatidyl inositol are important phosphoglycerides found in most cell membranes. They are particularly abundant in brain tissue.

sphingolipids

Sphingolipids are found in a number of tissues, especially in the brain and other nerve tissues. As the name implies, sphingolipids contain a compound called *sphingosine*. Different fatty acids are bound to the sphingosine group. A representative sphingolipid is shown:

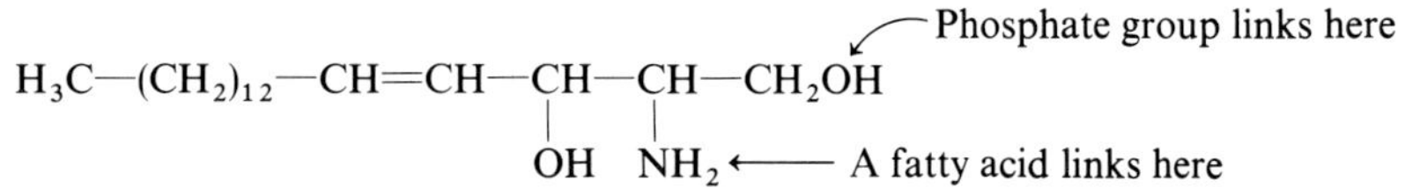

$$H_3C{-}(CH_2)_{12}{-}CH{=}CH{-}\underset{\displaystyle OH}{\underset{|}{CH}}{-}\underset{\displaystyle NH_2}{\underset{|}{CH}}{-}CH_2OH$$

Phosphate group links here (at CH_2OH); A fatty acid links here (at NH_2)

Sphingosine

$$H_3C{-}(CH_2)_{12}{-}CH{=}CH{-}\underset{\displaystyle OH}{\underset{|}{CH}}{-}\underset{\displaystyle NH{-}\overset{\displaystyle O}{\overset{\|}{C}}{-}R}{\underset{|}{CH}}{-}CH_2{-}O{-}\overset{\displaystyle O}{\overset{\|}{\underset{\displaystyle O^-}{\underset{|}{P}}}}{-}O{-}(CH_2)_2{-}\overset{\displaystyle CH_3}{\overset{|}{\underset{\displaystyle CH_3}{\underset{|}{{}^+N}}}}{-}CH_3$$

Fatty acid (at R)

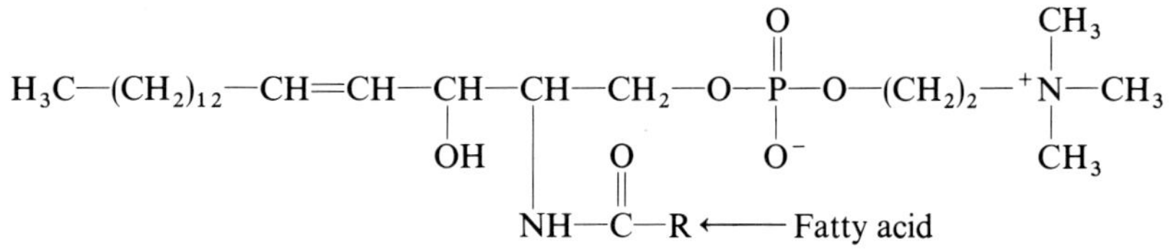

Sphingomyelin

13.6 DISEASES OF PHOSPHOLIPID AND SPHINGOLIPID METABOLISM

RESPIRATORY DISTRESS SYNDROME

Respiratory distress syndrome (RDS) with the ensuing hyaline membrane disease is one of the main causes of death in premature infants. The disease results from the inability of immature lungs to synthesize and secrete adequate amounts of lecithin (lung surfactant) into the alveolar air spaces. Lecithin apparently stabilizes the alveolar air spaces and keeps them from collapsing upon exhalation. Clinicians have been able to assess whether RDS may occur in an infant by measuring the lecithin to sphingomyelin (L/S) ratio in amniotic fluid or by using the foam stability test (FST). Both tests monitor the amount of lecithin, thus providing a basis for judging whether a premature infant has, or will develop, respiratory distress syndrome.

GLYCOSPHINGOLIPID STORAGE DISEASE

glycolipids

An important class of lipids are the **glycolipids** (lipids with attached sugar residues). Glycolipids contain a sphingosine backbone, a fatty acid, and a sugar residue (usually galactose or glucose). They do not contain a phosphate

ceramide

cerebroside

group and, hence, are not phospholipids. When sphingosine is associated with a fatty acid, the resulting compound is called a **ceramide**. If a sugar residue is bound to the sphingosine, the resulting compound is called a **cerebroside**. The generalized structures for these substances are shown in the following diagram:

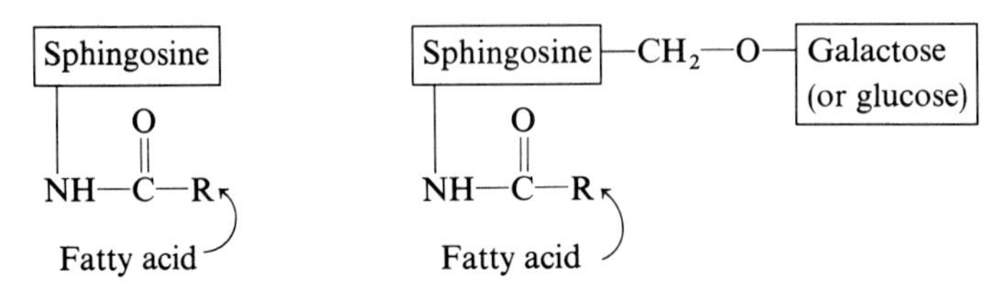

A ceramide (not a *glyco*lipid) A cerebroside (and, therefore, a *glyco*lipid)

glycosphingolipids

A more complex subgroup of the cerebrosides are the **glycosphingolipids**. These compounds contain the basic ceramide structure, but have additional sugars attached to the galactose unit in straight or branched chains.

Ceramide——Sugar——Sugar——Sugar

A typical glycosphingolipid, *globoside*, is found in the red blood cell.

Ceramide——Glucose——Galactose——Galactose
| N-acetyl-galactosamine

Many glycosphingolipids occur in the body and serve a number of

TABLE 13–3 The Glycosphingolipid Storage Diseases.

Disease	Enzymatic Defect	Remarks
1. Fabry's disease	α-galactosidase	Skin lesions, pain in the extremities, death usually in the fourth decade.
2. Gaucher's disease	β-glucosidase	Hepatosplenomegaly (enlargement of liver and spleen), frequently fatal; no known treatments.
3. Krabbe's disease	Galactocerebrosidase	Mental retardation, psychomotor retardation, progressive spasticity, invariably fatal.
4. Tay-Sach's disease (type I)	Hexosaminidase A	Red spot in retina, mental retardation, severe psychomotor retardation, blindness, invariably fatal; especially prevalent among Northern European Jews.

Source: Adapted from Bhagavan, N. V., *Biochemistry*, 2nd ed. (Philadelphia: J. B. Lippincott, 1978), pp. 860–861.

functions. Some glycosphingolipids are important in transmitting nerve impulses; others are involved in blood group and organ specificity. They are normally made and degraded by a variety of enzymes. The degradative enzymes are called *cerebrosidases*. The cerebrosidases may be deficient in people suffering from certain inherited diseases. If there is such a deficiency, the glycosphingolipids are not degraded at a proper rate and tend to accumulate in various tissues, often with disastrous effects. For example, in *Tay-Sachs Disease*, afflicted individuals have a deficiency of the enzyme *hexosaminidase A* (see Table 13–3). As a result of this defect, ganglioside G_{M2} accumulates in brain tissue according to the following reaction:

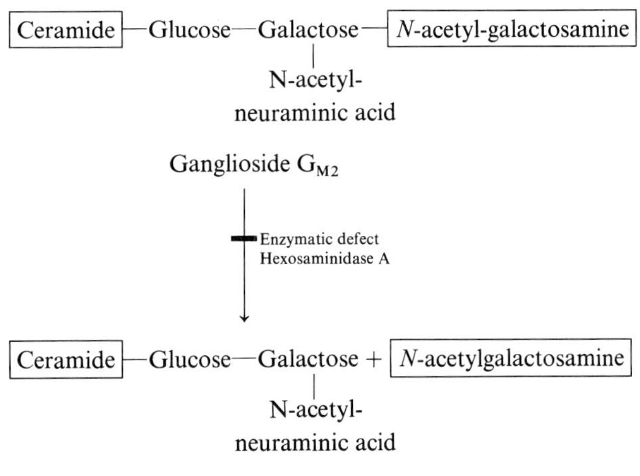

Mental retardation is one of the principal manifestations of these disease states. Unfortunately, clinical treatments are generally nonexistent. Table 13–3 lists a number of glycosphingolipid storage diseases. *Fabry's disease* results from a deficiency in α-galactosidase, *Gaucher's disease* results from a deficiency in β-glucosidase, and *Tay-Sach's disease* (as we have already mentioned) results from a deficiency in hexosaminidase A.

13.7 WAXES

waxes

Waxes are esters of long-chain alcohols and fatty acids.

$$H_3C{-}(CH_2)_n{-}CH_2{-}O{-}\overset{\overset{\displaystyle O}{\|}}{C}{-}(CH_2)_n{-}CH_3$$

Alcohol hydrocarbon chain (n units long) — Ester link — Fatty acid hydrocarbon chain

Each hydrocarbon chain is usually very long. Waxes are found in the protective coatings of feathers, fur, skin, leaves, and fruits. *Sebum*, a secretion of the *sebaceous glands* of the skin, contains many different waxes. It keeps the skin soft and prevents dehydration. The skin of the fetus is also protected by a waxy substance called the *vernix caseosa*.

13.8 TERPENES

terpenes

Terpenes are classified as nonsaponifiable lipids. They are constructed from the 5-carbon *isoprene* unit, as shown below:

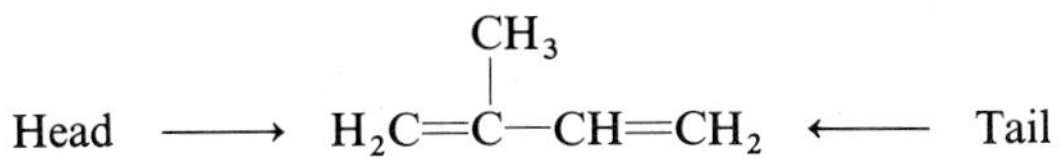

$$\text{Head} \longrightarrow H_2C{=}\underset{}{\overset{\displaystyle CH_3\atop|}{C}}{-}CH{=}CH_2 \longleftarrow \text{Tail}$$

Isoprene unit

Individual isoprene units can be joined together in two basic arrangements: (1) head-to-tail (the most common arrangement) or (2) tail-to-tail.

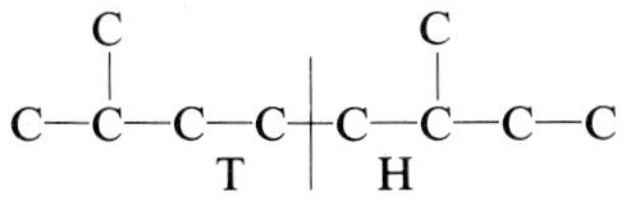

$$\begin{array}{c} C\text{–}\overset{C}{\overset{|}{C}}\text{–}C\text{–}\underset{T}{C}\,|\,\underset{H}{C}\text{–}\overset{C}{\overset{|}{C}}\text{–}C\text{–}C \\ \text{head-to-tail} \end{array} \qquad \begin{array}{c} C\text{–}\overset{C}{\overset{|}{C}}\text{–}C\text{–}\underset{T}{C}\,|\,\underset{T}{C}\text{–}C\text{–}\overset{C}{\overset{|}{C}}\text{–}C \\ \text{tail-to-tail} \end{array}$$

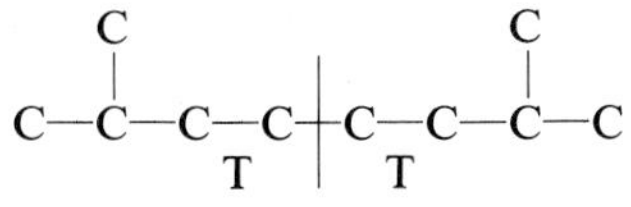

Longer structures of straight chains, branched chains, or cyclic compounds can result from these bonding arrangements. Terpenes are synthesized by plants cells and serve as the oils, fragrances, and pigments of many plants and trees. Representative terpenes are shown below:

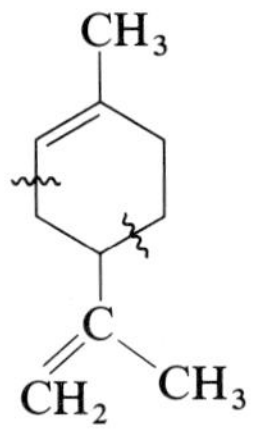

Limonene (from lemons)

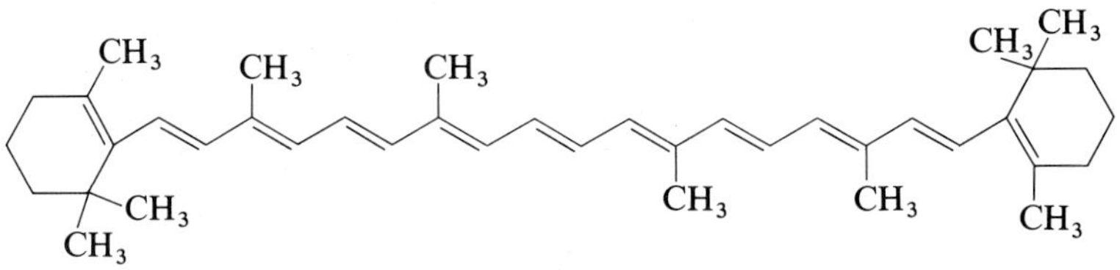

β-Carotene (from carrots)

Animals also make terpene compounds such as *squalene*, the precursor (parent compound) of *cholesterol.*

Squalene

fat-soluble vitamins

The **fat-soluble vitamins** (*vitamins A, D, E, and K*) are substances required in our diets and necessary for normal health. These vitamins are also terpene derivatives (except for Vitamin D, which is considered a steroid derivative). The biochemistry of the fat-soluble vitamins will be discussed in detail in Chapter 20, Vitamins and Nutrition.

Vitamin A_1 (retinol)

Vitamin E (α-Tocopherol)

Vitamin K_2 ($n = 6$ to 10)

13.9 STEROIDS

steroids

All **steroids** are derivatives of a 4-ring hydrocarbon system and belong to the class of nonsaponifiable lipids.

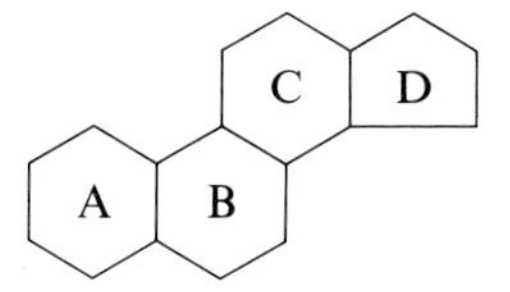

Steroid ring system

Individual steroids differ in (1) the number and position of substituents on the ring system, (2) the number and position of double bonds, and (3) the stereochemistry of the ring system. Cholesterol, the bile salts, the steroid hormones, and the steroid derivative vitamin D all belong to this class of lipids.

cholesterol

One of the most important steroids is **cholesterol**, a 27-carbon compound derived from squalene according to the following scheme:

$$H_3C{-}\overset{\overset{\displaystyle O}{\|}}{C}{-}S{-}CoA \longrightarrow \text{Squalene (30 C)}$$

Acetyl-CoA

Cyclization reaction

Lanosterol (30 C)

CH_3 CH_3 CH_3 CH_3 CH_3

Can form esters ⟶ HO
with fatty acids here

Cholesterol (27 C)

Recently, cholesterol has received a great deal of publicity regarding its role in the initiation of atherosclerosis. Some of this attention is warranted. However, one should note that the body normally synthesizes 1.5–2.0 grams of cholesterol per day (mainly in the liver) and that cholesterol is a very important precursor to a number of other steroids, such as the bile salts and steroid hormones. A normal diet contributes about 0.3 grams of cholesterol per day to the body. Atherosclerosis involves the accumulation of cholesterol esters and other lipids on the connective tissue of the arterial walls, leading to cholesterol plaque formation. If plaque formation is extensive, important blood vessels are occluded, and the blood supply is diminished or completely shut off. If a blood vessel on the heart muscle is blocked, a **myocardial infarction** (heart attack) might occur. If a blood vessel within the brain is blocked, a **stroke** might occur. It has been suggested that a correlation exists between elevated blood cholesterol levels and atherosclerosis. Unfortunately, convincing evidence is still lacking, and the issue is clouded by the complexity of the disease. It has been suggested that the initial injury to the arterial walls may be due to the presence of excessive amounts of *atherogenic lipoproteins* (low density lipoproteins, LDL), which contain large amounts of cholesterol and cholesterol esters. We will deal with this matter in more detail in Chapter 14, Lipid Metabolism.

myocardial infarction
stroke

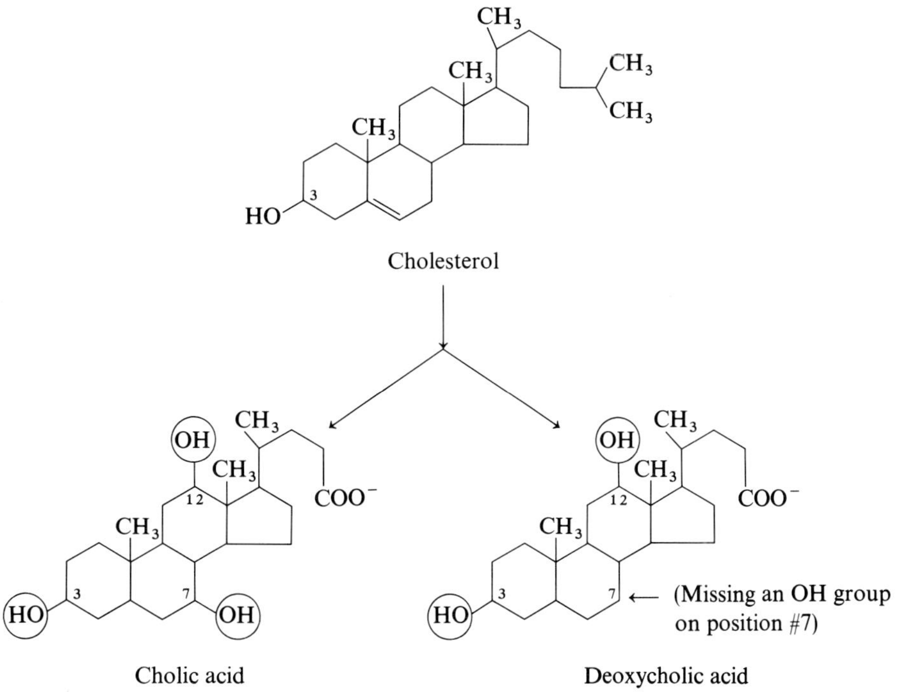
CH3
HO
3
Cholesterol
OH
CH3
CH3
12
COO⁻
CH3
HO
3
7
OH
Cholic acid
OH
CH3
CH3
12
COO⁻
CH3
HO
3
7
(Missing an OH group on position #7)
Deoxycholic acid

FIGURE 13–2 Representative bile salts derived from cholesterol.

As mentioned previously, cholesterol is the precursor to a number of *bile salts*. Bile salt biosynthesis occurs primarily in the liver (and also in the intestine). A simplified pathway is shown in Figure 13–2. These bile salts have more hydroxyl groups than the parent cholesterol molecule. Also notice that the side-chain has been modified. As a result of these modifications, the bile salts are rendered amphipathic, as shown in Figure 13–3. These amphipathic molecules can aggregate into micelles along with lecithin. Because of this ability, the lecithin/bile salt micelles can render soluble normally insoluble dietary lipids, such as cholesterol. This is done by incorporating them into the interior hydrophobic region of the micelle (see Chapter 14, Lipid Metabolism). Thus, excess cholesterol can be excreted

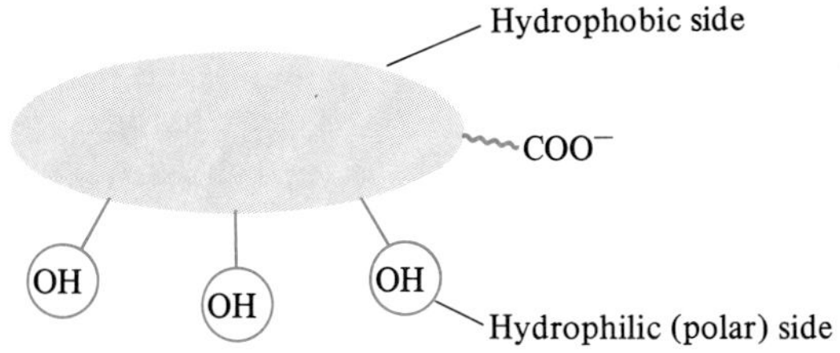
Hydrophobic side
COO⁻
OH
OH
OH
Hydrophilic (polar) side

FIGURE 13–3 An idealized diagram of a bile salt. The side view demonstrates the amphipathic nature of the bile salt.

into the intestine and eliminated in the feces. When bile salt/lecithin micelles cannot adequately incorporate all of the cholesterol, the cholesterol will precipitate out of solution and form *cholesterol gallstones*. This condition may be due to (1) increased synthesis of cholesterol by the liver, (2) increased dietary intake of cholesterol or lipids, (3) decreased bile salt synthesis, or (4) an injured gallbladder. After the gallstones form (usually on the walls of the gallbladder), they can be dislodged from the gallbladder and lodge in the cystic duct, in the common duct, or in the Sphincter of Oddi, causing a painful *gallbladder attack*. This condition is remedied by surgical intervention in which either the gallbladder is removed or it is scraped clean of the offending gallstones. *Cholestectomy* (gallbladder removal) is the most common surgical procedure performed in the United States (400 000 cases/year, 5000–8000 deaths/year, and over $1 billion/year in hospital costs).

steroid hormones

Steroid hormones are another important derivative of cholesterol. The structures and functions of important steroid hormones are shown in Table 13–4. A detailed discussion of steroid hormone function is found in Chapter 19, Hormones and Hormone Action.

TABLE 13–4 Some Important Steroid Hormones.

Structure and Name		Biological Function(s)
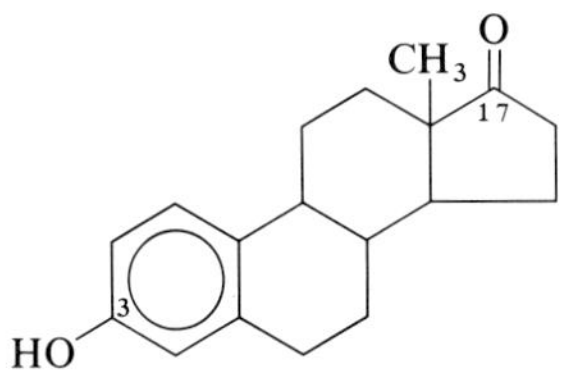 CH_3, O, 17, 3, HO	Estrone (an estrogen)	Estrogens promote the development, maturation, and function of the female reproductive organs. The proliferative phase of the uterine cycle is caused by the cyclic production of estrogen. They also suppress follicle-stimulating hormone (FSH) secretion and promote luteinizing hormone (LH) secretion.
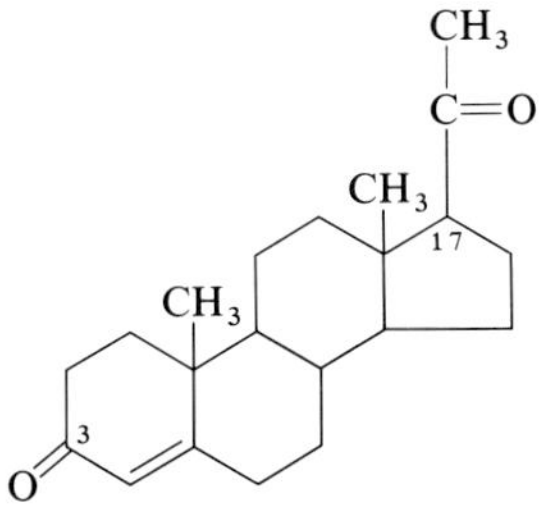 CH_3, C=O, CH_3, 17, CH_3, 3, O	Progesterone	In the nonpregnant female, the corpus luteum is the principal source of progesterone. The hormone prepares the uterus for the implantation of the ovum. It decreases uterine muscular contractions, stimulates the development of the breasts, and suppresses ovulation.
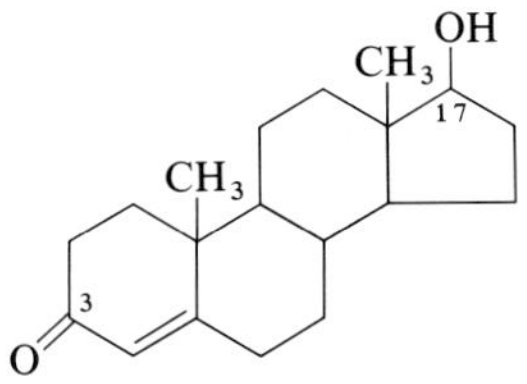 OH, CH_3, 17, CH_3, 3, O	Testosterone	Testosterone is synthesized by the testes. In males, it is responsible for the primary and secondary sexual characteristics; in females, for the development of body hair at puberty. The hormone causes an increase in skeletal muscle mass, stimulates bone growth, and increases the basal metabolic rate (BMR).

Source: Adapted from Bhagavan, N. V., *Biochemistry*, 2nd ed. (Philadelphia: J. B. Lippincott, 1978), pp. 1237–1241.

13.10 PROSTAGLANDINS

prostaglandins

The **prostaglandins** are 20-carbon fatty acids containing a 5-membered ring. In structure, they resemble a hairpin. The carbon atoms are numbered from 1 to 20, starting from the carboxyl group.

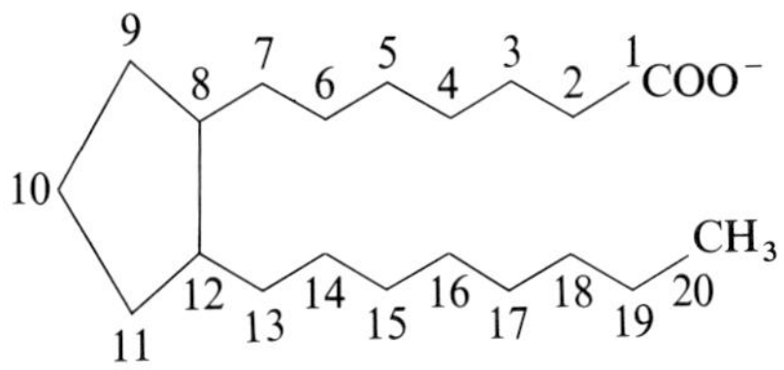

Generalized prostaglandin structure

The prostaglandins were first isolated from the *prostate* gland—hence the name. They are also found in seminal plasma, and recent research has demonstrated that nearly every organ in the body synthesizes this very important class of lipids. Prostaglandins can be defined as chemical messengers (or modulators), and regulate a number of important physiological functions. Table 13–5 summarizes a few of the prostaglandins and their diverse biological functions.

TABLE 13–5 Characteristics of Prostaglandin E (PGE) and Prostaglandin F_α (PGF_α).

Prostaglandin	Target Organ	Physiological Effect
PGE	Cardiovascular system	1. Increased cardiac output. 2. Dilation of blood vessels with decreased arterial blood pressure.
PGF	Cardiovascular system	The effects of PGF are the opposite of those described for PGE (that is, PGF constricts blood vessels and causes hypertension).
PGE_1	Kidneys	Increases plasma flow, urinary volume, and electrolyte excretion.
PGE	Brain	1. Produces sedation (in the cerebral cortex). 2. In the medulla, both PGE and PGF can produce either excitation or inhibition.
PGE (and PGF)	Reproductive system	Increased uterine contractions, leading to the expulsion of the fetus.

Note: PGE =

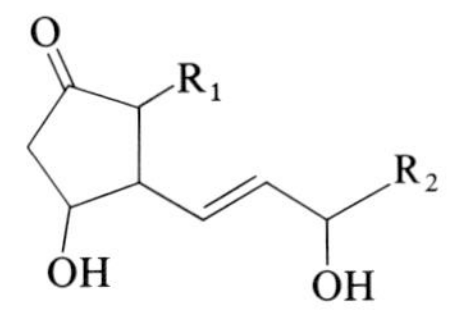

PGF_α =

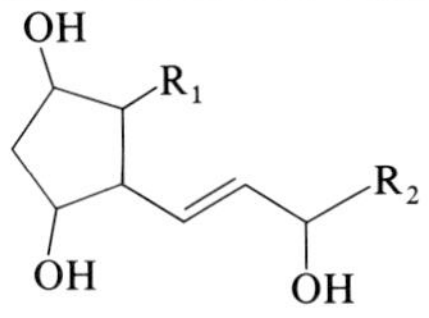

Source: Adapted from Bhagavan, N. V., *Biochemistry*, 2nd ed. (Philadelphia: J. B. Lippincott, 1978), pp. 818–30.

13.11 BIOLOGICAL MEMBRANES

cell membrane

A **cell membrane** is simply a chemical interface (a lipid bilayer) between two different aqueous environments. In essence, cell membranes are the structures that surround cells and hold them together. In addition, the cellular organelles (such as the nucleus, mitochondrion, and lysosomes) are surrounded by membranes. The endoplasmic reticulum is actually composed of a membrane.

The functions of the cell membrane are many and varied. As we have said, they separate different aqueous environments so that competing metabolic reactions can proceed without adversely affecting each other, and also to keep various substances from mixing. Biological membranes act as *permeability barriers* between the cell and the outside environment or between various compartments within the cell. Generally polar substances (and ions) do not readily pass across cell membranes. Therefore, specific *transport proteins* within the membrane serve to transport substances across the membrane. (We will discuss transport mechanisms in more detail in a moment.) Second, membranes provide physical support for a number of proteins and enzymes. In doing so, they may also control protein or enzyme activity. Third, they provide an alternative physical environment (a lipid/hydrophobic environment) for certain chemical reactions that might not take place in the normal aqueous environment of the cell. Fourth, they provide a means by which cells can communicate with each other. For example, as two cells come in contact with each other, each cell might modify certain components on the other cell's surface, thus exchanging chemical information.

Most cell membranes contain about 60 percent lipid and 40 percent protein. Some atypical membranes (such as the inner mitochondrial membrane) contain only about 20 percent lipid and 80 percent protein. The types of lipids present within a particular membrane will depend on the type of cell and the species of the plant or animal. However, phospholipids (phosphatidylcholine and phosphatidylethanolamine), sphingomyelin, and cholesterol predominate in most membranes. A variety of different proteins are associated with the membrane lipids. Precisely *how* the lipids and proteins are organized (and, thus, how the typical membrane is constructed) has been the subject of a great deal of research. Until recently, our understanding of the chemical structure and properties of cell membranes was rather limited. A more complete picture has emerged from the work of a number of dedicated biochemists (S. Singer, G. Nicolson, and M. Bretscher). The present model, the fluid-mosaic model, is diagrammed in Figure 13–4. The phospholipids (along with cholesterol and sphingomyelin) are organized in a **lipid bilayer** in which the hydrophobic chains extend toward the inside of the bilayer and the hydrophilic groups (the phosphate groups and other polar groups) are located on the outside, in contact with water. The proteins are embedded in the membrane; some float in the lipid bilayer like icebergs in the sea. There are two main classes of membrane-bound proteins: (1)

lipid bilayer

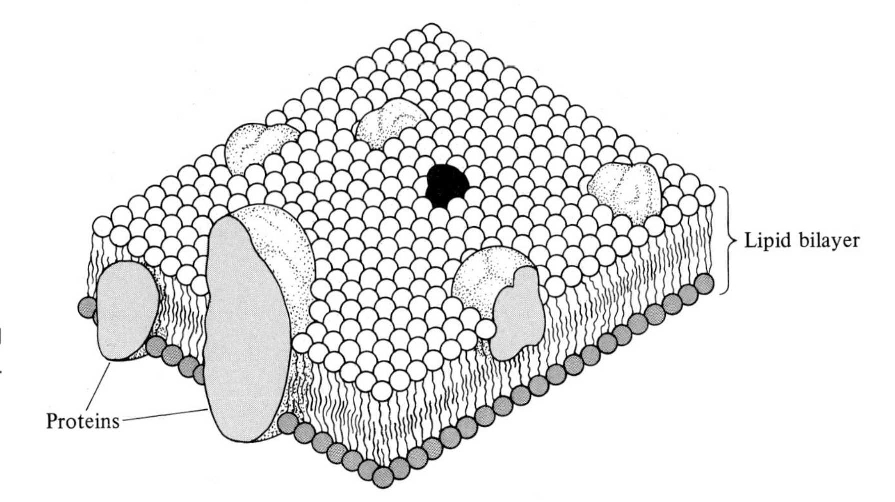

FIGURE 13–4
The fluid-mosiac model of membrane structure. (Adapted from S. J. Singer and G. L. Nicolson, *Science* 175 (1972): 720–31.)

extrinsic (peripheral)
intrinsic (integral)

extrinsic (peripheral) proteins and (2) **intrinsic (integral)** proteins. Extrinsic proteins are loosely associated with one side of the membrane, usually via electrostatic attractions with the electrically charged phospholipids. The intrinsic proteins are firmly embedded in the membrane by hydrophobic attractions between the hydrocarbon chains of the membrane lipids and various hydrophobic amino-acid R groups of the protein. Only intrinsic proteins are depicted in Figure 13–4. The proteins and lipids *are free to move laterally within the bilayer* like dancers on a crowded dance floor—hence the name fluid mosaic. Note that membrane lipids are usually unsaturated. Because of this feature, **fluidity** of the membrane is favored. The proteins and lipids will *not* flip-flop (that is, a specific lipid on one side of the bilayer will not flip over to the other side of the bilayer). Because flip-flop is not favored, membrane **asymmetry** is maintained. Thus, a specific population of lipids and proteins resides on one side of the bilayer, while a different population resides on the other. Referring to Figure 13–4, you will observe that some proteins reside on the outer surface of the membrane, others on the inner (cytoplasmic) surface, while still others extend completely across the membrane (the latter are called *transmembrane proteins*). Figure 13–4 is only a representation of what may actually exist. However, the model is based on sound experimental evidence derived from the membranes of red blood cells. Other cell types or cellular organelles may possess membranes arranged in a slightly different manner. For example, the inner mitochondrial membrane contains considerably more protein than the plasma membrane of the red blood cell; thus, the inner mitochondrial membrane may also be less fluid.

fluidity

asymmetry

Recent research indicates that *membrane-bound proteins* have many and varied functions, and that their organization within the membrane may be critical for their biological function. For example, some proteins serve as *transport proteins* and pump ions or small-molecular-weight substances

passive (facilitated)
active (facilitated)

across the membrane into the cell. There are two basic types of transport systems: (1) **passive (facilitated)** transport and (2) **active (facilitated)** transport. The term "facilitated" indicates that a carrier protein is required to transport a specific molecule across the membrane. Passive transport does not require energy input, as it usually proceeds down a concentration gradient from a high concentration to a low concentration. Active transport, on the other hand, requires energy input in the form of ATP to pump substances up a concentration gradient. An idealized model of active membrane transport is summarized in Figure 13–5.

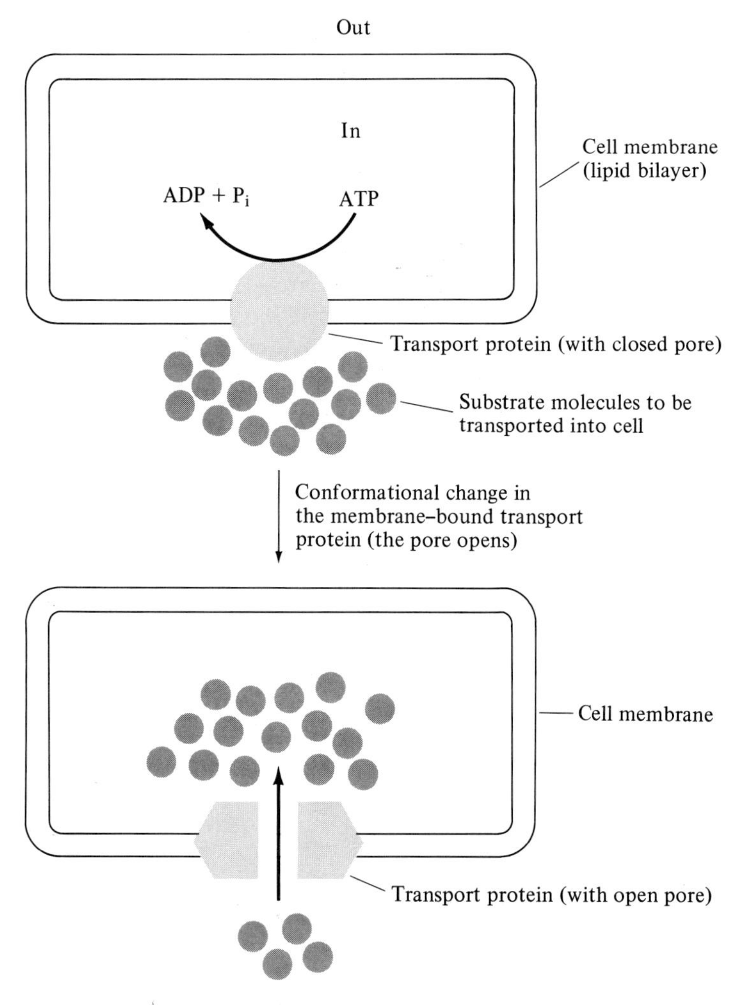

FIGURE 13–5 Active transport across a membrane. Active transport is an energy-dependent process in which ATP is used.

SUMMARY

Lipids serve as structural components for cellular membranes and are a form of cellular fuel. Some serve as hormones.

There are two main classes of lipids: (1) the saponifiable lipids (neutral lipids, phospholipids, sphingolipids, and waxes), and (2) the nonsaponifiable lipids (terpenes, steroids, and prostaglandins).

Fatty acids are long-chain hydrocarbons with a carboxylic acid functional group. They are composed of even numbers of carbon atoms, and the hydrocarbon chain can be saturated or unsaturated. They serve as the main structural component for the saponifiable lipids and can participate in ester links with alcohol groups. Fatty acids can associate with each other to form micelles.

Neutral lipids are composed of a glycerol backbone with one, two, or three fatty acids esterified to the hydroxyl groups of the glycerol molecule. The resulting lipids are called monoglycerides, diglycerides, and triglycerides, respectively. Triglycerides are an energy-rich cellular fuel.

A phospholipid is a molecule composed of a glycerol backbone with two fatty acids esterified to the #1 and #2 carbon hydroxyl groups. The #3 carbon hydroxyl group is involved in a phosphoester linkage with a phosphate group. Other substituents can bond with the phosphate group, forming a wide variety of compounds. Phospholipids are important structural components of most cell membranes. Sphingolipids, which are derived from sphingosine, are found in brain and nerve tissue.

Respiratory distress syndrome is one of the main causes of death in premature infants. It results from the inability of immature lungs to synthesize and secrete lecithin, a lung surfactant. Glycosphingolipid storage diseases result from a deficiency in specific cerebrosidases that break down the glycosphingolipids.

Waxes are esters of long-chain alcohols and fatty acids. They are found in the protective coatings of feathers, fur, skin, leaves, and fruit.

Terpenes are constructed from isoprene units and serve as plant oils, fragrances, and pigments. The fat-soluble vitamins A, E, and K are terpene derivatives.

All steroids are derivatives of a 4-ring hydrocarbon skeleton. Cholesterol, an important steroid, is synthesized from squalene. The bile salts and steroid hormones, in turn, are derived from cholesterol. Cholesterol (or lipoproteins containing cholesterol) may be involved in atherosclerosis. Gallbladder bile renders dietary lipids and excess cholesterol soluble.

The prostaglandins are 20-carbon fatty acids containing a 5-membered ring. They serve as chemical messengers and regulate a variety of physiological functions, such as blood flow, smooth muscle contraction, and the like.

Biological membranes are a chemical interface, or permeability barrier, separating two different chemical environments. They are the structures that surround cells and cellular organelles, holding them together. They also provide physical support for membrane-bound proteins and an alternative (hydrophobic) environment for certain chemical reactions. Most cell membranes are composed of a lipid bilayer in which intrinsic proteins are embedded in the proportions 60 percent lipid and 40 percent protein. Both lipids and proteins are free to move laterally within the membrane; thus, the membrane is thought to be fluid. Membrane-bound proteins (extrinsic and intrinsic) serve many functions.

REVIEW QUESTIONS

1. We introduced a number of important concepts in this chapter. Define (or briefly explain) each of the terms below.
   - **a.** Saponifiable lipids
   - **b.** Fatty acid
   - **c.** Ester link
   - **d.** Hydrophobic
   - **e.** Amphipathic
   - **f.** Micelle
   - **g.** Triglyceride
   - **h.** Phospholipid
   - **i.** Unsaturation
   - **j.** Sphingolipid
   - **k.** Terpene
   - **l.** Steroid hormone

m. Prostaglandin
n. Fluid-mosaic membrane model

2. Which of the following statements about cellular fatty acids are *not* correct?
a. They are usually straight-chain aldehydes.
b. The hydrocarbon portion of the chain can be saturated or unsaturated.
c. They usually contain chains with even numbers of carbon atoms.
d. The chains are usually 3 to 10 carbons in length.
e. When the chain is unsaturated, the double bond between the carbon atoms is usually *cis*.

3. Write the chemical structure for the fatty acid with the following shorthand notation: (20:1).

4. Write a complete chemical structure for a triglyceride containing myristic, palmitic, and oleic acids.

5. a. Draw the complete chemical structure for phosphatidyl serine (assume that palmitic and palmitoleic acids are the fatty acids esterified to glycerol). (*Note*: The serine OH group binds to the phosphate group.)
b. What is the net electrical charge on the molecule at pH 7?

6. a. What are the essential chemical features of the cholesterol molecule?
b. What are the main chemical differences between cholesterol, cholic acid, and deoxycholic acid?
***c.** What are the main chemical differences between the steroid hormones shown in Table 13–4? What modifications were required to the cholesterol molecule?

***7.** Prostaglandins probably bind to specific prostaglandin-binding proteins within the cells of the target tissue. Suggest important chemical features of these binding proteins—specifically, features of the prostaglandin binding site on the protein.

SUGGESTED READING

General Reference

Bhagavan, N. V. *Biochemistry*, 2nd ed., Ch. 8, pp. 807–72, 909–31; Ch. 10, pp. 1026–54. Philadelphia: J. B. Lippincott, 1978.

Lehninger, A. L. *Biochemistry*, 2nd ed., Chs. 11 and 28. New York: Worth, 1975.

Montgomery, R., Dryer, R. L., Conway, T. W., and Spector, A. A. *Biochemistry: A Case-Oriented Approach*, 3rd ed., Ch. 8, pp. 336–47; Ch. 10. St. Louis: C. V. Mosby, 1980.

Glycosphingolipid Storage Diseases

Bhagavan, N. V. *Biochemistry*, 2nd ed., pp. 857–68. Philadelphia: J. B. Lippincott., 1978.

Membranes

Bretscher, M. S. "Membrane Structure: Some General Principles." *Science* 181 (1973): 622–29.

Capaldi, R. A. "A Dynamic Model of Cell Membranes." *Scientific American* 230 (1974): 26.

Singer S. J., and Nicolson G. L., "The Fluid Mosaic Model of the Structure of Membranes." *Science* 175 (1972): 720–31.

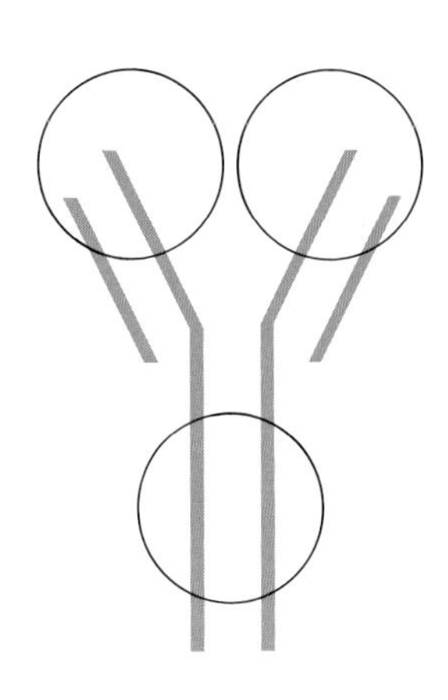

Chapter 14
Lipid Metabolism

14.1 INTRODUCTION

As we showed in the previous chapter, lipids have several important biological functions. Certain lipids serve as structural components of biological membranes, others are vitamins (the fat-soluble vitamins), and still others serve as chemical messengers (the steroid hormones). Certain classes of lipids serve as storage and transport forms of metabolic fuel. These lipids play an extremely important role in cellular metabolism, as they are an energy-rich fuel and large amounts can be stored in specialized fat-storing tissues (adipose tissue). In fact, lipids (in the form of triglycerides) contain about 2.5 times more chemical energy, weight for weight, than do carbohydrates and proteins. Lipids provide up to 40 percent of the total fuel requirement of a person on a normal diet.[1]

In this chapter, we shall discuss how the human body absorbs, transports, and stores dietary lipids. Lipid malabsorption diseases and abnormal lipoprotein biochemistry will also be discussed. The specific enzymatic steps involved in lipid degradation will be noted and compared to the steps involved in lipid biosynthesis. The role of lipids in the development of coronary heart disease will also be presented. Finally, the biosynthesis of cholesterol and phospholipids will be discussed.

14.2 LIPID ABSORPTION AND TRANSPORT

Nearly 40 percent of the total calorie intake of the average American is fat. In fact, many Americans ingest too much fat. As a consequence of this dietary imbalance, excessive fat can accumulate in the adipose tissue of these individuals. The absorption, transport, and storage of dietary lipids is summarized in Figure 14–1.

Because lipids are hydrophobic and, therefore, are insoluble in blood, our bodies require rather complex absorption and transport mechanisms for dealing with them. Consider, for a moment, the magnitude of the problem. How would you design a mechanism by which very insoluble lipids can be absorbed and transported through the aqueous cellular environment and into the blood stream?

After the ingestion of a meal containing lipids (in the form of triglycerides), the stomach contents enter the small intestine. It is here that

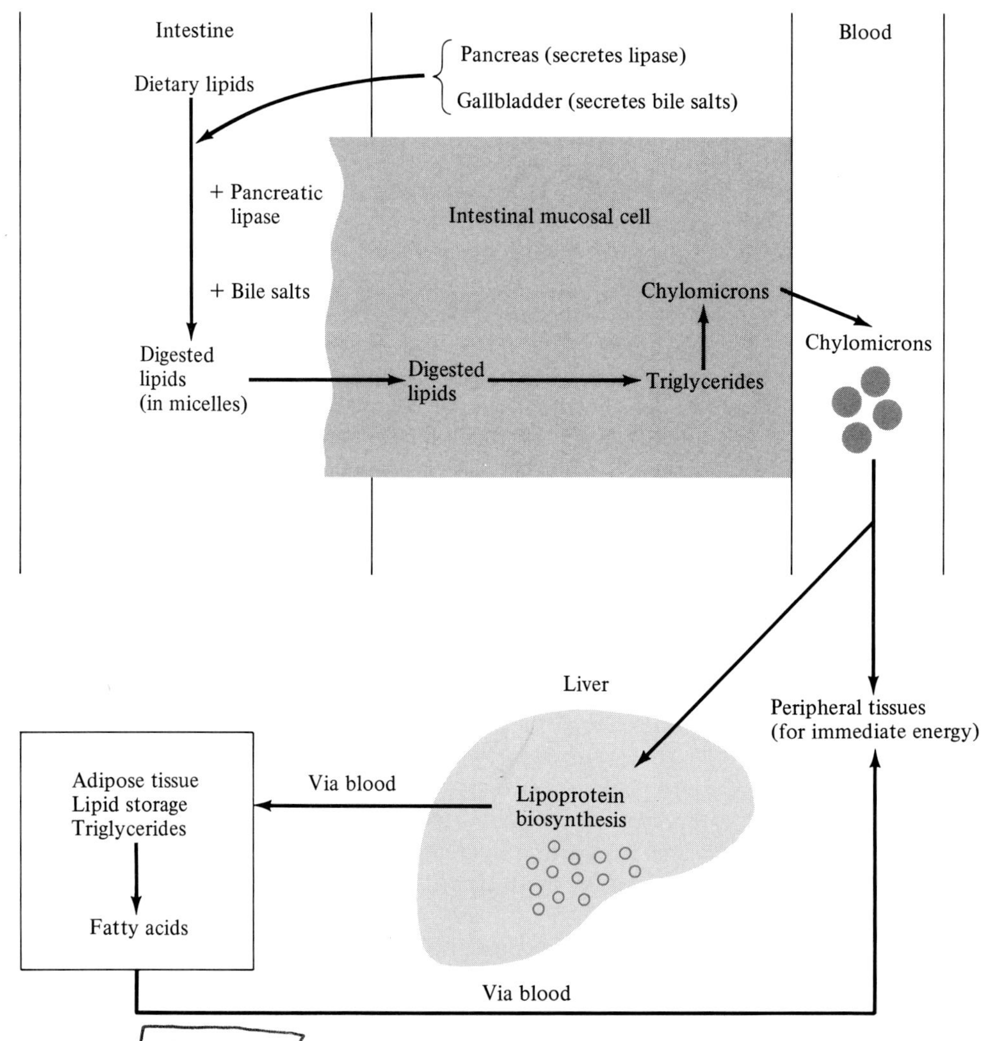

FIGURE 14–1 A summary of lipid absorption, transport, and storage. (Adapted from N. Tietz (ed.), *Fundamentals of Clinical Chemistry*, 2nd ed. (Philadelphia: W. B. Saunders), 1976, p. 484.)

peristalsis

lipid absorption actually takes place. Intestinal **peristalsis** (the mechanical mixing of the intestinal contents) emulsifies the lipids, breaking up the large lipid droplets into smaller structures. In response to the entrance of the stomach contents into the small intestine, the gallbladder and pancreas secrete *gallbladder bile and pancreatic juice*, respectively, into the intestinal lumen (the inner channel of the intestine). Pancreatic juice contains the enzyme *pancreatic lipase*, which degrades dietary triglycerides into monoglycerides and free fatty acids. Gallbladder bile contains bile salts and lecithin, which further emulsify the insoluble and partially digested lipids. If you recall, bile salts and lecithin form micelles (see Chapter 13, Lipids and Membranes). Dietary lipids are rendered soluble in the interior hydrophobic region of these micelles. Figure 14–2 summarizes lipase and bile salt action on dietary lipids.

chylomicrons

The intestinal mucosal cell can now absorb the micelle, along with the partially digested dietary lipids. Within the mucosal cells, the fatty acids and monoglycerides are reassembled into triglycerides. The reformed triglycerides (along with phospholipids, cholesterol esters, and specific proteins) aggregate into large spherical structures called **chylomicrons** (also known as lipoproteins). One can think of a chylomicron as a very large micelle in which the triglycerides are dissolved in the interior hydrophobic region of the structure. The phospholipids and proteins serve as a hydrophilic shell surrounding the hydrophobic core. Thus, chylomicrons serve as a means of transporting the hydrophobic triglycerides in an aqueous environment. Once formed, the chylomicrons travel into the lymphatic system and, finally, into the blood stream. It has been well established that the chylomicron content of blood and lymph increases dramatically after fat ingestion. For example, the turbidity of blood serum increases following the ingestion of a fatty meal, a phenomenon known as *postprandial lipemia*.

The chylomicrons are either (1) transported to the peripheral tissues for immediate energy production, or (2) transported to the liver for use in hepatic (liver) *lipoprotein* biosynthesis. The liver is responsible for the production of three main subclasses of lipoproteins: *very low density lipoproteins (VLDL), low density lipoproteins (LDL), and high density lipoproteins (HDL).* As their names imply, each class of lipoprotein differs from the others in density (in gm/ml). The chylomicrons and VLDLs contain a large percentage of lipids and a low percentage of proteins. Therefore, these substances are very low in density (remember that fats and oils are less dense than water and, therefore, "float" on water). In contrast, low density and high density lipoproteins contain less lipid and more protein and, therefore, are more dense than water. The physical properties of the lipoprotein classes are summarized in Table 14–1.

adipose tissue

After synthesis, the lipoproteins are transported into the **adipose tissue** (via the blood stream) for storage. Adipose tissue is present under the skin, around deep blood vessels, around the organs in the abdominal cavity, and in skeletal muscles. In normal adults, the tissue contains approximately 15 kilograms of triglycerides and, thus, is considered a major body organ.[2]

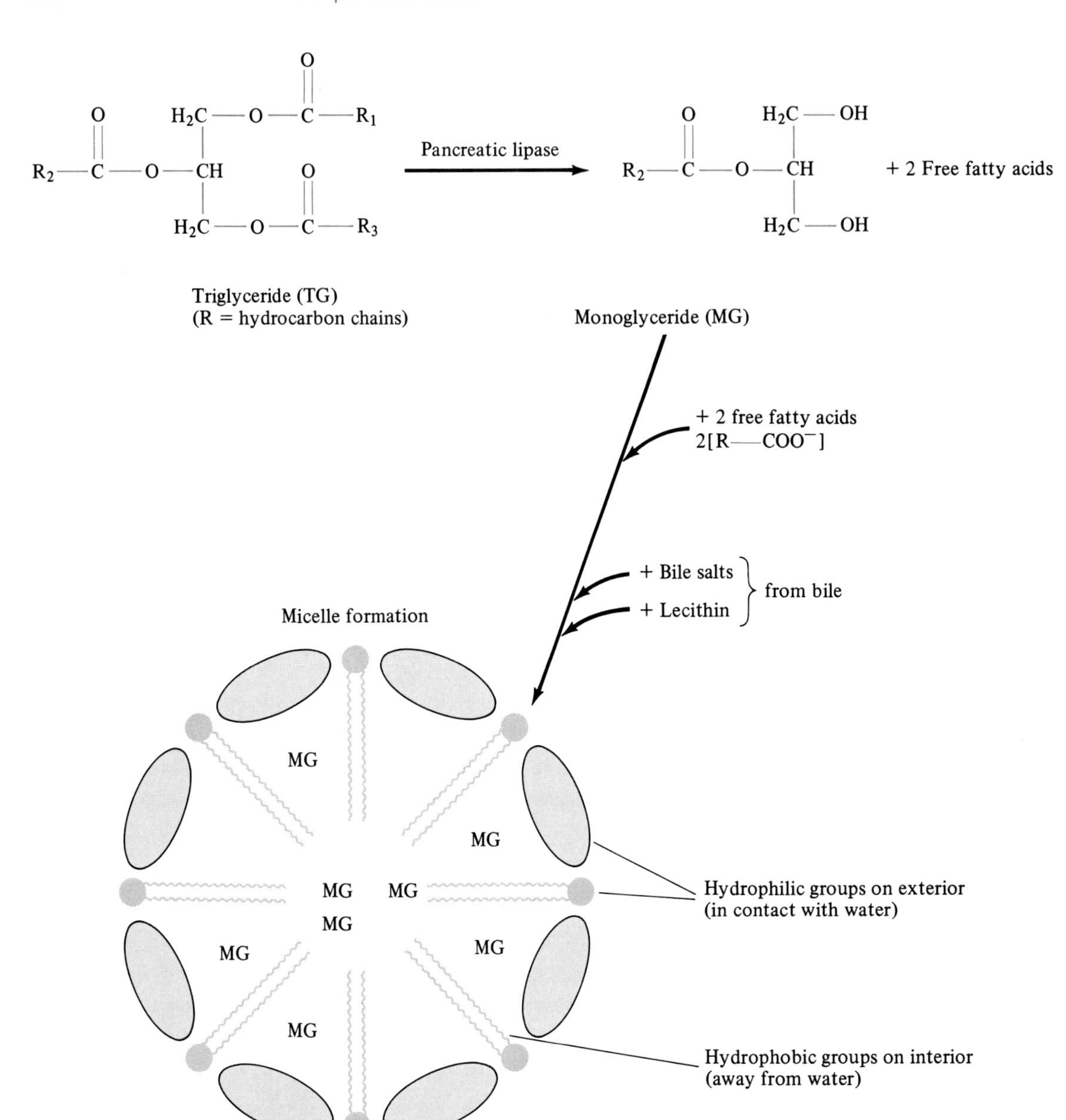

FIGURE 14–2 Triglyceride degradation and emulsification.

The tissue is very metabolically active and is composed of specialized fat cells in which the triglycerides are stored. Figure 14–3 shows a typical adipose tissue cell. Notice the very large fat vacuole that contains the triglycerides. This vacuole occupies a large percentage of the cytoplasm volume.

TABLE 14–1 The Major Lipoprotein Classes.

Lipoprotein Class	Density (gm/ml)	Triglyceride (percent)	Cholesterol (percent)	Protein (percent)
Chylomicron	<0.95	85	5	2
VLDL	0.95–1.006	56	15	10
LDL	1.006–1.063	10	60	25
HDL	1.063–1.21	1–5	20	50

Note: The chylomicrons and the VLDL are largely composed of triglycerides; the LDL and HDL particles contain less triglyceride but more protein. The LDL particle contains a large percentage of cholesterol (as free cholesterol and as cholesterol esters). See Section 14.4, on coronary heart disease, regarding this interesting fact.

Finally, it should be noted that when we need energy, the adipose tissue is mobilized (chemically signaled into action) and degrades triglycerides into free fatty acids (FFA). These free fatty acids are then transported from the adipose tissue to the peripheral tissues for energy production. Free fatty acids are normally bound to *serum albumin*, a protein, for transport through the blood stream.

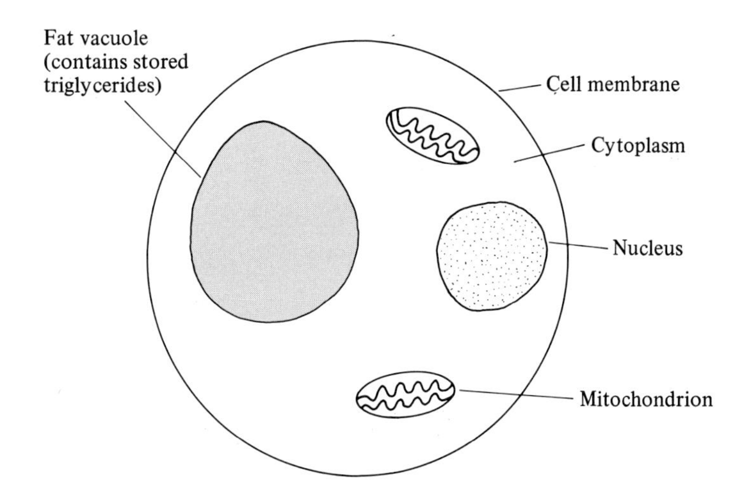

FIGURE 14–3
An adipose tissue cell. Notice the very large fat vacuole, containing stored triglycerides.

14.3 LIPID MALABSORPTION DISEASES

When dietary lipids *are not absorbed* by the small intestine, excessive amounts of lipids will be lost into the feces. This condition is referred to as *steatorrhea.* Disease states that lead to steatorrhea are called **lipid malabsorption** diseases. Defective lipid absorption can result from (1) defective triglyceride degradation in the small intestine, (2) defective intestinal mucosal cell function, leading to impaired formation of triglycerides and chylomicrons, or (3) ab-

lipid malabsorption

normal connections between the organs involved in the lipid absorption and transport processes.

A defect in intestinal triglyceride degradation may result from either a pancreatic lipase deficiency or a bile salt deficiency. A defect in pancreatic lipase activity could result from (1) a malfunctioning or damaged pancreas, (2) obstruction of the pancreatic ducts (by stones or abnormal growths), or (3) a genetic defect in the synthesis of the lipase itself. Whatever may be the problem, the pancreatic lipase is either not being produced by the pancreas, or it is not being secreted into the small intestine. Thus, dietary triglycerides are not degraded into monoglycerides and free fatty acids, and instead pass unchanged through the body. Should a defect occur in hepatic bile salt production or secretion, micelles could not form. This condition would impair absorption by the intestinal mucosal cells.

Defective mucosal cell function leads to impaired resynthesis of triglycerides and synthesis of chylomicrons. Thus, although triglyceride degradation, micelle formation, and associated absorption might occur within the intestinal lumen, chylomicrons could not form. This condition would produce a low blood chylomicron level.

Finally, abnormal connections between organs (resulting from inherited defects) could impair lipid absorption. For example, an abnormal connection between the urinary tract and the lymphatic system of the small intestine would result in the excretion of fatty (milk-white) urine. This

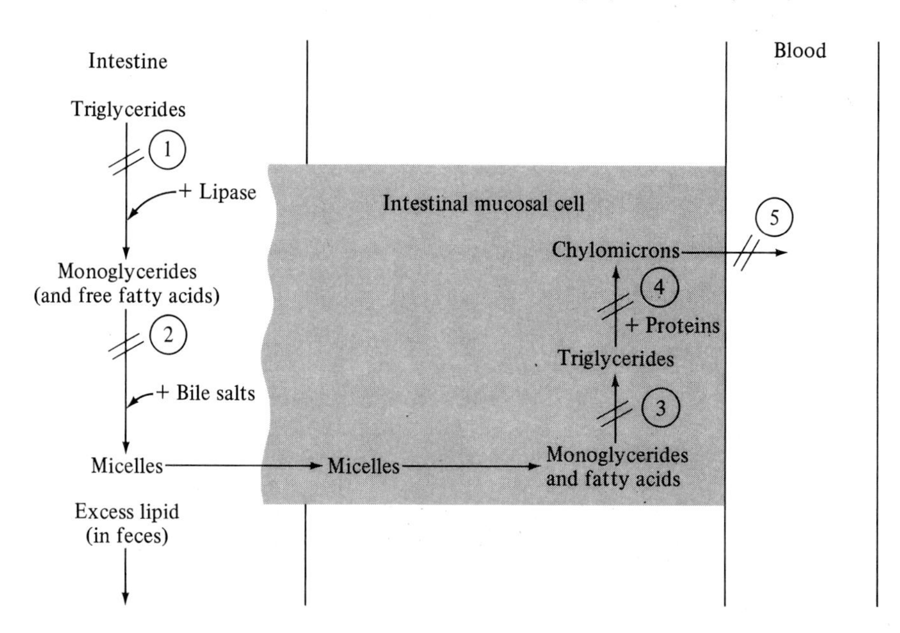

FIGURE 14–4 A summary of defective lipid absorption and transport. Comments: (1) Defective pancreatic lipase activity or secretion. (2) Defective bile salt formation or secretion, leading to defective micelle formation. (3) Defective triglyceride synthesis. (4) Defective chylomicron formation. (5) Defective anatomy.

chyluria condition is known as **chyluria**, and would produce a low blood chylomicron level.

One should note that defects in intestinal fat absorption can also impair the uptake of the fat-soluble vitamins (vitamins A, D, E, and K). A vitamin deficiency will result, thus producing secondary disease states. For example, if vitamin K absorption is impaired, a vitamin K deficiency will result, leading to impaired blood clotting. Figure 14–4 summarizes defects in lipid absorption.

14.4 ABNORMAL LIPOPROTEIN CHEMISTRY AND HEALTH

HYPERLIPOPROTEINEMIAS

As noted in a previous section, the body is responsible for the synthesis of four main classes of lipoproteins: chylomicrons, very low density lipoproteins, low density lipoproteins, and high density lipoproteins. These four classes vary in density, in size, and in lipid/protein composition. They can be separated from each other by size and density or by electrical charge. Abnormal lipoprotein levels are associated with certain disease states. Individuals afflicted with the disease **hyperlipoproteinemia** have elevated serum levels of one or more lipoproteins. There are five subclasses of hyperlipoproteinemias (noted as Types I–V); all result from inherited defects in lipoprotein metabolism. Table 14–2 summarizes the hyperlipoproteinemias and describes the genetic defect.

hyperlipoproteinemia

HDL CHOLESTEROL AND CORONARY HEART DISEASE

Referring to Table 14–2 (the classes of lipoproteins), one should note that the serum low density lipoproteins (LDL) contain a very large percentage

TABLE 14–2 The Hyperlipoproteinemias.

Type	Serum Triglycerides	Serum Cholesterol	Serum Lipoprotein	Genetic Defect(s)
I	Increased	Slightly increased	Increased chylomicron	Deficiency in lipoprotein lipase
II[a]	Normal	Increased	Increased VLDL and LDL	Unknown
III	Increased	Increased	Increased VLDL	Unknown
IV	Increased	Slightly increased	Increased VLDL; few or no chylomicrons	Impaired VLDL metabolism
V	Increased	Increased	Increased chylomicron and VLDL	Multiple genetic defects in in lipoprotein metabolism

[a] Most common of the five types.

TABLE 14–3 HDL Cholesterol and Risk Factor for Coronary Heart Disease

Seum HDL Cholesterol (mg/100 ml)	Risk Factor	
	Men	Women
30	1.82	—
40	1.22	1.94
50	0.82	1.25
60	0.55	0.80
70	—	0.52

Note: The risk factor is a unitless number that indicates the relative risk for coronary heart disease as compared to the national average. For example, if a man has an HDL cholesterol level of 30 mg/100 ml, he is 1.82 times as likely to succumb to coronary heart disease as the national average.
Source: Adapted from "HDL: The Heart of the Matter," Worthington Diagnostics Technical Bulletin, Worthington Diagnostics, Freehold, New Jersey 07728, 1978, p. 2.

of cholesterol and cholesterol esters. Cellular uptake of cholesterol and cholesterol esters can result in atherosclerosis. This probably occurs because the LDL particle binds to arterial cell membranes, with the subsquent deposition of cholesterol onto the arterial wall surface resulting in cholesterol plaque formation. High density lipoproteins (HDL) apparently interfere with the binding of the LDL particles, thus preventing the deposition of cholesterol onto the arterial cell wall.

Recent research (the Framington study) has demonstrated that an *increased level of serum HDL cholesterol* (the cholesterol in the HDL particle) is apparently beneficial to human health (see Table 14–3). The *higher* the HDL cholesterol level (the more cholesterol in the HDL particle), the less risk there is of the individual succumbing to coronary heart disease. In essence, the HDL particle entraps the cholesterol in the interior, hydrophobic region of the particle, thus preventing the cholesterol from forming plaques on the arterial cell membranes.

14.5 LIPID CATABOLISM

Several enzymatic steps must occur in order to mobilize and degrade reserve lipids. These steps are summarized in Figure 14–5.

STEPS 1 AND 2: MOBILIZATION AND TRANSPORT

Since lipids are stored in the adipose tissue, they must first be mobilized from this tissue and transported to the peripheral tissues for further de-

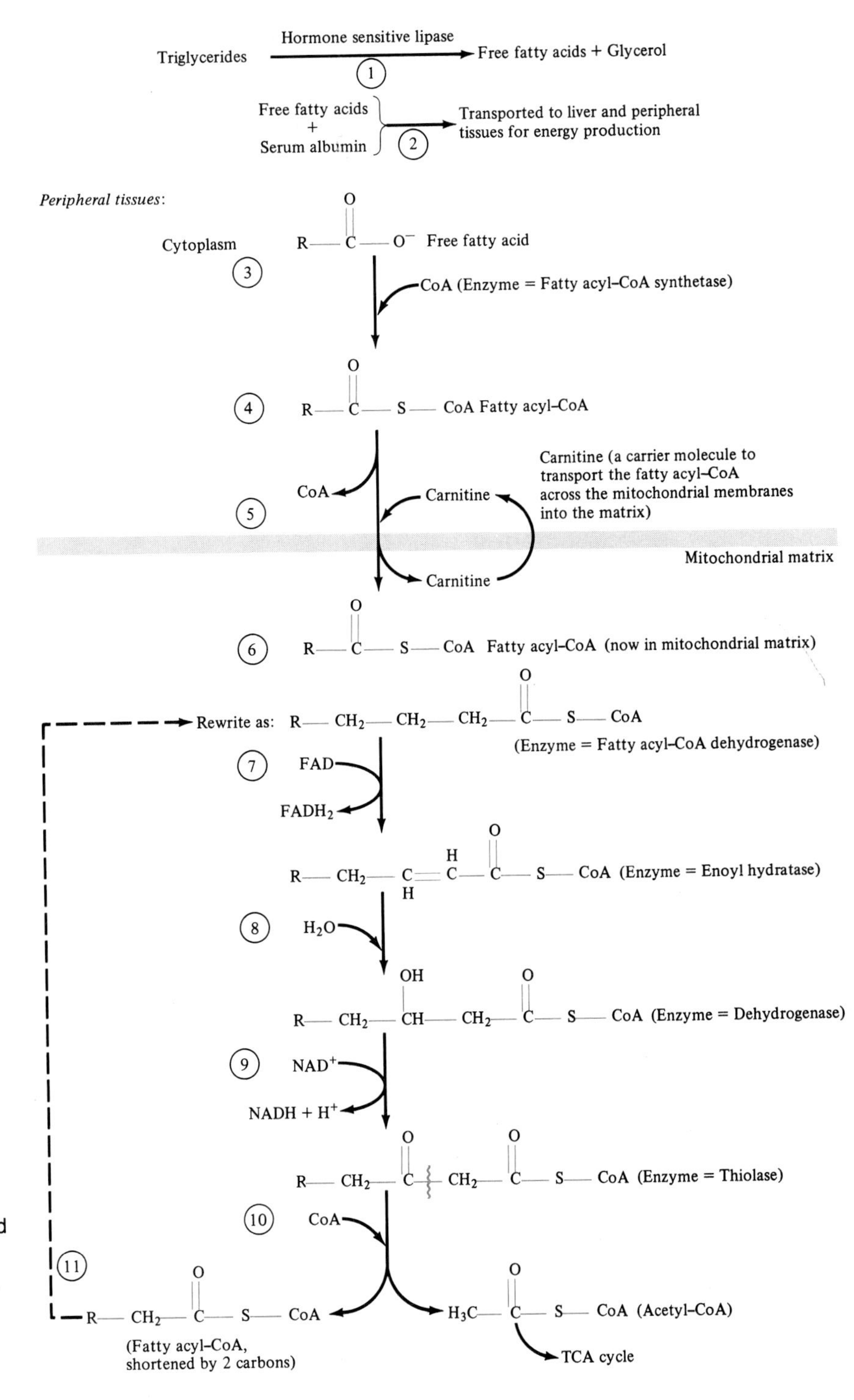

FIGURE 14–5
A summary of fatty acid degradation (α-oxidation). (Adapted from N. V. Bhagavan, *Biochemistry*, 2nd ed. (Philadelphia: J. B. Lippincott, 1978), p. 878.)

gradation and ATP production. Triglyceride mobilization occurs when *hormone sensitive lipase* degrades the triglycerides into glycerol and free fatty acids. The free fatty acids are subsequently bound to *serum albumin* for transport from the adipose tissue to the peripheral tissues. The hormone sensitive lipase is activated by *protein kinase* (see Chapter 10). The protein kinase activity is, in turn, regulated by the hormone epinephrine—hence the name. *Hormone insensitive lipase*, also present in adipose tissue, is not under hormonal control.

After the fatty acids are transported in the bloodstream to the peripheral tissues, several additional catabolic steps must occur. These include (1) transport of the fatty acid into the cell, (2) activation of the fatty acid in the cytoplasm, (3) transport of the activated fatty acid (fatty acyl-CoA) into the mitochondrial matrix, (4) oxidation of the fatty acid intermediates to yield $FADH_2$ and NADH, and (5) the sequential cleavage of the fatty acid chain to yield acetyl-CoA. The electrons of $FADH_2$ and NADH enter the electron transport system and drive ATP synthesis. Finally, the acetyl-CoA formed during fatty acid degradation enters the tricarboxylic acid cycle and generates additional NADH and $FADH_2$ molecules for ATP synthesis.

STEP 3: ACTIVATION

Upon transport of the fatty acid into the cytoplasm of the cell, the fatty acid must be activated by CoA and ATP. This is referred to as the priming step, and is catalyzed by the enzyme *acyl-CoA synthetase* according to the following reaction:

$$\underset{\text{Free fatty acid}}{\text{R-COO}^-} + \text{ATP} + \text{CoA-SH} \xrightarrow{\text{Acyl-CoA synthetase}} \underset{\substack{\text{Fatty acyl-CoA} \\ \text{(a thioester)}}}{\text{R}-\overset{\overset{\text{O}}{\|}}{\text{C}}-\text{S}-\text{CoA}} + \text{AMP} + \text{PP}_i$$

The term "acyl" refers to the long hydrocarbon chain of the fatty acid. The reaction favors the formation of fatty acyl-CoA. This is because ATP undergoes *pyrophosphate cleavage* to yield AMP and inorganic pyrophosphate (PP_i). The PP_i is, in turn, cleaved into two molecules of inorganic phosphate (P_i) by the enzyme *inorganic pyrophosphatase*, favoring product formation. In essence, two high-energy phosphate bonds are utilized to activate one fatty acid molecule. Small-, medium-, and long-chain acyl-CoA synthetases exist, each specific for a given range of fatty-acid chain length.

STEP 4: CARRIER FORMATION

Since long-chain fatty acids cannot easily cross the inner mitochondrial membrane, a carrier molecule is required to transport the fatty acid across the membrane and into the matrix of the mitochondrion. This molecule is called *carnitine*. The formation of *fatty acyl carnitine* is catalyzed by the enzyme *carnitine acyl transferase* according to the following reaction:

$$\underset{\text{Carnitine}}{H_3C-\overset{CH_3}{\underset{CH_3}{\overset{+}{N}}}-CH_2-\underset{OH}{CH}-CH_2-COO^-} + \underset{\text{Fatty acyl-CoA}}{R-\overset{O}{\overset{\|}{C}}-S-CoA} \xrightarrow{\text{Carnitine acyl-transferase}} \underset{\text{Fatty acyl carnitine}}{H_3C-\overset{CH_3}{\underset{CH_3}{\overset{+}{N}}}-CH_2-\underset{O-\overset{O}{\overset{\|}{C}}-R}{CH}-CH_2-COO^-}$$

Fatty acid links here (at the OH of carnitine)

STEP 5: TRANSPORT INTO MITOCHONDRION

The fatty acyl carnitine molecule is transferred from the cytoplasm into the mitochondrion via the carnitine carrier molecule. This *compartmentalization* means that acetyl-CoA, $FADH_2$, and NADH are all formed and trapped in the mitochondrial matrix, *where they can be used for ATP production by mitochondrial enzymes.*

STEP 6: FORMATION OF MITOCHONDRIAL FATTY ACYL-CoA

Once the fatty acyl carnitine molecule has entered the mitochondrion, the carnitine is hydrolyzed from the fatty acid and fatty acyl-CoA is reformed. The carnitine then diffuses back into the cytoplasm for reuse. The fatty acyl-CoA is next degraded by a series of mitochondrial enzymes that ultimately split off one acetyl-CoA molecule, thus shortening the fatty acyl-CoA by two carbons.

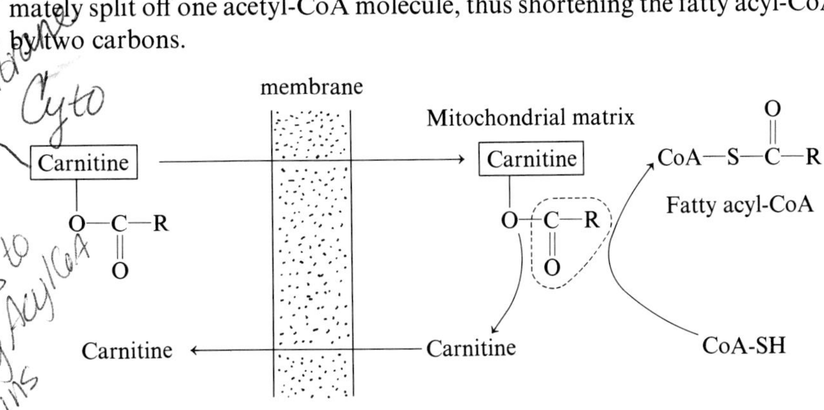

STEP 7: FIRST DEHYDROGENATION

The enzyme *fatty acyl-CoA dehydrogenase* removes two hydrogens (and therefore two electrons) from the fatty acid carbons. Since electrons are being removed, this process is referred to as oxidation. Because the β-carbon atom of the fatty acid is being oxidized (that is, electrons are being removed

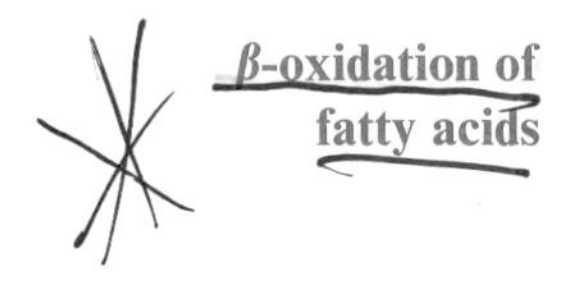

β-oxidation of fatty acids

from it), the entire process is often called the **β-oxidation of fatty acids**. The cofactor FAD is the hydrogen/electron acceptor; thus, $FADH_2$ is formed. The electrons of $FADH_2$ enter the electron transport system and produce ATP.

$$\underset{\text{Fatty acyl-CoA}}{R{-}\underset{\beta}{CH_2}{-}\underset{\alpha}{CH_2}{-}\overset{O}{\overset{\|}{C}}{-}S{-}CoA} \xrightarrow[\text{FAD} \quad \text{FADH}_2]{\text{Fatty acyl-CoA dehydrogenase}} \underset{\text{Enoyl-CoA}}{R{-}\underset{\beta}{CH}{=}\underset{\alpha}{CH}{-}\overset{O}{\overset{\|}{C}}{-}S{-}CoA}$$

STEP 8: HYDRATION

The next step involves the addition of a water molecule across the carbon-carbon double bond by the enzyme *enoyl-CoA hydratase*.

$$\underset{\text{Enoyl-CoA}}{R{-}CH{=}CH{-}\overset{O}{\overset{\|}{C}}{-}S{-}CoA} \xrightarrow[H_2O]{\text{Enoyl-CoA hydratase}} \underset{\text{3-Hydroxyl-acyl-CoA}}{R{-}\underset{3}{\overset{OH}{\overset{|}{CH}}}{-}\underset{2}{CH_2}{-}\underset{1}{\overset{O}{\overset{\|}{C}}}{-}S{-}CoA}$$

STEP 9: SECOND DEHYDROGENATION

A second dehydrogenation reaction (oxidation) next occurs, catalyzed by the enzyme *hydroxyacyl-CoA dehydrogenase*. Two hydrogens (and, therefore, two electrons) are removed; NAD^+ is the electron acceptor. The NADH formed from this reaction transfers its electrons to the electron transport system for ATP synthesis.

$$\underset{\text{3-Hydroxyacyl-CoA}}{R{-}\underset{3}{\overset{OH}{\overset{|}{CH}}}{-}\underset{2}{CH_2}{-}\underset{1}{\overset{O}{\overset{\|}{C}}}{-}S{-}CoA} \xrightarrow[NAD^+ \quad NADH + H^+]{\text{3-Hydroxy acyl-CoA dehydrogenase}} \underset{\text{3-Ketoacyl-CoA}}{R{-}\underset{3}{\overset{O}{\overset{\|}{C}}}{-}\underset{2}{CH_2}{-}\underset{1}{\overset{O}{\overset{\|}{C}}}{-}S{-}CoA}$$

STEP 10: CLEAVAGE

The ketoacyl-CoA intermediate is now ready for *thiolytic cleavage*, catalyzed by the enzyme *thiolase*. In this reaction, one acetyl-CoA molecule is cleaved off, leaving the fatty acid molecule two carbons shorter. Since coenzyme A (CoA) is also employed in this reaction, a new fatty acyl-CoA molecule is formed. A large standard free-energy change ($\Delta G^{\circ\prime} = -6.7$ kcal/mol) is observed for this very spontaneous reaction; therefore, cleavage is favored.

$$\underset{\text{3-Ketoacyl-CoA}}{R{-}\overset{O}{\overset{\|}{C}}{\wr}CH_2{-}\overset{O}{\overset{\|}{C}}{-}S{-}CoA} \xrightarrow[\text{CoA-SH}]{\text{Thiolase}} \underset{\text{Fatty acyl-CoA}}{R{-}\overset{O}{\overset{\|}{C}}{-}S{-}CoA} + \underset{\text{Acetyl-CoA}}{H_3C{-}\overset{O}{\overset{\|}{C}}{-}S{-}CoA}$$

STEP 11: SEQUENTIAL DEGRADATION

The remaining fatty acyl-CoA molecule (now shorter by two carbons) is recycled through the dehydrogenase/hydration/dehydrogenation/thiolase steps (steps 7–10), yielding additional $FADH_2$, NADH, and acetyl-CoA molecules. Thus, the very long hydrocarbon chain of a fatty acid molecule is *sequentially degraded*, two carbons at a time, via the β-oxidation pathway. For example:

Fatty acyl-CoA degradation via β-oxidation

$$H_3C{-}CH_2 \wr CH_2{-}CH_2 \wr CH_2{-}CH_2 \wr CH_2{-}\overset{O}{\overset{\|}{C}}{-}S{-}CoA$$

β-oxidation

1 acetyl-CoA	1 acetyl-CoA	1 acetyl-CoA	1 acetyl-CoA
+ $FADH_2$	+ $FADH_2$	+ $FADH_2$	+ $FADH_2$
+ NADH	+ NADH	+ NADH	+ NADH

Table 14–4 summarizes the number of ATP molecules that can be generated from one fatty acid molecule. As you can see, the number is quite large.

TABLE 14–4 How Many ATP Molecules can be Generated from One Triglyceride Molecule?

Assumption: We are degrading tripalmitin, which consists of 3 molecules of palmitic acid linked to a glycerol backbone. Each molecule of palmitic acid can be degraded to 8 molecules of acetyl-CoA, which can enter the TCA cycle. (The production of ATP from the glycerol backbone is ignored in this calculation.)

Reaction	Yield of ATP/7 cycles of β-oxidation
1st dehydrogenation = 7 $FADH_2$ (2 ATP/$FADH_2$)	14
2nd dehydrogenation = 7 NADH (3 ATP/NADH)	21
8 acetyl-CoA produced (12 ATP/acetyl-CoA)	96
	131
Initial priming step = 2 high-energy bonds (ATP $\rightarrow$ AMP + $PP_i \rightarrow 2\ P_i$)	−2
	129 ATP/palmitic acid

Since there are 3 palmitic acid molecules per tripalmitin molecule,

Net yield of ATP/tripalmitin = 3 × 129 = 387 ATP

Compare this figure with the number of ATP molecules generated for each glucose molecule during aerobic glycolysis. Now can you see why lipids are considered a very energy-rich fuel?

Finally, it should be noted that *unsaturated fatty acids* can also be degraded to yield acetyl-CoA. An *isomerase* catalyses the change in location of the double bond within the fatty acid molecule so that β-oxidation can proceed unhindered.

14.6 LIPID ANABOLISM

fatty acid biosynthesis

Fatty acid biosynthesis (lipid anabolism) takes place primarily in the liver and in adipose tissue. It occurs when *excess* carbohydrates and amino acids are present and when these substances are not needed for ATP production. Since both carbohydrates and amino acids are degraded to acetyl-CoA, the body can utilize this *excess* acetyl-CoA for fatty acid biosynthesis.

Fatty acids are synthesized from acetyl-CoA by a complex sequence of enzymatic steps, summarized in Figure 14–6. One should note that the biosynthesis of fatty acids *is not* simply a reversal of the fatty acid degradation pathway. Different enzyme systems and cofactors are employed, and the pathway is segregated in the cytoplasm of the cell. (Anabolic pathways are usually segregated from catabolic pathways within the cell.) Table 14–5 summarizes the differences between fatty acid biosynthesis and fatty acid degradation.

The carbon atoms of a fatty acid are ultimately derived from acetyl-CoA. This acetyl-CoA is normally formed in the mitochondrion during the degradation of carbohydrates or amino acids. Since the acetyl-CoA cannot pass out of the mitochondrion into the cytoplasm, it must first be converted into citrate, which can then be transported out of the mitochondrion. The citrate (now in the cytoplasm) is cleaved by the *citrate cleavage enzyme* to yield *cytoplasmic acetyl-CoA* and oxaloacetate (Figure 14–7).

STEP 1: MALONYL-CoA SYNTHESIS

Malonyl-CoA is formed from acetyl-CoA and bicarbonate (HCO_3^-) by the enzyme *acetyl-CoA carboxylase*. This cytoplasmic enzyme utilizes the

TABLE 14–5 A Summary of the Differences between Fatty Acid Biosynthesis and Fatty Acid Degradation.

Property	Biosynthesis	Degradation
Location	Cytoplasm	Mitochondrial matrix
Enzymes	Multienzyme complex	Separate enzymes
Acyl group carrier	Acyl-carrier protein (ACP-SH)	Coenzyme A (CoA-SH)
Cofactors	NADPH used	NADH (and $FADH_2$) made

FIGURE 14–6 (opposite) A summary of fatty acid biosynthesis.

Step 1: Malonyl–CoA synthesis:

$$\underset{\text{Acetyl–CoA}}{H_3C-\overset{\overset{\displaystyle O}{\|}}{C}-S-CoA} + \boxed{HCO_3^-} \xrightarrow[\text{ATP} \quad \text{ADP} + P_i]{\text{Acetyl–CoA carboxylase}} \underset{\text{Malonyl–CoA}}{\boxed{^-OOC}-CH_2-\overset{\overset{\displaystyle O}{\|}}{C}-S-CoA}$$

Step 2: The fatty acid synthetase complex and priming reaction:

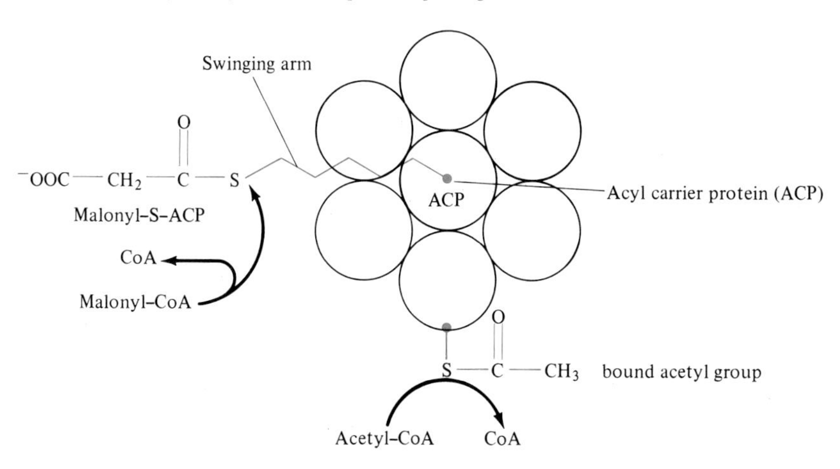

Step 3: Condensation and reduction reactions:

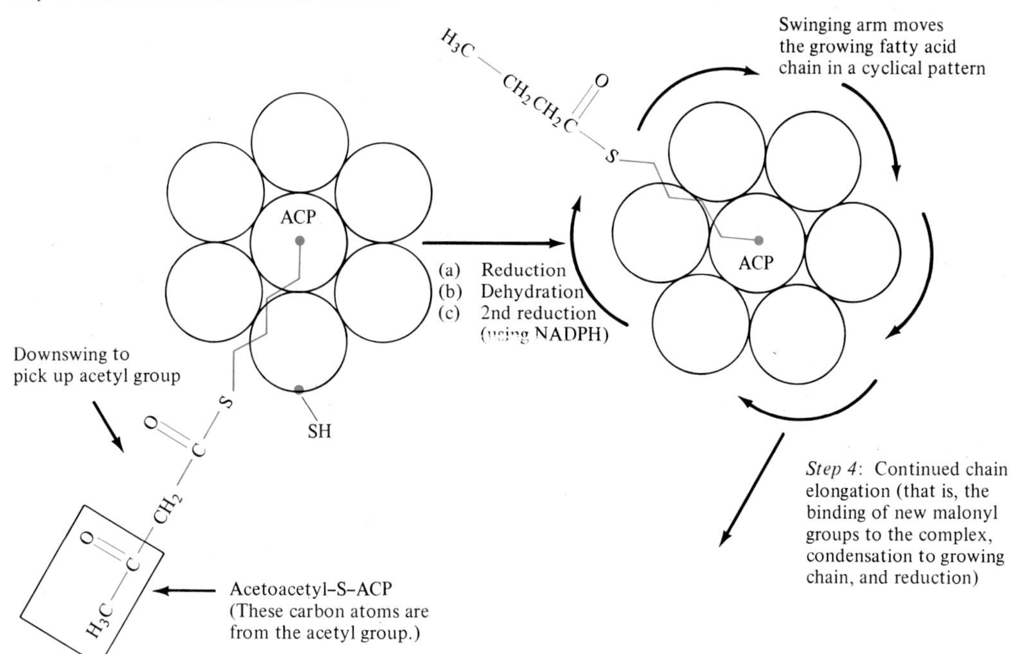

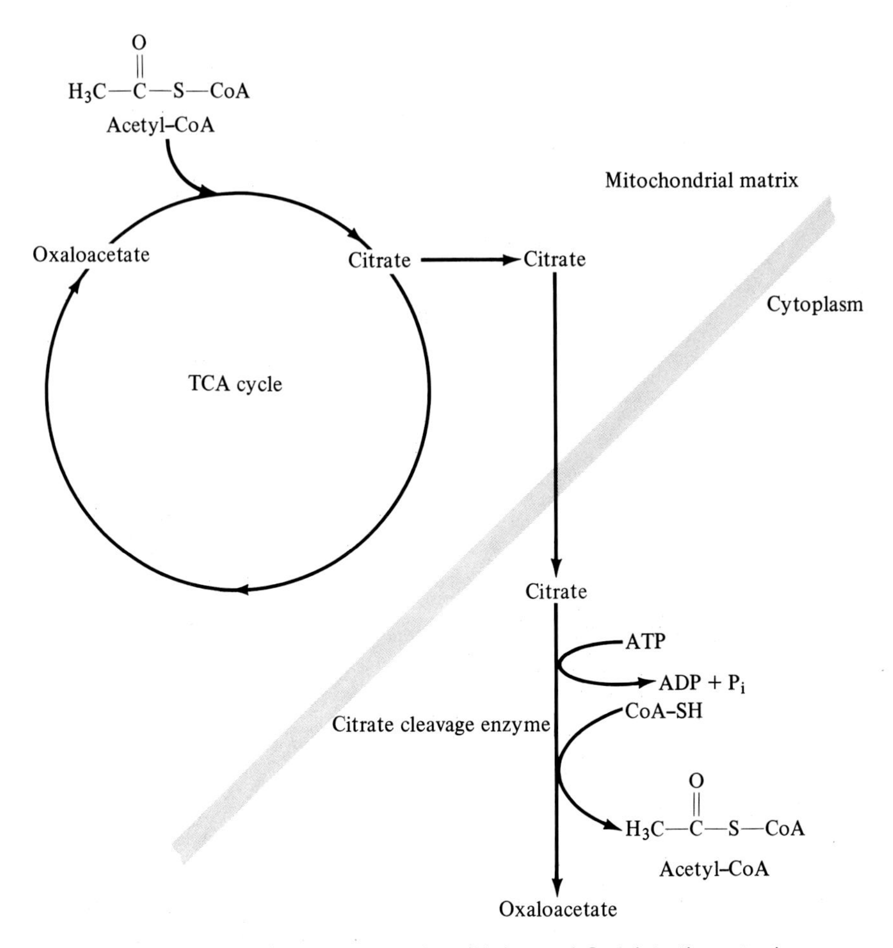

FIGURE 14–7 The transport of mitochondrial acetyl-CoA into the cytoplasm.

carboxylation reaction

vitamin cofactor biotin and catalyzes a **carboxylation reaction** (the addition of CO_2 to a substrate). In this reaction, the CO_2 group is added to the methyl group of acetyl-CoA. Acetyl-CoA carboxylase is a regulatory enzyme and responds to positive (citrate) and negative (fatty acyl-CoA) regulatory effectors. This step is considered the rate-limiting reaction in fatty acid biosynthesis.

$$H_3C-\overset{O}{\overset{\|}{C}}-S-CoA + HCO_3^- \xrightarrow[\text{ATP} \rightarrow \text{ADP} + P_i]{\text{Acetyl-CoA carboxylase}} {}^-OOC-CH_2-\overset{O}{\overset{\|}{C}}-S-CoA$$

Cytoplasmic acetyl-CoA — Malonyl-CoA (^-OOC from HCO_3^-)

The swinging arm of the acyl carrier protein

$$\text{CH}-\text{CH}_2-\text{O}-\underset{\text{O}^-}{\overset{\text{O}}{\overset{\|}{\text{P}}}}-\text{O}-\text{CH}_2-\underset{\text{CH}_3}{\overset{\text{CH}_3}{\text{C}}}-\overset{\text{OH}}{\text{CH}}-\overset{\text{O}}{\overset{\|}{\text{C}}}-\text{NH}-\text{CH}_2-\text{CH}_2-\overset{\text{O}}{\overset{\|}{\text{C}}}-\text{NH}-\text{CH}_2-\text{CH}_2-\text{SH}$$

Polypeptide chain of ACP

Malonyl binds here

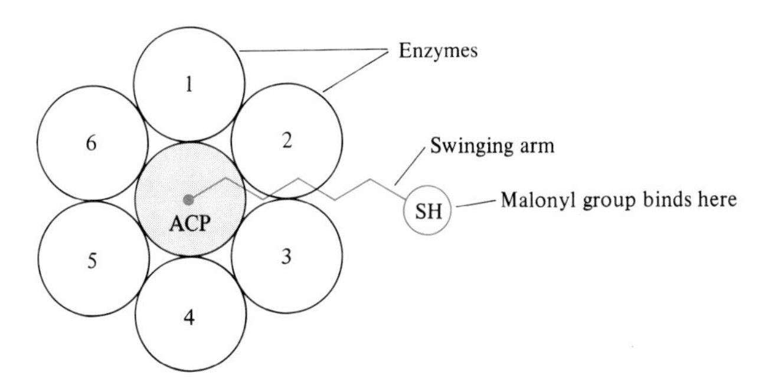

The fatty-acid synthetase complex

FIGURE 14–8 The fatty-acid synthetase complex and the "swinging arm."

STEP 2: THE FATTY-ACID SYNTHETASE COMPLEX AND THE PRIMING REACTION

fatty-acid synthetase complex

The **fatty-acid synthetase complex** (mammalian) is a multienzyme complex found in the cytoplasm of the cell. It contains six different enzymes (2 enzymes covalently-linked in subunit A and 4 enzymes covalently-linked in subunit B) and has a molecular weight of 400,000 daltons. Within the fatty-acid synthetase complex is a protein entitled *acyl carrier protein* (*ACP*). A sulfhydryl derivative of *pantothenic acid* is covalently bound to the polypeptide chain of the ACP and serves as a swinging arm carrying malonyl units and other fatty acid intermediates during fatty acid biosynthesis (Figure 14–8). If you recall, pantothenic acid (a vitamin) is also part of Coenzyme A (see Chapter 11, The Tricarboxylic Acid Cycle). The swinging arm of the fatty-acid synthetase complex functions much like the swinging arm of the pyruvate dehydrogenase complex; it binds various intermediates and moves them from one enzyme to another.

An acetyl-CoA molecule is first bound to the ACP swinging arm and transferred to one of the enzymes of the fatty acid complex. This is called the *priming reaction.* The ACP swinging arm next picks up a molecule of malonyl-CoA. When either malonyl-CoA or acetyl-CoA bind to the ACP swinging arm, the CoA is released into the cytoplasm.

STEP 3: CONDENSATION AND HYDROGENATION REACTIONS

Once the fatty-acid synthetase complex has been primed with bound malonyl and acetyl groups, a condensation reaction occurs in which the acetyl group is transferred to the malonyl-S-ACP, releasing the free carboxyl group as CO_2 (a decarboxylation reaction). The decarboxylation reaction drives condensation; hence, chain elongation is favored. The molecule of CO_2 contains the carbon atom that was introduced as HCO_3^- in the acetyl-CoA carboxylase reaction. The product of the condensation reaction is called *acetoacetyl-S-ACP* and is now four carbons in length (Figure 14–9).

The acetoacetyl-S-ACP group is next moved by the swinging arm to the active sites of the other enzymes of the complex. The acetoacetyl-S-ACP undergoes reduction (the addition of hydrogens), dehydration (the removal of a water molecule), and a second reduction reaction. The product of this series of reactions is called *butyryl-S-ACP* (Figure 14–10). The proton/electron donor for the reduction steps is the coenzyme **NADPH** (*n*icotinamide *a*denine *d*inucleotide *p*hosphate, reduced form).

NADPH

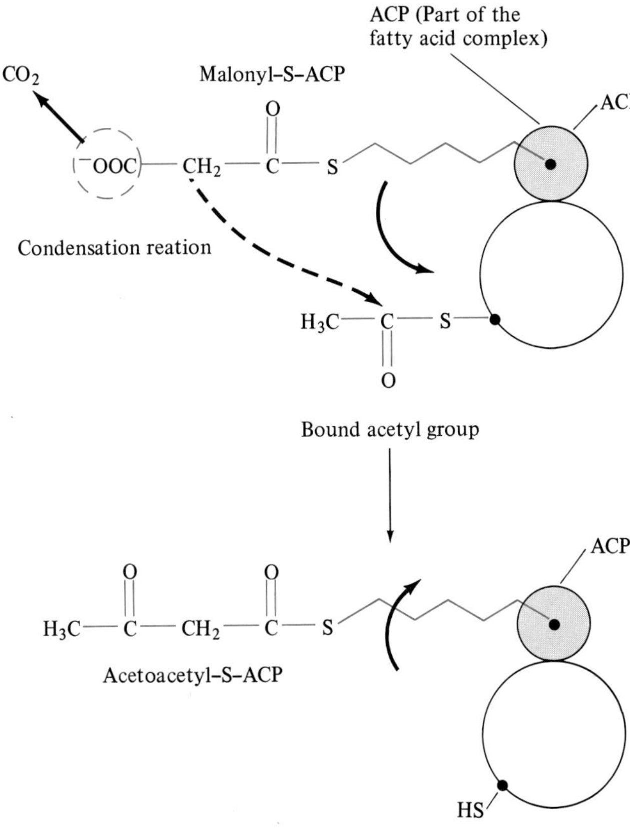

ACP (Part of the fatty acid complex)

CO_2

Malonyl–S–ACP

ACP

^{-}OOC—CH_2—C(=O)—S

Condensation reation

H_3C—C(=O)—S

Bound acetyl group

ACP

H_3C—C(=O)—CH_2—C(=O)—S

Acetoacetyl–S–ACP

HS

FIGURE 14–9
The condensation reaction between malonyl-S-ACP and the bound acetyl group.

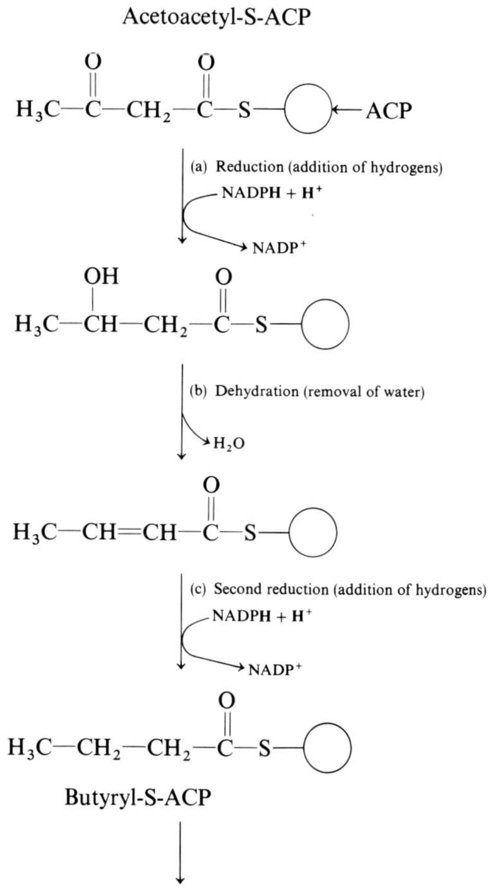

Acetoacetyl-S-ACP

$$H_3C-\overset{O}{\overset{\|}{C}}-CH_2-\overset{O}{\overset{\|}{C}}-S-\bigcirc \leftarrow ACP$$

(a) Reduction (addition of hydrogens)

$NADPH + H^+ \rightarrow NADP^+$

$$H_3C-\overset{OH}{\overset{|}{CH}}-CH_2-\overset{O}{\overset{\|}{C}}-S-\bigcirc$$

(b) Dehydration (removal of water)

$\rightarrow H_2O$

$$H_3C-CH{=}CH-\overset{O}{\overset{\|}{C}}-S-\bigcirc$$

(c) Second reduction (addition of hydrogens)

$NADPH + H^+ \rightarrow NADP^+$

$$H_3C-CH_2-CH_2-\overset{O}{\overset{\|}{C}}-S-\bigcirc$$

Butyryl-S-ACP

↓

(Ready for condensation with a second molecule of malonyl-S-ACP and a repeat of above steps.)

FIGURE 14–10
A summary of the reduction and dehydration steps of fatty acid synthesis.

STEP 4: CHAIN ELONGATION

Once the 4-carbon intermediate is formed, the fatty acyl chain can be further elongated with repeated cycles of the fatty-acid synthetase complex. Specifically, condensation with another malonyl group will occur, followed by decarboxylation, reduction, dehydration, and a second reduction. Thus, the fatty acid chain is elongated until a length of 16 carbons is reached. Once a 16-carbon chain is formed, the fatty acid is cleaved from the complex to yield *fatty acyl-CoA*. Further elongation of the chains, or the formation of unsaturated fatty acids, is done by other enzyme systems.

ESSENTIAL FATTY ACIDS

Humans (and other mammals) cannot synthesize certain unsaturated fatty acids, such as *linoleic acid* and *linolenic acid*. Because these substances must

essential fatty acids

be obtained in our diets, they are called **essential fatty acids**. The structures of linoleic acid and linolenic acid are shown in the following diagram:

Linoleic acid (18:2) $H_3C(CH_2)_4CH{=}CH{-}CH_2{-}CH{=}CH{-}(CH_2)_7{-}COO^-$

Linolenic acid (18:3)

$$H_3C{-}CH_2{-}CH{=}CH{-}CH_2{-}CH{=}CH{-}CH_2{-}CH{=}CH(CH_2)_7COO^-$$

Notice that these fatty acids have a double bond located between the terminal methyl group and the sixth carbon from that group. Mammals cannot introduce double bonds in these regions.

TRIGLYCERIDE BIOSYNTHESIS

Once fatty acyl-CoA molecules have been formed and separated from the fatty-acid synthetase complex, mono-, di-, and triglyceride formation can occur according to the following reaction scheme:

Glycerol phosphate ($H_2C{-}OH$, $HO{-}C{-}H$, $H_2C{-}O{-}$Ⓟ) + $R{-}C({=}O){-}CoA$ (1st fatty acyl-CoA) —Enzyme→ CoA + $H_2C{-}O{-}C({=}O){-}R_1$, $HO{-}CH$, $H_2C{-}O{-}$Ⓟ

+ $R{-}C({=}O){-}S{-}CoA$ (2nd fatty acyl-CoA) —Enzyme→ CoA + $H_2C{-}O{-}C({=}O){-}R_1$, $R_2{-}C({=}O){-}O{-}C{-}H$, $H_2C{-}O{-}$Ⓟ

+ $R{-}C({=}O){-}S{-}CoA$ (3rd fatty acyl-CoA) —Enzyme→ CoA, P_i + A triglyceride ($H_2C{-}O{-}C({=}O){-}R_1$, $R_2{-}C({=}O){-}O{-}C{-}H$, $H_2C{-}O{-}C({=}O){-}R_3$)

PHOSPHOLIPID BIOSYNTHESIS

cytidine triphosphate (CTP)

The phospholipids are synthesized from *diglycerides* and a unique intermediate called **cytidine triphosphate (CTP)**. *Phosphatidylethanolamine*, *phosphatidylserine*, and *phosphatidylcholine* are representative examples (see Chapter 13, Lipids and Membranes). We will only consider the biosynthesis of phosphatidylethanolamine.

Phosphatidylethanolamine is synthesized according to a reaction scheme in which *ethanolamine* is phosphorylated by ATP and a kinase, yielding *phosphoethanolamine*. The phosphoethanolamine then reacts with cytidine triphosphate (CTP) to yield cytidine diphosphate ethanolamine. Cytidine triphosphate is similar to ATP in that it has a high-energy triphosphate "tail" capable of undergoing hydrolysis with the release of free energy.

$$HO{-}CH_2{-}CH_2{-}NH_2 \xrightarrow[\text{ATP} \to \text{ADP}]{\text{Ethanolamine kinase}} {}^{-}O{-}\overset{\overset{\displaystyle O}{\|}}{\underset{\underset{\displaystyle O^-}{|}}{P}}{-}O{-}CH_2{-}CH_2{-}NH_2$$

Ethanolamine

Phosphoethanolamine

$${}^{-}O{-}\overset{\overset{\displaystyle O}{\|}}{\underset{\underset{\displaystyle O^-}{|}}{P}}{-}O{-}CH_2{-}CH_2{-}NH_2 \xrightarrow[\boxed{\text{CTP}} \to \text{PP}_i]{\text{Phosphoethanolamine cytidyltransferase}}$$

Phosphoethanolamine

Cytidine *trip*hosphosphate (similar to ATP)

$$H_2N{-}CH_2{-}CH_2{-}O{-}\overset{\overset{\displaystyle O}{\|}}{\underset{\underset{\displaystyle {}^-O}{|}}{P}}{-}O{-}\overset{\overset{\displaystyle O}{\|}}{\underset{\underset{\displaystyle {}^-O}{|}}{P}}{-}O{-}CH_2\text{(ribose, OH OH)}{-}\boxed{\text{Cytosine}}$$

From phosphoethanolamine

Cytidine diphosphate ethanolamine
(CDP-ethanolamine)

Finally, CDP-ethanolamine reacts with a diglyceride to yield phosphatidylethanolamine:

$$\boxed{\text{CDP-ethanolamine}} + \begin{array}{l} \quad\quad\quad\quad\ \ H_2C{-}O{-}\overset{\overset{\displaystyle O}{\|}}{C}{-}R_1 \\ R_2{-}\overset{\overset{\displaystyle O}{\|}}{C}{-}O{-}CH \\ \quad\quad\quad\quad\ \ H_2C{-}OH \end{array} \xrightarrow[\to \text{CMP}]{\text{Phosphatidyl ethanolamine transferase}}$$

A diglyceride

$$\longrightarrow \begin{array}{l} \quad\quad\quad\quad\ \ H_2C{-}O{-}\overset{\overset{\displaystyle O}{\|}}{C}{-}R_1 \\ R_2{-}\overset{\overset{\displaystyle O}{\|}}{C}{-}O{-}CH \\ \quad\quad\quad\quad\ \ H_2C{-}O{-}\overset{\overset{\displaystyle O}{\|}}{\underset{\underset{\displaystyle O^-}{|}}{P}}{-}O{-}CH_2{-}CH_2{-}NH_2 \end{array}$$

Phosphatidylethanolamine

14.7 CHOLESTEROL BIOSYNTHESIS

Normally, we synthesize about 1.5–2.0 gm of cholesterol per day, for the most part in the liver (1.0–1.5 gm/day). As we mentioned previously (Chapter 13, Lipids and Membranes), cholesterol is used in constructing biological membranes and is required for synthesizing the bile salts and the steroid hormones.

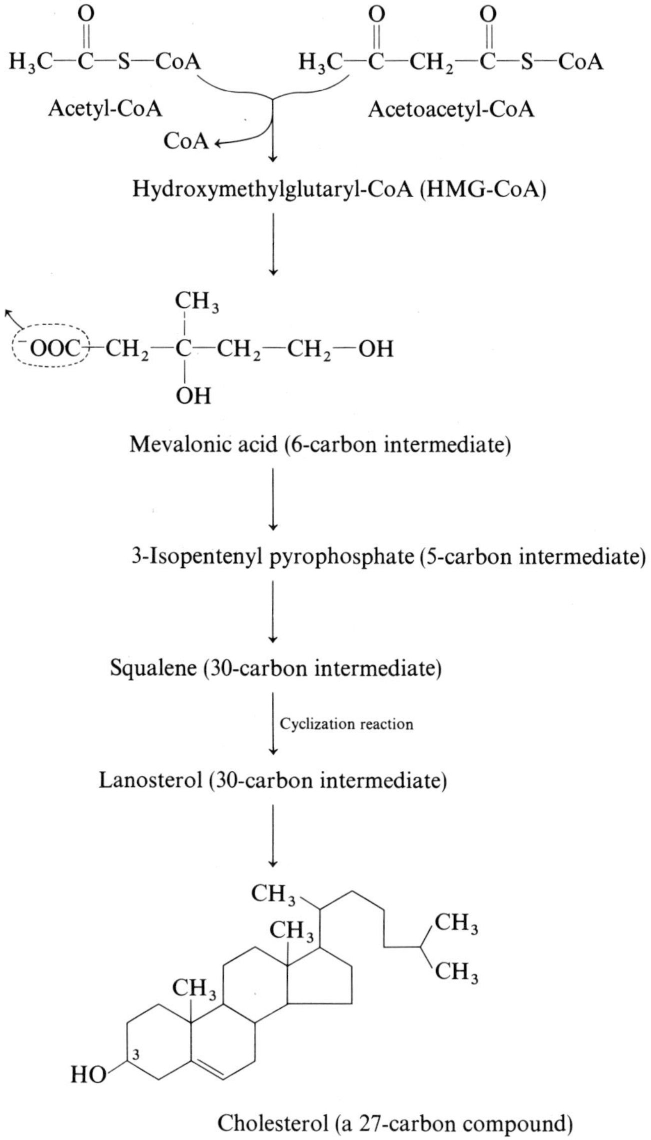

$H_3C{-}C({=}O){-}S{-}CoA$ Acetyl-CoA

$H_3C{-}C({=}O){-}CH_2{-}C({=}O){-}S{-}CoA$ Acetoacetyl-CoA

CoA ←

↓

Hydroxymethylglutaryl-CoA (HMG-CoA)

↓

$^{-}OOC{-}CH_2{-}C(CH_3)(OH){-}CH_2{-}CH_2{-}OH$

Mevalonic acid (6-carbon intermediate)

↓

3-Isopentenyl pyrophosphate (5-carbon intermediate)

↓

Squalene (30-carbon intermediate)

↓ Cyclization reaction

Lanosterol (30-carbon intermediate)

↓

CH_3 CH_3 CH_3 CH_3 CH_3 HO 3

Cholesterol (a 27-carbon compound)

FIGURE 14–11
A summary of cholesterol biosynthesis.

The synthesis of cholesterol is rather complex, involving as it does 25 separatc cnzymatic steps. Because of this complexity, we have simplified the pathway considerably (see Figure 14–11).

The complex 4-ring system of the cholesterol molecule is ultimately constructed from acetate molecules (acetyl-CoA). Specifically, acetyl-CoA and acetoacetyl-CoA undergo condensation to form *hydroxymethylglutaryl-CoA*, from which, in turn, *mevalonic acid* is formed. Once mevalonic acid has been formed, this 6-carbon compound is converted to a 5-carbon *isoprenoid* unit (*3-isopentenyl pyrophosphate*). This conversion is accomplished by 3 separate phosphorylation steps using ATP, along with a decarboxylation reaction (the removal of one carbon as CO_2). *Squalene*, a branched 30-carbon chain, is next constructed from six 5-carbon isoprenoid units. Squalene is converted into *lanosterol* via a unique cyclization reaction in which the four rings are closed via a series of electron-pair shifts catalyzed by the enzyme *squalene monooxygenase*. Other enzymes modify the lanosterol molecule to form cholesterol, a 27-carbon compound.

14.8 THE RELATIONSHIP OF LIPID METABOLISM TO OTHER METABOLIC PATHWAYS

Lipid and carbohydrate metabolism are closely interrelated and under constant regulation in most organs. It must be stressed that in a normal individual on a well-balanced diet, both lipid and carbohydrate metabolism are *in balance*. Certain organs utilize both lipids and carbohydrates for their energy needs, but in varying amounts. The utilization rate of a specific metabolic fuel often depends on the physiologic state of the individual. For example, resting skeletal muscle primarily utilizes free fatty acids for ATP production. However, during muscular activity, carbohydrate oxidation (via aerobic or anaerobic glycolysis) also occurs to meet the increased energy demands. Heart muscle tissue utilizes free fatty acids, carbohydrates, and even ketone bodies for its normal functioning. The brain is unique in that it cannot utilize free fatty acids for energy production, but must depend on glucose as its main fuel. During periods of severe metabolic stress (as in starvation), the brain can utilize **ketone bodies** as well as glucose for its energy needs.

ketone bodies

Ketone bodies (*acetoacetic acid* and *β-hydroxybutyric acid*) are produced by the liver during starvation or in the disease diabetes mellitus. During starvation, for example, carbohydrate intake is obviously low or nonexistent, glycogen stores are low, and lipid mobilization and degradation is consequently very high. The capacity of the liver to degrade the excess *acetyl-CoA* is exceeded, and the liver forms ketone bodies from the excess. These ketone bodies can be utilized by the peripheral tissues and by the brain as an *alternative energy* source during periods of low carbohydrate intake. Excess ketone bodies are either excreted into the urine or broken down to acetone, producing the characteristic "acetone breath" of a ketotic individual. This metabolic state is known as **ketosis**. If blood ketone body concentrations reach too high a level, **ketoacidosis** can result, lowering the

ketosis
ketoacidosis

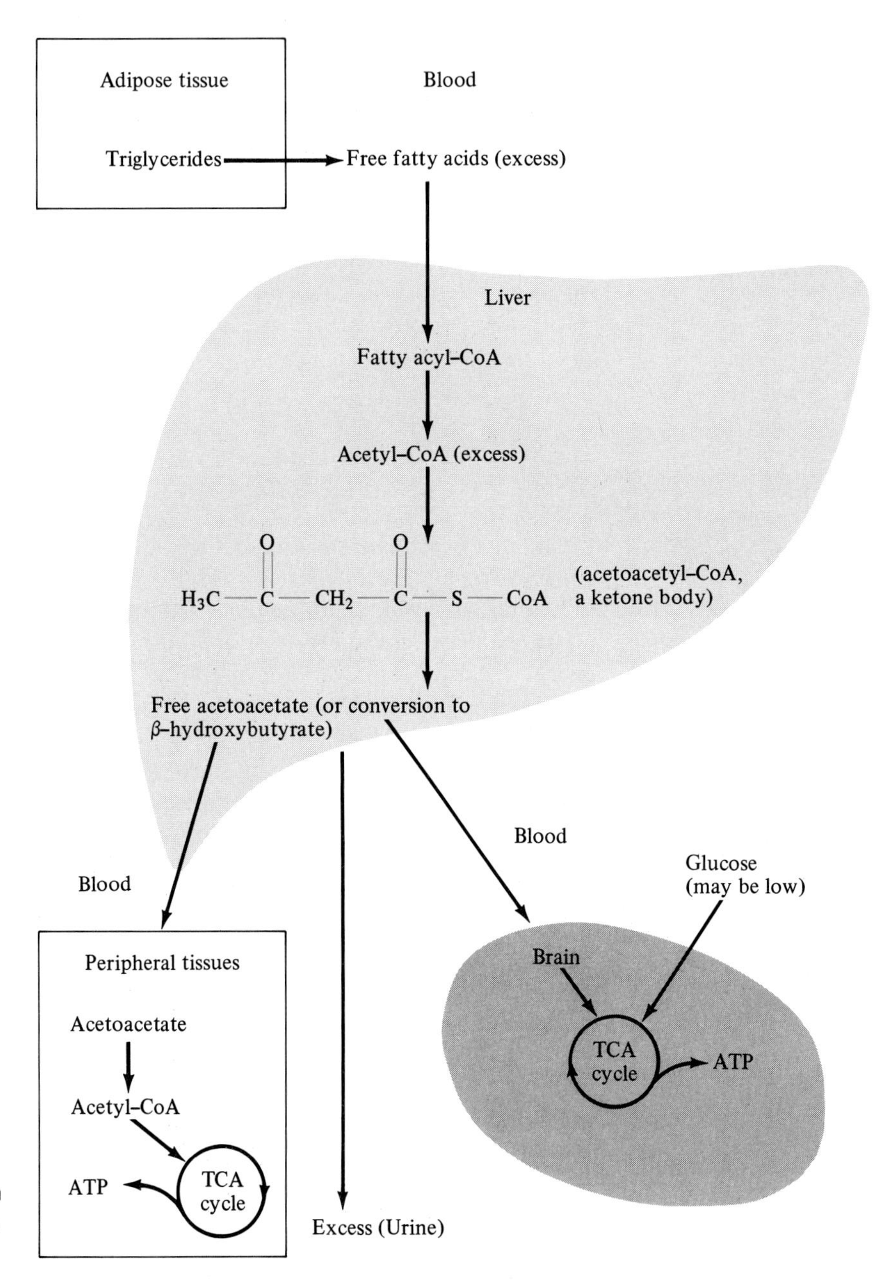

FIGURE 14–12
Ketone body formation during starvation or in diabetes mellitus.

blood pH. Ketone body formation and utilization is summarized in Figure 14–12.

SUMMARY

Lipids have several important functions and serve as an energy-rich cellular fuel. Large amounts can be stored in the adipose tissue.

Lipid emulsification (aided by pancreatic lipase, bile salts, and peristalsis) produces micelles, which can be absorbed by the intestinal mucosal cell. In the intestinal mucosal cell, triglycerides are resynthesized and chylomicrons are formed. Chylomicrons transport dietary lipids through the blood stream. In the liver, chylomicrons are modified to form three types of hepatic lipoproteins, VLDL, LDL, and HDL. These substances transport the triglycerides to the adipose tissue for storage.

When dietary lipids are not absorbed by the body, an excessive amount of lipid will be lost into the feces (steatorrhea), and a low blood chylomicron level will result. Lipid malabsorption can result from either (1) defective triglyceride degradation in the small intestine, (2) defective intestinal mucosal-cell function leading to impaired triglyceride and chylomicron formation, or (3) abnormal connections between the organs involved in the lipid absorption and transport processes.

Abnormally elevated levels of blood lipoproteins are associated with the hyperlipoproteinemias. High density lipoproteins (HDL) are important in preventing cholesterol from being deposited on arterial cell membranes. High levels of serum HDL cholesterol are apparently beneficial to human health, indicating as they do a lower susceptibility to coronary heart disease.

Triglycerides are degraded into free fatty acids and glycerol by hormone sensitive (and hormone insensitive) lipases within the adipose tissue. The free fatty acids and glycerol are transported to the peripheral tissues, where β-oxidation and the sequential cleavage of the fatty acid chain occurs, yielding acetyl-CoA, $FADH_2$ and NADH (see Figure 14–5, p. 259). Large numbers of ATP molecules are thus produced during the degradation of the fatty acid molecule.

Fatty acids are synthesized from acetyl-CoA via a pathway different from the β-oxidation pathway. Fatty acid biosynthesis occurs in the cytoplasm of the cell and employs a large multienzyme complex known as the fatty-acid synthetase complex. Initially, malonyl-CoA is formed from acetyl-CoA and bicarbonate (HCO_3^-). Both malonyl-CoA and acetyl-CoA are used to prime the fatty-acid synthetase complex. The acetyl group is condensed with malonyl-S-ACP, accompanied by decarboxylation, reduction, dehydration, and further reduction. Thus, a 4-carbon chain is formed and hydrogenated. The chain can be elongated further by the complex, up to 16-carbon atoms in length. The newly synthesized fatty acid is then cleaved from the complex as fatty acyl-CoA. Further elongation of the chain or the formation of unsaturated fatty acids is done by other enzyme systems. Humans cannot synthesize certain unsaturated fatty acids, such as linoleic acid and linolenic acid.

Once appropriate fatty acids are made within the cell, mono-, di-, and triglycerides can be synthesized using glycerol phosphate and molecules of fatty acyl-CoA.

Phospholipids can be synthesized from a diglyceride, cytidine triphosphate (CTP), and some substance (X) such as ethanolamine or serine.

Humans synthesize about 1.5–2.0 gm cholesterol/day, primarily in the liver. The complex, 25-step pathway was summarized in Figure 14–11, p. 272. Cholesterol is constructed from a large number of acetyl-CoA molecules. Mevalonic acid, 3-isopentenyl pyrophosphate, and squalene are intermediates. Six 5-carbon (isoprenoid) units constructed from mevalonic acid are used to construct the 30-carbon squalene molecule. Squalene is converted into lanosterol via a cyclization reaction, and lanosterol, in turn, is converted to cholesterol via several additional steps.

Lipid metabolism is closely interrelated with carbohydrate metabolism in the normal individual. Most organs utilize both free fatty acids and carbohydrates for their energy needs. The brain, however, cannot use free fatty acids and must rely on glucose (or, during starvation, ketone bodies) for its energy needs. During periods of severe metabolic stress, as in starvation, some of the fatty acids are degraded to ketone bodies for energy production.

REVIEW QUESTIONS

1. We introduced a number of important concepts in this chapter. Define (or briefly explain) each of the following terms:
   a. Micelle
   b. Chylomicron
   c. Lipoprotein
   d. Lipid malabsorption disease
   e. Pancreatic lipase deficiency
   f. Chyluria
   g. Hyperlipoproteinemia
   h. HDL cholesterol
   i. β-Oxidation
   j. Carnitine
   k. Fatty-acid synthetase complex
   l. Malonyl-CoA
   m. Acyl carrier protein (ACP)
   n. NADPH
   o. Essential fatty acid
   p. Cytidine triphosphate (CTP)
   q. Mevalonic acid
   r. Squalene
   s. Ketone body
2. A number of enzyme reactions were presented in this chapter. Write out the reaction catalyzed by the enzymes given below, giving the structures of reactants and products. Note whether any vitamin cofactors are employed in the reaction.
   a. Pancreatic lipase
   b. Hormone sensitive lipase
   c. Fatty acyl-CoA dehydrogenase
   d. Thiolase
   e. Acetyl-CoA carboxylase
   f. Fatty-acid synthetase complex (general scheme)
3. A patient was admitted to General Hospital. Laboratory analyses revealed (1) a low blood chylomicron level, (2) fatty stools, and (3) normal bile salt production. What diseases are indicated by these results? What biochemical abnormalities might explain these results?
4. How many ATP molecules are produced when the 18-carbon lipid *stearic acid* is completely degraded by the β-oxidation pathway?
5. * What are the similarities between fatty acid biosynthesis and degradation? (*Hint*: Compare the individual enzymatic reactions for both pathways.)
6. What vitamin cofactors are involved in fatty acid biosynthesis? List them, and explain the reactions in which they participate.
7. Describe the biochemical pathways that are affected during periods of prolonged starvation. Explain in detail.
8. * A laboratory mouse was fed ^{14}C pyruvate in which the radioactive label appears on carbon #3.

$$\underset{3}{H_3\overset{*}{C}}-\underset{2}{\overset{\overset{O}{\|}}{C}}-\underset{1}{COO^-}$$

Pyruvate

Where would the label appear in acetoacetyl-CoA? Draw structures.

SUGGESTED READING

Bhagavan, N. V. *Biochemistry*, 2nd ed., Ch. 8, pp. 852–57, 872–979. Philadelphia: J. B. Lippincott, 1978.

Lehninger, A. L. *Biochemistry*, 2nd ed., Chs. 20 and 24. New York: Worth, 1975.

Montgomery, R., Dryer, R.L., Conway, T. W., and Spector, A. A. *Biochemistry: A Case-Oriented Approach*, 3rd ed., Ch. 8, pp. 349–91; Ch. 10, pp. 446–86. St. Louis: C. V. Mosby, 1980.

Newsholme, E. A., and Start, C. *Regulation in Metabolism*, Chs. 5 and 7. New York: John Wiley, 1973.

NOTES

1. Lehninger, A. L. *Biochemistry*, 2nd ed. (New York: Worth, 1975), pp. 543, 836.
2. Lehninger, *Biochemistry*, 2nd ed., p. 836.

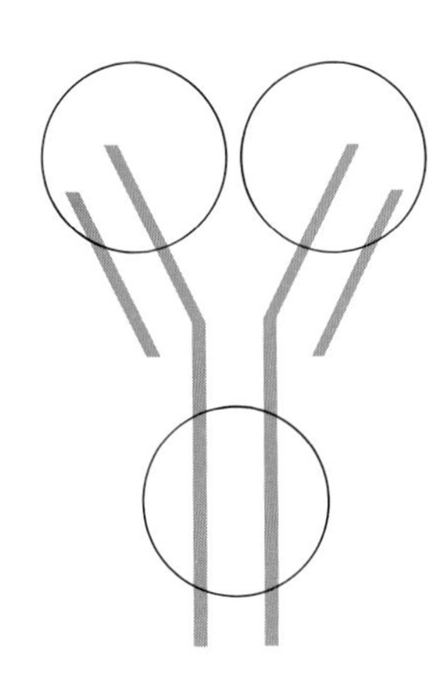

Chapter 15
Amino Acid Metabolism

15.1 INTRODUCTION

In Chapter 3 (Amino Acids and Proteins), we described the 20 amino acids that constitute the structural building blocks of proteins. Amino acids not only can be used as structural building blocks for the proteins of the body, but surplus amino acids can be used to produce energy. Once the α-amino group has been removed from the amino acid, the remaining carbon skeleton can be enzymatically transformed into either *acetyl-CoA, pyruvate, or a TCA cycle intermediate such as oxaloacetate, α-ketoglutarate, or succinyl-CoA*. These products of amino acid catabolism, along with those from carbohydrates and fats, fuel the TCA cycle and aid in producing NADH for ATP synthesis. In addition, because amino acids are degraded into either acetyl-CoA, pyruvate, or TCA cycle intermediates, these degradation products can be used by the cell as precursor molecules for a number of other cellular constituents. For example, acetyl-CoA can be used to make fatty acids, and pyruvate can be used to make glucose.

Amino acid metabolism, as a whole, is extremely complex, because we must consider the degradation and biosynthesis of 20 different amino acids. Some of the pathways are rather simple and only employ one or two enzymes; others can employ as many as 10 separate steps. We will greatly simplify this complex subject by summarizing both amino acid catabolism and anabolism in a few pages of text.

In this chapter, we will discuss how proteins are degraded into free amino acids and how the amino acids are transported into the blood stream. Next, we will discuss how the cell removes the α-amino group from the amino acid and disposes of the nitrogen via the urea cycle. A brief summary of amino acid catabolism (degradation) and amino acid anabolism (biosynthesis) will also be presented.

15.2 PROTEIN DIGESTION AND AMINO ACID TRANSPORT

The recommended dietary allowance (RDA) for protein was established in 1979 by the Food and Nutrition Board of the National Research Council. This board established that 0.6 g protein per Kg of body weight (ideal weight) per day was the amount of high-quality protein needed for optimal health. This amounts to around 45 g/day of high-quality protein for a 70 Kg

1. Food enters the stomach.

↓

2. The presence of food stimulates the release of gastric juice, increasing acidity (H^+).

3. Pepsinogen (Inactive) —[H^+]→ Pepsin (Active at pH 1–2)

↓

4. Active pepsin digests some proteins in the stomach.

↓

5. Acidic stomach contents enter the small intestine and are neutralized to pH 7–8.

↓

6. Stomach contents entering the small intestine stimulate the release of pancreatic juice, which contains zymogen forms of the proteolytic enzymes (along with intestinal enteropeptidase, an enzyme).

↓

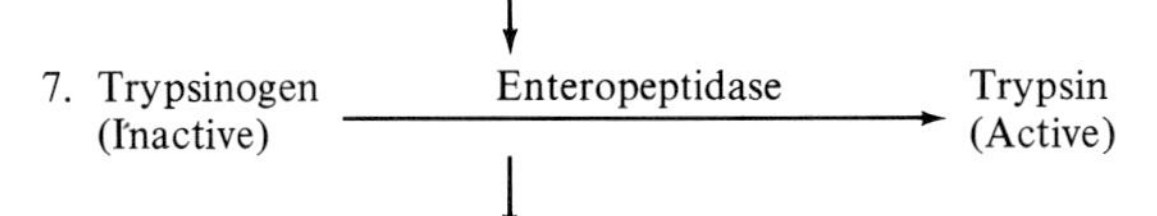

↓

8. Once trypsin has been activated, it activates other zymogens:

↓

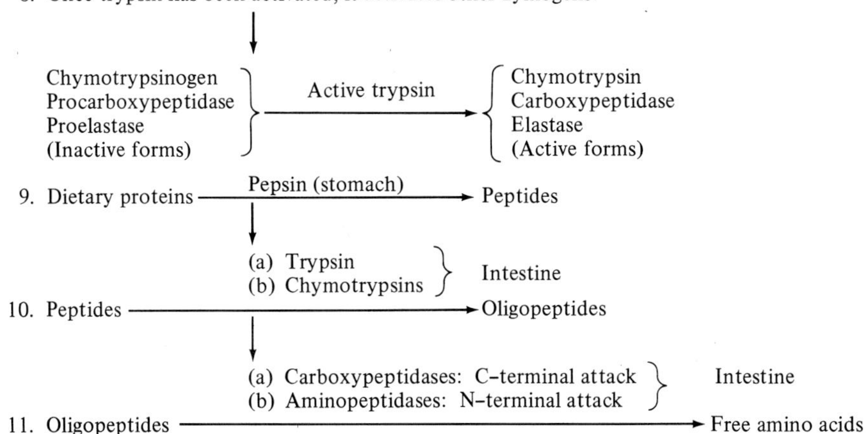

FIGURE 15–1
The activation of proteolytic enzymes and the digestion of proteins.

adult male. Many Americans consume significantly greater amounts of protein, in excess of 100 g/day. The role of dietary protein in human nutrition will be discussed in greater detail in Chapter 20 (Nutrition).

When protein is consumed, a number of biochemical events occur by which the protein is hydrolyzed into free amino acids. The sequence of events is summarized in Figure 15–1. A number of **pancreatic zymogens** (see Chapter 6, Enzymes) are activated by the enzyme *trypsin*. These proteolytic enzymes catalyze the hydrolysis of peptide bonds at specific points within the protein, thus degrading large protein molecules into small peptides and free amino acids. Various proteolytic enzymes, with their specificities, are listed in Table 15–1.

pancreatic zymogens

It should be noted that gastrointestinal tract physiology is under *hormonal control*. A textbook on medical physiology should be consulted for more specific details on the different hormones that regulate the GI tract.

Once proteins have been hydrolyzed into free amino acids, these amino acids are transported from the intestinal lumen into the intestinal mucosal cell and, finally, into the blood stream for transport to other regions of the body. Intestinal transport of amino acids is rather complex and will be summarized.

TABLE 15–1 The Specificities of Gastrointestinal Tract Proteolytic Enzymes.

Enzyme	Prefers to Cleave Peptide Bonds in Which
Trypsin	R_1 = Arg or Lys R_2 = any amino acid residue
Chymotrypsin	R_1 = Aromatic amino acid (Phe, Tyr, Trp) R_2 = any amino acid residue
Carboxypeptidase A	R_1 = any amino acid residue R_2 = any C-terminal residue except Arg, Lys, or Pro
Elastase	R_1 = Neutral (uncharged) residues R_2 = any amino acid residue
Aminopeptidase	R_1 = Most N-terminal residues of a polypeptide chain R_2 = any residue except Pro
Pepsin	R_1 = Trp, Phe, Tyr, Met, Leu R_2 = any amino acid residue

Note:

Peptide bond cleaved

$$H_2N\cdots NH-\underset{R_1}{CH}-\overset{O}{\overset{\|}{C}}\downarrow NH-\underset{R_2}{CH}-\overset{O}{\overset{\|}{C}}\cdots C(=O)OH$$

Source: Adapted from Bhagavan, N. V., *Biochemistry*, 2nd ed. (Philadelphia: J. B. Lippincott, 1978), p. 561.

The 20 amino acids are transported across the intestinal mucosal cell membranes by *specific* transport (carrier) systems. Five transport systems have been identified, each of which is specific for certain amino acids and will only transport amino acids *similar in structure and electrical charge*.

System 1: Small neutral amino acids
System 2: Large neutral amino acids
System 3: Positively charged amino acids
System 4: Negatively charged amino acids
System 5: The imino acid *proline*

In addition, small oligopeptides are transported by these systems. Amino acid transport is an energy-dependent process requiring ATP hydrolysis.

Another transport system, *not* present within mucosal cells, is known as the *γ-glutamyl cycle*. This transports amino acids from the blood into the many organs of the body. The γ-glutamyl cycle is shown in Figure 15–2. A membrane-bound enzyme called *γ-glutamyl transferase* transports amino acids across the cell membrane into the cell. The enzyme utilizes a substance

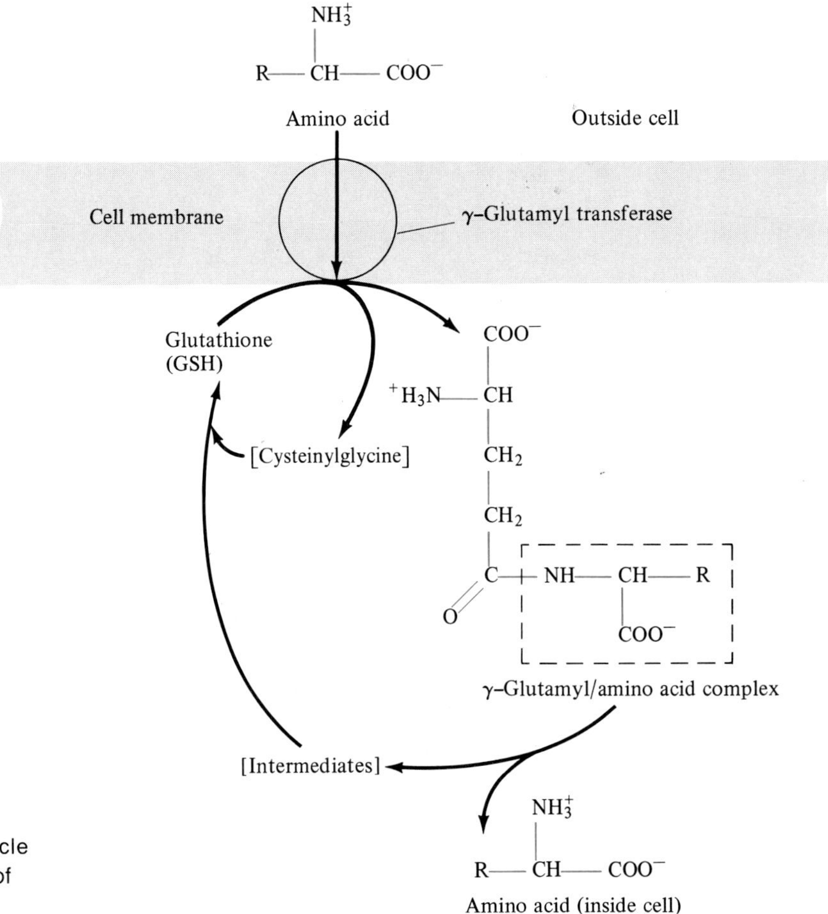

FIGURE 15–2
The γ-glutamyl cycle for the transport of amino acids.

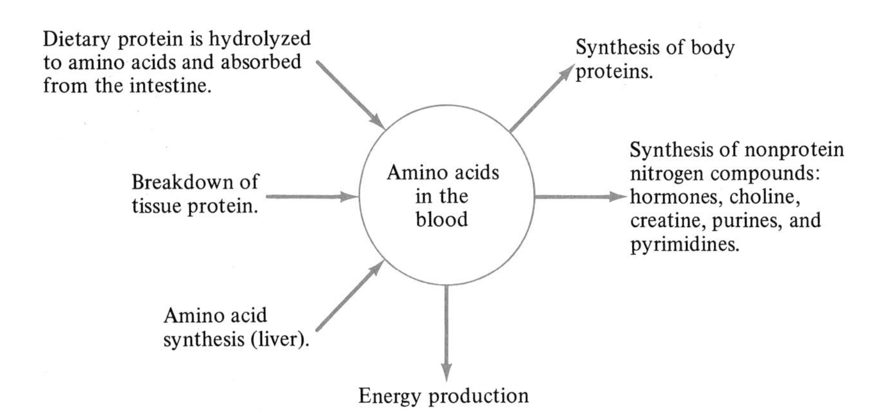

FIGURE 15–3
Generalized scheme of protein and amino acid metabolism. (Redrawn from N. V. Bhagavan, *Biochemistry*, 2nd ed. (Philadelphia: J. B. Lippincott, 1978), p. 544.)

called *glutathione* (*GSH*). The generalized structure of glutathione is as follows:

$$\boxed{\gamma\text{-Glutamyl}}\text{—cysteine—glycine}$$
(Glutamic acid)

Glutathione donates the γ-glutamyl (glutamic acid) residue used during amino acid transport. The amino acid is bound to the γ-glutamyl residue, and this γ-glutamyl/amino acid complex is transported into the cell cytoplasm. Once inside the cell, the amino acid is liberated from the γ-glutamyl residue, which is then reused to make glutathione.

Although we have only dealt with the fate of dietary proteins, our bodies can also reuse amino acids from tissue proteins. In fact, the tissues of the body are constantly "turning over." Tissue proteins are synthesized, have a given lifetime, and are then degraded back into constituent amino acids. In addition, new amino acids are synthesized (primarily in the liver). Figure 15–3 summarizes the sources and utilization of amino acids.

15.3 NITROGEN DISPOSAL: DEAMINATION AND TRANSAMINATION

Before amino acids can be completely degraded, the α-amino group of the amino acid (and amino groups on some R groups) must be hydrolyzed from the amino acid. The nitrogen atom of the α-amino group will be referred to as an *amino group nitrogen*. The detached amino groups are either (1) reused in other biosynthetic pathways or (2) disposed of (in humans, in the form of *urea*,

$$H_2N\text{—}\overset{\overset{\displaystyle O}{\|}}{C}\text{—}NH_2).$$

deamination

A number of different **deamination** reactions occur within the cell.

1. OXIDATIVE DEAMINATION

In this reaction, the amino group is removed by *amino acid oxidase* as free ammonia (NH_3). It should be noted that at physiological pH values, ammonia (NH_3) actually exists in solution as the ammonium ion (NH_4^+). The reaction is oxidative in nature and requires an electron/hydrogen acceptor, such as a flavin cofactor.

$$\underset{\alpha\text{-Amino acid}}{R-\overset{NH_3^+}{\overset{|}{C}H}-COO^-} \xrightarrow[\text{Flavin (Ox)} \rightarrow \text{Flavin-}H_2 \text{ (Red)}, \; NH_4^+]{\text{D-Amino acid oxidase and L-Amino acid oxidase}} \underset{\alpha\text{-Keto acid}}{R-\overset{O}{\overset{||}{C}}-COO^-}$$

(Flavin = FMN or FAD)

Glutamate dehydrogenase also catalyzes an oxidative deamination reaction (of L-glutamate).

$$\underset{\text{L-Glutamic acid}}{^-OOC-(CH_2)_2-\overset{NH_3^+}{\overset{|}{C}H}-COO^-} \xrightarrow[NAD^+ \rightarrow NADH + H^+]{\text{Enzyme}, \; NH_4^+} \underset{\alpha\text{-Ketoglutaric acid}}{^-OOC-(CH_2)_2-\overset{O}{\overset{||}{C}}-COO^-}$$

2. DEAMINATION BY DEHYDRATION

Serine and threonine can be deaminated by removing a water molecule from the amino acid.

$$\underset{\text{Serine}}{HO-CH_2-\overset{NH_3^+}{\overset{|}{C}H}-COO^-} \xrightarrow[H_2O]{\text{Enzyme}, \; NH_4^+} \underset{\text{Pyruvate}}{H_3C-\overset{O}{\overset{||}{C}}-COO^-}$$

Specific *amino acid dehydratases* catalyze this type of reaction.

3. TRANSAMINATION

transaminases

One of the most important types of deamination reactions is that catalyzed by the class of enzymes known as the **transaminases**. The transaminases catalyze the transfer of the α-amino group of an amino acid to an α-keto acid, thus forming a new α-amino acid. In essence, the transaminase reaction is both a *deamination* reaction (of the first amino acid) and an *amination* reaction (of the α-keto acid). The reversible transaminase reaction is shown in the following diagram:

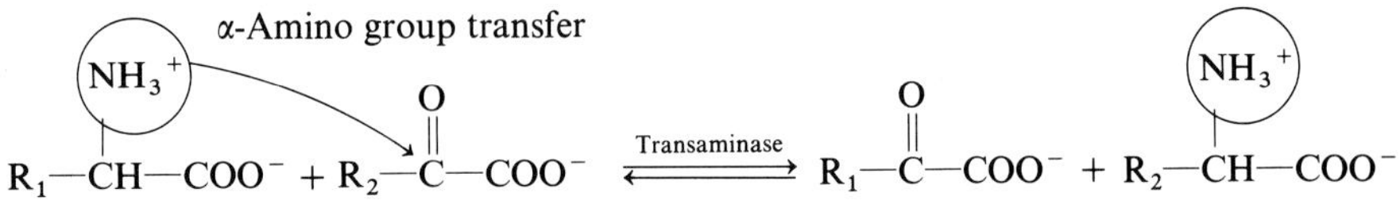

α-Amino group transfer

$$R_1{-}\underset{\underset{NH_3^+}{|}}{CH}{-}COO^- + R_2{-}\overset{\overset{O}{\|}}{C}{-}COO^- \xrightleftharpoons{\text{Transaminase}} R_1{-}\overset{\overset{O}{\|}}{C}{-}COO^- + R_2{-}\underset{\underset{NH_3^+}{|}}{CH}{-}COO^-$$

α-Amino Acid (1) α-Keto Acid (2) α-Keto Acid (1) α-Amino Acid (2)

A number of transaminases are present in various tissues. *Aspartate aminotransferase*, (*AST*)—also known as *glutamate-oxaloacetate transaminase*, (*GOT*)—and *alanine aminotransferase* (*ALT*) are important transaminases. Aspartate aminotransferase catalyzes the transfer of the α-amino group of aspartic acid to α-ketoglutaric acid (an α-keto acid and a TCA cycle intermediate). Glutamic acid is thus formed by aminating α-ketoglutaric acid. Deaminating aspartic acid yields oxaloacetate (OAA, also a TCA cycle intermediate). The reaction is reversible.

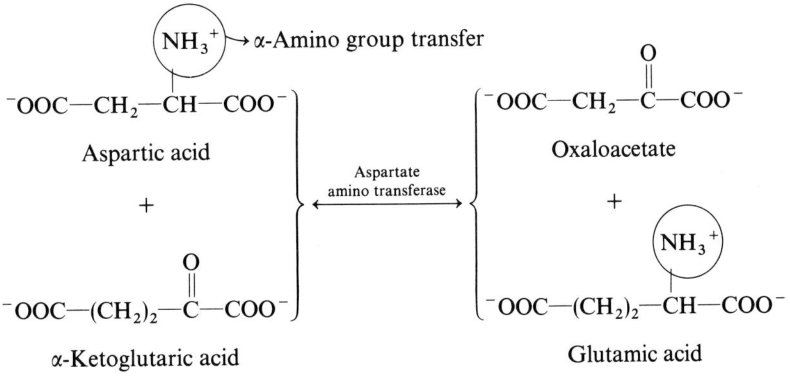

α-Amino group transfer

$$\left\{\begin{array}{c} {}^-OOC{-}CH_2{-}\underset{\underset{NH_3^+}{|}}{CH}{-}COO^- \\ \text{Aspartic acid} \\ + \\ {}^-OOC{-}(CH_2)_2{-}\overset{\overset{O}{\|}}{C}{-}COO^- \\ \alpha\text{-Ketoglutaric acid} \end{array}\right\} \xleftrightarrow{\text{Aspartate amino transferase}} \left\{\begin{array}{c} {}^-OOC{-}CH_2{-}\overset{\overset{O}{\|}}{C}{-}COO^- \\ \text{Oxaloacetate} \\ + \\ {}^-OOC{-}(CH_2)_2{-}\underset{\underset{NH_3^+}{|}}{CH}{-}COO^- \\ \text{Glutamic acid} \end{array}\right\}$$

pyridoxal phosphate (PLP)

All transaminases require **pyridoxal phosphate (PLP)** as an enzyme cofactor. Pyridoxal phosphate is derived from the vitamin pyridoxine (vitamin B_6). The structures of pyridoxine and pyridoxal phosphate are as follows:

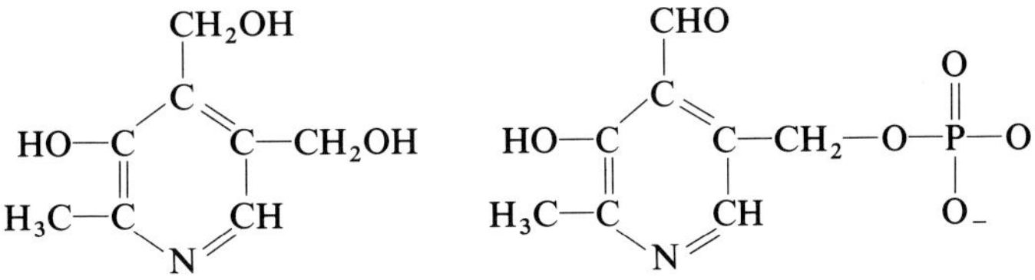

Pyridoxine: ring C bearing CH_2OH; $HO{-}C$, $C{-}CH_2OH$; $H_3C{-}C$, CH; N

Pyridoxine

(Vitamin B_6)

Pyridoxal phosphate: ring C bearing CHO; $HO{-}C$, $C{-}CH_2{-}O{-}\overset{\overset{O}{\|}}{\underset{\underset{O^-}{|}}{P}}{-}O^-$; $H_3C{-}C$, CH; N

Pyridoxal phosphate

(PLP)

Pyridoxal phosphate is bound to the active site of the transaminase and binds the α-amino group of the incoming amino acid. Upon reaction, a Schiff base is formed (the —HC=N— bonding structure is characteristic of a Schiff base).

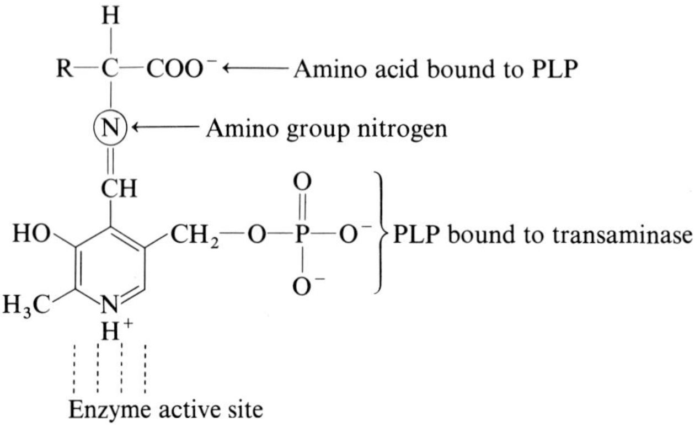

R—C(H)—COO^- ← Amino acid bound to PLP

(N) ← Amino group nitrogen

CH

HO, H_3C, N, H^+, CH_2—O—P(=O)(—O^-)—O^- } PLP bound to transaminase

Enzyme active site

Aspartate aminotransferase (AST) and alanine aminotransferase (ALT) are of particular clinical importance. These enzymes exist in both the cytoplasm of the cell and in the mitochondria. A number of organs contain varying amounts of these enzymes, as shown in Table 15–2. When a particular organ system is injured, resulting in cell lysis and death, both AST and ALT will be released from the cells into the blood stream. Thus, elevated levels of these enzymes will be detected in the blood via an assay specific for transaminases. For example, since heart muscle tissue contains very high levels of AST, a myocardial infarction (heart attack) will produce

TABLE 15–2 The Activities of Transaminases in Various Organ Systems.

Organ	AST	ALT
Heart	156	7
Liver	142	44
Skeletal muscle	99	5
Kidney	91	19
Serum	0.02	0.02

Note: AST = Aspartate aminotransferase (also known as glutamate-oxaloacetate transaminase, GOT); ALT = Alanine aminotransferase. The units stated above are × 10^{-4}/g of wet tissue. *Source*: Bhagavan, N. V., *Biochemistry*, 2nd ed. (Philadelphia: J. B. Lippincott, 1978), p. 584.

elevated levels of AST in the serum. Peak levels will occur between 24 and 36 hours following the attack, and decrease after 4 to 5 days. The magnitude of the serum enzyme activity reflects the severity of the attack. The situation is analogous to that for serum lactate dehydrogenase (LDH) isoenzymes (see Chapter 9, Carbohydrate Metabolism). Elevated transaminase levels are also produced in other disease states (Table 15–3).

TABLE 15–3 Clinical Conditions and Serum Transaminase Levels.

Clinical Condition	Transaminase Level
1. Myocardial infarction	Elevated AST
2. Infectious hepatitis (viral)	Elevated AST and ALT
3. Renal Infarction	Elevated AST
4. Progressive muscular dystrophy	Elevated AST
5. Other muscle disease states (and injuries affecting the muscles—crushing of skeletal muscles; in automobile and industrial accidents)	Elevated AST

15.4 THE EXCRETION OF EXCESS NITROGEN: THE UREA CYCLE

To review, the α-amino groups of amino acids are either: (1) removed as free ammonia (NH_3 or NH_4^+) via specific deamination reactions or (2) transferred to α-keto acids via transamination reactions. *Ammonia* (actually, the NH_4^+ ion) is highly *toxic* to cells; therefore, it must be converted into a more innocuous substance (urea) before it can be excreted from the body.

First, ammonia is primarily carried through the blood stream as glutamine or alanine. The synthesis of glutamine is shown in the following reaction:

$$\underset{\text{Glutamic acid}}{^{-}OOC{-}(CH_2)_2{-}\overset{\overset{NH_3^+}{|}}{CH}{-}COO^-} \xrightarrow[NH_4^+,\ ATP \rightarrow ADP + P_i]{\text{Glutamine synthetase}} \underset{\text{Glutamine}}{\overset{O}{\underset{NH_2}{C}}{-}(CH_2)_2{-}\overset{\overset{NH_3^+}{|}}{CH}{-}COO^-}$$

Thus, *glutamine* and *alanine* serve as carriers of ammonium ions.

Second, glutamine is deaminated by *glutaminase* (found in liver or kidney cells), yielding free ammonia (NH_3^+) and glutamic acid.

$$\underset{\underset{(NH_2)}{/}}{\overset{\overset{O}{\|}}{C}}-(CH_2)_2-\overset{\overset{NH_3^+}{|}}{CH}-COO^- \xrightarrow[\searrow (NH_4^+) \rightarrow \text{Urea synthesis (liver)}]{\text{Glutaminase (liver)}} {}^-OOC-(CH_2)_2-\overset{\overset{NH_3^+}{|}}{CH}-COO^-$$

The free ammonia is used by the liver for synthesizing urea. Finally, the urea is transported out of the liver to the kidneys for excretion in the urine.

Figure 15–4 summarizes the major steps involved in excreting amino acid nitrogen.

Urea is formed in the liver via a series of enzymes that constitute the urea cycle

urea cycle. This important metabolic pathway, first elucidated by H. A. Krebs and K. Henseleit in 1932, is also known as the Krebs-Henseleit cycle. The structure of urea is as follows:

$$(H_2N)-\overset{\overset{O}{\|}}{C}-(NH_2)$$

Urea

Notice that the molecule is rather simple in structure and possesses two amino groups. These amino groups come from the catabolism of amino acids.

The urea cycle is shown in Figure 15–5. Three important points about the pathway should be stressed: (1) The enzymes of the pathway are segregated. Some reactions take place in the cytoplasm, others in the mitochondrial matrix. (2) The process is energy-dependent, meaning that it requires the expenditure of energy in the form of ATP. (3) The process is directly linked to the TCA cycle and utilizes TCA cycle intermediates.

The initial reaction of the urea cycle is catalyzed by the enzyme *carbamoyl phosphate synthetase*, and occurs within the mitochondrial matrix. The enzyme condenses ammonia (NH_3), carbon dioxide (CO_2), and water (H_2O) to form an unstable compound called *carbamoyl phosphate*. This irreversible reaction requires the expenditure of 2 ATP molecules, and is summarized in the following equation:

$$(NH_3) + CO_2 + H_2O + 2\ ATP \xrightarrow[\searrow 2\ ADP + P_i]{\text{Carbamoyl phosphate synthetase}} \boxed{H_2N-\overset{\overset{O}{\|}}{C}}-O-\overset{\overset{O}{\|}}{\underset{\underset{O^-}{|}}{P}}-O^-$$

NH_3 from: Glutamine, Glutamic Acid

Carbamoyl group

Carbamoyl phosphate

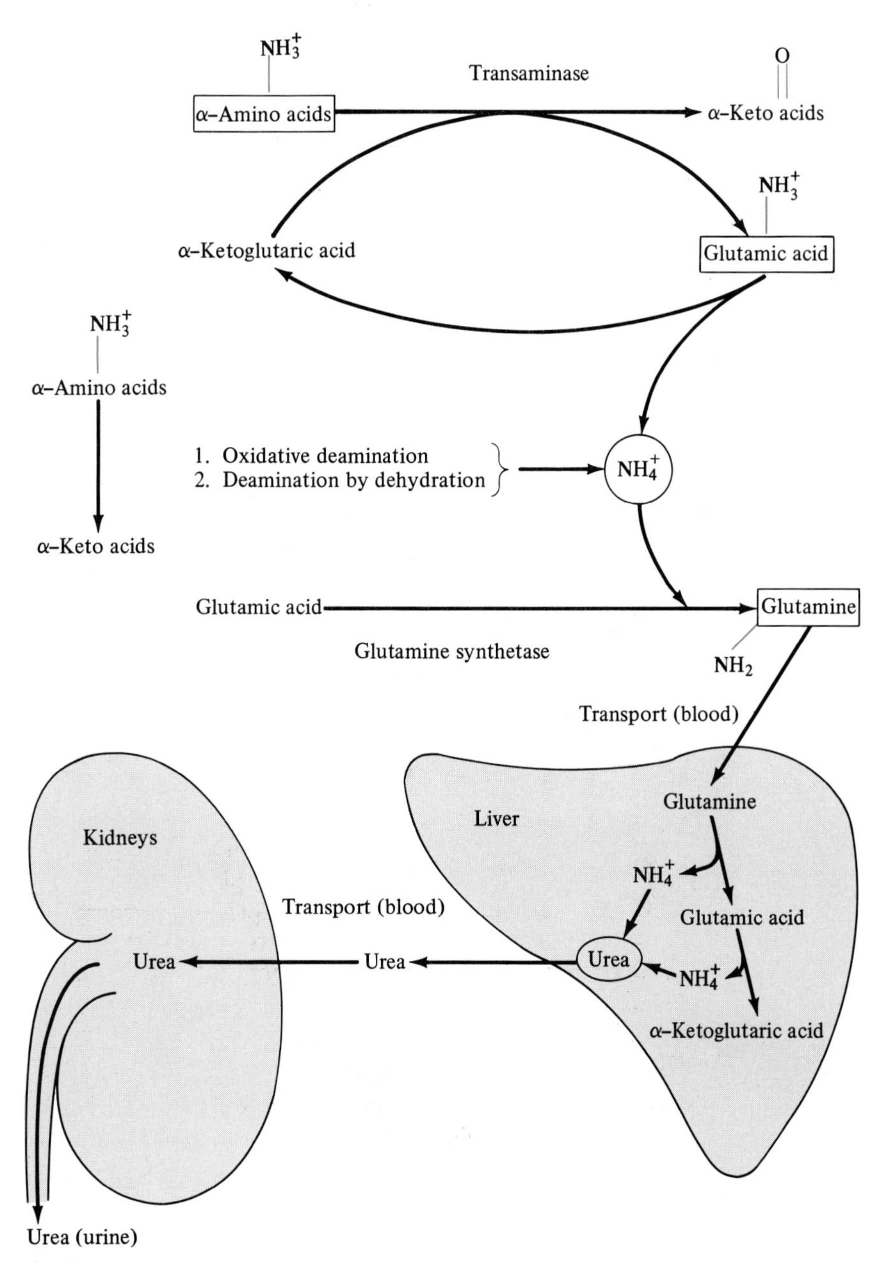

FIGURE 15–4
A summary of the excretion of amino acid nitrogen.

The free ammonia came from glutamine (via the glutaminase reaction) and from glutamic acid (via oxidative deamination, catalyzed by glutamate dehydrogenase).

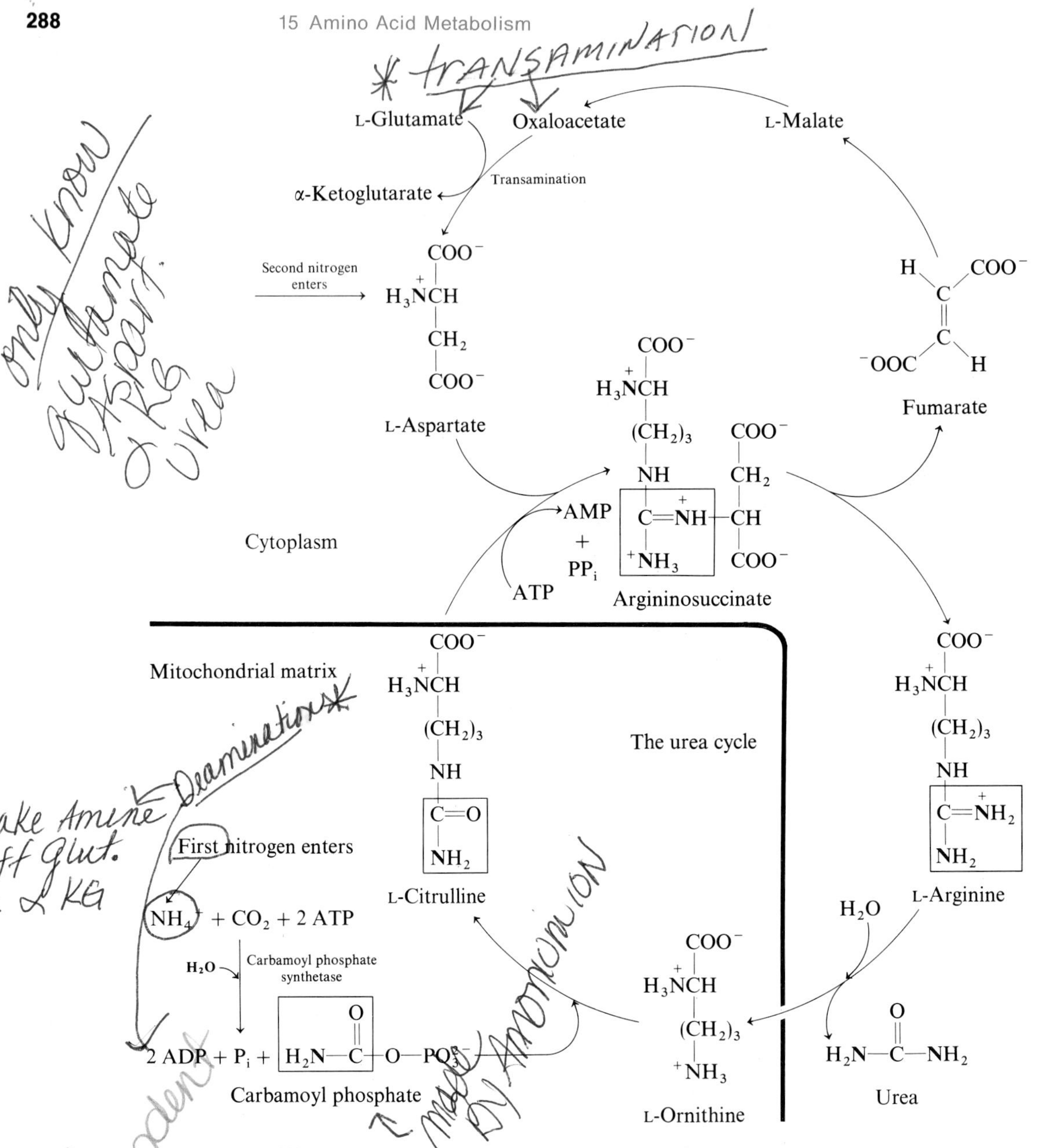

FIGURE 15–5 The urea cycle. The details of the urea cycle are explained in the text.

Next, the carbamoyl group (from carbamoyl phosphate) condenses with ornithine to form *citrulline*. This reaction is catalyzed by the enzyme *ornithine carbamoyl transferase*. Ornithine is initially generated in the cytoplasm and transported into the mitochondrion. Once citrulline is formed in the mitochondrion, it is immediately transported back into the cytoplasm.

Cytoplasmic citrulline condenses with aspartic acid to form *arginino-succinate*. This reaction is catalyzed by the enzyme *argininosuccinate synthetase* and requires the expenditure of an ATP molecule (pyrophosphate cleavage). The aspartic acid used in this step carries the second amino group nitrogen. This nitrogen was transferred to aspartic acid from glutamic acid via aspartate aminotransferase according to the following reaction:

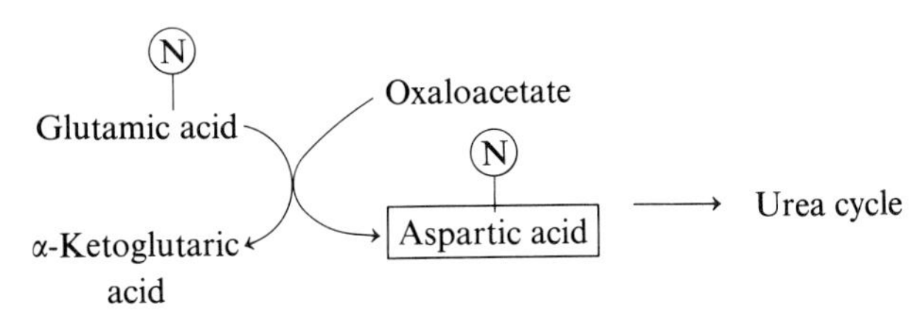

Next, argininosuccinate is cleaved by *argininosuccinase* to yield arginine and *fumarate* (a TCA cycle intermediate). (It should be noted that arginine now carries *two* amino groups, one from free ammonia in the mitochondrion and the other from aspartic acid.)

Arginase catalyzes the *hydrolysis* of urea from arginine and yields ornithine. Urea is transported out of the liver and passes in the blood stream to the kidneys for excretion in the urine. Ornithine is transported from the cytoplasm of the cell into the mitochondrion for reaction with carbamoyl phosphate. A total of 4 ATP molecules are required to synthesize 1 molecule of urea.

The urea cycle is linked to the TCA cycle because fumarate is produced at the argininosuccinase step. The fumarate formed in this step is converted to malate (by fumarase), and the malate, in turn, is converted to oxaloacetate (by malate dehydrogenase). The oxaloacetate can be converted to aspartic acid (by a transaminase), or used to fuel the TCA cycle.

AMMONIA TOXICITY AND HYPERAMMONEMIA

We have stated that ammonia is extremely toxic to cells. The toxic effects of ammonia are attributed to the *reductive amination* of α-ketoglutarate (catalyzed by glutamate dehydrogenase), yielding glutamic acid.

$$\alpha\text{-Ketoglutaric acid} \xrightarrow[\text{NH}_4^+ \text{ (excess)} \quad \text{NADH} \rightarrow \text{NAD}^+ \text{ (reduction)}]{\text{Glutamate dehydrogenase}} \text{Glutamic acid}$$

Thus, α-ketoglutaric acid would be depleted and the TCA cycle severely impaired. The production of NADH and ATP would drop significantly. Brain cells are particularly susceptible to an inhibition of cellular respiration; thus, excess ammonia **(hyperammonemia)** especially affects the central nervous system. Symptoms of ammonia toxicity include blurred vision, slurred speech, weakness, tremors, seizures, and coma, leading to death.

hyperammonemia

Hyperammonemia is a clinical term describing the situation in which the blood ammonia level is very elevated. Normal levels of ammonia range between 40 and 80 μg/100 ml blood. Hyperammonemia is caused by a number of serious diseases, including inherited defects of the urea cycle enzymes (that is, a partially or completely inactive enzyme), and extensive liver damage resulting in the inability of the liver to detoxify ammonia.

Genetic defects of one of the urea cycle enzymes cause a partial or complete defect in a particular enzyme of the cycle. This blocks the cycle, leading to the accumulation of ammonia. Coma and death usually result if a *complete* block occurs. Mental retardation associated with psychomotor retardation, seizures, vomiting, and lethargy result if a *partial* block occurs.

Severe damage to the liver will obviously interfere with the ability of the liver to detoxify ammonia. Alcohol abuse (leading to cirrhosis of the liver), ingestion of various toxic chemicals (such as carbon tetrachloride), and certain viral infections (that damage hepatic cells) all can cause serious liver damage and lead to hyperammonemia.

Treatment of hyperammonemia may include: (1) reduction of protein intake, (2) division of the daily protein intake into smaller portions (to prevent ammonia buildup), or (3) treatment with antibiotics (to inhibit GI tract bacteria that produce large amounts of ammonia).

BLOOD UREA NITROGEN AND KIDNEY FUNCTION

Urea is normally excreted out of the body by the kidneys. A steady state level of urea is normally present in the blood; 15–39 mg urea/100 ml blood (7–18 mg of "urea nitrogen"/100 ml blood) are considered normal levels.

*B*lood *U*rea *N*itrogen (BUN)

The term **"*B*lood *U*rea *N*itrogen" (BUN)** is used to express the actual amount of nitrogen present in the blood.

Increased BUN values are a reflection of either: (1) increased protein/amino acid catabolism (leading to increased urea synthesis and excretion by the liver), (2) dehydration, (3) kidney malfunction (inability of the kidneys to excrete urea), or (4) obstruction of the urinary tract by stones or abnormal growth. *Decreased BUN* levels are observed during: (1) protein malnutrition, (2) pregnancy, or (3) rehydration (massive water intake/infusion).

15.5 AMINO ACID CATABOLISM

As we stated in the introduction, amino acid metabolism is extremely complex. Amino acid catabolic and anabolic pathways often involve many separate enzymatic steps; therefore, a detailed discussion of them is beyond the scope of this book. We will instead summarize the main amino acid catabolic and anabolic pathways, and present a few representative and interesting pathways to illustrate the genetic basis of certain disease states.

In the previous section, we discussed the steps by which the amino groups are removed from the α-amino acids. Let us now turn our attention to the "carbon skeletons" of the amino acids. To summarize, the 20 amino acids are degraded into either (1) pyruvate, (2) acetyl-CoA, (3) acetoacetyl-CoA (which is degraded into acetyl-CoA), and (4) various TCA cycle intermediates (α-ketoglutaric acid, succinyl-CoA, fumarate, and oxaloacetate). Each of these substances directly feeds into and fuel the TCA cycle. It is, therefore, obvious that amino acid catabolism is very important in generating NADH for ATP production. Figure 15–6 summarizes the catabolic fate of each of the 20 amino acids.

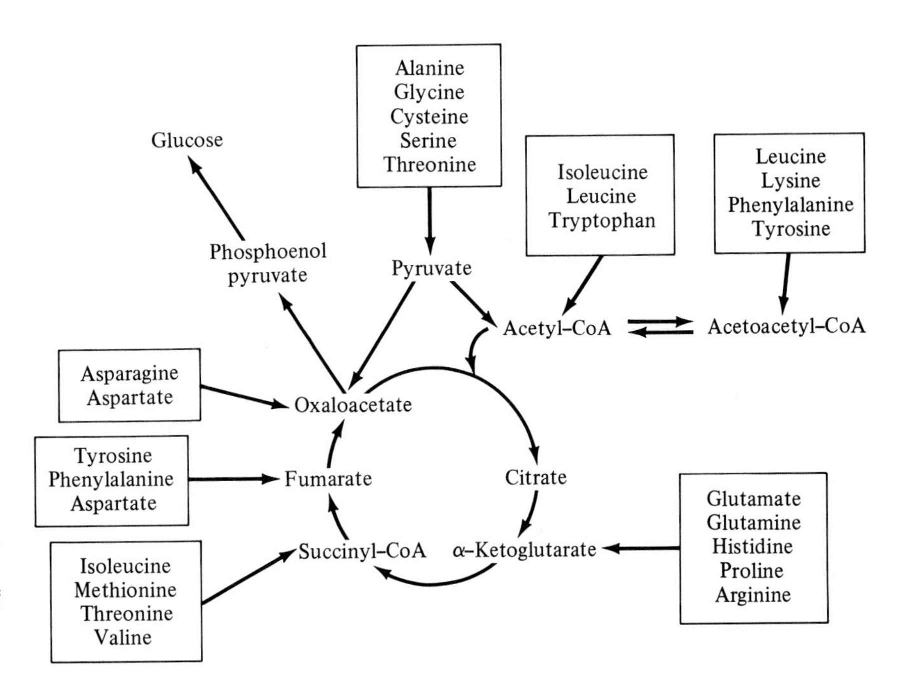

FIGURE 15–6
Amino acid catabolism. The catabolic fate of the carbon skeletons of the 20 amino acids is summarized.

Amino acids that are degraded into pyruvate or a TCA cycle intermediate can be used to make glucose; such amino acids are termed **glucogenic** amino acids. Amino acids that are degraded into acetoacetyl-CoA and acetyl-CoA can be used to make ketone bodies (and also lipids); such amino acids are termed **ketogenic** amino acids. Table 15–4 summarizes a catabolic classification of the 20 amino acids. It should be noted that leucine is the only amino acid that is *solely ketogenic*. Phenylalanine, tryosine, tryptophan, lysine, and isoleucine are *both* ketogenic and glucogenic; the remaining 14 amino acids are solely glucogenic.

glucogenic

ketogenic

During periods of starvation (when carbohydrate intake is low or nonexistent), fats are mobilized and fatty acids are oxidized to generate ATP. Ketone bodies (see Chapter 14, Lipid Metabolism) are also formed and are utilized by the brain for its energy needs. In addition, protein and

TABLE 15–4 A Classification of the Amino Acids Based upon Their Catabolic Fate.

Solely Ketogenic	Ketogenic and Glucogenic	Solely Glucogenic
Leucine	Isoleucine	Alanine
	Lysine	Arginine
	Phenylalanine	Aspartic acid
	Tryptophan	Asparagine
	Tyrosine	Cysteine
		Glutamic acid
		Glutamine
		Glycine
		Histidine
		Methionine
		Proline
		Serine
		Threonine
		Valine

amino acid catabolism is *increased* dramatically. Proteins from the digestive tract and from the skeletal muscle system are degraded into amino acids for energy production. A significant portion of the glucogenic amino acids are used for gluconeogenesis to maintain an adequate blood glucose level for the brain. In essence, the organs of the body are mobilized during starvation to maintain blood glucose at a relatively constant level for proper brain function.

AROMATIC AMINO ACID METABOLISM

The pathway for the degradation of *phenylalanine* and *tyrosine* will be discussed in detail (see Figure 15–7).

Phenylalanine is first converted into tyrosine by the enzyme *phenylalanine 4-monooxygenase*. The reaction *hydroxylates* the aromatic ring, and requires oxygen and NADPH (as the hydrogen/electron carrier). Next, *tyrosine aminotransferase* removes the α-amino group, converting tyrosine into the α-keto acid *4-hydroxyphenylpyruvic acid*. α-Ketoglutaric acid is the α-amino group acceptor. Glutamic acid is also produced in the reaction. *4-Hydroxyphenylpyruvic acid dioxygenase* catalyzes the second hydroxylation of the aromatic ring, along with a decarboxylation reaction producing *homogentisic acid*. These hydroxylation reactions are necessary to *break* the aromatic ring. A dioxygenase catalyzes the rupture of the aromatic ring, yielding *4-maleylacetoacetate* (*cis* isomer). An isomerase catalyzes the formation of the *trans* isomer (*4-fumarylacetoacetate*). Next, *fumaryl aceto-*

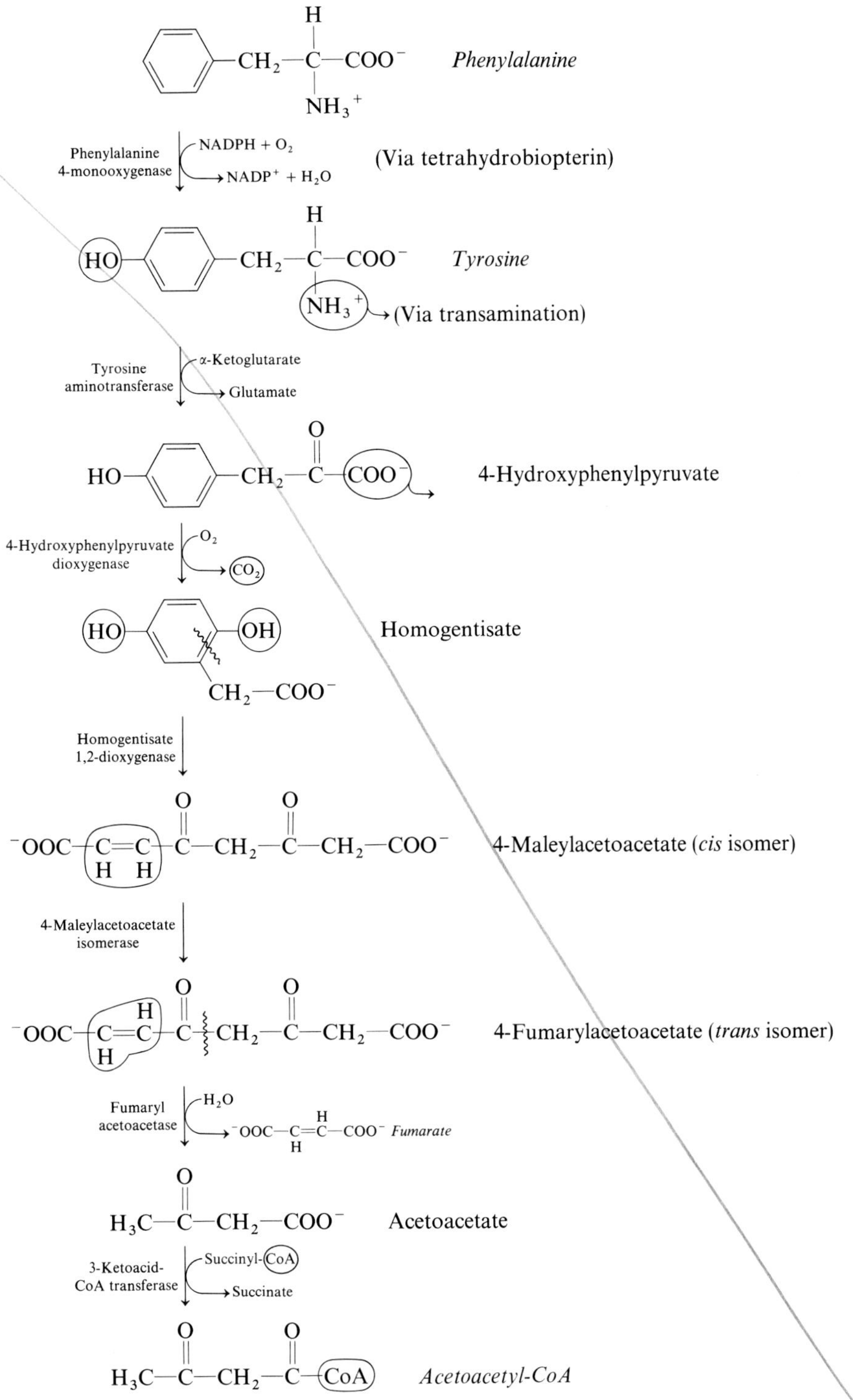

Phenylalanine

Phenylalanine 4-monooxygenase ($NADPH + O_2$ → $NADP^+ + H_2O$) (Via tetrahydrobiopterin)

Tyrosine

(Via transamination)

Tyrosine aminotransferase (α-Ketoglutarate → Glutamate)

4-Hydroxyphenylpyruvate

4-Hydroxyphenylpyruvate dioxygenase (O_2 → CO_2)

Homogentisate

Homogentisate 1,2-dioxygenase

4-Maleylacetoacetate (*cis* isomer)

4-Maleylacetoacetate isomerase

4-Fumarylacetoacetate (*trans* isomer)

Fumaryl acetoacetase (H_2O → $^-OOC—CH=CH—COO^-$ *Fumarate*)

Acetoacetate

3-Ketoacid-CoA transferase (Succinyl-CoA → Succinate)

Acetoacetyl-CoA

FIGURE 15–7
The catabolism of phenylalanine and tyrosine. The carbon skeletons of both Phe and Tyr are degraded into fumarate and acetoacetyl-CoA.

acetatase catalyzes the hydrolysis of 4-fumarylacetoacetate into *fumarate* and *acetoacetate*. The fumarate can enter the TCA cycle, and its carbons can be used in synthesizing glucose (gluconeogenesis). Thus, part of both phenylalanine and tyrosine are glucogenic. Acetoacetate is converted into acetoacetyl-CoA (using succinyl-CoA as CoA donor) as the last step in the pathway. Acetoacetyl-CoA can be cleaved into two molecules of acetyl-CoA, which can then enter the TCA cycle. Alternatively, the acetoacetate portion of the original amino acid molecule can be used in synthesizing ketone bodies and is, therefore, ketogenic.

INHERITED DISEASES OF AROMATIC AMINO ACID CATABOLISM

phenylketonuria
alkaptonuria

Phenylketonuria (PKU) and **alkaptonuria** are two inherited diseases associated with abnormal aromatic amino acid metabolism. In phenylketonuria, there is an inherited *deficiency* of the enzyme phenylalanine 4-monooxygenase (also known as phenylalanine hydroxylase). As a result, phenylalanine cannot be converted to tyrosine. The entire phenylalanine catabolic pathway is blocked by this enzyme defect, leading to the accumulation of phenylalanine and phenylalanine metabolites in the tissues and blood (Figure 15–8).

Increased levels of phenylalanine and phenylalanine metabolites damage developing brain cells, causing severe mental retardation (IQ = 20 in untreated, institutionalized individuals). Other symptons include defective myelination of the nerves, hyperactive reflexes, and abnormal brain weight. The lifespans of untreated individuals afflicted with PKU are significantly shorter than normal (50 percent are dead by age 20 and 75 percent are dead by age 30). It is estimated that 1 out of every 20 000 new-

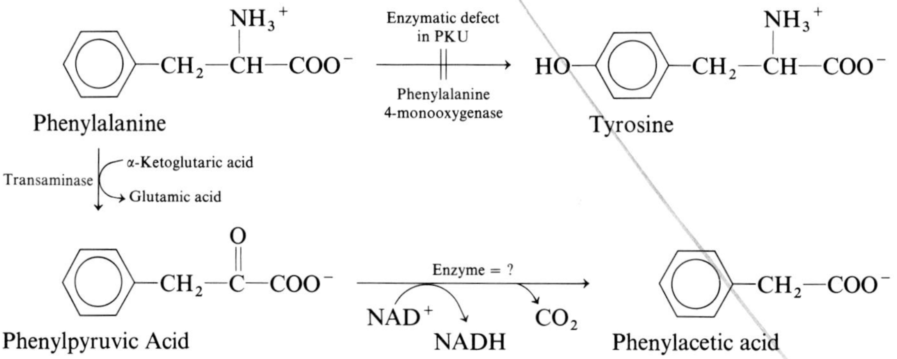

FIGURE 15–8 The enzymatic defect in phenylketonuria (PKU). A defect in phenylalanine 4-monooxygenase causes increased levels of phenylalanine, phenylpyruvic acid, and phenylacetic acid in the blood and urine.

borns is afflicted with PKU. The mechanism of damage is presently unknown. The disease can be detected very easily by analyzing blood or urine samples for phenylalanine or one of its metabolites. Generally, the test is run on newborns 6–14 days of age. Normal concentrations of phenylalanine range between 1.2 and 3.4 mg/100 ml of blood. Although very sensitive tests for phenylalanine are now used in the laboratory, the *diaper test* has been used as a screening test for PKU in the past. A solution of ferric chloride ($FeCl_3$) is placed on a urine-soaked diaper. If a green or blue-green color is produced, a very high concentration of *phenylpyruvic acid* (a metabolite of phenylalanine) is present. The Guthrie test (based on a microbial assay) is also used to screen for PKU; it is, in fact, superior to the diaper test in detecting PKU.

The disease can be treated by restricting the amount of phenylalanine in the diet of the infant starting 2–3 weeks after birth. Some phenylalanine must still be allowed in the diet, because this essential amino acid is required for constructing various proteins. Strict dietary management is not as critical after the age of six years, because the brain has matured by this age and is not as susceptible to the toxic effects of phenylalanine or its metabolites.

Finally, it should be noted that phenylalanine and tyrosine are essential precursor molecules to a number of other important important substances. For example, *epinephrine*, *thyroxine* (a hormone that controls the basal metabolic rate), and *melanin* (skin pigment) are all derived from tyrosine. Defects in skin and hair pigmentation are characteristic clinical features of PKU. Apparently, the high phenylalanine level inhibits the production of melanin in the skin.

Alkaptonuria, another (fortunately benign) genetic disease, is the result of a deficiency of the enzyme *homogentisic acid oxidase*. *Homogentisic acid* accumulates in the urine of afflicted individuals. Upon exposure to air, the urine turns black because degradation products of homogentisic acid have formed and polymerized.

Phe ⟶ HO–(ring)–OH (with CH_2—COO^-) —Alkaptonuria—╫⟶ 4-Maleylacetoacetate

Homogentisic acid

↘ Urine ($+O_2 \rightarrow$ Dark polymerization products)

Many other defects in amino acid metabolism are known to occur, and produce a variety of disease states with rather unusual clinical symptoms. Most are very serious (cause mental retardation) and are often lethal.

TABLE 15–5 The Essential and Nonessential Amino Acids.

Nonessential	Essential
Alanine	Arginine[a]
Asparagine	Histidine[b]
Aspartic acid	Isoleucine
Cysteine	Leucine
Glutamic acid	Lysine
Glutamine	Methionine
Glycine	Phenylalanine
Proline	Threonine
Serine	Tryptophan
Tyrosine	Valine

[a] Required by growing children only.
[b] Required by infants; only recently established as required by adults.

15.6 AMINO ACID ANABOLISM

The 20 amino acids are synthesized in various organisms through somewhat complex metabolic pathways; therefore, we will only summarize amino acid biosynthesis in this section. Adult humans can only make 11 of the 20 amino acids, the so-called *nonessential amino acids*. The 9 amino acids that we cannot make (and, therefore, must acquire in a balanced diet) are called the **essential amino acids**. Table 15–5 lists the essential and non-essential amino acids.

essential amino acids

All 20 amino acids can be constructed from only a few precursor molecules: two TCA cycle intermediates (α-ketoglutaric acid and oxalo-acetate), and various carbohydrate metabolites (3-phosphoglycerate, pyruvate, phosphoenolpyruvate, erythrose 4-phosphate, and ribose 5-phosphate). Figure 15–9 summarizes the biosynthesis of the 20 amino acids from these precursor molecules. Not all of the 20 amino acids can be synthesized by every organism; remember, humans cannot make 9 of them (see Table 15–5).

The nonessential amino acids, by and large, are easier to synthesize than the essential amino acids. Figure 15–10 summarizes the biosynthesis of six nonessential amino acids. It should be noted that five of these can be made from either oxaloacetate or α-ketoglutarate in a few enzymatic steps. This illustrates the amazing simplicity and economy of the cell. Alanine is synthesized from pyruvate. One should also note how amination and transamination reactions are used in these interconversions. The syntheses of serine, glycine, cysteine, arginine, and tyrosine are not depicted in Figure 15–10.

α-Ketoglutarate
Glutamate
Glutamine Proline Arginine

Oxaloacetate
Aspartate
Asparagine Methionine Threonine Lysine
Isoleucine

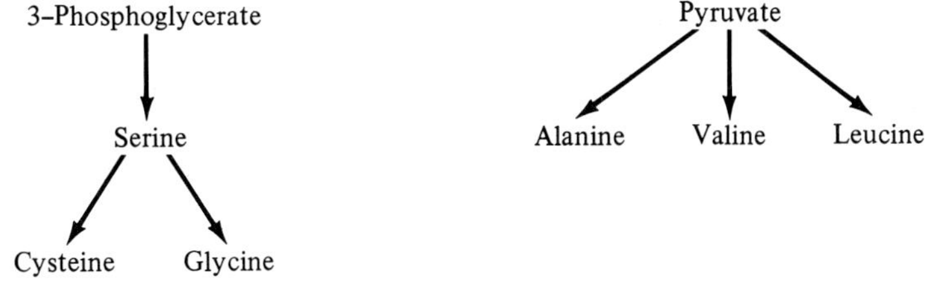

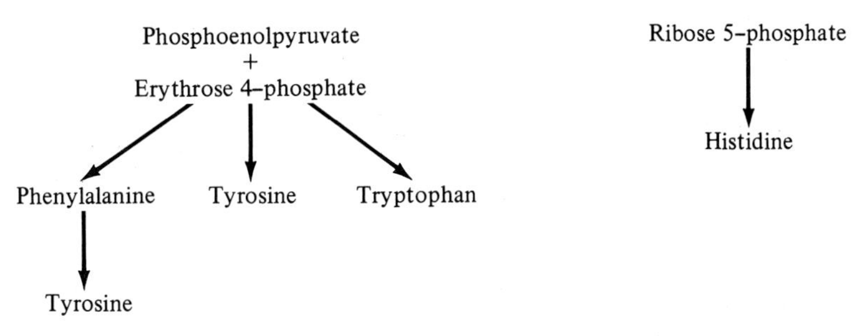

FIGURE 15–9 A summary of amino acid biosynthesis

SUMMARY

Amino acids can be degraded to obtain energy, or the carbon skeletons can be used to make other biological molecules.

Dietary proteins are completely digested into small peptides and amino acids by activated pancreatic proteolytic enzymes (Figure 15–1, p. 278). These enzymes catalyze the hydrolysis of peptide bonds at specific points (Table 15–1, p. 279). Amino acids are transported from the intestine and into the blood stream via five specific transport systems. The γ-glutamyl cycle transports amino acids from the blood stream into tissue cells using glutathione (GSH).

Oxidative deamination, deamination by dehydration, and transamination (amino group transfer) are three enzymatic reactions responsible for removing the α-amino group from amino acids.

Transaminases catalyze the transfer of the α-amino group of an amino acid to an α-keto acid, forming a new α-amino acid. Transaminases require pyridoxal phosphate for activity. Aspartate amino-

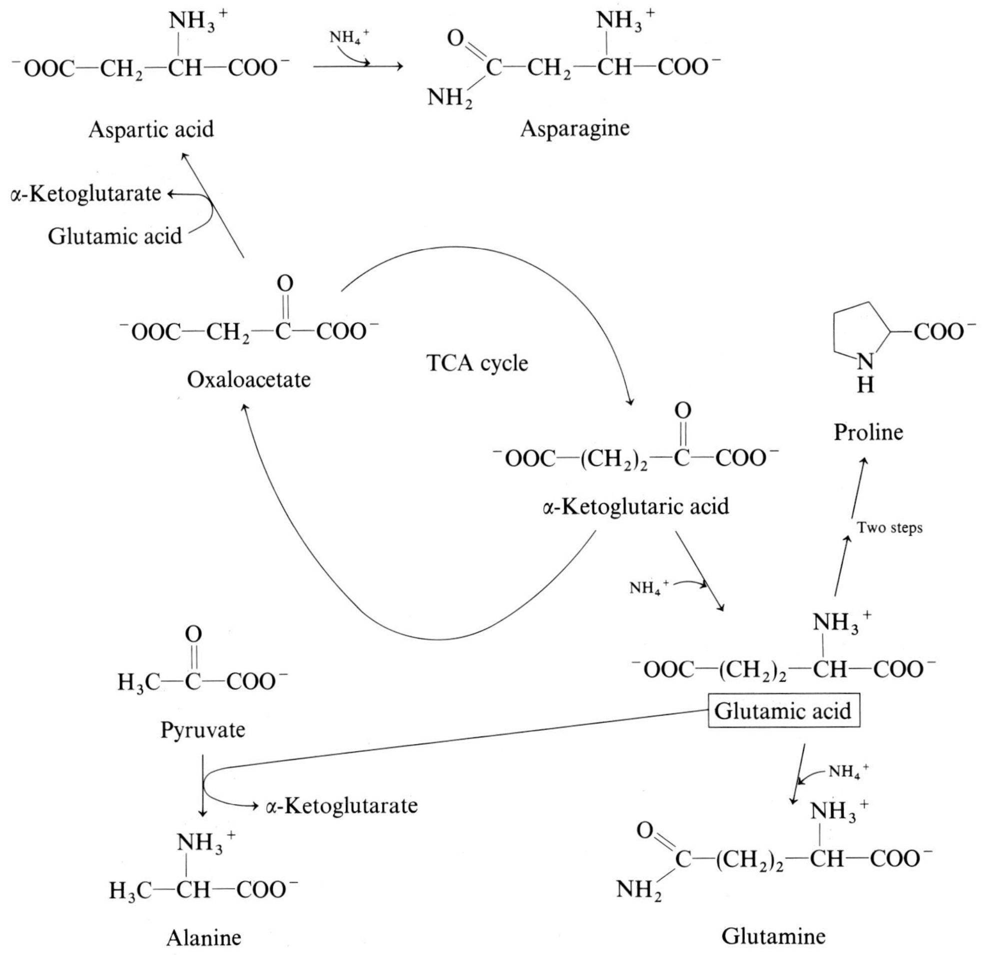

$^{-}OOC{-}CH_2{-}CH(NH_3^+){-}COO^-$ $\xrightarrow{NH_4^+}$ $H_2N{-}C(=O){-}CH_2{-}CH(NH_3^+){-}COO^-$

Aspartic acid → Asparagine

α-Ketoglutarate ← Glutamic acid

$^{-}OOC{-}CH_2{-}C(=O){-}COO^-$

Oxaloacetate

TCA cycle

$^{-}OOC{-}(CH_2)_2{-}C(=O){-}COO^-$

α-Ketoglutaric acid

NH_4^+

$^{-}OOC{-}(CH_2)_2{-}CH(NH_3^+){-}COO^-$

Glutamic acid

Two steps

Proline (COO^-, N, H)

NH_4^+

$H_2N{-}C(=O){-}(CH_2)_2{-}CH(NH_3^+){-}COO^-$

Glutamine

$H_3C{-}C(=O){-}COO^-$

Pyruvate

α-Ketoglutarate

$H_3C{-}CH(NH_3^+){-}COO^-$

Alanine

FIGURE 15–10 A summary of the biosynthesis of six nonessential amino acids. The biosyntheses of serine, glycine, cysteine, arginine, and tyrosine are not shown in this diagram.

transferase (AST) and alanine aminotransferase (ALT) are transaminases of clinical importance.

Ammonia (NH_3 or NH_4^+) is toxic to cells and must be transported out of the cells through the blood stream to the liver as either glutamine or alanine. The liver transforms the ammonia into urea, which is then transported to the kidneys for excretion (Figure 15–4).

Urea is formed in the liver via the urea cycle. The urea cycle has parts segregated in the cytoplasm and in the mitochondrial matrix. It is an energy dependent process, and is linked to the TCA cycle. The cycle is depicted in Figure 15–5, p. 288. Amino group nitrogens come from (1) deamination reactions and (2) from a transamination reaction that generates aspartic acid. The cycle generates arginine. Urea is hydrolyzed from arginine and then transported out of the liver cell into the blood stream for excretion by the kidneys.

Ammonia toxicity is probably caused by the reductive amination of α-ketoglutaric acid, which depletes the TCA cycle of this intermediate and inhibits cellular respiration. Hyperammonemia occurs when the level of ammonia in the bloodstream is very high. Hyperammonemia is caused by inherited defects of the urea cycle enzymes or by extensive liver damage (cirrhosis). Blood urea nitrogen (BUN) was discussed and a variety of disease states producing increased or decreased BUN levels was presented.

The 20 amino acids are degraded into (1) pyruvate, (2) acetyl-CoA, (3) acetoacetyl-CoA, and (4) various TCA cycle intermediates (α-ketoglutaric acid, succinyl-CoA, fumarate, and oxaloacetate). Figure 15–6 summarizes the catabolic fate of the 20 amino acids. Amino acids that are degraded into pyruvate or into a TCA cycle intermediate can be used to make glucose (are glucogenic); amino acids that are degraded into acetyl-CoA and acetoacetyl-CoA can be used to make ketone bodies (are ketogenic). Starvation causes increased amino acid catabolism in which the carbon skeletons of the glucogenic amino acids are used to make glucose, keeping the blood glucose level constant for normal brain function.

The aromatic amino acids phenylalanine and tyrosine are degraded into acetoacetyl-CoA and into fumarate. They are, therefore, both ketogenic and glucogenic (see Figure 15–7, p. 293). Phenylketonuria (PKU) is an inherited disease in which the enzyme phenylalanine 4-monooxygenase is deficient, blocking the conversion of phenylalanine into tyrosine. As a result, increased levels of phenylalanine (and other metabolic intermediates) accumulate in the blood, damaging brain cells and causing severe mental retardation. Alkaptonuria results from a deficiency in the enzyme homogentisic acid oxidase. Homogentisic acid accumulates in the blood and urine; homogentisic acid metabolites in urine turn black upon exposure to oxygen.

Adult humans can only synthesize 11 out of the 20 amino acids. These 11 are called the nonessential amino acids. The 9 essential amino acids must be acquired in a well-balanced diet (Table 15–5, p. 296).

The synthesis of the 20 amino acids is summarized in Figure 15–9, p. 297. α-Ketoglutarate, oxaloacetate, 3-phosphoglycerate, pyruvate, phosphoenolpyruvate, erythrose 4-phosphate, and ribose 5-phosphate are the only precursor molecules needed to synthesize the 20 amino acids. The biosynthesis of 6 nonessential amino acids was presented in Figure 15–10.

A number of the nonessential amino acids can be interconverted via amination and transamination reactions.

REVIEW QUESTIONS

1. We have presented a number of important concepts and terms in this chapter. Define (or briefly explain the following terms:
   - **a.** Zymogen (see Chapter 6)
   - **b.** Trypsin
   - **c.** γ-Glutamyl cycle
   - **d.** Glutathione (GSH)
   - **e.** Oxidative deamination
   - **f.** Transamination
   - **g.** Pyridoxal phosphate
   - **h.** Urea
   - **i.** Urea cycle
   - **j.** Carbamoyl phosphate
   - **k.** Hyperammonemia
   - **l.** Ammonia toxicity
   - **m.** BUN
   - **n.** Glucogenic
   - **o.** PKU
   - **p.** Essential amino acid
   - **q.** Nonessential amino acid

2. We have presented a number of enzymatic reactions in this chapter. Write the complete reaction for each of the following enzymes:
   - **a.** Trypsin (show the structure of a hypothetical protein substrate)
   - **b.** Elastase (show the structure of a hypothetical protein substrate)
   - **c.** Glutamate dehydrogenase
   - **d.** Aspartate aminotransferase (AST)
   - **e.** Alanine aminotransferase (ALT)
   - **f.** Glutamine synthetase
   - **g.** Glutaminase
   - **h.** Carbamoyl phosphate synthetase
   - **i.** Phenylalanine 4-monooxygenase

3. Would increased or decreased rates of protein degradation be observed if:
   - **a.** significantly less acid than normal is excreted by the stomach?
   - **b.** the intestinal contents remained acidic (pH = 2)?
   - **c.** the intestine did not produce any enterokinase?

* **4.** Amino acid transport system 2 is a membrane-bound protein responsible for transporting

large, neutral amino acids. What are some essential physical-chemical features of the amino acid binding site of transport system 2?

5. Draw the complete chemical structure of glutathione (GSH).

*6. Suggest possible enzyme reactions for converting pyridoxine to pyridoxal phosphate.

7. Very high levels of both AST and ALT transaminases were detected in a clinical blood sample. In addition, a high serum LDH level was found, and electrophoresis revealed a very high level of LDH_5 (M_4) isoenzyme (see Chapter 9). Which organ system is probably the source of these enzymes? What disease states might be responsible for their release into the bloodstream?

*8. Suppose 2 molecules of alanine are degraded according to the following equation:

$$\text{Alanine} \xrightarrow[-NH_3]{} \text{Pyruvate} \xrightarrow{CO_2} \text{Acetyl-CoA} \longrightarrow \text{TCA cycle}$$

a. How many ATP molecules are generated if the carbon skeletons of alanine are fed into the TCA cycle?
b. How many ATP molecules are required to convert the 2 nitrogen atoms of the 2 alanine molecules into urea?
c. What is the *net* production of ATP in this reaction?

9. Look at the structure of the urea cycle intermediates. L-Ornithine is similar in structure to which amino acid?

10. In the text, we stated that ammonia toxicity was caused by the depletion of the TCA cycle intermediate α-ketoglutaric acid. What other substance would also be depleted, and why would its depletion cause serious problems?

11. a. Why would dehydration lead to increased BUN levels?
b. Why would pregnancy cause decreased BUN levels?

*12. Assume that phenylalanine is radioactively labeled at position 4 in the aromatic ring, as follows:

$$\overset{*}{4}\text{-}C_6H_5\text{-}CH_2\text{—}\underset{}{\overset{NH_3^+}{|}}{CH}\text{—}COO^-$$

In which degradation product (fumarate or acetoacetyl-CoA) will the label appear? In what position?

SUGGESTED READING

Bhagavan, N. V. *Biochemistry*, 2nd ed., Ch. 6. Philadelphia: J. B. Lippincott, 1978.

Grisolia, S., Baguena, R., and Major, F. *The Urea Cycle*. New York: John Wiley, 1976.

Lehninger, A. L. *Biochemistry*, 2nd ed., Chs. 21 and 25. New York: Worth, 1975.

Meister, A. *Biochemistry of the Amino Acids*, 2nd ed., vols. 1 and 2. New York: Academic Press, 1965.

Stanbury, J. O., Wyngaarden, J. B., and Fredrickson, D. S., eds. *The Metabolic Basis of Inherited Disease*, 3rd ed. New York: McGraw-Hill, 1972.

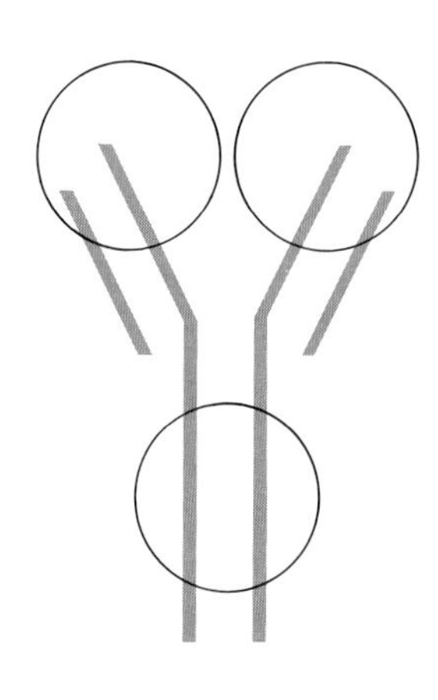

Chapter 16
Nucleic Acids

16.1 INTRODUCTION

Nucleic acids are a very important class of biological molecules. They are information storage molecules and, therefore, contain all of the genetic information necessary for the growth, development, and reproduction of living organisms. Nucleic acids were first isolated and studied by Friedrick Miescher in the later part of the 19th century. Miescher isolated a substance (which he called *nuclein*) from the nuclei of pus cells. This substance contained phosphate and was acidic; hence, it was given the name nucleic acid. Further analysis revealed that this substance was constructed from sugar molecules, phosphate, and organic molecules, called purines and pyrimidines, that contained nitrogen. It was not until the middle of the 20th century that the true biological significance of nucleic acids was fully appreciated. Biology has been revolutionized by the discovery and elucidation of the role of these substances in living systems.

Before we can discuss how cells store and retrieve the information coded in nucleic acids and how they synthesize proteins, we must first describe the chemistry and structure of nucleotides and nucleic acids. A brief discussion of nucleotide biosynthesis and degradation will also be presented. The chemistry and structure of deoxyribonucleic acid (DNA) and ribonucleic acid (RNA) will be discussed. The mechanism of DNA replication and the mechanism by which genetic information is retrieved is presented in Chapter 17.

16.2 NUCLEOTIDE STRUCTURE AND NOMENCLATURE

nucleotides
deoxyribonucleic acid (DNA)
ribonucleic acid (RNA)

Nucleotides are the structural components of a number of very important biological molecules. For example, **deoxyribonucleic acid (DNA)** and **ribonucleic acid (RNA)**, the molecules that store and transfer genetic information, are polymers of nucleotides. Adenosine triphosphate (ATP) is an extremely important nucleotide, because the cell stores and transfers chemical energy via this compound. A number of coenzymes (NAD^+, FAD, and coenzyme A) are derived from the adenine nucleotide. Finally, cyclic nucleotide derivatives (example, cAMP) serve as mediators of hormone action.

purine
pyrimidine

A nucleotide is a compound that is constructed from a *base* containing nitrogen (in the form of a **purine** or a **pyrimidine**), a sugar, and one or more phosphate groups.

There are two major classes of nitrogenous bases: the purines and the pyrimidines. *Adenine* (*A*) and *guanine* (*G*) are the major purines; *cytosine* (*C*), *thymine* (*T*), and *uracil* (*U*) are the major pyrimidines. The structures of these bases are shown:

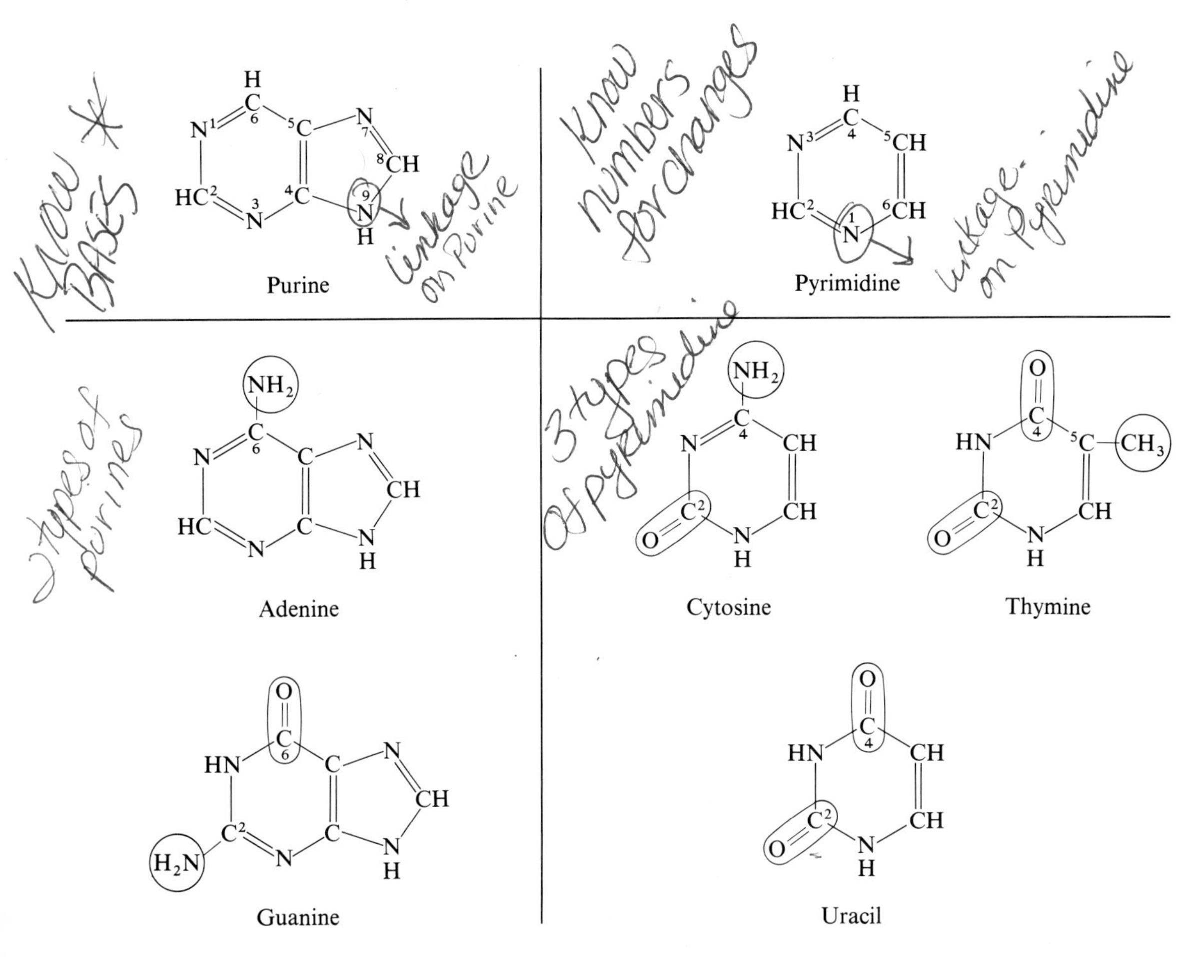

The DNA molecule contains:

Purine	*Pyrimidine*
Adenine (A)	Thymine (T)
Guanine (G)	Cytosine (C)

whereas RNA molecules contain: adenine, guanine, cytosine, and uracil. Notice that thymine occurs only in DNA and uracil occurs only in RNA.

Nucleotides are constructed from two types of sugars: D-*ribose* and D-*2-deoxyribose*. The structures of these two important pentoses are shown. A hydrogen atom has replaced the OH group in the 2′ position in D-2-deoxyribose. *Ribo*nucleic acid is made from ribose; *deoxyribo*nucleic acid is constructed from D-2-deoxyribose.

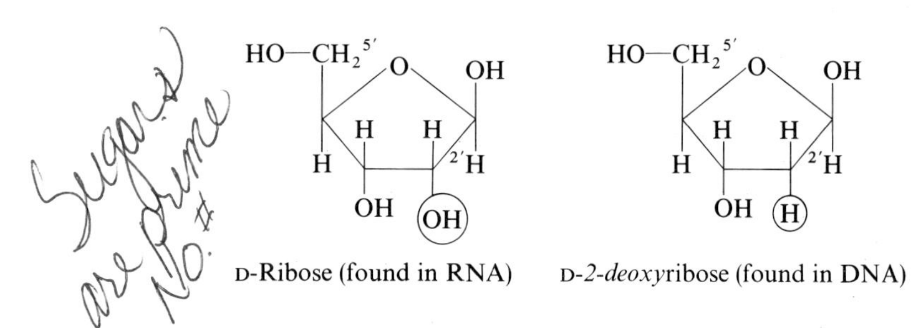

D-Ribose (found in RNA) D-2-*deoxy*ribose (found in DNA)

nucleoside

A **nucleoside** is constructed from a purine or pyrimidine base linked to either a ribose sugar or a deoxyribose sugar. The sugar is bonded in β-configuration to the #1 nitrogen of the pyrimidine base or to the #9 nitrogen of the purine base. Thus, *adenosine*, *guanosine*, *cytidine*, and *uridine* are the major *ribonucleosides* (see structures).

Adenosine Guanosine Uridine Cytidine

Deoxyadenosine, *deoxyguanosine*, *deoxycytidine*, and *deoxythymidine* are the major *deoxyribonucleosides* (see structures). Deoxyuridine is not naturally found in nucleic acids, as uracil is normally linked to ribose to form uridine.

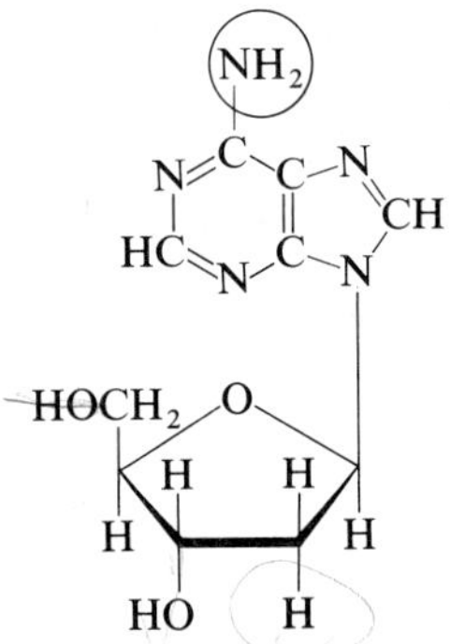

Deoxyadenosine 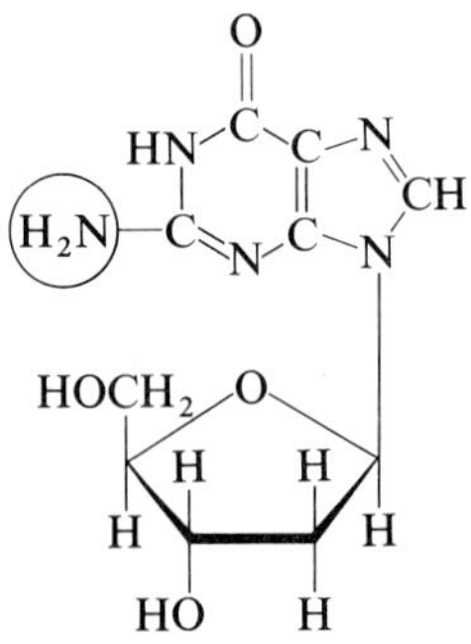 Deoxyguanosine 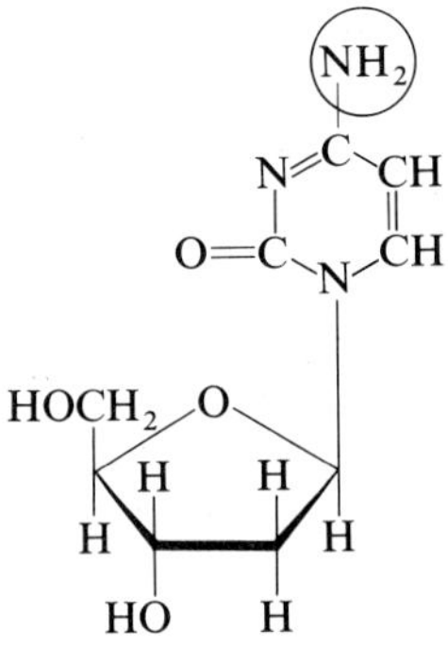 Deoxycytidine 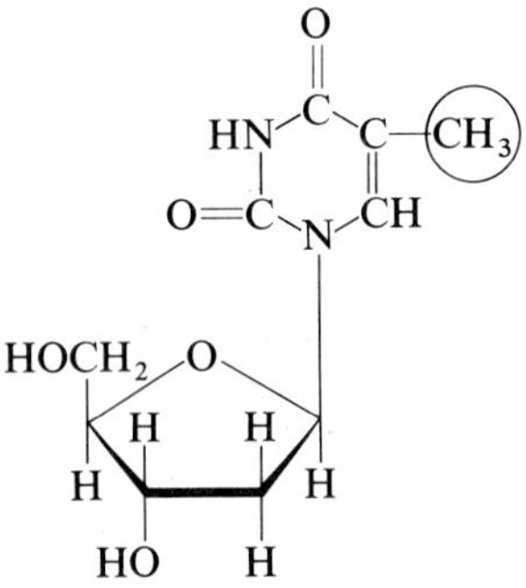 Deoxythymidine

nucleotide

A **nucleotide** contains one or more phosphate groups usually esterified to the 5′ carbon as shown in the following structures.

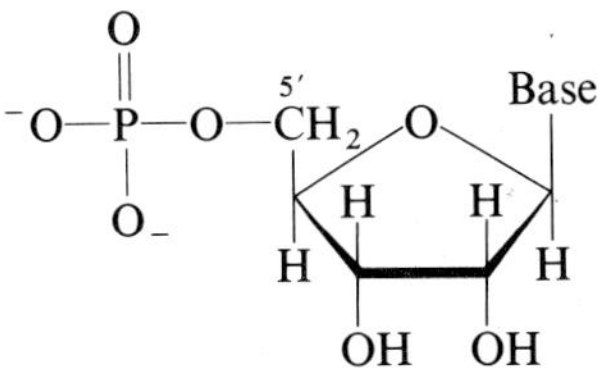

Ribonucleoside 5′-phosphate 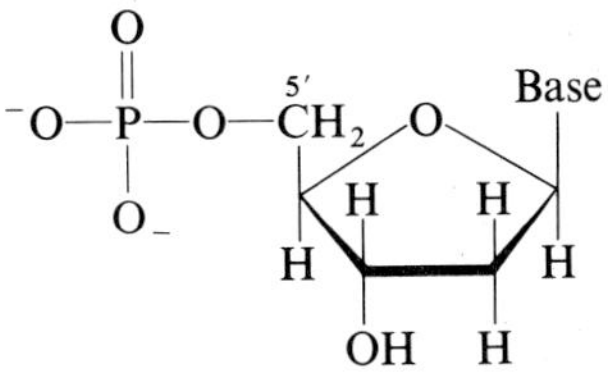 Deoxyribonucleoside 5′-phosphate

Table 16–1 summarizes the nomenclature of the major nucleotides.

TABLE 16–1 Nucleoside and Nucleotide Nomenclature.

Base	Ribonucleoside	Ribonucleotide (5′-monophosphate)[a]
Adenine (A)	Adenosine	Adenylate (AMP)
Guanine (G)	Guanosine	Guanylate (GMP)
Uracil (U)	Uridine	Uridylate (UMP)
Cytosine (C)	Cytidine	Cytidylate (CMP)
Base	**Deoxyribonucleoside**	**Deoxyribonucleotide[b] (5′-monophosphate)**
Adenine (A)	Deoxyadenosine	Deoxyadenylate (dAMP)
Guanine (G)	Deoxyguanosine	Deoxyguanylate (dGMP)
Thymine (T)	Deoxythymidine	Deoxythymidylate (dTMP)
Cytosine (C)	Deoxycytidine	Deoxycytidylate (dCMP)

[a] If one phosphate group is esterified to the 5′ position of the sugar, the resulting nucleotide is called 5′-monophosphate.

[b] The small "d" prefix denotes the deoxyribose sugar.

Nucleoside monophosphate esters have been defined and appropriate structures shown. Nucleoside diphosphate and triphosphate esters are also present in the cell. If one phosphate group is esterified to the 5′ position, the resulting nucleotide is called *nucleoside 5′-monophosphate* (*NMP*). If two phosphate groups are so esterified, the compound is called a *nucleoside 5′-diphosphate* (*NDP*); if three, a *nucleoside 5′-triphosphate* (*NTP*). The structures and nomenclature of the adenosine mono-, di-, and triphosphates are shown in the following diagram:

Adenosine 5′-monophosphate (AMP)
Adenosine 5′-diphosphate (ADP)
Adenosine 5′-triphosphate (ATP)

Finally, it should be noted that cyclic derivatives of monophosphate nucleotides control various biological functions. For example, cyclic adenosine monophosphate (*cyclic AMP* or *cAMP*) contains a phosphate group esterified to the 5′ and 3′ positions of the ribose ring (adenosine 3′, 5′-monophosphate).

Cyclic AMP
(adenosine 3′,5′-mono
phosphate)

16.3 PURINE AND PYRIMIDINE METABOLISM

PURINE BIOSYNTHESIS

The purine ring system is constructed from a number of precursor molecules by a very interesting biosynthetic pathway. The origin of the individual atoms of the purine ring is shown in Figure 16–1. Glycine, CO_2, and derivatives of *tetrahydrofolate* (*FH_4*) (see Chapter 20) contribute the carbon atoms, while glutamine and aspartic acid contribute the nitrogen atoms.

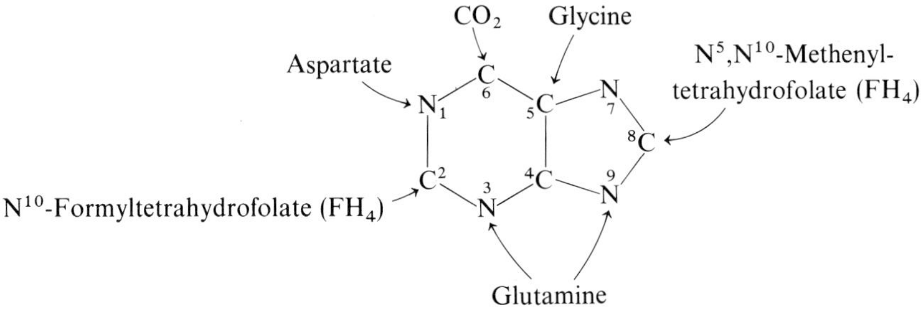

CO_2 Glycine
Aspartate
N^5,N^{10}-Methenyl-tetrahydrofolate (FH_4)
N^{10}-Formyltetrahydrofolate (FH_4)
Glutamine

FIGURE 16–1 The origin of the atoms of the purine ring system.

The biosynthesis of the purine ring system and the major purine nucleotides is shown in Figure 16–2.

Interestingly, the purine ring system is *assembled while attached to ribose phosphate* as shown in Figure 16–2. If you recall, ribose 5-phosphate is synthesized by the pentose phosphate pathway. Ribose 5-phosphate is phosphorylated by ATP to yield *5-phosphoribosyl-1-pyrophosphate* (*PRPP*). Hydrolysis of the pyrophosphate group (in a later reaction) serves to drive that reaction to completion. Next, the purine ring system is assembled—first the 5-membered ring, then the 6-membered ring. Glutamine contributes an amino group that displaces pyrophosphate from the 1′ position of the ribose ring. Additional reactions take place—condensation of glycine (contributing carbons 4 and 5 and nitrogen 7 of the ring), addition of carbon 8 by methenyl-FH_4, addition of nitrogen #3 by glutamine, and ring closure. The 6-membered ring is constructed next. A CO_2 molecule contributes carbon #6, an aspartic acid contributes nitrogen #1 (from the amino group of aspartic acid), and formyl-FH_4 contributes carbon #2, followed by ring closure. *Inosine 5′-monophosphate* (*IMP*) (also known as *inosinate*) is the final product of the pathway.

Inosinate is the precursor to *adenylate* (*AMP*) and *guanylate* (*GMP*). An aspartic acid molecule contributes an amino group to the #6 carbon of the inosinate purine ring, forming adenylate. Alternatively, the #2 carbon

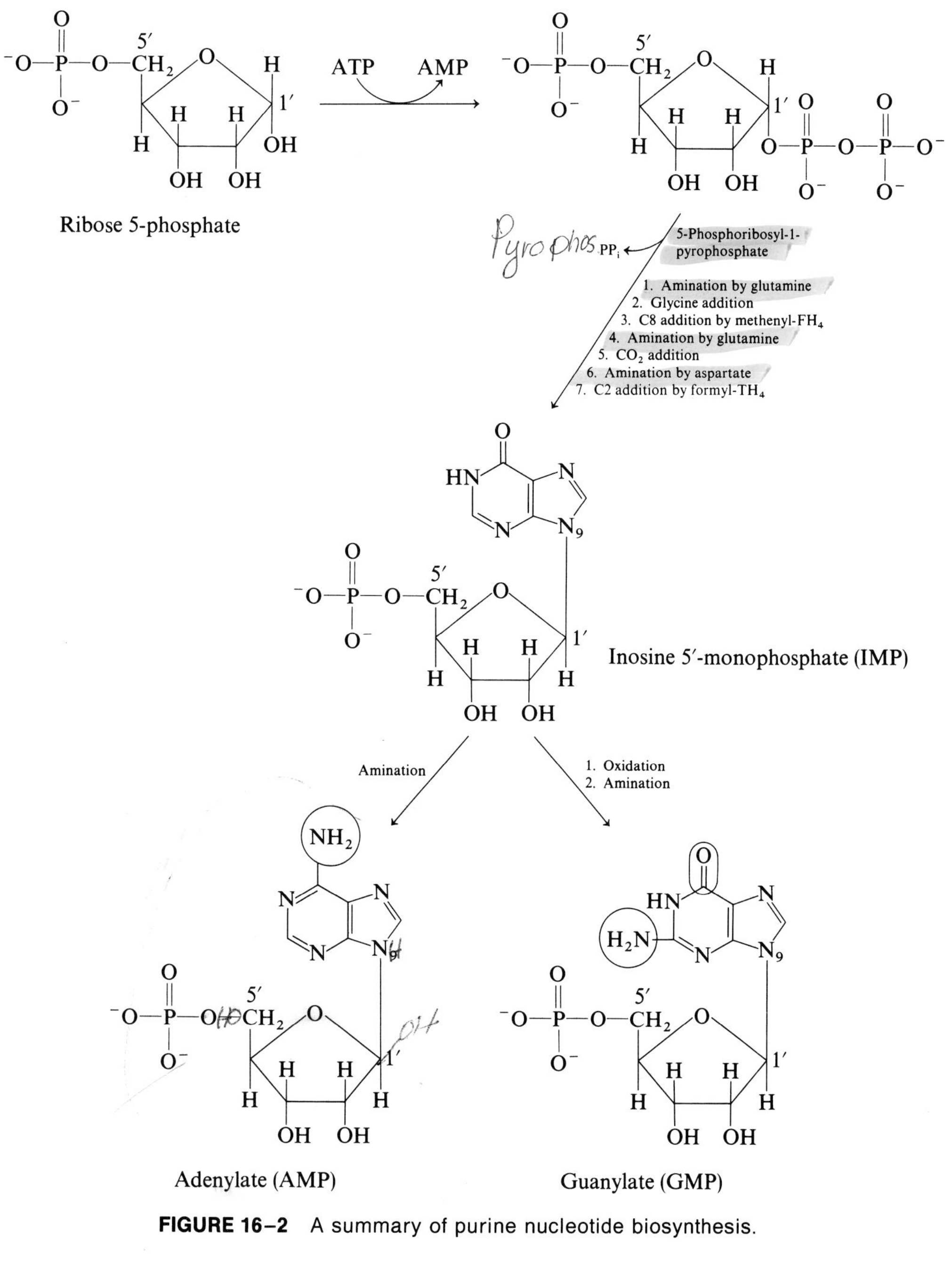

Ribose 5-phosphate

ATP AMP

5-Phosphoribosyl-1-pyrophosphate

PP_i

1. Amination by glutamine
2. Glycine addition
3. C8 addition by methenyl-FH_4
4. Amination by glutamine
5. CO_2 addition
6. Amination by aspartate
7. C2 addition by formyl-TH_4

Inosine 5′-monophosphate (IMP)

Amination

1. Oxidation
2. Amination

Adenylate (AMP)

Guanylate (GMP)

FIGURE 16–2 A summary of purine nucleotide biosynthesis.

of the inosinate ring is oxidized and then aminated (by glutamine) to form guanylate (GMP).

PYRIMIDINE BIOSYNTHESIS

Pyrimidine nucleotides are synthesized in a different manner from purine nucleotides. Specifically, the pyrimidine ring is *first constructed* and then attached to the ribose phosphate, forming the nucleotide. Pyrimidine nucleotide biosynthesis is shown in Figure 16–3.

Carbamoyl phosphate and *aspartic acid* are condensed by *aspartate transcarbamoylase* to form *N-carbamoylaspartate.* N-Carbamoylaspartate is cyclized and oxidized to form *orotate.* Orotate next reacts with PRPP to form *orotidylate.* Orotidylate is then decarboxylated (that is, a CO_2 group is removed) to form *uridylate (UMP).*

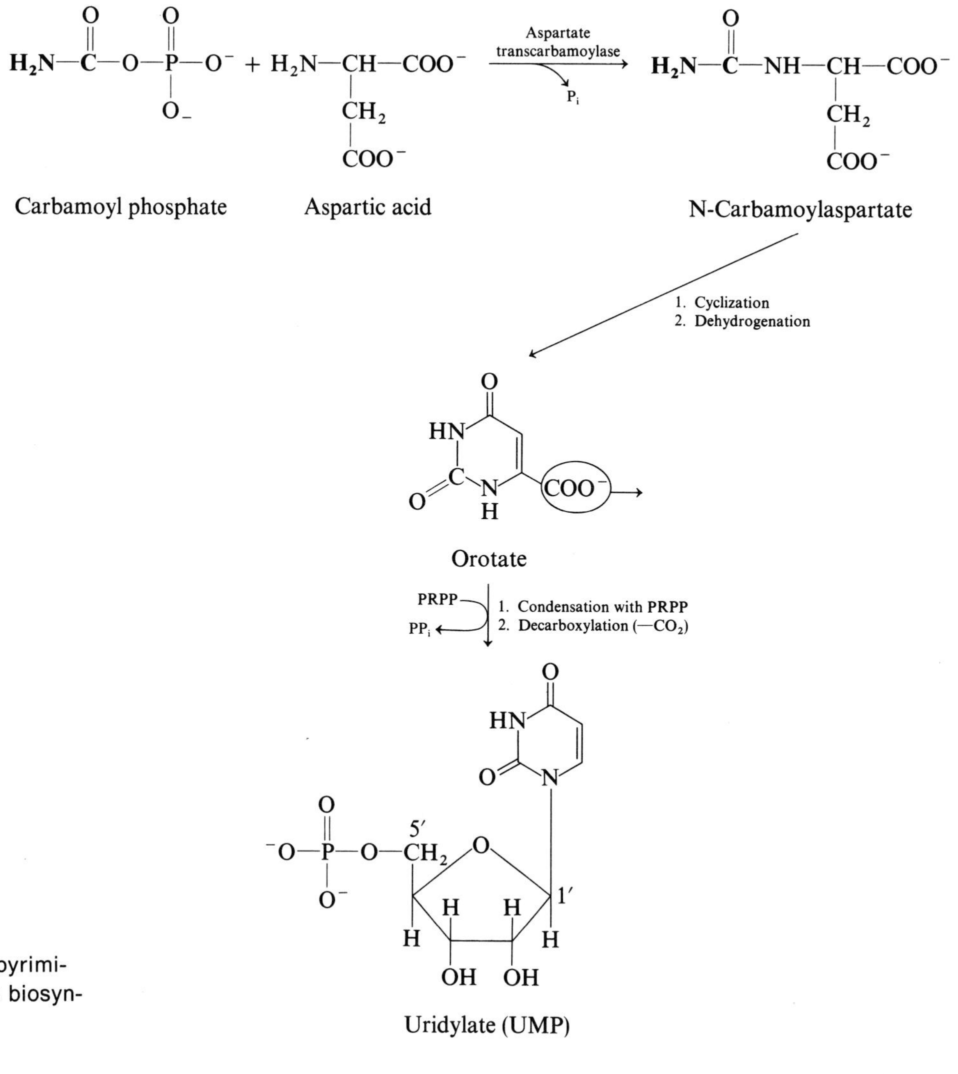

Carbamoyl phosphate + Aspartic acid —(Aspartate transcarbamoylase; P_i)→ N-Carbamoylaspartate

N-Carbamoylaspartate —(1. Cyclization 2. Dehydrogenation)→ Orotate

Orotate —(PRPP → PP_i; 1. Condensation with PRPP 2. Decarboxylation ($-CO_2$))→ Uridylate (UMP)

FIGURE 16–3
A summary of pyrimidine nucleotide biosynthesis.

The other pyrimidine nucleotides are constructed from UMP via the following reactions:

1. Phosphorylation of UMP to yield UTP:

$$\text{UMP} \xrightarrow{\text{ATP} \;\; \text{ADP}} \text{UDP} \xrightarrow{\text{ATP} \;\; \text{ADP}} \text{UTP}$$

2. Formation of *cytidine triphosphate* (*CTP*) from UTP:

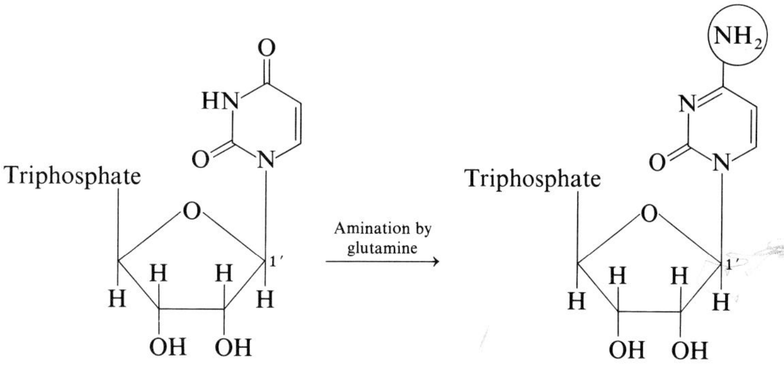

Triphosphate

O HN O N O H H H H 1′ OH OH

UTP

Amination by glutamine

Triphosphate

NH_2 N O N O H H H H 1′ OH OH

CTP

3. Formation of *deoxythymidylate* (*dTMP*) from deoxyuridylate (dUMP):

a. Reduction of uridylate diphosphate (UDP) to deoxyuridylate diphosphate (dUDP) (the deoxyribonucleotides are derived from this general reduction pathway):

$$\text{UDP} \xrightarrow{\text{NADPH} + \text{H}^+ \;\; \text{NADP}^+} \text{dUDP}$$

(The pentose sugar now lacks an OH group in the #2 position)

b. Formation of deoxythymidylate (dTMP):

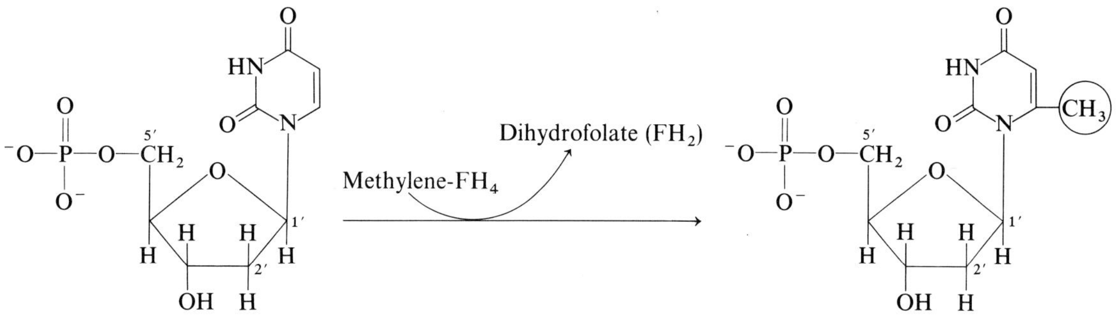

$^-O-P(=O)(O^-)-O-CH_2$ (5′) O HN O N O H H H H 1′ 2′ OH H

dUMP

Methylene-FH_4 → Dihydrofolate (FH_2)

$^-O-P(=O)(O^-)-O-CH_2$ (5′) O HN O N CH_3 O H H H H 1′ 2′ OH H

dTMP

PURINE AND PYRIMIDINE CATABOLISM

Dietary nucleic acids are digested in the intestine into mononucleotides via pancreatic nucleases. *Ribonucleases* degrade RNA into constitutive mononucleotides; *deoxyribonucleases* degrade DNA into its constitutive deoxymononucleotides. *Nucleoside phosphorylases* degrade the nucleotides into the free base (purine or pyrimidine) and phosphorylated sugar (ribose 1-phosphate or deoxyribose 1-phosphate). The sugars can be reused by the cell.

The degradation of purines leads to the production of *uric acid*, as shown in Figure 16–4. Adenine is degraded into *hypoxanthine*, whereas guanine is degraded into *xanthine*. Hypoxanthine is transformed into xanthine by xanthine oxidase, and xanthine is then converted to uric acid.

Humans degrade purines into uric acid, which is excreted in the urine. Other mammals and reptiles and fish further degrade uric acid into either *allantoin* or urea. In humans, the *excess* production of uric acid causes an increase in the level of uric acid in the blood. The excess uric acid can crystallize and precipitate in the synovial fluid of the joints, leading to a painful condition known as **gout** characterized by severe arthritis. Exactly what causes the excess production of uric acid is largely not understood. A defi-

gout

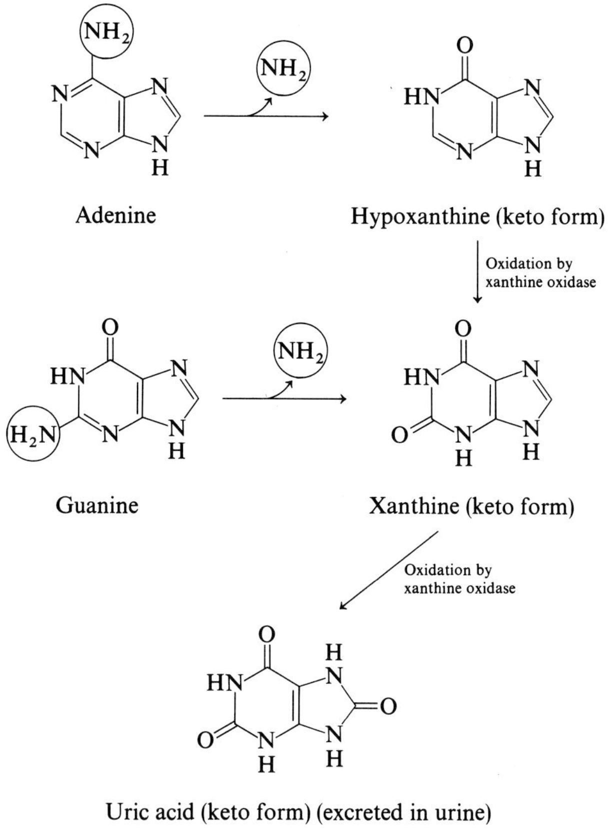

Adenine

Hypoxanthine (keto form)

Oxidation by xanthine oxidase

Guanine

Xanthine (keto form)

Oxidation by xanthine oxidase

Uric acid (keto form) (excreted in urine)

FIGURE 16–4
A summary of purine catabolism.

ciency in an enzyme that converts hypoxanthine back into inosinate (a salvage pathway) may account for the excess uric acid levels, as may a *deficiency* of glucose 6-phosphatase. Although the latter enzyme is part of a different pathway, an indirect metabolic relationship can be established. A deficiency in glucose 6-phosphatase activity would lead to an increase in glucose 6-phosphate level. This sugar would be diverted into the pentose phosphate pathway and lead to increased ribose 5-phosphate production. Increased PRPP synthesis would occur and, in turn, stimulate purine synthesis, leading to overproduction of uric acid (via purine catabolism). It should be noted that foods high in nucleic acids may aggravate the condition; therefore, dietary restrictions are often imposed on patients with gout.

Finally, pyrimidines are catabolized in the liver to ammonia, carbon dioxide, malonate, and methyl-malonate (pathway not shown).

16.4 THE STRUCTURE OF DEOXYRIBONUCLEIC ACID

deoxyribonucleic acid (DNA)

genetic code

Simply defined, **deoxyribonucleic acid (DNA)** is a long double-stranded polymer composed of a very large number of deoxyribonucleotides. The *sequence* of purine and pyrimidine bases constitute a code (often called the **genetic code**). In effect, the DNA molecule contains codes for all of the proteins, enzymes, and nucleic acids necessary for the duplication, growth, development, and normal functioning of the cell. Put another way, the nucleus of each and every cell in your body contains deoxyribonucleic acid, and it is the information in this molecule that defines each of the biochemical, physiological, and physical features that constitute "you".

The structure of deoxyribonucleic acid was not understood until 1953, when Dr. J. Watson and Dr. F. Crick published their historic paper that beautifully described the structure of the DNA molecule. This concise paper (only 900 words) was published in *Nature*, a British scientific journal. Its publication was a landmark event in biology.

phosphodiester linkages

The essential structural and chemical features of the DNA molecule will be summarized. Figure 16–5 shows the chemical structure of a portion of one strand of a DNA molecule. A phosphate sugar (deoxyribose) backbone is readily apparent in the diagram. As the name implies, this backbone serves a structural role and contains no genetic information. The individual phosphate sugar residues are joined by **phosphodiester linkages** between the 5′ and 3′ carbons. A phosphate group linked to the 3′ carbon is attached to the 5′ carbon of the next deoxyribose unit. Notice that the top of the structure is a 5′ phosphate group, while the bottom is an open link at the 3′ position. Thus, the phosphate deoxyribose backbone exhibits *directionality* (noted as 5′ → 3′). The individual purine and pyrimidine bases are linked to the 1′ position of the deoxyribose units. The *sequence of the purine and*

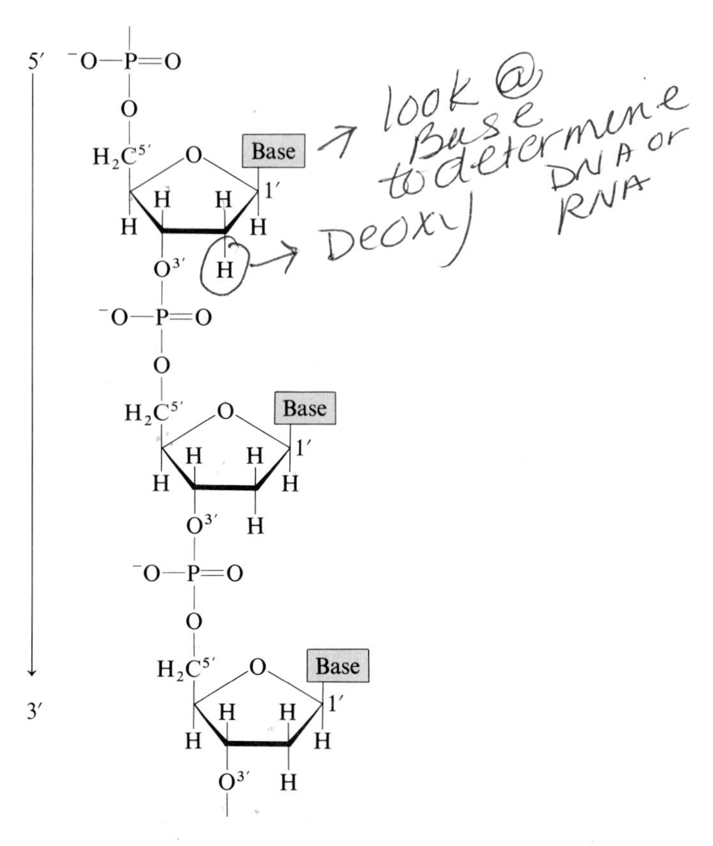

FIGURE 16–5
The structure of a portion of the DNA chain.

genetic code

pyrimidine bases is the important feature. This sequence is the **genetic code**. A hypothetical base sequence is shown in Figure 16–6.

Chemical analysis of DNA isolated from various sources has revealed that the cells of each living species contain specific quantities of each of the four bases. In addition, a *unique ratio* of purines to pyrimidines was found. The amount of adenine (A, purine) was equal to the amount of thymine (T, pyrimidine), and the amount of guanine (G, purine) was equal to the amount of cytosine (C, pyrimidine). This relationship is summarized in Table 16–2. From this information (along with physical measurements obtained by a technique called X-ray diffraction analysis), Watson and Crick postulated that adenine could only hydrogen bond with thymine (A=T; two H-bonds) and that guanine could only hydrogen bond with cytosine (G≡C; 3 H-bonds). The hydrogen-bonding relationships are shown in Figure 16–7.

The X-ray diffraction studies indicated that the DNA molecule was a *helix*. Watson and Crick used this data, along with the manner in which the purine/pyrimidine bases are paired, to formulate a possible structure. They proposed that the DNA molecule is constructed from two very long strands of deoxyribonucleotides wound about each other in a *right-handed double helix*. The phosphate-deoxyribose backbone of the two strands is hydrophilic; therefore, it is located on the *outside* of the double helix, in contact with

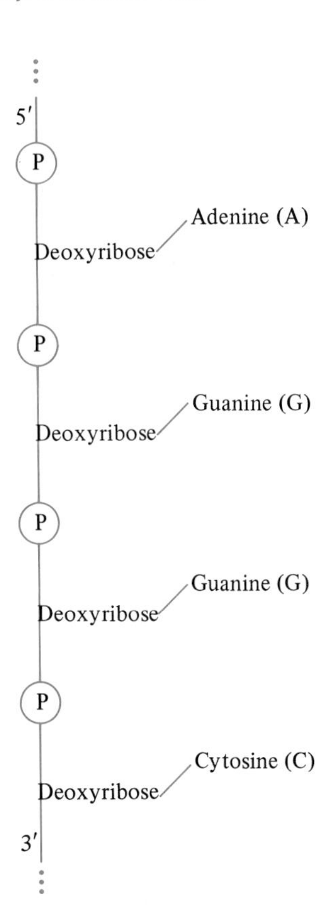

FIGURE 16–6
A hypothetical base sequence. This base sequence would be read from the 5′ position and could be abbreviated as 5′—A—G—G—C—3′. The base sequence of the DNA within a cell may be millions of nucleotides in length.

water. The hydrophobic purine-pyrimidine base pairs are located on the *inside* of the helix, away from water. The base pairs are perpendicular to the long axis of the helix. As we mentioned previously, each strand exhibits directionality (5′ → 3′). Watson and Crick proposed that the two strands within the double helix are **antiparallel**—that is, the orientation of one strand is *opposite* to the orientation of the other strand. Stated more simply, the chains run in opposite directions, as shown in Figures 16–7 and 16–8. The two strands are held together by the hydrogen bonds connecting the thousands of purine/pyrimidine base pairs (see Figure 16–7). Finally, the X-ray diffraction data gave important clues about the dimensions of the molecule. One complete turn of a strand about the helix axis requires ten base

antiparallel

TABLE 16–2 The Composition of Human DNA.

Nucleotide Base	Composition (percent)	Base Ratio
Adenine (purine)	30.9	1.05 (A/T)
Thymine (pyrimidine)	29.4	
Guanine (purine)	19.9	1.00 (G/C)
Cytosine (pyrimidine)	19.8	

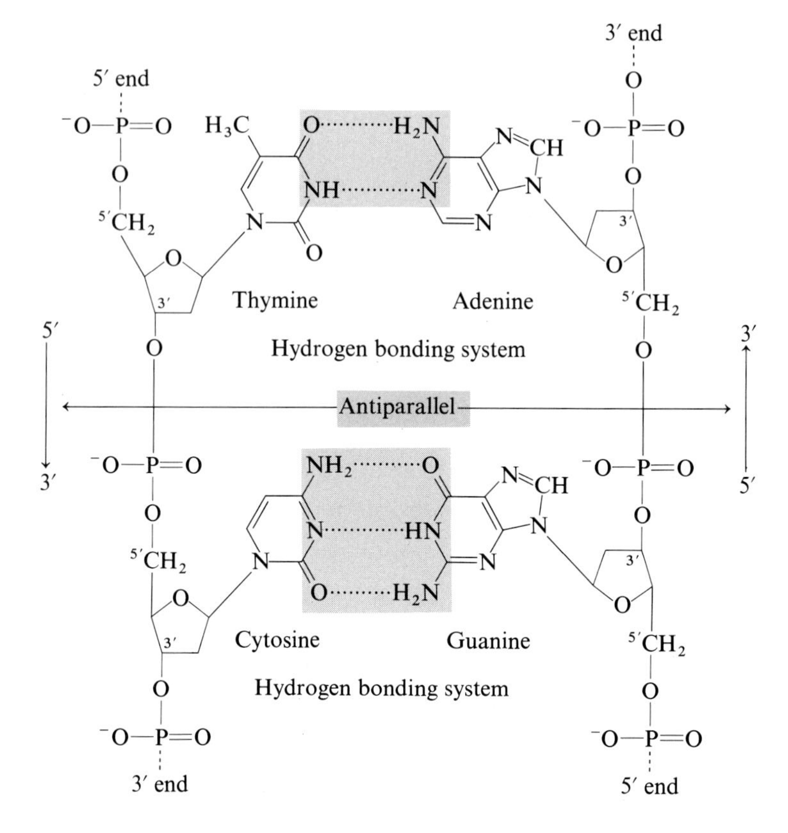

FIGURE 16–7 Hydrogen-bonding relationships between the purine and pyrimidine base pairs of a DNA molecule. Note the antiparallel orientation of the individual strands. (Adapted from E. E. Conn and P. K. Stumpf, *Outlines of Biochemistry*, 4th ed. (New York: John Wiley, 1976), p. 131.)

pairs and 34 Å. The molecule is 20 Å in diameter and many thousands of Å in length. The structure of the DNA molecule is shown in Figures 16–8 and 16–9.

complementary

An important point is that the strands are **complementary**. Because adenine (A) can only hydrogen bond with thymine (T) and guanine (G) can only bond with cytosine (C), the base sequence of one strand defines the base sequence in the opposite (complementary) strand of the double helix. To illustrate, consider the following base sequence:

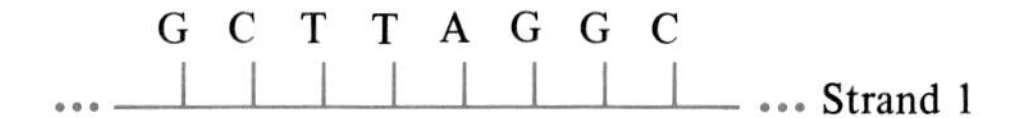

The base sequence of the other strand must be:

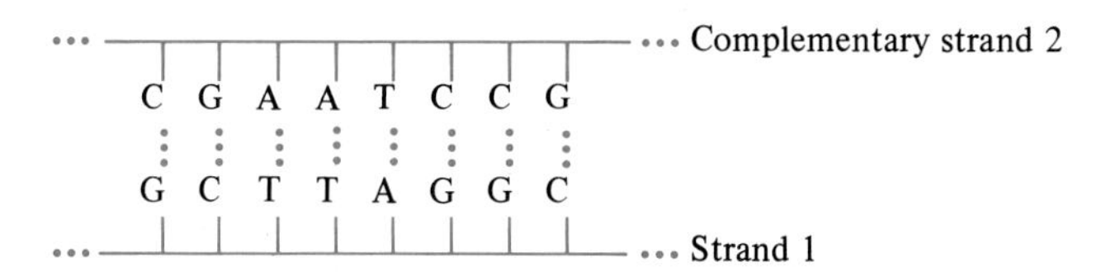

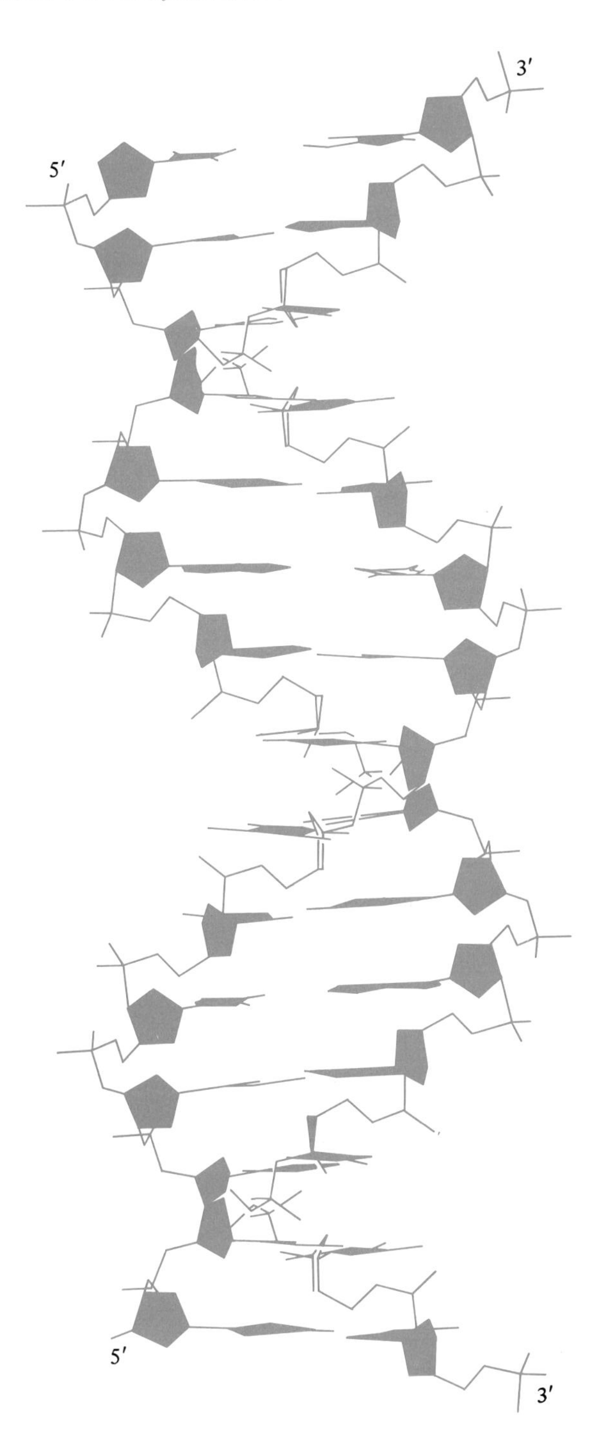

FIGURE 16–8
The structure of the DNA molecule. The phosphate deoxyribose backbone is prominent (the deoxyribose units appear as dark pentagons). The purine-pyrimidine base pairs appear as thin horizontal lines.

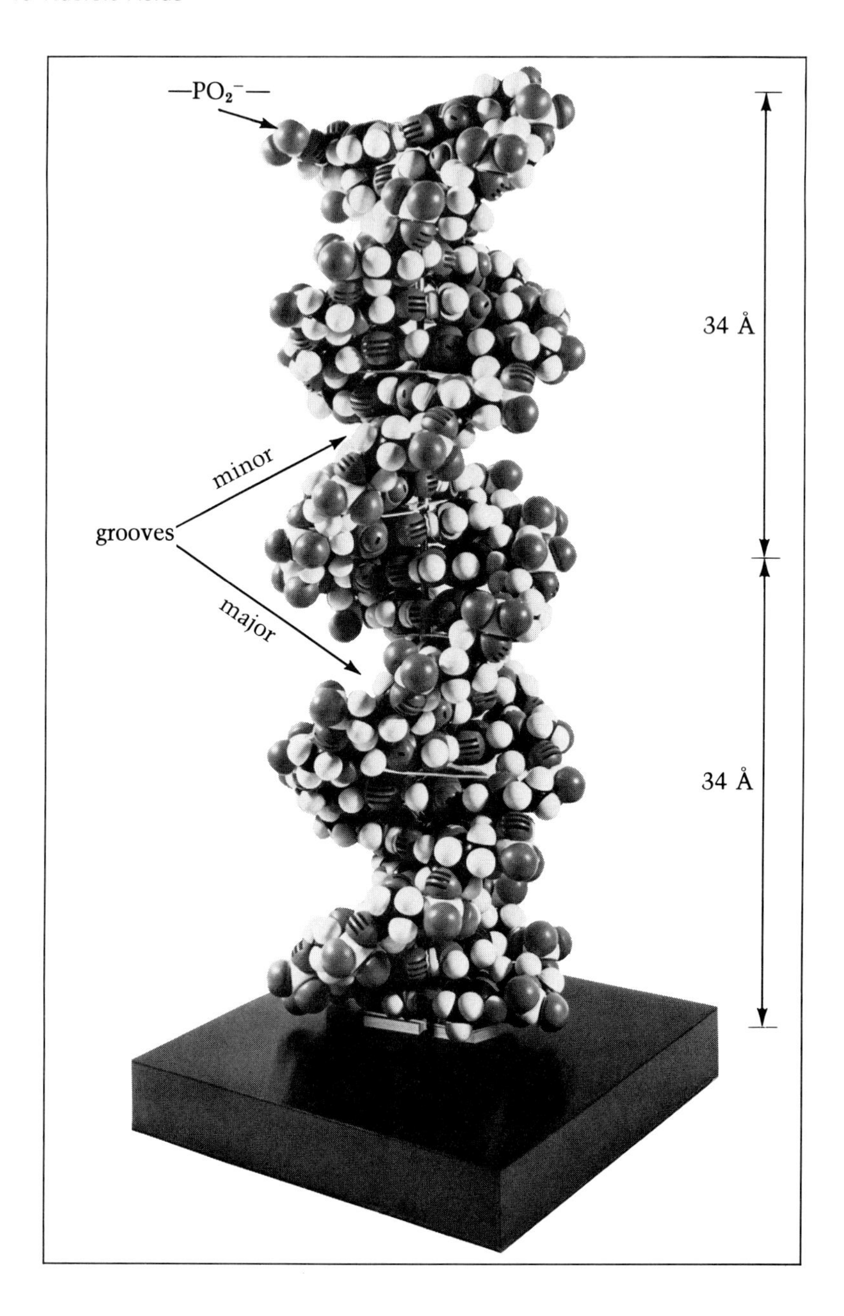

FIGURE 16–9
A space-filling model of the DNA molecule. Notice the presence of grooves in the double helix. One complete turn of the helix involves 10 base pairs and requires 34 Å. Photo courtesy of the Ealing Corporation, Natick, Massachusetts.

As we shall see in the next chapter, complementary strands are a critical feature of DNA replication—the production of new "daughter" strands of DNA containing the original base sequence of the "parent" strands.

Finally, we should mention the organization of DNA in the cell nucleus. In *prokaryotes* (bacteria) the DNA double helix is actually a very large coiled circle. In *eukaryotes* (animal cells) the DNA double helix exists as long threads coiled into a compact bundle. These bundles (along with associated proteins) constitute the chromosomes of the cell nucleus.

16.5 THE STRUCTURE OF AND CLASSES OF RIBONUCLEIC ACID

ribonucleic acid (RNA)

Ribonucleic acid (RNA) is a very long polymer of *ribonucleotides* linked via 3′, 5′-phosphodiester bonds. This nucleic acid is constructed from D-ribose instead of D-2-deoxyribose. The molecule consists of a single strand containing adenine and guanine (the purines) and cytosine and uracil (the pyrimidines). The bases are joined to the individual ribose units at the 1′ position. Notice that RNA does not contain the base thymine. The structure of part of an RNA molecule is shown in Figure 16–10.

Although RNA molecules are single-stranded, regions of an RNA molecule can hydrogen bond into double-helices in the form of *hairpin loops*. Purine bases hydrogen bond with complementary pyrimidine bases (A=U and G≡C) in these regions as shown in the following diagram:

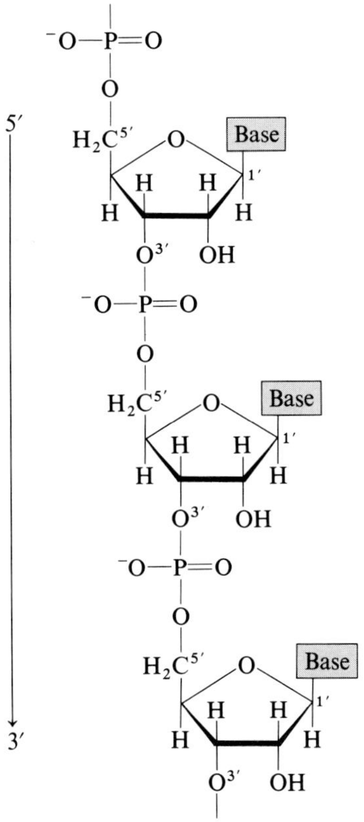

FIGURE 16–10
The structure of a portion of the RNA chain.

```
                       C
                     U   C
                    U     G
                     G—C
                     |  |
                     G—C
                     |  |
                     C—G
                     |  |
                     A—U
                     |  |
                     A—U
                     |  |
                     U—A
                    /    \
···—G—G—C—G—A—G—G—/      \G—G—U—A—G—C—G—···
```

A single-stranded molecule of RNA can develop a number of hairpin loops and, thus exhibit a complicated 3-dimensional structure (see Figure 16–11).

There are three major classes of RNA molecules: *ribosomal RNA* (*rRNA*), *messenger RNA* (*mRNA*), and *transfer RNA* (*tRNA*). Table 16–3 shows the amounts of each class of RNA found in an *E. Coli* (bacterium) cell. Ribosomal RNA (rRNA) is found in the greatest concentration within the cell because this type of RNA is the major structural component of the **ribosome**, the organelle on which protein synthesis takes place. Different sizes of rRNA have been isolated and extensively studied. Messenger RNA (mRNA) is a short-lived form of RNA that carries the chemical message

ribosome

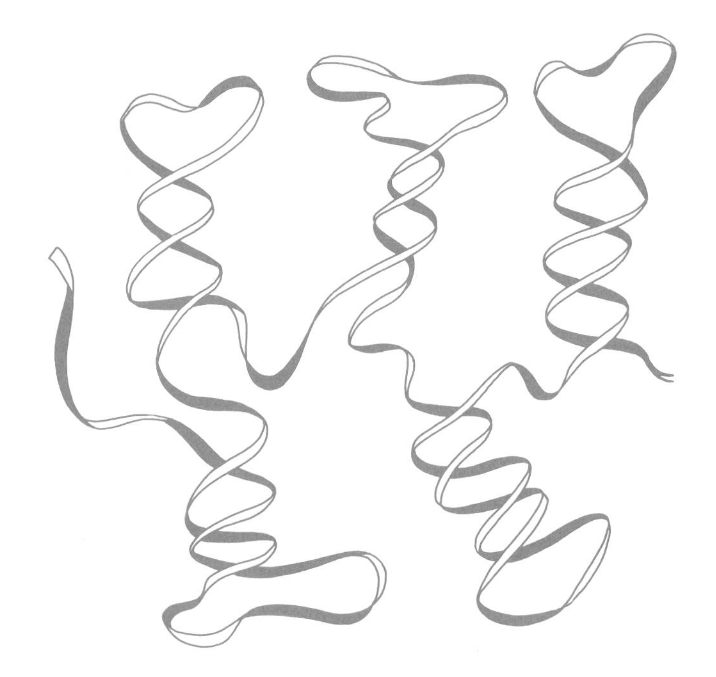

FIGURE 16–11
A hypothetical molecule of RNA exhibiting single-strand coiling and 3-dimensional structure.

TABLE 16–3 The Different Forms of RNA in *E. Coli*.

Class of RNA	Amount in Cell (percent)	Number of RNA Subtypes	Number of Nucleotides
Ribosomal RNA (rRNA)	82	3	~120 ~1700 ~3700
Transfer RNA (tRNA)	16	60	73–93
Messenger RNA (mRNA)	2	Many	75–3000

from DNA within the nucleus to the ribosome in the cytoplasm prior to protein synthesis. Finally, transfer RNA (t RNA) molecules are short RNA molecules (73–93 nucleotides in length) that bind and transfer amino acids during protein synthesis. Cells contain many different tRNA molecules, each specific for a particular amino acid. A more detailed discussion of the chemistry and function of these classes of RNA molecules is found in the next two chapters.

SUMMARY

Nucleic acids—(deoxyribonucleic acid (DNA) and ribonucleic acid (RNA)—carry all of the chemical information needed for the growth, development, and reproduction of living organisms.

Nucleotides are constructed from a purine or pyrimidine base, a sugar, and one or more phosphate groups. Adenine and guanine are the major purines, cytosine, thymine, and uracil are the major pyrimidines. Ribonucleic acid (RNA) is constructed from D-ribose, deoxyribonucleic acid (DNA) is constructed from D-2-deoxyribose. Nucleoside mono-, di-, and triphosphate esters were defined (with structures).

The following are required to make the purine ring system: glycine, CO_2, derivatives of tetrahydrofolate (FH_4), and the amino acids glutamine and aspartic acid. Ribose 5-phosphate is first converted into 5-phosphoribosyl-1-pyrophosphate (PRPP), after which the purine ring system is synthesized.

Pyrimidines are synthesized from carbamoyl phosphate and aspartic acid, which form orotate. Uridylate (UMP) is made from orotidylate.

Cytidine triphosphate (CTP) is synthesized from UTP; deoxythymidylate (dTMP) is made from deoxyuridylate (dUMP).

The catabolism of the purine bases adenine and guanine involves deamination and oxidation reactions that form xanthine, which is converted to uric acid for excretion. Increased uric acid levels produce gout, characterized by painful arthritis and the precipitation of uric acid crystals in the joints.

The deoxyribonucleic acid (DNA) molecule is constructed from two polymeric strands of deoxyribonucleotides linked together via 3′,5′-phosphodiester bonds. The two strands wind about each other in the form of a double helix. The strands run in opposite directions—that is, are antiparallel. The base adenine (a purine) can hydrogen bond with thymine (a pyrimidine), and guanine (a purine) can hydrogen bond with cytosine (a pyrimidine). Genetic information is contained in the sequence of bases in the DNA molecule. The base pairs reside on the interior of the DNA helix, away from water. Each strand is complementary—that is, the base sequence of one strand is duplicated in the complementary sequence of the other strand.

Ribonucleic acids (RNA) are very long, single-stranded polymers of ribonucleotides connected via 3′,5′-phosphodiester links. Adenine, guanine, cytosine, and uracil are the four major bases. The RNA molecule does not exist as a double helix; however, portions of the molecule can form helical regions in which complementary base pairs are involved in hydrogen-bonding associations. There are three major classes of RNA: ribosomal RNA (rRNA), which is found in ribosomes; messenger RNA (mRNA), which carries the genetic message; and transfer RNA (tRNA), which binds and transfers amino acids during protein synthesis.

REVIEW QUESTIONS

1. We have introduced a number of important terms and concepts in this chapter. Define (or briefly explain) each of the following terms:
   - **a.** Purine base
   - **b.** Pyrimidine base
   - **c.** Deoxyribose
   - **d.** Nucleoside
   - **e.** Nucleotide
   - **f.** Nucleic acid
   - **g.** Adenylate (AMP)
   - **h.** Inosinate (IMP)
   - **i.** Orotate
   - **j.** Uric acid
   - **k.** DNA (the double helix)
   - **l.** Antiparallel
   - **m.** Complementary
   - **n.** RNA
   - **o.** mRNA
   - **p.** tRNA
2. Draw the complete chemical structures for:
   - **a.** Guanosine mono-, di-, and triphosphates.
   - **b.** Cytidine mono-, di-, and triphosphates.
3. Draw the chemical structure of cyclic GMP.
4. Why is a pyrophosphate group first bound to the ribose 5-phosphate molecule *before* purine ring biosynthesis?
5. **a.** Which amino acids are required for purine biosynthesis?
   b. From which TCA cycle intermediates do these amino acids arise?

*6. From what you know about allosteric enzymes, suggest a possible mechanism by which inosinate (IMP), AMP, and GMP regulate nucleotide biosynthesis.

7. Briefly compare and contrast purine and pyrimidine biosynthesis.
8. Draw the *complete* chemical structure for these segments of DNA:
   a. 5′—A—G—3′ **b.** 5′—G—C—T—A—3′
9. Draw the complementary strands for the following segments of deoxyribonucleic acids:
   a. 5′—A—A—T—G—C—A—T—C—3′
   b. 5′—G—G—C—T—T—A—C—C—G—A—T—A—G—C—3′
10. A portion of a DNA molecule contained the following base sequence:

    5′—A—G—G—C—T—A—G—G—C—C—C—A—A—G—G—C—T—A—3′

    Draw the complementary base sequence for the corresponding mRNA molecule.

SUGGESTED READING

Bhagavan, N. V. *Biochemistry*, 2nd ed., Ch. 5, pp. 371–412, 492–543. Philadelphia: J. B. Lippincott, 1978.

Lehninger, A. L. *Biochemistry*, 2nd ed., Chs. 12, 26, 31, and 32. New York: Worth, 1975.

Montgomery, R., Dryer, R. L., Conway, T. W., and Spector, A. A. *Biochemistry: A Case-Oriented Approach*, 3rd ed., Ch. 11. St. Louis: C. V. Mosby, 1980.

Stryer, L. *Biochemistry*, 2nd ed., Chs. 22 and 24. San Francisco: W. H. Freeman, 1981.

Watson, J. D. *Molecular Biology of the Gene*, 3rd ed. Menlo Park, CA: W. A. Benjamin, 1976.

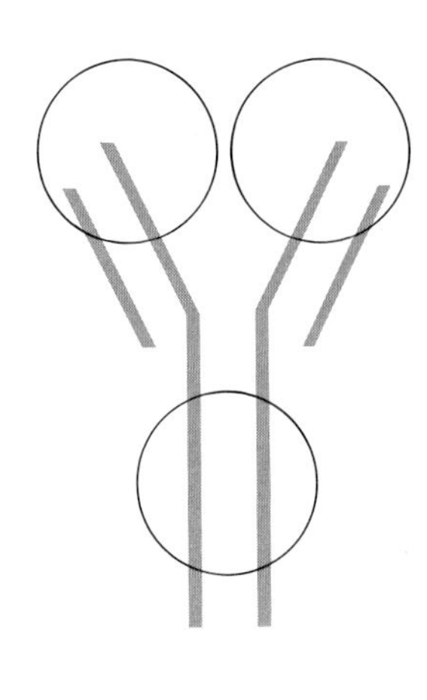

Chapter 17

The Storage and Transmission of Genetic Information

17.1 INTRODUCTION

The structure of nucleic acids (DNA and RNA) was introduced in the previous chapter. In this chapter, we will discuss how the genetic information in these molecules is stored and transmitted. Our insight into these chemical processes is very recent, and incomplete. The reader should consult the references cited in the Suggested Reading section for more information about this important—and very interesting—aspect of biochemistry.

flow of genetic information

Basically, this chapter deals with the **flow of genetic information** from one type of biological molecule to another. This flow of information can be represented as follows:

$$\text{DNA} \xrightarrow{\text{Transcription}} \text{mRNA} \xrightarrow{\text{Translation}} \text{Protein}$$

The information contained in the DNA molecule (the base sequence) is transmitted by the process of transcription to *messenger RNA* (*mRNA*). Messenger RNA contains a *complementary sequence* of bases constructed using the DNA base sequence as a template. A specific 3-base sequence, a codon, in the mRNA molecule corresponds to a particular amino acid. Therefore, a series of 3-base mRNA sequences defines the final amino acid sequence of the protein. Previous discussions of protein structure (Chapter 3, Amino Acids and Proteins) stressed that the amino acid sequence of a protein defines the 3-dimensional shape of the molecule. The biological function of the protein is largely defined by its shape and chemical properties. Therefore, the *3-dimensional shape of a protein is ultimately defined by the*

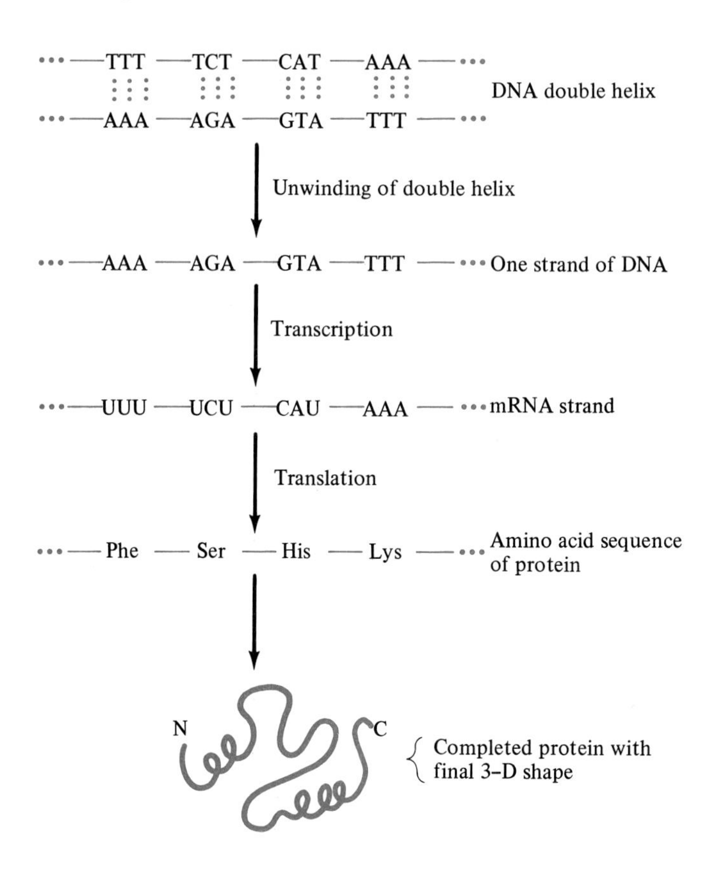

FIGURE 17–1
The flow of genetic information from DNA to protein.

nucleotide base sequence of the DNA molecule. The flow of genetic information is diagrammed in more detail in Figure 17–1.

In this chapter, we will first discuss how the DNA molecule is *replicated* to form subsequent generations of daughter strands. We will then discuss the process of **transcription** by which mRNA is synthesized and processed. (The rather complicated process of protein synthesis **(translation)** will be described in Chapter 18.) We will then discuss DNA repair and recombinant DNA technology.

transcription
translation

17.2 THE REPLICATION AND REPAIR OF DNA

The replication of the DNA molecule is a complex and fascinating phenomenon. How can DNA precisely replicate itself—that is, how is the base sequence (the genetic code) transferred from one generation of DNA molecules to another? To answer, DNA replication is dependent upon the *unique structure* of the DNA molecule (see Chapter 16, section 16.4). The concept of complementary strands beautifully explains this important point. During replication, the double helix of the DNA molecule is unwound and the individual strands separated from each other. Thus, the individual base pairs within each strand are free and no longer hydrogen bonded to their complementary base. During DNA replication, new bases are inserted to form a new strand. For example, if part of one strand contains a free adenine residue, a thymine nucleotide will join (via hydrogen bonds) to that residue. This process continues until a new (daughter) strand is constructed using the old (parent) strand of DNA as a **template** to define the base sequence. Thus, the new double helix molecule contains one strand from the original (parent) DNA molecule and one from the new (daughter) strand. This process is called **semiconservative replication**, and is diagrammed in Figure 17–2.

template

semiconservative replication

At least 15 different proteins and enzymes are involved in DNA replication. This complexity is probably necessary to ensure that the genetic

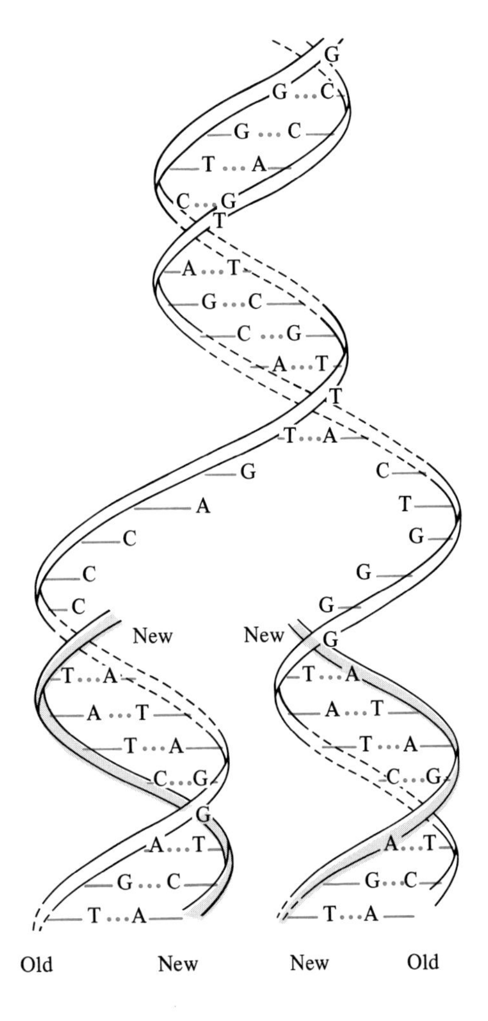

FIGURE 17–2
Semiconservative replication of the DNA molecule. When the DNA molecule replicates, the double helix unwinds and exposes the individual strands of the helix. New and complementary strands are synthesized and rewound, forming daughter DNA molecules. (Adapted from J. D. Watson, *The Molecular Biology of the Gene,* 3rd ed. (Menlo Park, CA: W. A. Benjamin, 1965), p. 267.)

code is *accurately copied* during replication. It has been estimated that the cell makes only one copying error for each 10^9 to 10^{10} base pairs! In addition, the cell contains enzymes that can repair a damaged or defective DNA molecule.

In 1956, Dr. A. Kornberg and colleagues discovered an enzyme that could synthesize new strands of DNA. They named the enzyme *DNA polymerase* (it was later renamed DNA polymerase I). The enzyme requires one strand of the DNA double helix, which acts as a template for the addition of new bases. Also required for DNA synthesis are the *deoxyribonucleoside triphosphates* (dATP, dGTP, dTTP, and dCTP), an RNA primer, and Mg^{2+}. The enzyme catalyzes the sequential addition of mononucleotide units to the free 3′ position of the new and growing strand of DNA. The direction of synthesis is $5' \rightarrow 3'$. Pyrophosphate (PP_i) cleavage from the incoming dNTP provides sufficient free energy to form the new 3′,5′-phosphodiester bond. DNA polymerase can add about 1000 nucleotides per minute to the replicating DNA molecule. This reaction is shown in Figure 17–3. Other DNA polymerases (DNA polymerases II and III) also exist within the cell and carry out additional enzymatic functions via similar mechanisms. Current research has shown that *DNA polymerase III* is probably the "real" DNA polymerase responsible for the synthesis of the new DNA strand; DNA polymerase I is only responsible for removing RNA "primers" and for filling in gaps within the new DNA strand (see Figure 17–6, page 327).

Interestingly, DNA polymerase I can catalyze two additional reactions: (1) *$3' \rightarrow 5'$ exonuclease* activity and (2) *$5' \rightarrow 3'$ exonuclease* activity. The $3' \rightarrow 5'$ exonuclease catalyzes the sequential hydrolysis of nucleotides, starting at the 3′ end of the strand and proceeding towards the 5′ end. The enzyme requires a free nucleotide (at the 3′ end) that is *not* part of the double helix. This exonuclease removes a *wrong nucleotide* that may have been added to the growing DNA strand. Thus, the enzyme exhibits *editing* properties—it checks the accuracy of the newly formed base sequence. The $5' \rightarrow 3'$ exonuclease catalyzes the cleavage of a 3′,5′-phosphodiester bond in a strand that is still part of the DNA double helix. In effect, this enzyme catalyzes the *excision of a section of DNA* as part of the repair process.

During semiconservative replication, the original (parent) DNA molecule is separated into two antiparallel strands, as shown in Figure 17–4. It was originally postulated that one strand is synthesized in the $5' \rightarrow 3'$ direction, while the other strand is synthesized in the $3' \rightarrow 5'$ direction (Figure 17–4). However, DNA polymerase I (and III) can only synthesize a new DNA strand in the $5' \rightarrow 3'$ direction.

Reiji Okazaki discovered that newly synthesized DNA was synthesized in *small fragments* (later named *Okazaki fragments*) and that these fragments were joined together by *DNA ligase* to form a completed strand of DNA. This resolved the controversy over the direction in which new DNA strands

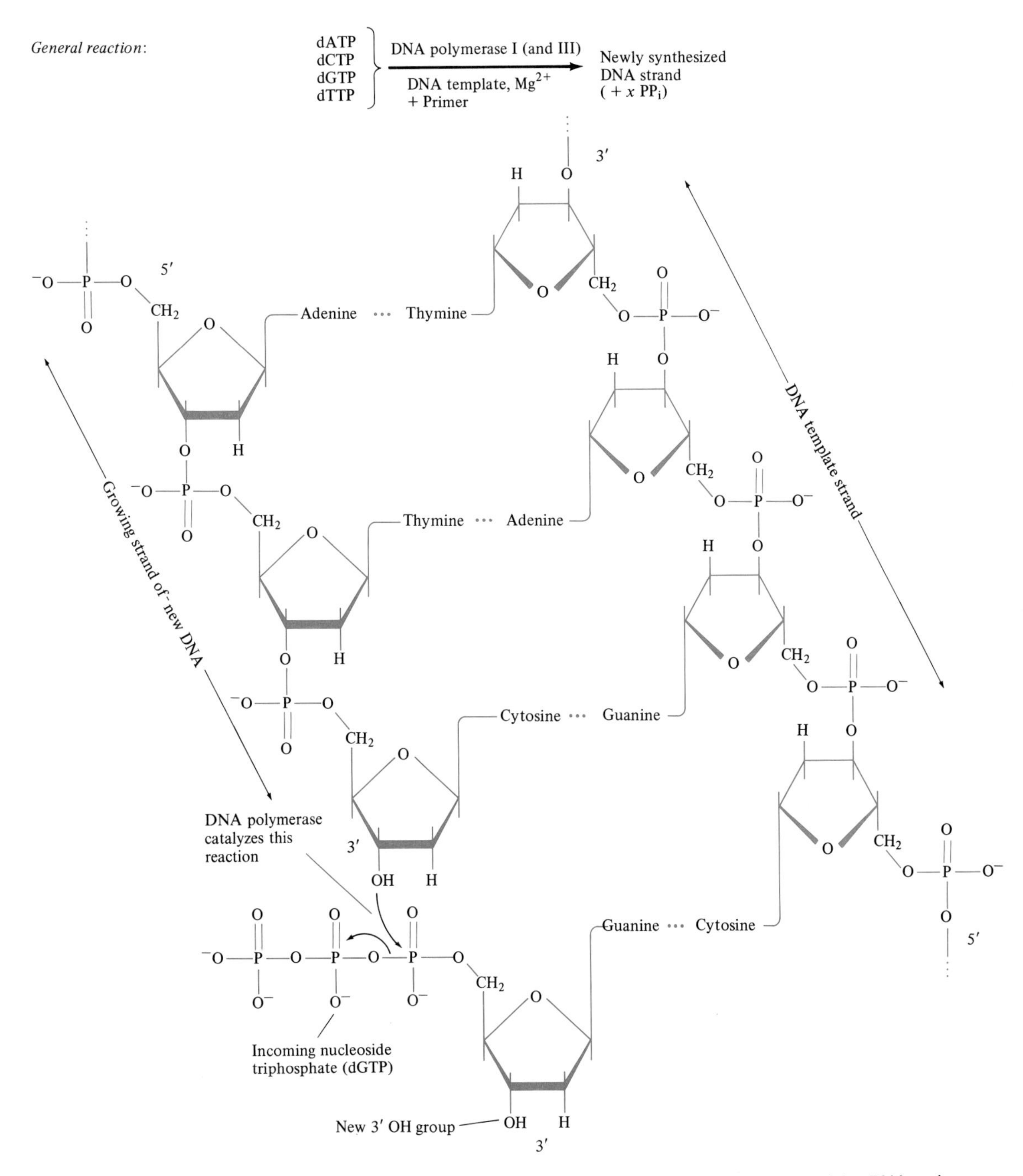

FIGURE 17–3 The synthesis of a new DNA strand catalyzed by DNA polymerase I (and III). One strand of DNA serves as a template for the insertion of the new strand.

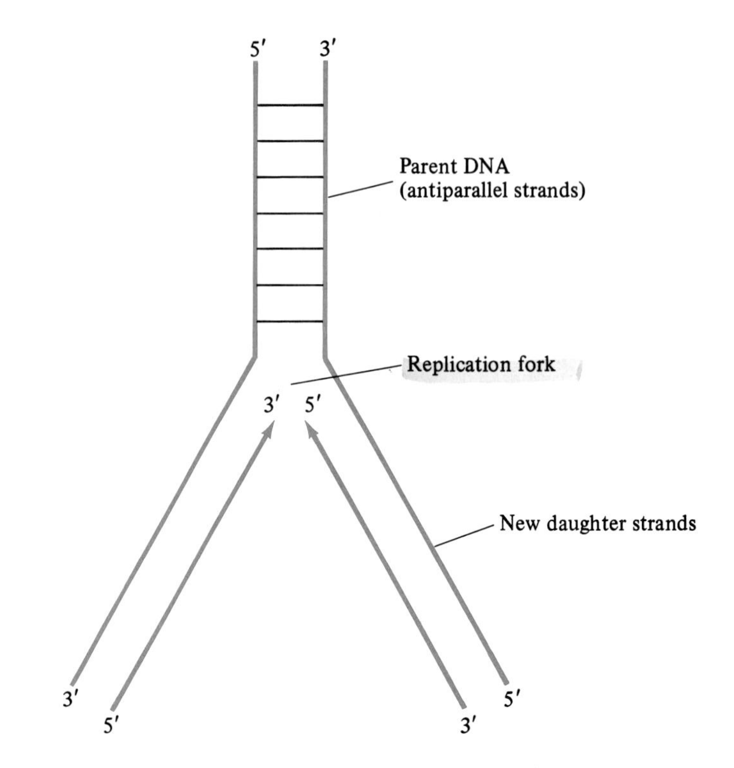

FIGURE 17–4
Original hypothesis regarding semiconservative replication of DNA. One strand would be synthesized in the $5' \rightarrow 3'$ direction; the other would be synthesized in the $3' \rightarrow 5'$ direction.

are synthesized. As shown in Figure 17–5, one strand (the *leading strand*) is synthesized in the normal $5' \rightarrow 3'$ direction. Okazaki fragments are also synthesized in the $5' \rightarrow 3'$ direction (by DNA polymerase III, present at the replication fork). These fragments are then joined by *DNA ligase* to yield an intact chain (the *lagging strand*). The entire process is called **discontinuous replication**, because the lagging strand is synthesized in discontinuous fragments.

discontinuous replication

The individual steps involved in DNA replication are briefly summarized here. (See Figure 17–6).

STEP 1: UNWINDING OF THE DNA DOUBLE HELIX

Since the base pairs of the DNA molecule are inside the double helix, the helix must be first unwound to expose the individual strands and, thus, provide a *template* for the synthesis of the new strands. The unwinding process is not completely understood, but it must involve an *unwinding protein* (rep protein, helicase). Other proteins (*prepriming* proteins and *DNA B* protein) open up and stabilize the unwound strands (Figure 17–7).

The leading strand is synthesized primarily as one unit by DNA polymerase III (in the $5' \rightarrow 3'$ direction). The lagging strand is synthesized in a more complicated fashion, as described in steps 2–5 (see Figure 17–6).

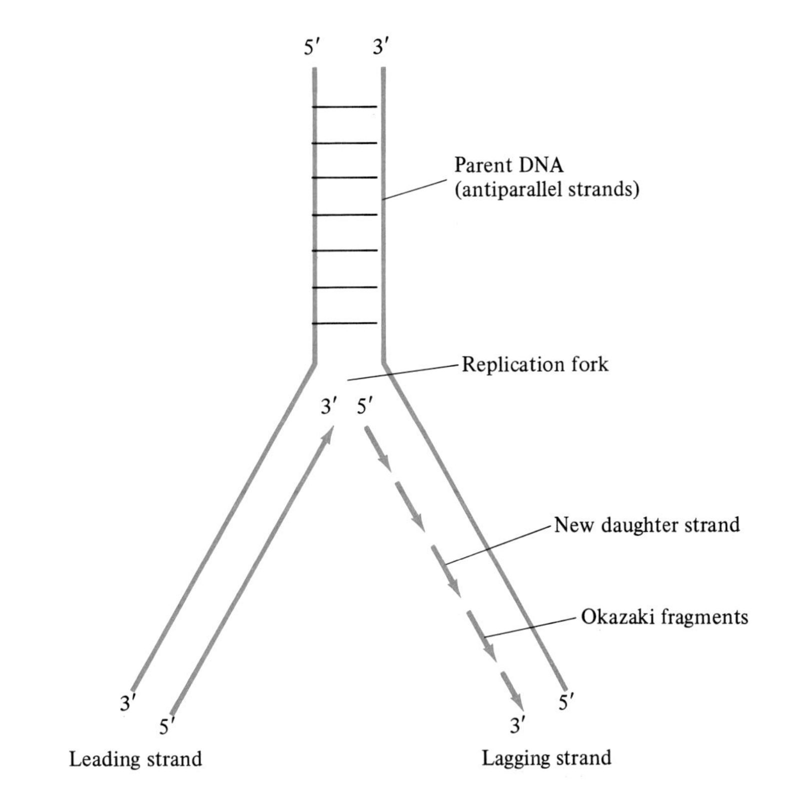

FIGURE 17–5
Current concept of the mechanism by which DNA is replicated. The leading strand is synthesized in the $5' \rightarrow 3'$ direction. The lagging strand is synthesized piecemeal—also in the $5' \rightarrow 3'$ direction—after which the Okazaki fragments are joined by DNA ligase.

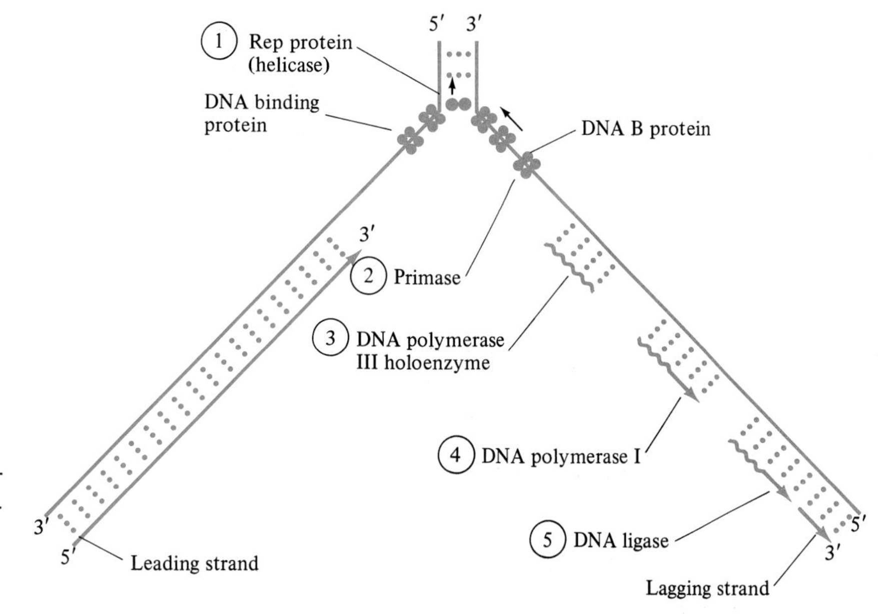

FIGURE 17–6
A diagram of the enzymatic steps that occur during DNA replication. (Adapted from A. Kornberg, *DNA Replication* (San Francisco: W. H. Freeman), 1980.)

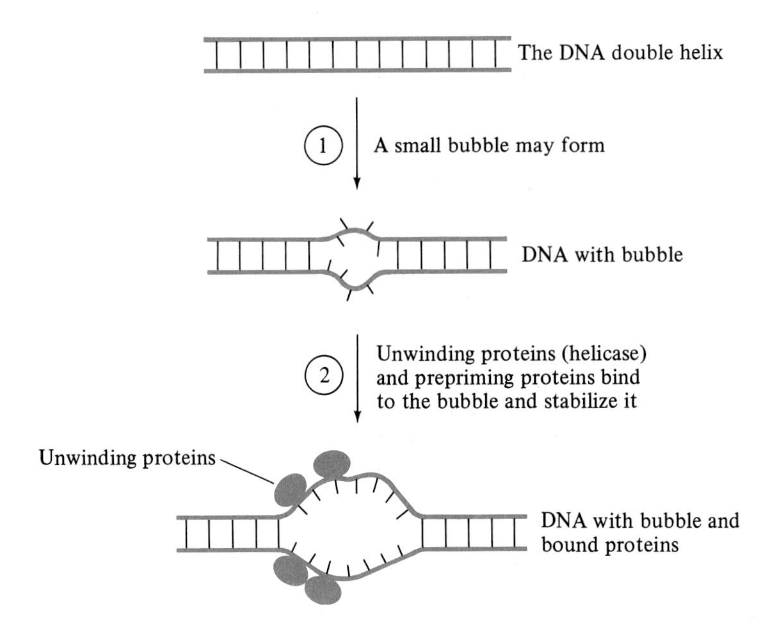

FIGURE 17–7 A suggested mechanism by which the DNA double helix is unwound. (Adapted from A. L. Lehninger, *Biochemistry,* 2nd ed. (New York: Worth, 1975), p. 909.)

STEP 2: FORMATION OF THE RNA PRIMER STRAND

Interestingly, a short RNA chain 10–30 nucleotides long must be synthesized if DNA is to be synthesized. This short RNA chain serves as the *primer* for DNA synthesis. An enzyme called *primase* (also known as *DNA-directed RNA polymerase*) catalyzes the formation of the RNA chain. The RNA strand is complementary to the base sequence of the original strand of DNA.

STEP 3: SYNTHESIS OF THE NEW DNA STRAND

DNA Polymerase III next catalyzes the replication of the new DNA strand onto the RNA primer strand. The RNA primer strand is then hydrolyzed by DNA polymerase I, leaving only the new DNA strand. The $5' \rightarrow 3'$ exonuclease of DNA polymerase I catalyzes the removal of the RNA primer strand.

STEP 4: FILL GAPS

The gaps resulting from the removal of the RNA primer strands are filled in by DNA polymerase I.

STEP 5: JOIN SEGMENTS

Finally, DNA ligase joins the free 3′ OH group of one nucleotide unit with the 5′ phosphate group of the adjacent nucleotide unit, forming a 3′,5′-

phosphodiester linkage. Thus, the two newly synthesized segments of DNA are joined together as follows:

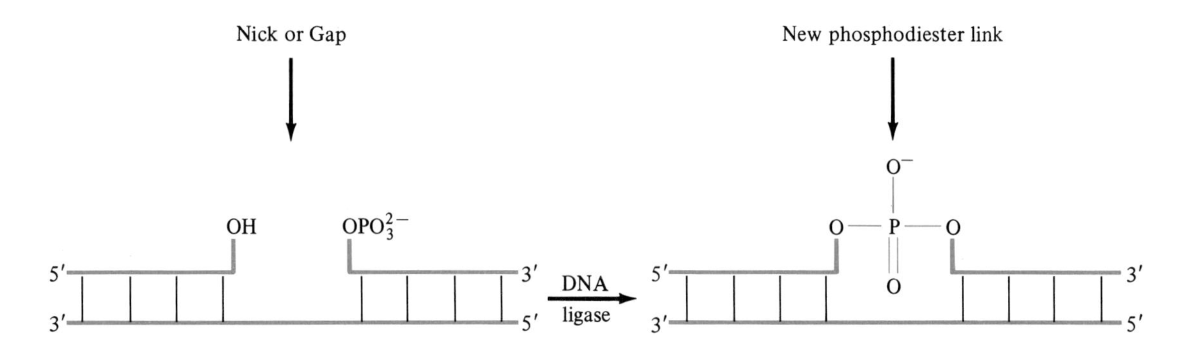

THE REPAIR OF DNA

A number of chemical and environmental factors (such as ionizing radiation) can damage DNA. Usually, the cell can repair this damaged DNA. An example of a specific repair mechanism will illustrate this point.

Ultraviolet (UV) radiation causes the formation of *thymine dimers*, as shown in the following reaction:

UV light

—deoxyribose—P—deoxyribose— → —deoxyribose—P—deoxyribose—

Adjacent thymine residues

Thymine *dimer*

This thymine dimer *prevents* the replication of the DNA molecule.

The region of the DNA molecule containing the thymine dimer can be repaired via an unusual *excision* mechanism, as shown in Figure 17–8. The presence of the thymine dimer causes a distortion in the structure of the DNA double helix. This distortion is detected by a specific *endonuclease* that produces a nick in the strand of DNA that contains the dimer. DNA polymerase I catalyzes the formation of a new strand of DNA using the other, intact strand of the DNA double helix as a template. The region of the old strand that contains the thymine dimer is next excised by the $5' \rightarrow 3'$ exonuclease activity of DNA polymerase I. Finally, DNA ligase forms a 3′,5′-phosphodiester linkage, thus joining the two segments of the DNA strand together.

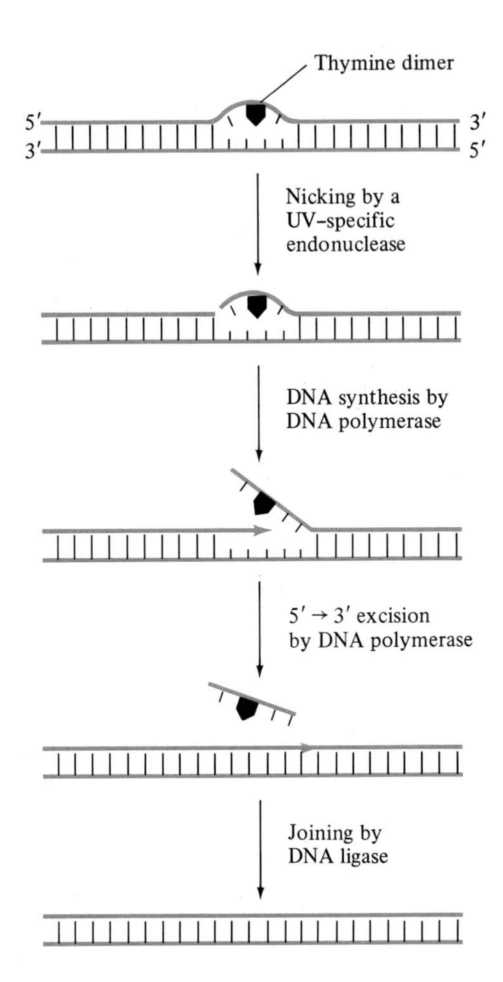

FIGURE 17–8
The excision-repair mechanism that removes thymine dimers from DNA. (Adapted from P. C. Hanawalt, *Endeavor* 31 (1972): 83.)

17.3 RECOMBINANT DNA

The 1970s were an explosively productive decade for biochemical research on DNA. New and powerful research techniques were developed during this period. Using these techniques, the sequence and structure of many genes were determined, and such new technologies as recombinant DNA and cloning were developed.

restriction endonucleases

The commercial availability of a new class of enzymes, the **restriction endonucleases**, was essential to the rapid progress in DNA research. Restriction endonucleases, which are present in prokaryotes, catalyze the hydrolysis of phosphodiester linkages at specific points within the nucleotide base sequence of the DNA strands. Research has shown that the base sequence of certain regions of DNA exhibit a remarkable symmetry, as shown in the following structure:

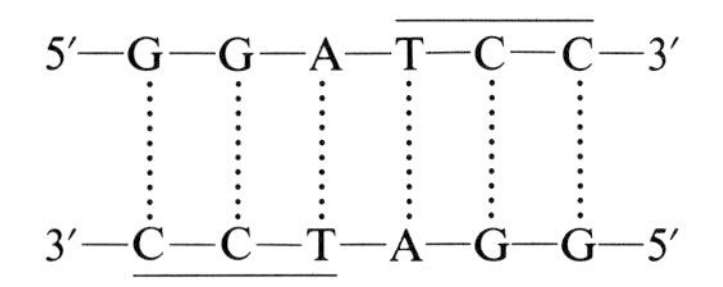

Notice that the sequence (read in the 5′ to 3′ direction) is repeated in the other strand, but in reverse order. Since the strands are antiparallel, the sequence reads the same in the 5′ to 3′ direction as it does in the other strand. These regions are called **palindromes**. A palindrome is defined as a word, sentence, phrase, or code that reads the same from left to right as it does from right to left. A specific restriction endonuclease recognizes a particular palindromic region of the DNA double helix, binds to that region, and cleaves specific phosphodiester linkages within the sequence. Table 17–1 lists a few palindromic sequences, along with the restriction endonucleases that recognize and cleave those sequences.

palindromes

The biological role of these bacterial endonucleases is to degrade *foreign DNA* that has entered the bacterial cell. If endonucleases are present within the bacterial cell, what prevents them from destroying the cell's own DNA? Bacterial DNA contains methylated bases in a certain sequence. Foreign DNA also contains methylated bases, but in a sequence different from that in bacterial DNA. Because of this, foreign DNA is susceptible to degradation by the bacterial endonucleases. A sequence of DNA containing methylated adenine residues is shown in the following diagram:

CH_3
3′—T—T—A—A—G—5′
5′—A—A—T—T—C—3′
CH_3

TABLE 17–1 The Specificities of a Few Restriction Endonucleases.

Endonuclease	Sequence Recognized and Cleaved
BamH I	5′—G↓G—A—T—C—C—3′ 3′—C—C—T—A—G↑G—5′
Eco R I	5′—G↓A—A—T—T—C—3′ 3′—C—T—T—A—A↑G—5′

Note: Each restriction endonuclease is identified by the abbreviated name of the bacterial species from which it was originally isolated. For example, *Eco* R I endonuclease was isolated from *E. Coli.* The arrows show where the endonuclease cleaves the phosphodiester linkages.

Because restriction endonucleases cleave DNA molecules at specific points offset along the two strands, the resulting fragments have single-stranded ends with free bases. These free bases can hydrogen bond with complementary bases and are, therefore, called *sticky ends* (see Figure 17–9).

Alternatively, sticky ends can be generated by enzymatically forming a tail of repeating deoxyadenosine (poly dA) or deoxythymidine (poly dT) nucleotide units.

5′·········dA—dA—dA—. . .—dA—OH—3′ ⟵ Poly dA tail, or sticky end

3′·········5′

A tail of poly dA can associate with a tail of poly dT, thus joining two different segments of DNA.

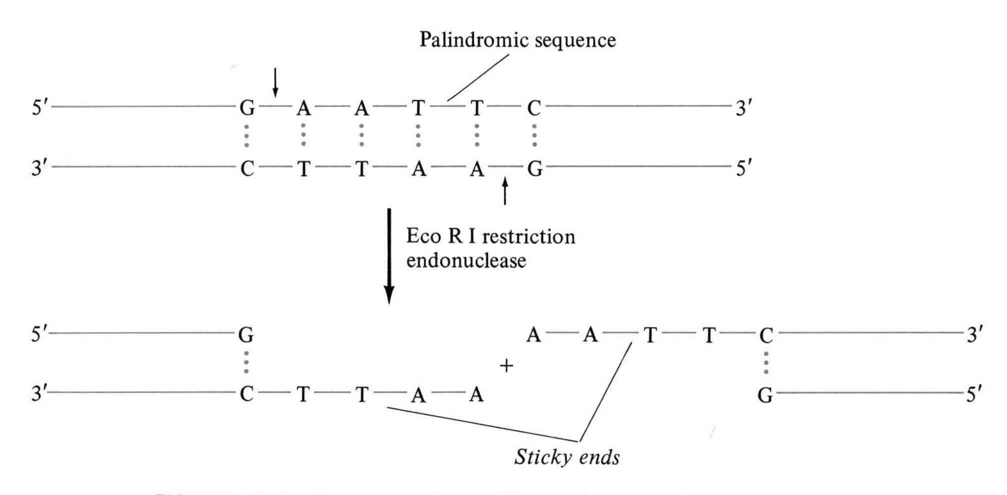

FIGURE 17–9 Fragmentation of DNA and the production of "sticky ends."

recombinant DNA

With this background information in mind, let us describe the **recombinant DNA** technique. Bacteria (certain strains of *E. coli*) have two types of DNA: *chromosomal DNA* (tightly coiled circular DNA) and *plasmid DNA* (coiled circular DNA, much smaller in size). Plasmid DNA is often called extrachromosomal DNA. In the recombinant DNA technique, plasmid DNA is separated from chromosomal DNA and treated with specific restriction endonucleases to fragment the DNA molecules and produce sticky ends. DNA from another organism is then isolated, purified, and also treated with restriction endonucleases. The two sets of treated molecules are then mixed together and the sticky ends are allowed to recombine with each other, forming complementary regions of double stranded DNA. The phosphate-deoxyribose backbone of the hydrogen bonded sticky ends is next covalently linked by DNA ligase. Thus, a closed circle of double-stranded DNA is formed, as shown in Figure 17–10. This *recombinant DNA*

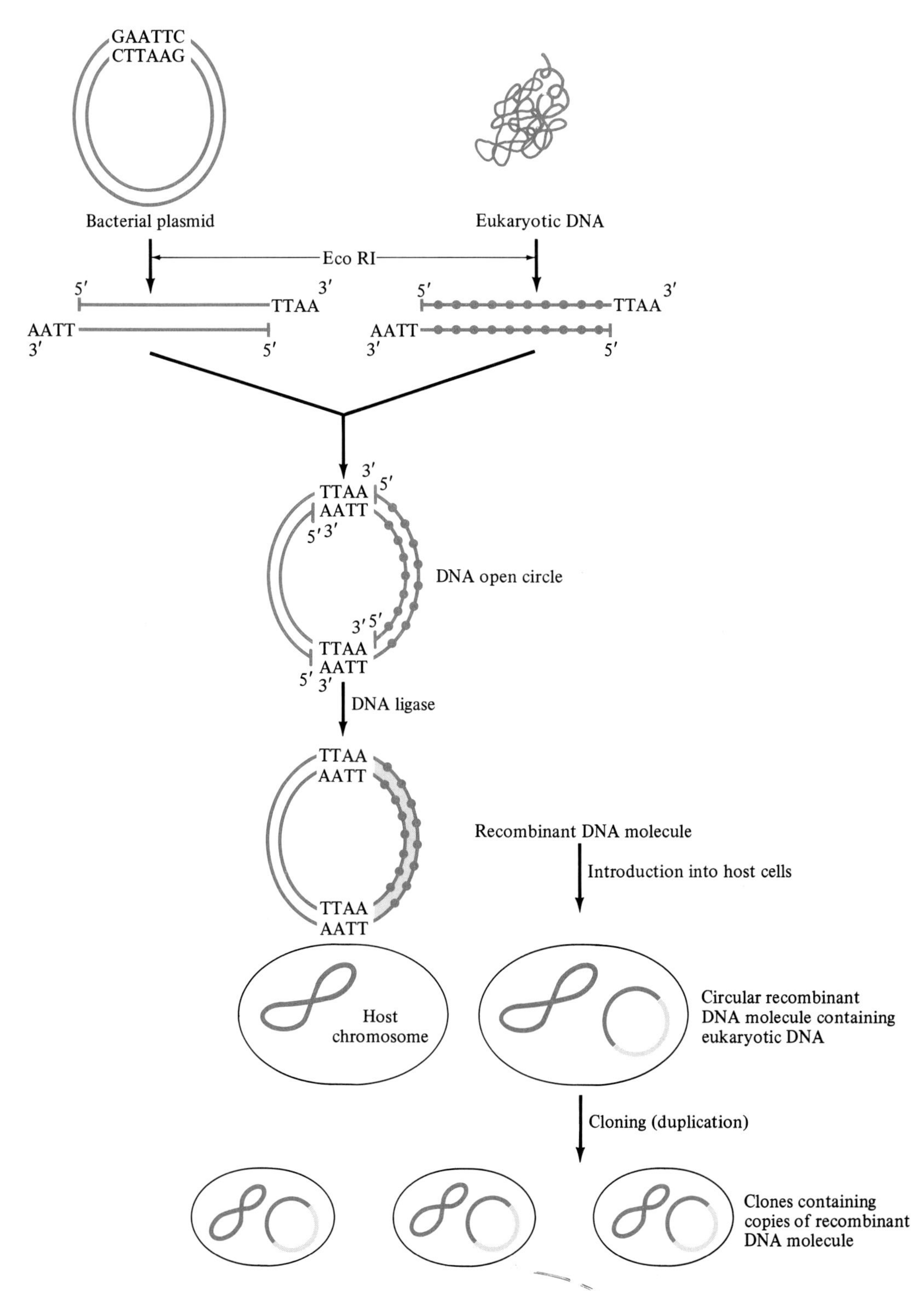

FIGURE 17–10 The recombinant DNA technique. (Adapted from F. B. Armstrong and T. P. Bennett, *Biochemistry* (New York: Oxford University Press, 1979), pp. 404–05.)

molecule is a hybrid molecule because it contains a portion of DNA that came from the bacterial plasmid DNA and a second portion that came from a different organism.

This procedure has incredible technological potential. It has already spawned a new arena of biochemical research and development called *genetic engineering*. For example, the portion of the DNA molecule that codes for human insulin can be attached to a bacterial plasmid DNA via recombinant DNA techniques. This recombinant DNA molecule is inserted into a host *E. Coli* bacterium. The host bacterium contains its own DNA and the enzymes necessary for DNA replication and gene expression. The host bacteria are allowed to reproduce, producing many copies of the recombinant DNA molecule containing the human insulin gene. These bacteria are then directed to produce insulin. This insulin can be isolated from the bacterial cells and nutrient media and purified for commercial use.

17.4 TRANSCRIPTION: THE SYNTHESIS OF RIBONUCLEIC ACID

The structure and classes of ribonucleic acid (RNA) were described in Chapter 16. Recall that RNA molecules are single strands of ribonucleotides linked together via 3′,5′-phosphodiester bonds. The RNA molecule is constructed from adenine, guanine, cytosine, and uracil bases, and hairpin loops can form when complementary base pairs make hydrogen bonds (A═U and G≡C).

RNA polymerase

The salient features of RNA synthesis are as follows: RNA synthesis requires an enzyme called **RNA polymerase** (prokaryotic), which catalyzes the initiation and elongation of the RNA chain. This enzyme requires one strand of the *double-stranded DNA molecule as the template*, the ribonucleoside triphosphates ATP, GTP, UTP, and CTP, Mg^{2+}, and no primer. The general reaction is shown in Figure 17–11. The RNA chain is synthesized in the 5′ → 3′ direction, as DNA chains are. The hydrolysis of pyrophosphate groups provides sufficient free energy to make 3′,5′-phosphodiester linkages. Ribosomal RNA (rRNA), messenger RNA (mRNA), and transfer RNA (tRNA) are all synthesized by RNA polymerase, and in the same manner.

The RNA polymerase molecule is a large, 5-subunit enzyme complex. It contains two alpha (α) subunits, one beta (β) subunit, one β' subunit, and one sigma (σ) subunit. The holoenzyme is symbolized as $(\alpha_2, \beta, \beta', \sigma)$. The *core enzyme* $(\alpha_2, \beta, \beta')$ catalyzes RNA chain synthesis; the sigma (σ) subunit *identifies* the initiation region on the DNA molecule.

RNA synthesis can be divided into three major phases: *initiation phase*, *elongation phase*, and *termination phase*. This sequence is summarized in Figure 17–12.

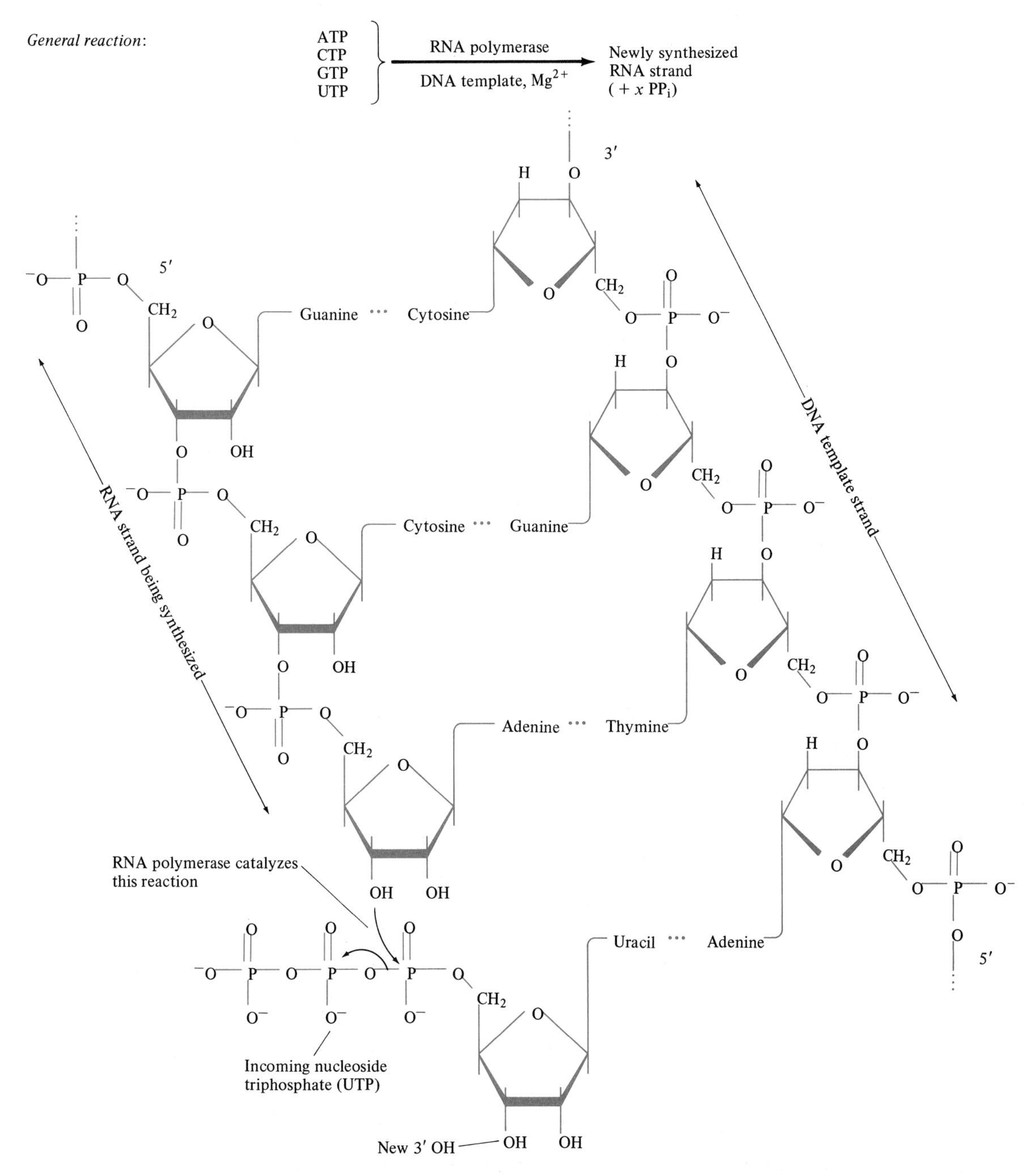

FIGURE 17–11 The synthesis of RNA, catalyzed by RNA polymerase. One strand of DNA serves as a template.

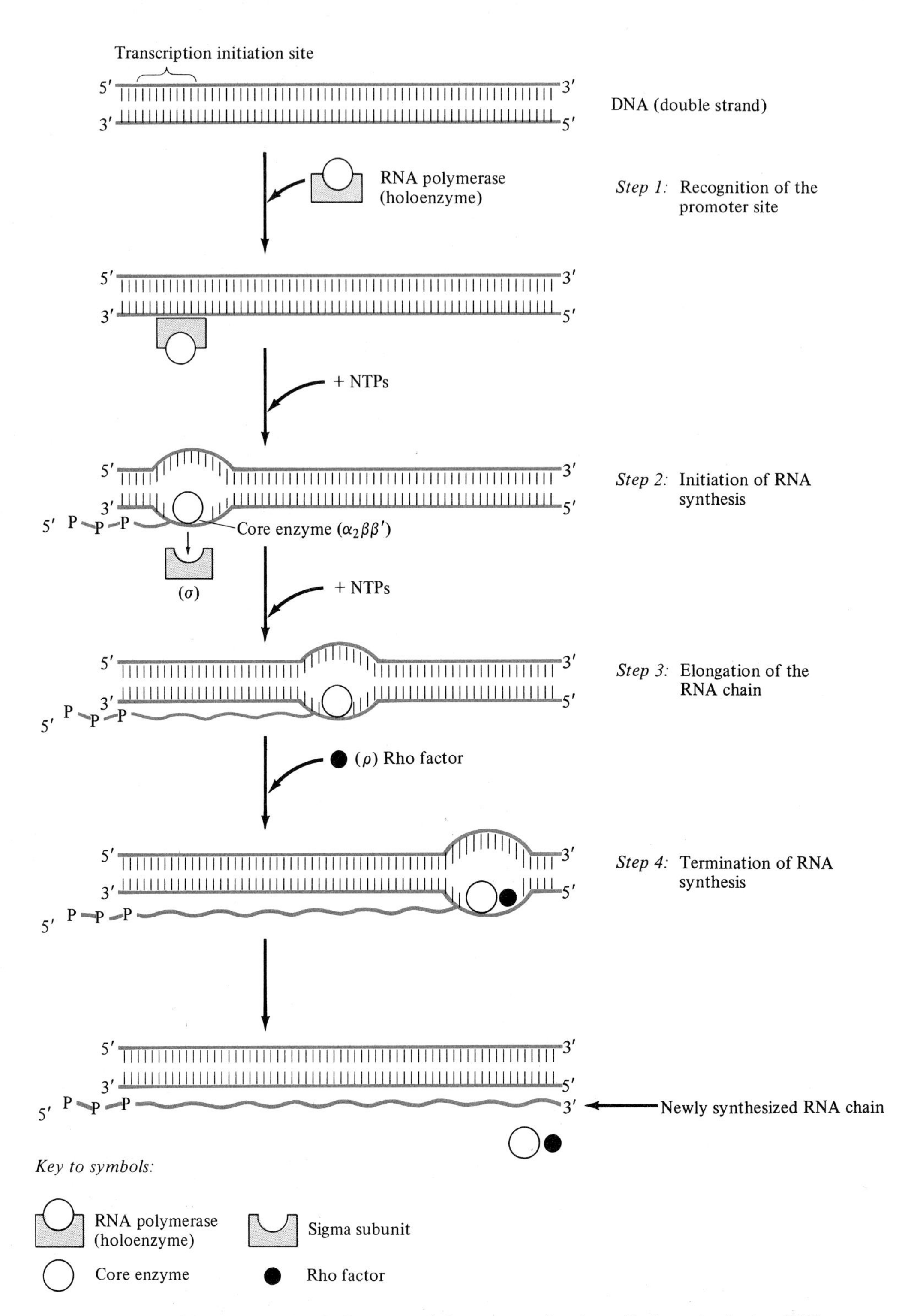

FIGURE 17–12 A diagram of the enzymatic steps that occur during RNA synthesis (transcription). (Adapted from F. B. Armstrong and T. P. Bennett, *Biochemistry* (New York: Oxford University Press, 1979), p. 423.)

STEP 1: RECOGNITION OF THE PROMOTER SITE

promoter regions

How does the RNA polymerase molecule know which regions of the DNA double helix are to be transcribed into RNA? Research has shown that specific **promoter regions** within the DNA base sequence bind the RNA polymerase molecule. These promoter regions are rich in adenine-thymine base pairs, as shown in the following sequence:

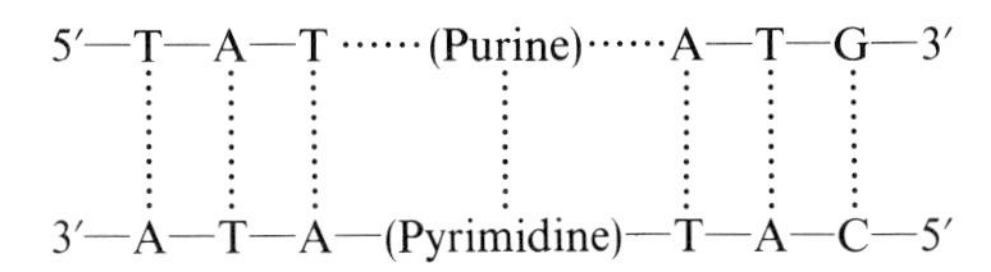

The promoter region on DNA

The sigma subunit of RNA polymerase recognizes the promoter region and, thus, directs the RNA polymerase to the *initiation site*. The sigma subunit also opens up the double helix, exposing the template strand of the DNA molecule for RNA synthesis.

STEP 2: INITIATION OF RNA SYNTHESIS

After the RNA polymerase holoenzyme has bound to the DNA molecule, the sigma subunit dissociates from the holoenzyme complex. The core enzyme now catalyzes the formation and growth of the new RNA chain. The 5′ end of the chain is either 5′—pppA or 5′—pppG (ppp indicates a triphosphate tail). The core enzyme catalyzes the sequential addition of ribonucleotides, in the process forming 3′,5′-phosphodiester linkages. Interestingly, only *one* strand of the DNA double helix is transcribed, the RNA polymerase molecule moving along this strand in the $3' \rightarrow 5'$ direction. The *new complementary RNA* strand is, thus, synthesized in the $5' \rightarrow 3'$ direction. It should be noted that, unlike DNA chain synthesis, RNA chain synthesis *does not* require a primer.

STEP 3: ELONGATION OF THE RNA CHAIN

The RNA chain is elongated by the successive addition of ribonucleotide units ($\sim$3000 ribonucleotide units per minute). As the RNA polymerase complex proceeds down the DNA helix, the helix unwinds, exposing the template strand. The newly synthesized RNA strand trails off to the side of the DNA helix. The DNA helix reforms after the RNA polymerase has transcribed a particular section (see Figure 17–12).

STEP 4: TERMINATION OF RNA SYNTHESIS

The DNA template contains a specific region called the *termination site*. In this region is a sequence with many guanine-cytosine base pairs, followed by a region with many adenine-thymine base pairs. It is thought that the RNA polymerase molecule recognizes this sequence and terminates RNA chain synthesis at this point. In certain cases, a protein called the *rho* (ρ) *factor* binds to the newly synthesized RNA chain and then displaces the RNA polymerase molecule from the DNA template.

POSTTRANSCRIPTIONAL MODIFICATION OF RNA

Prokaryotic and eukaryotic RNA generally undergoes additional enzymatic modifications *after* a particular RNA molecule has been transcribed and released from the DNA template/RNA polymerase complex. Eukaryotic RNA is subject to more extensive modification than prokaryotic RNA. There are three major types of RNA modifications: (1) cleavage of a large precursor RNA molecule (often noted as pre-RNA), (2) addition of poly-adenylate sequences to the 3′ end of the mRNA chain, and (3) enzymatic modifications of individual RNA nucleotide bases and sugar units.

Ribosomal RNA (eukaryotic rRNA) is derived from the cleavage of a large precursor RNA molecule (pre-rRNA), as shown in Figure 17–13. The precursor RNA molecule (noted as 45S RNA—the S denotes the size of the molecule) is processed into a 41S RNA molecule and a 24S RNA molecule. The 41S RNA undergoes further modifications to yield the 28S and 18S rRNA that are used in constructing the ribosome.

The addition of polyadenosine sequences is another type of post-transcriptional RNA modification. *Poly A polymerase* adds a sequence of poly adenosine residues to the 3′ end of the mRNA transcript (noted as poly A). Usually, between 50 and 200 adenosine nucleotides are present in

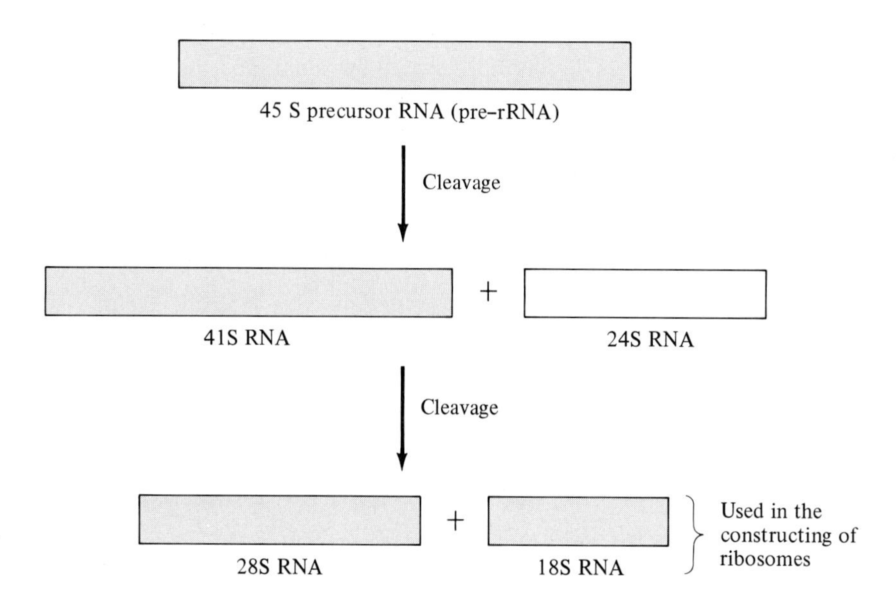

FIGURE 17–13 Posttranscriptional modification of eukaryotic ribosomal RNA.

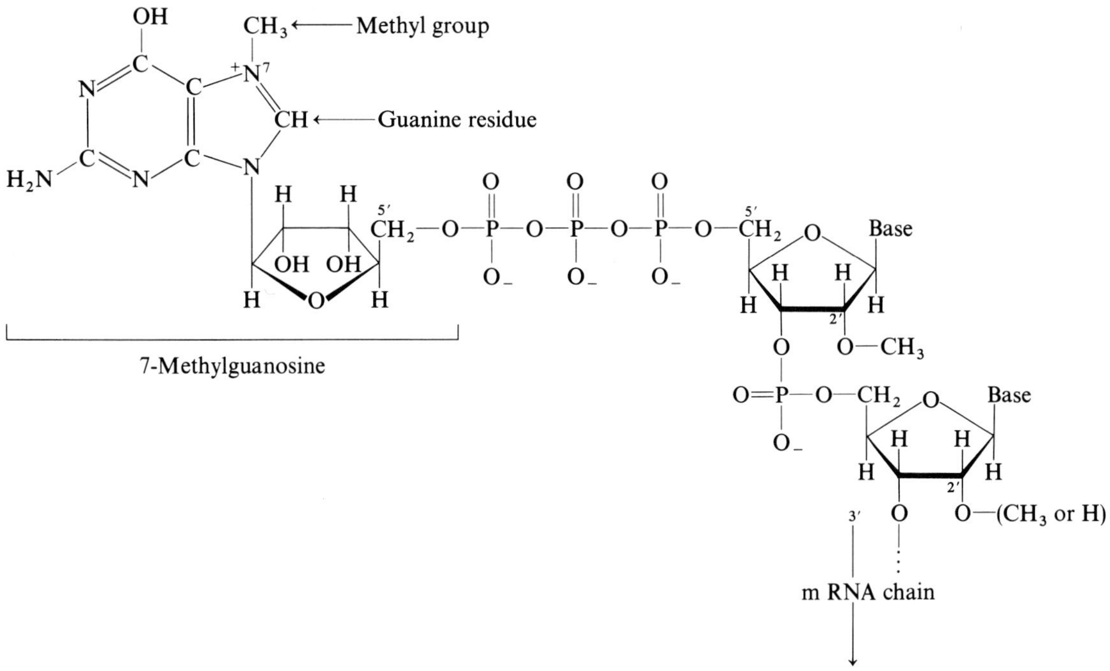

FIGURE 17–14 The 5′ "cap" at the 5′ end of eukaryotic mRNA.

this 3′-poly A tail. The poly A region may participate in transporting the mRNA from the nucleus to the cytoplasm.

The 5′ end of the mRNA molecule is also modified. This contains *7-methylguanosine* esterified to the 5′ triphosphate tail (Figure 17–14), a modification known as the *5′ cap*. The biological function of this cap is still unknown.

Finally, the base and sugar units of the RNA chain are often enzymatically modified to form derivatives. Transfer RNA (tRNA—see Chapter 18) contains a large number of unusual bases that are derived in this way.

SUMMARY

Genetic information flows from DNA to RNA (by transcription) and from RNA to proteins (by translation). The 3-dimensional shape of a protein is defined by the nucleotide base sequence of DNA.

The DNA molecule is duplicated via semiconservative and discontinuous replication. The DNA double helix is first unwound, forming a replication fork. The two separated strands serve as templates for the synthesis of two new strands, called the leading strand and the lagging strand. The leading strand is continuously replicated in the 5′ → 3′ direction by DNA polymerase. The lagging strand is discontinuously replicated by the following sequence of reactions: (1) DNA-directed RNA polymerase forms short RNA primer strands attached to the lagging strand. (2) Sections of DNA are replicated in the $5' \rightarrow 3'$ direction onto the RNA primer strands. (3) The RNA primer strands are removed. (4) The resulting sections of DNA (Okazaki fragments) are joined together by DNA ligase. Damaged sections of DNA can be repaired by excising the damaged region, synthesizing a new section, and joining this section to the undamaged DNA by DNA ligase.

Restriction endonucleases recognize palindromic sequences within DNA and hydrolyze the phosphodiester linkages between specific bases. This fragments the DNA molecule and produces sticky ends.

The recombinant DNA technique involves (1) isolating bacterial plasmid DNA and DNA from another organism, (2) treating these two forms of DNA with restriction endonucleases to produce sticky ends and (3) combining the DNA fragments to yield recombined DNA. This recombinant DNA is then treated with DNA ligase to form a closed circle of DNA, one section of which came from bacterial plasmid DNA and the other of which came from a different organism. The recombinant DNA can then be introduced into a host bacterium to produce more copies of the recombinant DNA molecule.

RNA is synthesized by RNA polymerase, which requires one strand of the DNA helix (as a template) and the ribonucleoside triphosphates (ATP, GTP, UTP, and CTP). RNA synthesis can be divided into three major phases: (1) the initiation phase, in which RNA polymerase binds to the DNA promoter region and the 5′ end of mRNA forms; (2) the elongation phase, in which the RNA chain is elongated by the successive addition of ribonucleotide units in the $5' \rightarrow 3'$ direction; and (3) the termination phase, in which the RNA polymerase molecule stops RNA chain synthesis in response to a specific terminal sequence. The RNA molecule next dissociates from the DNA chain and undergoes posttranscriptional modification. For example, precursor RNA molecules can be cleaved into shorter ones; sequences of RNA units can be added to the chain (for instance, the poly-A tail at the 3′ end); and individual nucleotide bases and sugar units within the RNA molecule can be chemically modified.

REVIEW QUESTIONS

1. We have presented a number of important concepts and terms in this chapter. Define (or briefly explain) each of the following terms:
   a. DNA
   b. DNA replication (semiconservative)
   c. Okazaki fragments
   d. Discontinuous replication
   e. RNA primer strand
   f. Thymine dimers
   g. Palindromic sequence
   h. Recombinant DNA
   i. Plasmid DNA
   j. Transcription
   k. Posttranscriptional modification
   l. 3′-Poly A tail (on mRNA)

2. We have discussed a number of important enzymes in this chapter. For each enzyme listed, describe what the enzyme does and note why it is important. If possible, diagram the reaction catalyzed by each enzyme.
   a. DNA polymerase I
   b. DNA polymerase III
   c. DNA ligase
   d. Restriction endonuclease
   e. RNA polymerase

3. Briefly describe the flow of genetic information.

* **4.** In detail, describe the principal steps of:
   a. DNA replication
   b. DNA transcription (including posttranscriptional modification)

* **5.** Referring to Question 4, compare and contrast the replication of DNA with the transcription of DNA to form RNA. What steps are involved in these processes? What enzymes? How are the steps the same? How do they differ?

6. Complete the following palindromic sequence:

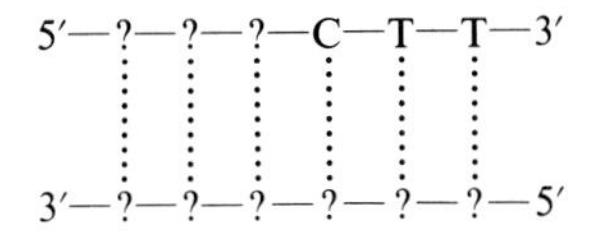

SUGGESTED READING

Bhagavan, N. V. *Biochemistry*, 2nd ed., Ch. 5, pp. 371–484. Philadelphia: J. B. Lippincott, 1978.

Kornberg, A. *DNA Replication*. San Francisco: W. H. Freeman, 1980.

Lehninger, A. L. *Biochemistry*, 2nd ed., Chs. 12, 31, and 32. New York: Worth, 1975.

Montgomery, R., Dryer, R. L., Conway, T. W., Spector, A. A. *Biochemistry: A Case-Oriented Approach*, 3rd ed., Chs. 11 and 12. St. Louis: C. V. Mosby, 1980.

Stryer, L. *Biochemistry*, 2nd ed., Chs. 24 and 25. San Francisco: W. H. Freeman, 1981.

Watson, J. D. *Molecular Biology of the Gene*, 3rd ed. Menlo Park, CA: W. A. Benjamin, 1976.

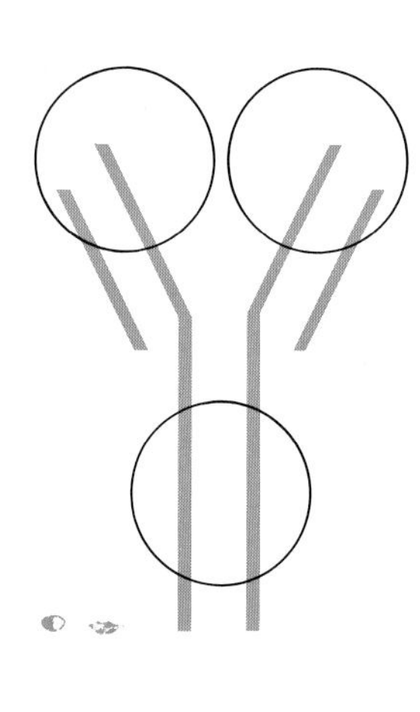

Chapter 18

Protein Synthesis: The Expression of Genetic Information

18.1 INTRODUCTION

In Chapter 17, we stated that the chemical information (the genetic code) in the DNA molecule is expressed as the complementary base sequence of a messenger RNA (mRNA) molecule and that this base sequence codes for the sequence of amino acids in the final protein product (see Figure 17–1). Thus, the correct reading of the genetic code establishes the correct sequence of amino acids in a specific protein molecule.

translation

In this chapter, we will describe the genetic code and how the code is read. The intricate enzymatic steps that occur during protein synthesis **(translation)** will be presented in detail. The structure of the ribosome and of transfer RNA (tRNA) molecules will be discussed. Finally, the structure, regulation, and expression of prokaryotic and eukaryotic genes will be examined.

18.2 THE GENETIC CODE

messenger RNA (mRNA)

Scientists have determined that protein synthesis occurs in the cytoplasm of the typical eukaryote. In 1961, F. Jacob and J. Monod hypothesized that an *intermediate polynucleotide carries the genetic message from DNA in the nucleus to the cytoplasm, where protein synthesis takes place.* According to Jacob and Monod, this intermediate **(messenger RNA, mRNA)** (1) is transcribed within the nucleus; (2) contains a base sequence complementary to the original DNA base sequence; (3) is transported out of the nucleus and into the cytoplasm; (4) binds to ribosomes (the cytoplasmic organelles where proteins are synthesized; and (5) allows its genetic message to be read by the ribosome to form a complete polypeptide chain.

genetic code

What is the **genetic code**, and how is this code read? Proteins are constructed from the 20 different amino acids; yet, mRNA is constructed from only 4 bases (adenine, guanine, cytosine, and uracil). Obviously, if each base corresponded to one amino acid, these bases could only code for 4 amino acids. If two adjacent bases (a doublet) coded for amino acids, only 16 combinations are possible ($4 \times 4 = 16$ combinations) and, thus, only 16 amino acids could be coded. However, if *three bases* coded for each amino acid, more than enough combinations are possible ($4 \times 4 \times 4 = 64$ combinations). Thus, it was concluded that each of the 20 amino acids is coded by a three-base (triplet) nucleotide sequence within the mRNA molecule. This triplet is called a **codon**. It should be noted that a complementary triplet would be found in the original DNA molecule (see Figure 17–1).

codon

It has been firmly established that the genetic information of the DNA double helix is read in a *triplet*, *degenerate*, *nonoverlapping*, *and commaless code*.[1] To explain further, a sequence of three nucleotides (the *triplet* or *codon*), codes for a particular amino acid. "*Degeneracy*" will be explained in a moment. The code is read sequentially, one codon at a time. An overlapping code is not resident within original code (that is, adjacent codons do not overlap). In addition, the code is read without commas or spaces, as is noted in the following diagram:

1. Codons in mRNA "read" sequentially, in triplets, and *without* "overlaps" or "commas":
   · · · *UUU UCU CAU AAA* · · ·
2. If "overlaps" possible:
   · · · *UUU UCU CAU AAA* · · · (original)
   · · · *UCU CUC UCA* · · · (with overlap)
3. If "commas" possible:
   · · · *UUU UCU CAU AAA* · · · (original)
   · · · UUU *U* CUC *A* UAA *A* · · · (with "commas")
   ↑ ↑ ↑
   "Comma"

The genetic code contains 64 triplet codons (remember, $4 \times 4 \times 4 = 64$), as shown in Table 18–1. In the table, the codes are listed as read in the $5' \rightarrow 3'$ direction (that is, AUC would be read as 5′—AUC—3′). There are 61 codons that code for the 20 amino acids. In addition, there are three *stop codons* (UAG, UAA, and UGA) that signal the end of a particular genetic message. They are like periods at the end of a sentence.

degenerate

One important feature of the genetic code is that it is **degenerate**. From Table 18–1, you will note that most of the amino acids (except for tryptophan and methionine) have more than one codon. For example, UUU and UUC *both* code for phenylalanine, and UCU, UCC, UCA, and UCG all code for serine. Reading from the 5′ end of each codon, positions 1 and 2 are

TABLE 18–1 The Genetic Code.

First Base	Second Base							
	U		C		A		G	
U	UU**U**	Phe	UC**U**	Ser	UA**U**	Tyr	UG**U**	Cys
	UU**C**	Phe	UC**C**	Ser	UA**C**	Tyr	UG**C**	Cys
	UU**A**	Leu	UC**A**	Ser	UA**A**	Term.	UG**A**	Term.
	UU**G**	Leu	UC**G**	Ser	UA**G**	Term.	UG**G**	Trp
C	CU**U**	Leu	CC**U**	Pro	CA**U**	His	CG**U**	Arg
	CU**C**	Leu	CC**C**	Pro	CA**C**	His	CG**C**	Arg
	CU**A**	Leu	CC**A**	Pro	CA**A**	Gln	CG**A**	Arg
	CU**G**	Leu	CC**G**	Pro	CA**G**	Gln	CG**G**	Arg
A	AU**U**	Ile	AC**U**	Thr	AA**U**	Asn	AG**U**	Ser
	AU**C**	Ile	AC**C**	Thr	AA**C**	Asn	AG**C**	Ser
	AU**A**	Ile	AC**A**	Thr	AA**A**	Lys	AG**A**	Arg
	AU**G**	Met	AC**G**	Thr	AA**G**	Lys	AG**G**	Arg
G	GU**U**	Val	GC**U**	Ala	GA**U**	Asp	GG**U**	Gly
	GU**C**	Val	GC**C**	Ala	GA**C**	Asp	GG**C**	Gly
	GU**A**	Val	GC**A**	Ala	GA**A**	Glu	GG**A**	Gly
	GU**G**	Val	GC**G**	Ala	GA**G**	Glu	GG**G**	Gly

Note: Of the 64 triplets (codons) in the genetic code, 61 are codes for the 20 amino acids and 3 are stop (terminal) codons. Codons are read in the $5' \rightarrow 3'$ direction. The third base (bold face) is variable.

generally constant for a particular amino acid, and the base in the *third position varies*. For example, the codon for phenylalanine contains either U or C in the third position. Since both UUU and UUC code for phenylalanine, phenylalanine would be inserted into the newly synthesized polypeptide chain if either UUU or UUC were read during protein synthesis. This third, variable base is called the *wobble base*.

At this point, we might ask how the genetic code is read during protein synthesis. The specific steps of protein synthesis will be discussed in detail in Section 18.4. Francis Crick suggested that an *adaptor molecule* brings the amino acid to the place where the polypeptide chain is being assembled. This adaptor molecule then binds to a specific codon within the mRNA molecule. Crick further hypothesized that at least 20 different adaptor molecules would be required, each specific for one of the 20 amino acids.

transfer RNA (tRNA)

Transfer RNA (tRNA) was later found to be this adaptor molecule.

18.3 TRANSFER RNA

The average tRNA molecule is about 75 nucleotides in length. The structure of *phenylalanine transfer RNA* (*Phe tRNA*), the tRNA molecule that binds and transfers phenylalanine, is shown in Figure 18–1. Transfer RNA molecules are usually depicted by this convenient cloverleaf representation. The tRNA molecule contains three major stems and hairpin loops (also

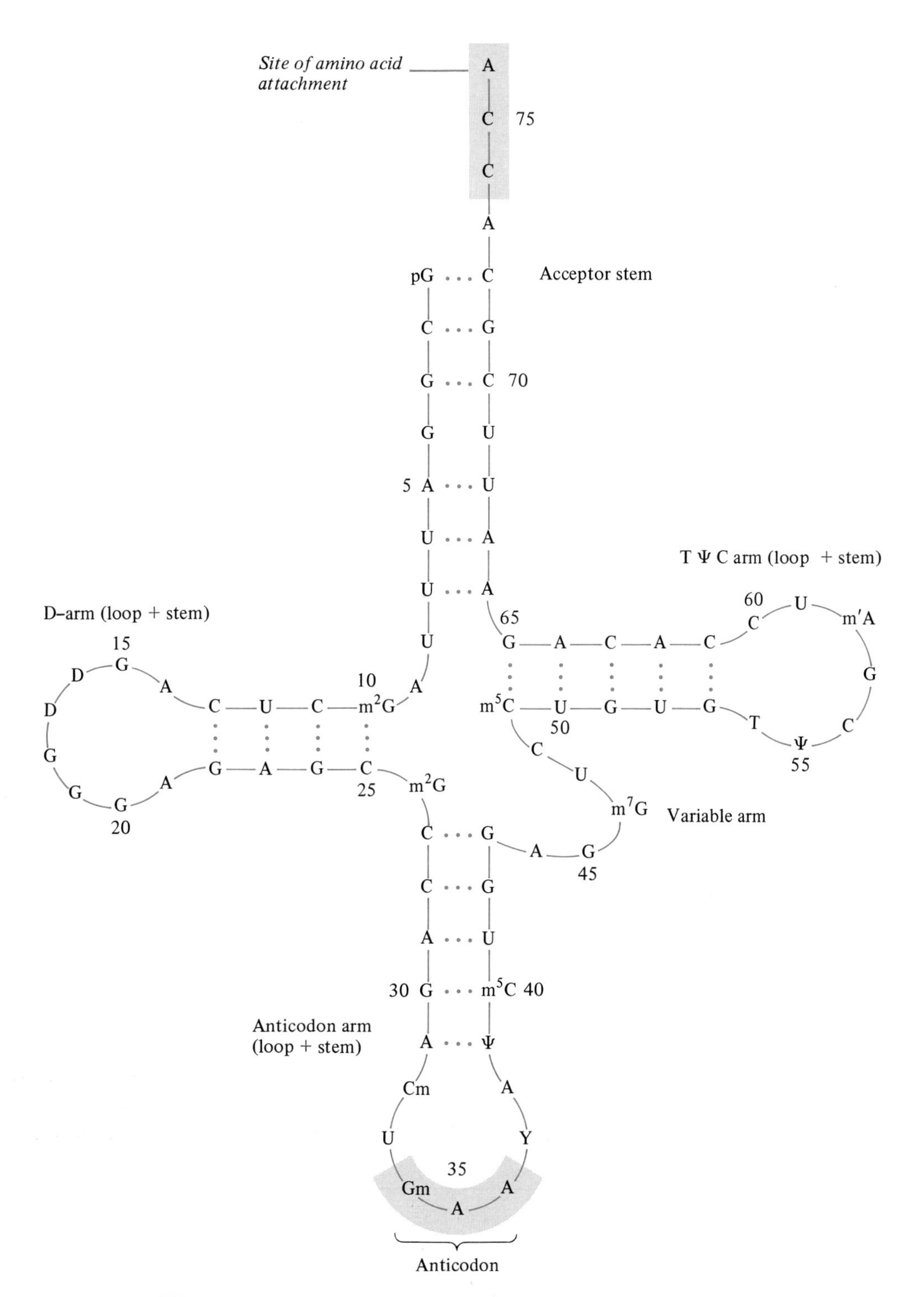

FIGURE 18–1 The nucleotide sequence of yeast phenylalanine transfer RNA (tRNA). All tRNA molecules are depicted using this cloverleaf representation. The stems are held together by hydrogen bonds; the loops are not bonded. A number of unusual nucleotide bases are shown: D = dihydrouridine; Ψ = pseudouridine; m^2G = dimethylguanosine; and Cm = methylcytidine (to name a few).

called arms). A fourth (minor) stem is found in some tRNA molecules. The arms form because specific base-pairs are hydrogen bonded. The bases within the loops are not hydrogen bonded. You will note that the upper (acceptor) arm contains a CCA base sequence terminating at the 3′ position. This sequence is constant in all tRNAs. The *specific amino acid* binds to the terminal adenosine residue. The lower arm and loop is called the **anticodon loop**, since this loop contains a specific 3-base sequence (anticodon) that is *complementary* to a 3-base sequence (codon) in the mRNA molecule. Thus, the anticodon will hydrogen bond to the codon region of mRNA during protein synthesis (see Section 18.4). The other arms and loops are important in determining the shape of the tRNA molecule. Finally, a number of unusual bases (dihydrouridine, pseudo uridine, dimethylguanosine) are found in the tRNA base sequence. The chemistry of these unusual bases is beyond the scope of this text.

anticodon loop

The cloverleaf representation shown in Figure 18–1 is used for convenience; it does not represent the *actual* shape of the tRNA molecule in 3 dimensions. Figure 18–2, illustrates the 3-dimensional shape of a tRNA molecule. The phosphate ribose backbone is shown by heavy lines in the figure; the individual base-pairs are drawn as "rungs" within this complicated and highly folded molecule.

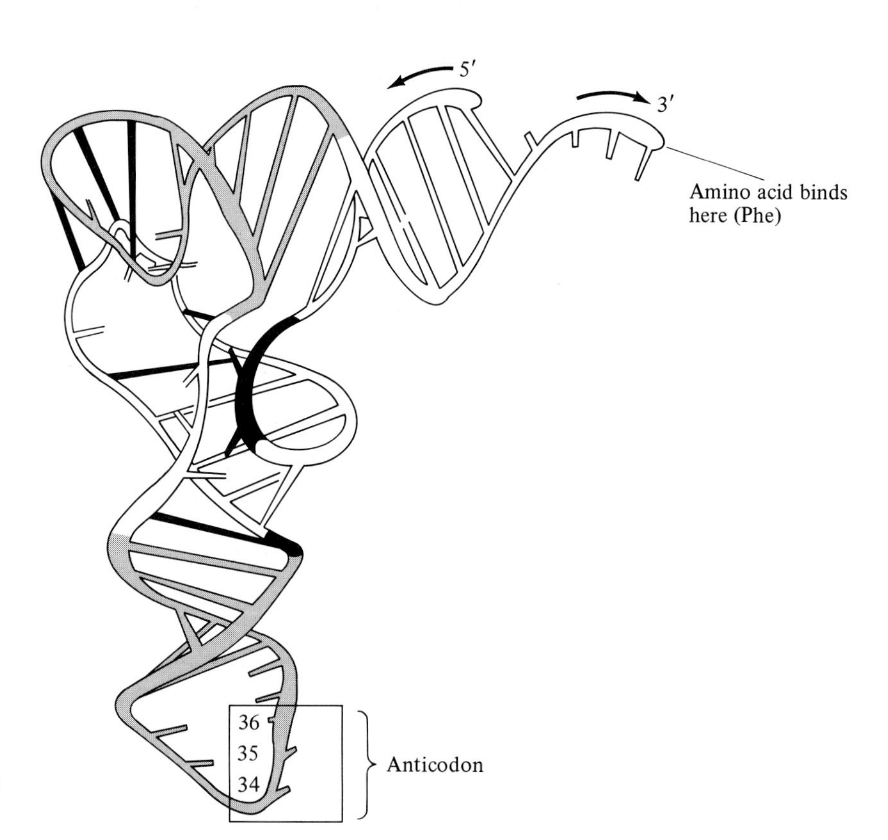

FIGURE 18–2
The three-dimensional structure of transfer RNA (tRNA). The anticodon loop is shown. This tRNA carries the amino acid phenylalanine and, therefore, is called phenylalanine-tRNA. (Adapted from S. H. Kim et al., *Science* 185 (1974): 435.)

18.4 PROTEIN SYNTHESIS: THE EXPRESSION OF THE GENETIC CODE

translation

The amino acid sequence of a protein molecule is defined from the sequence of codons within the mRNA molecule; thus, it is ultimately determined by the original base sequence of the DNA molecule. **Translation** is the process in which the codon sequence is read and the resulting polypeptide chain is assembled into an intact, functioning protein. Polypeptide chain assembly requires the following ingredients: (1) a mRNA molecule containing a specific genetic message to be translated; (2) a protein-synthesizing apparatus (the ribosome); (3) amino acids bound to tRNA molecules (aminoacyl-tRNAs); and (4) various enzymes and protein factors responsible for the *initiation*, *elongation*, and *termination* of polypeptide chain synthesis. A protein is assembled by the *sequential addition* of amino acids to the growing polypeptide chain, which is physically bound to the mRNA/ribosome complex. The mRNA is read from the $5' \rightarrow 3'$ direction; the polypeptide chain is synthesized from the amino terminal to the carboxyl terminal.

The major steps of protein synthesis include: (1) formation of "*activated*" *aminoacyl-tRNA* molecules (each of the 20 amino acids; must first be bound to a specific tRNA molecule); (2) initiation of polypeptide chain synthesis (formation of the mRNA/ribosome initiation complex), (3) elongation of the polypeptide chain; (4) termination of protein synthesis; and (5) *posttranslational modification* of the polypeptide chain, yielding the final, active protein product.

THE RIBOSOME

ribosomes

Ribosomes are the sites where protein synthesis takes place. They are located in the cytoplasm of both prokaryotes and eukaryotes. In eukaryotes, some ribosomes are bound to the outer membranes of the endoplasmic reticulum, forming what is known as the *rough endoplasmic reticulum* (*rough ER*). The structure and composition of a prokaryotic ribosome is shown in Figure 18–3.

Prokaryotic ribosomes are composed of two major subunits: a 50S subunit (constructed from 23S rRNA, 5S rRNA, and 34 different proteins) and a 30S subunit (constructed from 16S rRNA and about 21 different proteins). The "S" suffix defines the relative sizes of the subunits and RNA. Together, these two subunits form the complete 70S prokaryotic ribosome. Eukaryotic ribosomes are a little larger than their prokaryotic counterparts. The 80S eukaryotic ribosome is also constructed from two major subunits—a large 60S subunit and a smaller 40S subunit.

During the initiation phase of prokaryotic protein synthesis, the smaller ribosome subunit attaches to the mRNA chain first, followed by the larger subunit. This forms the *70S initiation complex*. A number of individual ribosomes (each conducting protein synthesis) are often bound to a

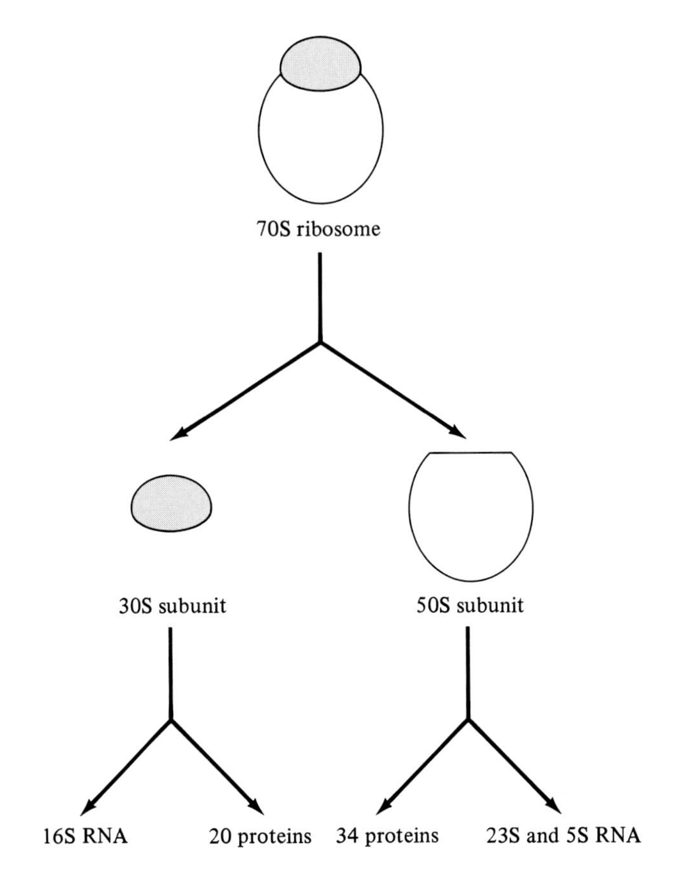

FIGURE 18–3
The structure and composition of the prokaryotic ribosome.

single mRNA chain, forming a *polyribosome* (or *polysome*). In Figure 18–4 is shown a polyribosome with five ribosomes carrying out protein synthesis in the 5′ → 3′ direction along the mRNA chain.

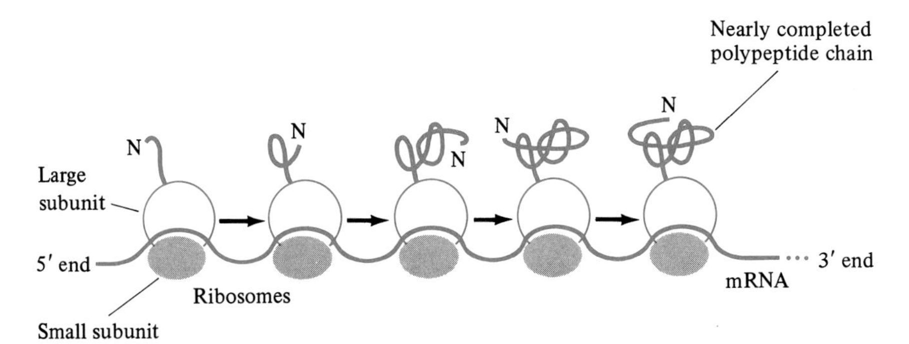

FIGURE 18–4 A polyribosome. The individual ribosomes attached to the mRNA molecule are involved in protein synthesis. The ribosomes move along the mRNA chain in the 5′ → 3′ direction. The ribosomes near the 5′ end of the chain have just begun protein synthesis; the ribosomes near the 3′ end are almost finished.

Step 1: Activation of Amino Acids: Formation of Aminoacyl-tRNA

Before an amino acid can be used in protein synthesis, it must first be bound to a tRNA molecule and, thus, activated. An *aminoacyl-tRNA synthetase* catalyzes the condensation of the amino acid to the 3′ terminal OH group of a specific tRNA molecule. An *ATP molecule* (and an Mg^{2+}) is required for the reaction (Figure 18–5). The synthetase catalyzes the formation of an *aminoacyl-adenylate intermediate.* This intermediate is actually bound to

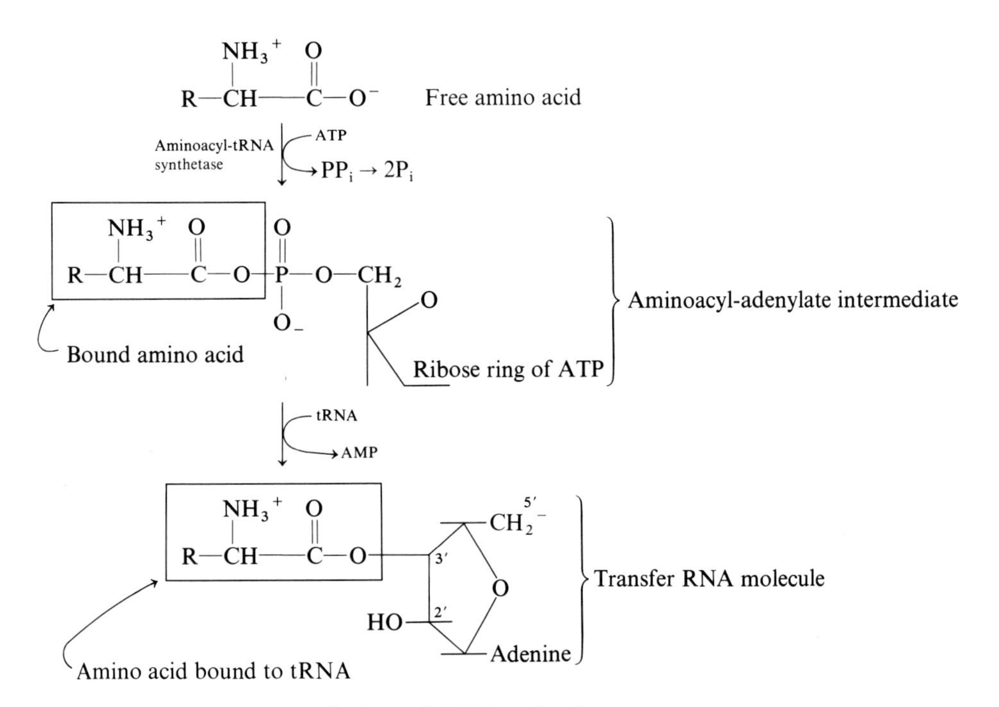

Alanine
Anticodon
Alanyl-tRNA

FIGURE 18–5
Formation of aminoacyl-tRNA.

the enzyme. The synthetase next transfers the amino acid to the 3′ OH group of the tRNA molecule, forming the aminoacyl-tRNA. The carboxyl group of the amino acid is linked to the 3′ OH group by an ester linkage. The reaction is driven to completion by the cleavage of a pyrophosphate (PP_i) group from ATP and subsequent hydrolysis to yield 2 molecules of inorganic phosphate (P_i). Different aminoacyl-tRNA synthetases, each specific for a particular amino acid and tRNA molecule catalyze the addition of each amino acid to the proper tRNA molecule.

PROTEIN SYNTHESIS IN THE PROKARYOTE

Step 2: Initiation of Polypeptide Chain Synthesis

An initiation complex is constructed from the following component: (1) the 30S (small) subunit of the ribosome; (2) the mRNA molecule; (3) an initiation tRNA (together with a bound derivative of methionine called *N-formylmethionine*); and (4) three initiation factors (proteins). The formation of *N-formylmethionyl tRNA* (*fMet-tRNA*^{fmet}) is shown in the following reaction:

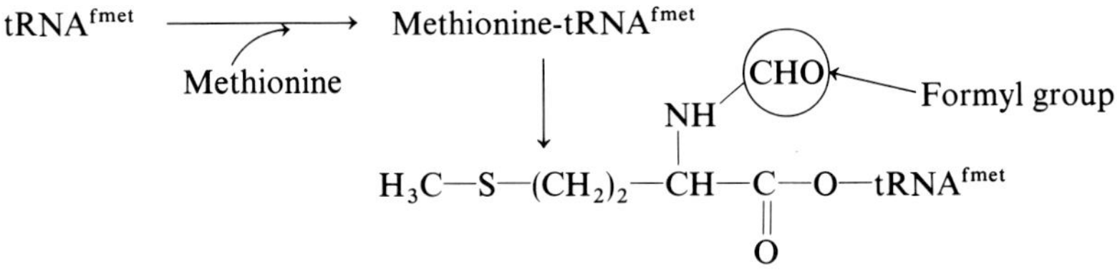

$$\text{tRNA}^{fmet} \xrightarrow{\text{Methionine}} \text{Methionine-tRNA}^{fmet}$$

$$\downarrow$$

$$\text{H}_3\text{C—S—(CH}_2)_2\text{—}\underset{\substack{|\\ \text{NH—CHO}}}{\text{CH}}\text{—}\underset{\substack{\|\\ \text{O}}}{\text{C}}\text{—O—tRNA}^{fmet}$$

(CHO: Formyl group)

N-Formylmethionyl-tRNA^{fmet}

The assembly of the initiation complex is shown in Figure 18–6. The 30S subunit combines with initiation factors and with the fMet-tRNA^{fmet}. A molecule of GTP is also bound at this point. The 30S/(fMet-tRNA)/initiation factor complex binds close to the 5′ end of the mRNA. The anticodon of the fMet-tRNA aligns itself with the AUG (start) codon on mRNA (the AUG codon is specific for the fMet-tRNA). Next, the 50S subunit binds, forming the complete 70S ribosome. The GTP is hydrolyzed into GDP and P_i and the initiation factors are released. The 70S initiation complex is now ready to carry out protein synthesis. The start signal (in mRNA) for protein synthesis is AUG and a hairpin loop within the mRNA molecule.

Step 3: Polypeptide Chain Synthesis and Elongation

The events associated with polypeptide chain synthesis and elongation are complicated; they require that one envision a couple of things happening in sequence and then repeated. Three major events take place: (1) the aminoacyl-tRNA binds to the ribosome and the codon is recognized; (2) a peptide bond forms; and (3) *translocation* (movement of the tRNA contain-

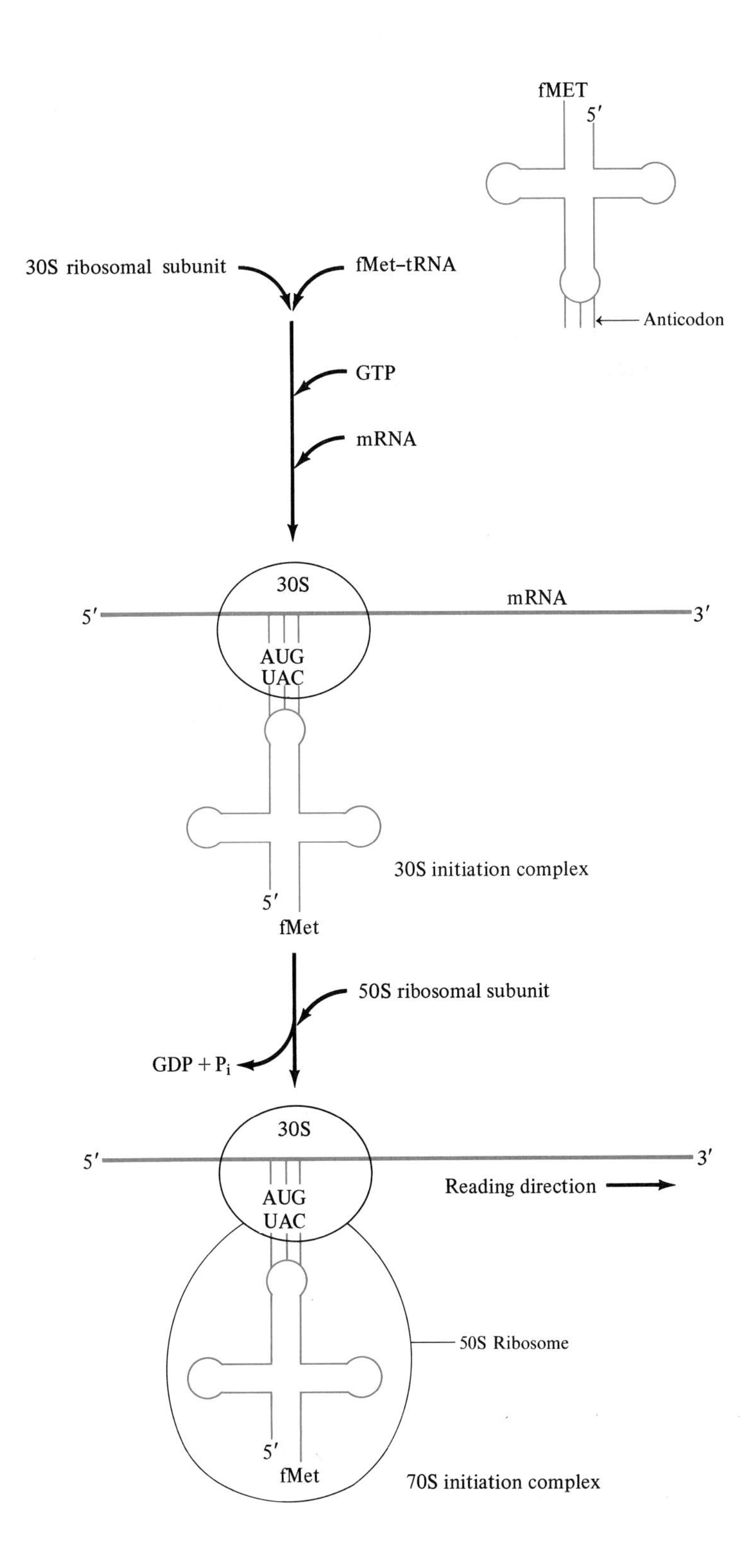

FIGURE 18–6
The assembly of the 70S initiation complex.

ing the bound peptide chain, and movement of the ribosome, so that the next codon can be read). These events are shown in Figure 18–7.

The large 50S subunit contains two distinct t*RNA* binding sites: the *P* (*peptidyl*) *site* and the *A* (*aminoacyl*) *site*. At the start of polypeptide chain formation, the fMet-tRNA is bound to the P (peptidyl) site and to the AUG codon on mRNA. An aminoacyl-tRNA molecule (containing an amino acid) next binds to the A (aminoacyl) site. The hydrolysis of a GTP molecule is required for binding. *Elongation factors* (proteins) are also required for this process. Assume now that the first amino acid to be added is alanine. An alanyl-tRNA, therefore, binds to the A (aminoacyl) site and aligns with the codon for alanine (GCA) on mRNA, as shown in Figure 18–7. Next, *peptidyl transferase*, an enzyme bound to the 50S subunit, catalyzes the formation of a peptide bond between the fMet and the amino group of the incoming alanine. In essence, the entire fMet group of fMet-tRNA is transferred to the free amino group of the bound alanine, as shown in Figure 18–7. The P (peptidyl) site now contains the free tRNA specific for fMet, while the A (aminoacyl) site contains a tRNA with a bound dipeptide, (in this case, fMet-Ala-tRNA).

In the next step, the fMet-Ala-tRNA is moved (*translocated*) from the A (aminoacyl) site to the P (peptidyl) site. The tRNA that no longer has a bound amino acid (in this case, the tRNA specific for fMet) is first released from the P site and diffuses away from the 70S ribosome complex. The fMet-Ala-tRNA is then bound into the P (peptidyl) site leaving the A (aminoacyl) site empty. An elongation factor (also called *translocase*) is required for this translocation reaction, as is the hydrolysis of a GTP molecule. During translocation, the 70S ribosome moves relative to the mRNA molecule, positioning a new codon in the A site ready to accept a new aminoacyl-tRNA molecule with an appropriate anticodon. In this example, the new codon is CAU (coding for histidine). Thus, the anticodon of a histidyl-tRNA molecule will bind to the CAU (His) codon, positioning the His-tRNA in the A site. The sequence of steps is thus repeated: binding of a new aminoacyl-tRNA, formation of a peptide bond, and translocation. In this way, the mRNA is read sequentially one codon at a time by the ribosome, which constructs the new polypeptide chain one amino acid at a time. As more amino acids are added to the polypeptide chain, the chain may begin to fold into a specific 3-dimensional shape (see Figure 18–4, p. 348).

Step 4: Termination of Polypeptide Chain Synthesis

The genetic code (see Table 18–1) contains three stop or termination, signals (UAA, UAG, and UGA). These stop codons (a typical mRNA molecule would contain one of the three) signal the end of the genetic message and, thus, the end of the polypeptide chain. *Release factors* (proteins) bind to the A site of the ribosome and recognize the stop codon. The peptidyl transferase (on the 50S subunit) then catalyzes the hydrolysis of the polypeptide chain from the final tRNA molecule bound to the P site, releasing the polypeptide chain from the ribosome/mRNA complex. The releasing factors dissociate from the ribosome and the ribosome itself dissociates into

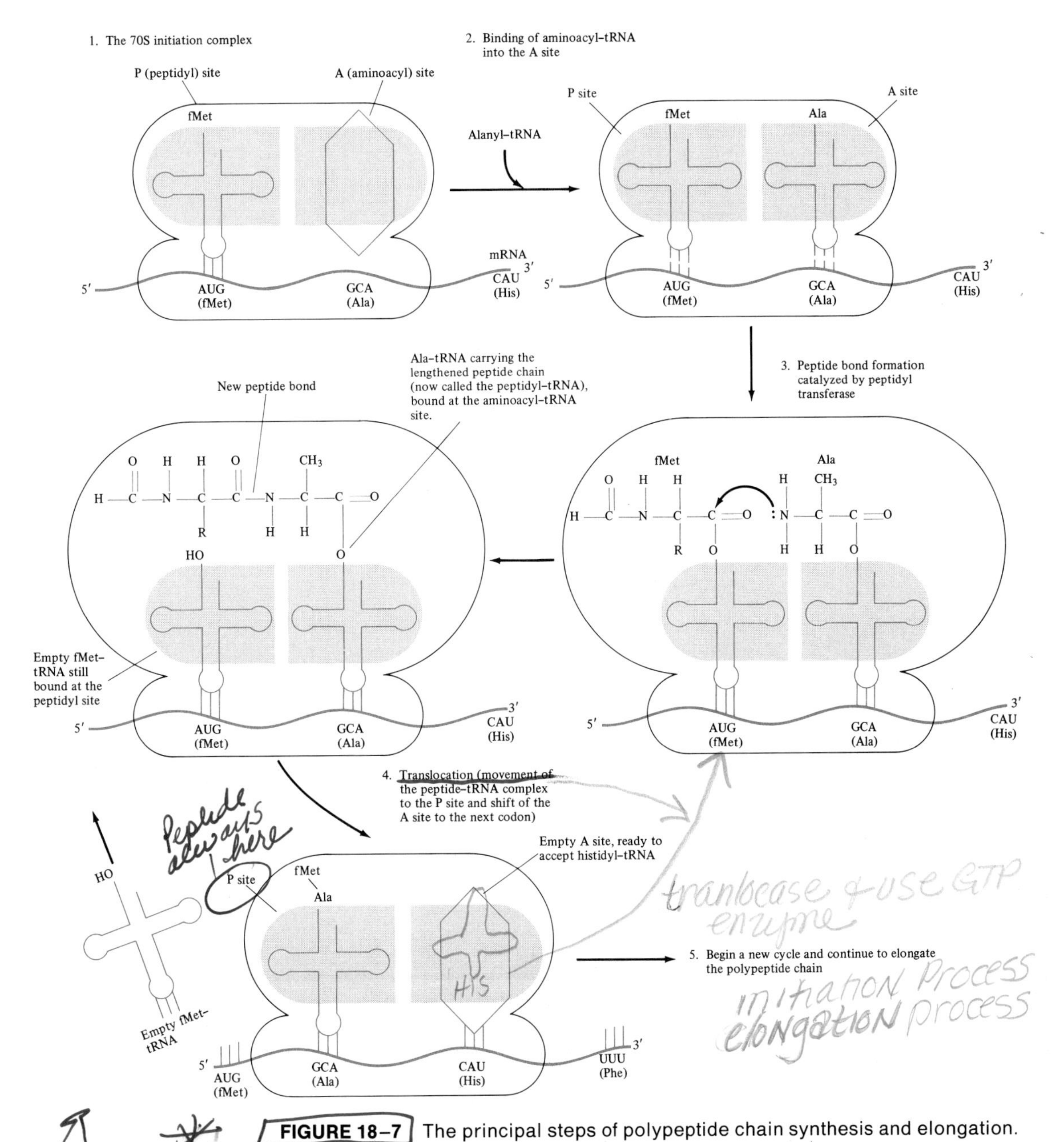

FIGURE 18–7 The principal steps of polypeptide chain synthesis and elongation.

the 30S and 50S subunits, which release from the mRNA molecule and diffuse away into the cytoplasm.

Step 5: Posttranslational Modification of Proteins:

The newly synthesized polypeptide chain may still require additional enzymatic steps **(posttranslational modification)** to form a biologically active protein. The N-formyl group and the methionine residue on the N-terminal

posttranslational modification

end of the polypeptide chain are enzymatically removed. Other covalent modifications include *intra-* and *inter*chain disulfide bond formation; glycosylation (attachment of one or more sugar residues, forming a glycoprotein); attachment of prosthetic groups; and proteolytic cleavage (see Chapter 6, Enzymes).

ANTIBIOTICS INHIBIT BACTERIAL PROTEIN SYNTHESIS

A number of antibiotics inhibit bacterial protein synthesis and are, therefore, very effective agents against certain strains of bacteria. These antibiotic drugs inhibit a specific enzyme associated with protein synthesis. A few representative antibiotics and their biological action are summarized in Table 18–2.

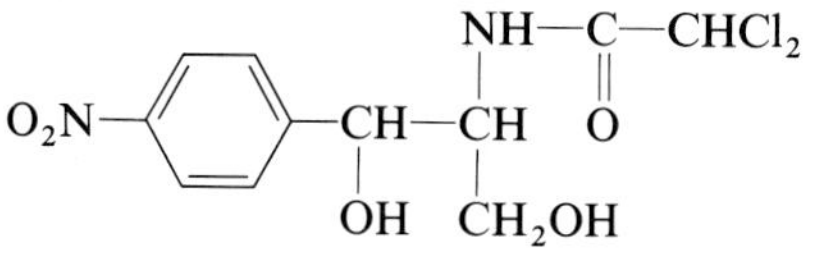

Chloramphenicol

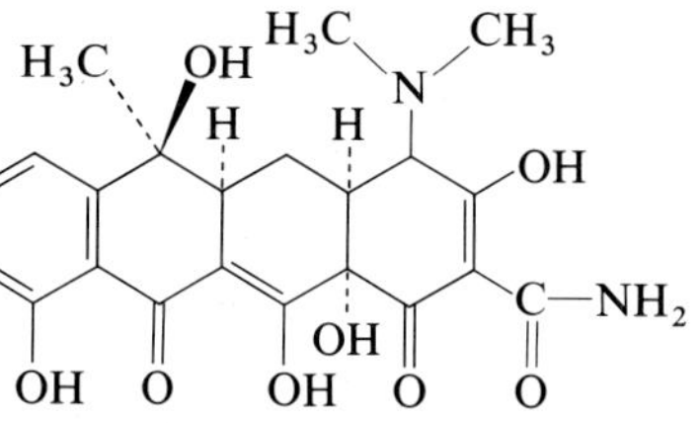

Tetracycline

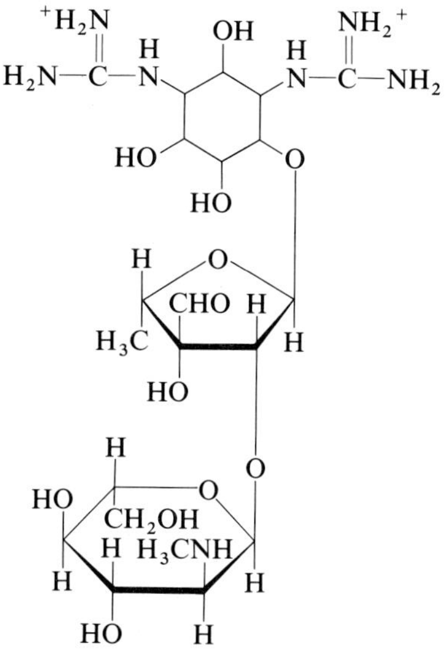

Streptomycin

TABLE 18–2 Antibiotic Inhibitors of Protein Synthesis.

Antibiotic	Biological Action
Chloramphenicol	Inhibits the peptidyl transferase of the 50S ribosomal subunit
Erythromycin	Binds to the 50S subunit and inhibits the translocation step.
Puromycin	Induces premature polypeptide chain termination
Streptomycin	Inhibits the initiation step of protein synthesis (and also causes the mRNA to be misread)
Tetracycline	Binds to the 30S subunit and inhibits the binding of aminoacyl-tRNA

18.5 THE STRUCTURE OF PROKARYOTIC AND EUKARYOTIC GENES

PROKARYOTIC GENE STRUCTURE AND THE REGULATION OF GENE EXPRESSION

operon

The elucidation of the structure of the *lac operon* was a landmark event in biochemistry. It not only gave insight into the structure of a specific set of genes in the DNA of *E. coli* (a prokaryote), but also provided important information on how the *expression of these genes is regulated.* An **operon** can be defined as a segment of DNA that contains an operator gene controlling the expression of closely spaced structural genes. The structural genes code for specific proteins or enzymes that, in turn, carry out particular biochemical processes. The *lac* operon is responsible for producing the enzymes required for lactose transport and metabolism. A diagram of the *lac* operon is shown in Figure 18–8.

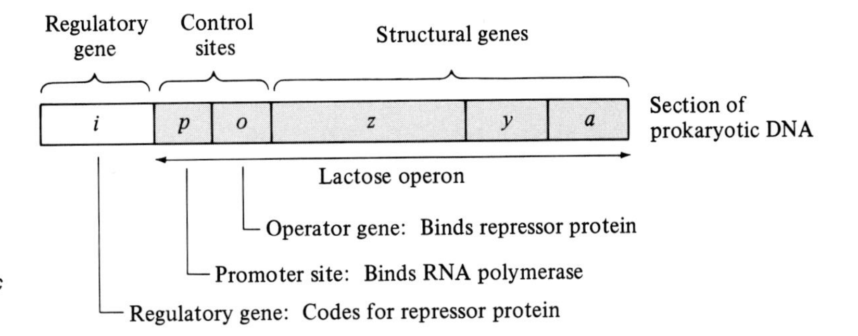

FIGURE 18–8
The structure of the *lac* operon.

lac operon

The ***lac* operon** is a section of the bacterial DNA molecule. It consists of three distinct parts: the *regulatory gene* (i), the *operator gene* (o), and the *structural genes* (z, y, and a). The regulatory gene codes for a *repressor protein* that binds to the operator gene (o). The operator gene is subdivided into two regions: the *promoter region* (p), which binds RNA polymerase, and the operator gene (o), which binds the repressor protein. The structural genes (z), (y), and (a) code for three separate enzymes involved with lactose metabolism. The z gene codes for *β-galactosidase*, the enzyme that hydrolyzes lactose into glucose and galactose for use in energy production.

Lactose + H_2O —(β-Galactosidase)→ Galactose + Glucose

The *y* gene codes for *galactoside permease*, the protein that mediates the transport of lactose into the cell. The *a* gene codes for *galactoside acetylase* (function unknown).

How, then, does the *E. Coli* cell mediate the transport and metabolism of lactose? When the bacterial cell has plenty of glucose, the *lac* operon is repressed—that is, it does not produce mRNA that can be expressed as new β-galactosidase and galactoside permease molecules. However, if glucose is depleted, then the cell must utilize an alternative energy source. The presence of *lactose* within the culture medium (and its diffusion into the cell) immediately initiate the production of galactoside permease and β-galactosidase. Once the levels of these enzymes increase, then lactose is transported more quickly into the cell and is hydrolyzed into glucose and galactose for energy. Thus, the production of these enzymes only takes

1. *Repressed state (noninduced):*

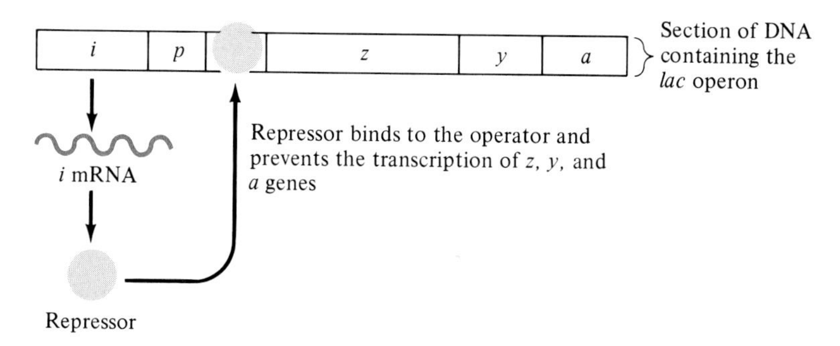

2. *Induced state* (in the presence of the inducer, lactose):

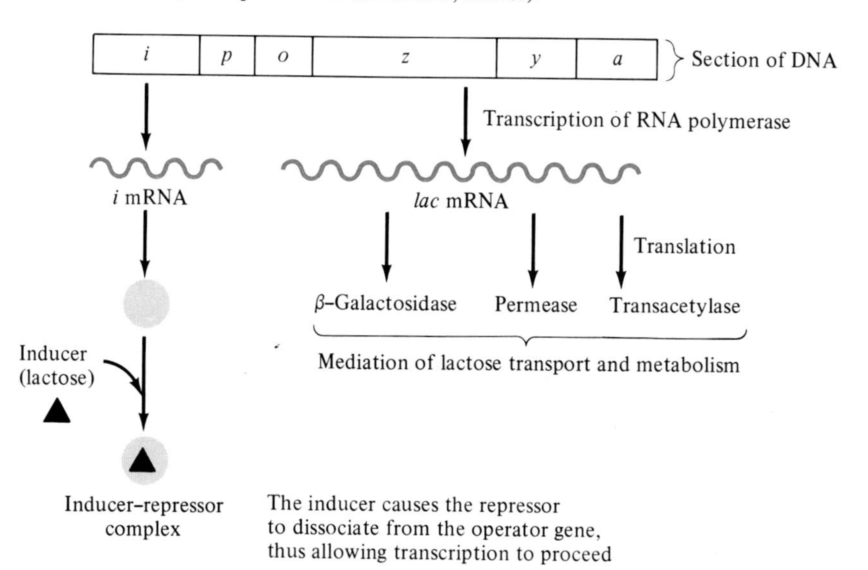

FIGURE 18–9
The regulation of the *lac* operon. (Adapted from L. Stryer, *Biochemistry*, 2nd ed. (San Francisco: W. H. Freeman, 1981), p. 672.)

place in the presence of an *inducer* (lactose) and only when the cell needs these enzymes.

How does the *lac* operon function, and how is it regulated? If the inducer is not present, the repressor protein is bound to the operator gene and prevents the transcription of the DNA by RNA polymerase. Thus, the operon does not produce the mRNA that codes for β-galactosidase, galactoside permease, and galactoside acetylase, and the level of these enzymes in the cell is low.

Induction of the *lac* operon takes place when the inducer (lactose) binds to the repressor protein and causes it to dissociate from the operator gene. The RNA polymerase bound to the promoter region can now transcribe the structural genes *z*, *y*, and *a*. The mRNA transcript coding for proteins *z*, *y*, and *a* is then used in protein synthesis, and these proteins are made. The levels of galactoside permease and β-galactosidase therefore increase and promote the uptake and initial metabolism of lactose. These events are summarized in Figure 18–9.

THE STRUCTURE OF THE EUKARYOTIC CHROMOSOME AND GENE

The genetic material of eukaryotic cells is more complex than that of prokaryotes. In addition, the eukaryotic DNA is organized in a more complex manner. There are a number of differences between eukaryotic and prokaryotic DNA. First, the prokaryotic DNA exists as a very large, tightly coiled circle, whereas eukaryotic DNA exists as a very long chain of double-stranded DNA tightly coiled into a chromosome. Prokaryotes do not have a nucleus and, therefore, the DNA and the enzyme systems responsible for its replication, transcription, and translation are present within the cytoplasm of the bacterium. The DNA of eukaryotes, on the other hand, is packaged within a nucleus and, thus, is separated from the cytoplasm. In eukaryotes, protein synthesis (translation) takes place in the cytoplasm, separated from the nuclear DNA. Finally, it should be noted that eukaryotic DNA is associated with basic, positively charged proteins called **histones**. Prokaryotic DNA does not have bound histones associated with it.

histones

Microscopic evaluation of eukaryotes (including the cells in the human body) has revealed that the cell nucleus contains structures called *chromosomes*. Human cells contain 23 chromosome pairs. Each chromosome pair differs from the others in size and shape and can be segregated into separate classes. How is the DNA organized within the eukaryotic chromosome?

Eukaryotic DNA is constructed from very long chains of double-stranded DNA. This DNA is *not* circular (as in prokaryotic DNA), nor is it branched or organized into shorter segments. The typical chromosome may contain millions or hundreds of millions of individual base pairs. Histone proteins are associated with the DNA double helix and form a repeating unit called a **nucleosome**. A diagram of the nucleosome is shown

nucleosome

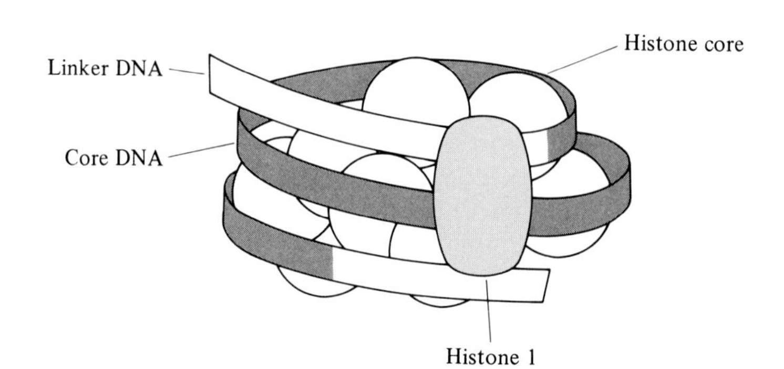

FIGURE 18–10 A suggested structure for the nucleosome. (Adapted from L. Stryer, *Biochemistry*, 2nd ed. (San Francisco: W. H. Freeman, 1981), p. 690.)

in Figure 18–10. The DNA double helix is wound around a core of histone proteins. A specific histone (the H 1 histone) is bound to the DNA chain on the outside of the nucleosome.

Individual nucleosomes are connected together via *linker DNA* segments, forming a string of connected nucleosomes. This nucleosome chain is probably organized into a compact helix, as shown in Figure 18–11. The nucleosome chain may then be wound into a complex 3-dimensional structure, as shown in Figure 18–12. This structure is the chromosome observed with the microscope.

At this point, we might ask, How are individual genes organized within the chromosome? This is not an easy question to answer. Research has shown that the eukaryotic gene is very complex. Our present under-

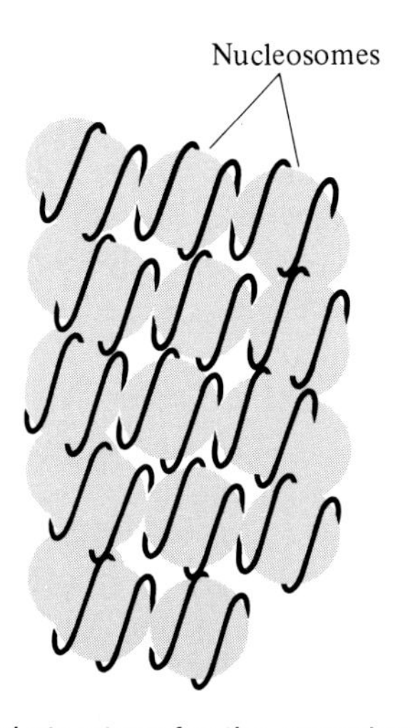

FIGURE 18–11 A suggested structure for the organization of nucleosomes. The spherical structures are the histone core; the dark line is the DNA double helix. (Adapted from L. Stryer, *Biochemistry*, 2nd ed. (San Francisco: W. H. Freeman, 1981), p. 691.)

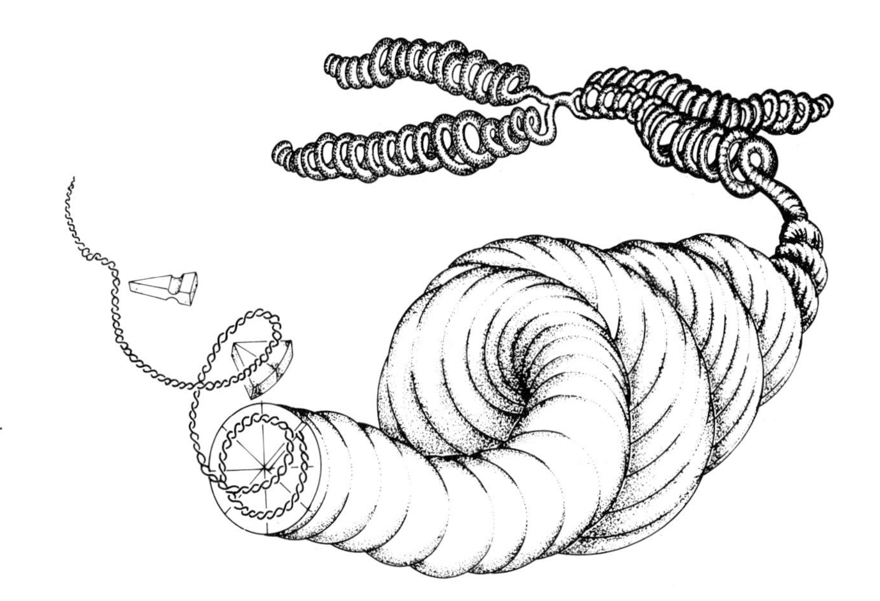

FIGURE 18–12
A model showing how DNA is tightly coiled into a compact chromosome. (Adapted from M. R. MacDonald, *The Origin of Johnny* (New York: Alfred A. Knopf, 1975), p. 73.)

standing of the structure and organization of eukaryotic genes can be summarized as follows:

satellite DNA

1. Research has shown that a large percentage of eukaryotic DNA consists of regions of *repeated* base sequences. This *repetitive DNA*, also known as **satellite DNA**, is primarily located near the *centromere* (the central region of the chromosome that links the two legs of the chromosome). It is hypothesized that these satellite DNA sequences help align the chromosome during mitosis and meiosis.[2]

2. The genes that code for ribosomal RNA (rRNA) and for the histones *are repeated* many times within the chromosome. Between 100 and several thousand copies of these genes are found in the typical eukaryotic chromosome. These copies are clustered closely together.

3. Most proteins other than histones are derived from so-called *single-copy genes*. That is, a single gene within the DNA molecule codes for a particular protein.

coding sequences

4. The structure of single-copy genes is rather complex because the gene is interrupted by repetitive sequences. A diagram of the eukaryotic gene coding for the β-chain of hemoglobin is shown in Figure 18–13. Two major types of base sequences have been established for this gene. The bases that code for the final amino acid sequence of the protein are called **coding sequences** (or *exons*). The intervening, repetitive bases that do *not* code for the amino acid sequence are called *introns*. A number of eukaryotic genes have been analyzed, and most contain these intron sequences. For example, the genes coding for the α- and β-chains of hemoglobin contain 2 introns. The ovalbumin gene contains 7 introns, the conalbumin gene contains 17 introns, and the α-collagen gene contains *52* introns! Apparently, both the exon and the intron sequences

are transcribed into mRNA. However, only the exon sequences are spliced together to yield the mature mRNA which is then used for the synthesis of the final protein product. This scheme is shown in Figure 18–13.

5. Certain sections of DNA are *mobile* and can be inserted into other sections of DNA. These mobile units are called *translocatable elements* (*transposons*).

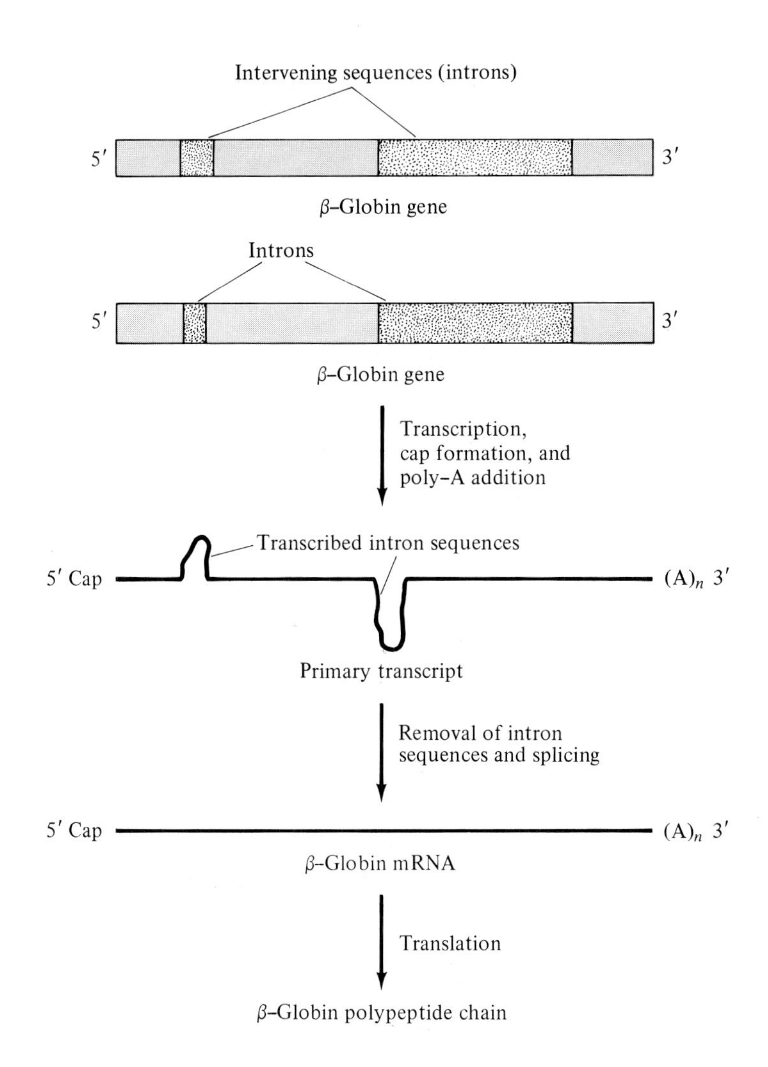

FIGURE 18–13
The structure of the eukaryotic β-globin gene. The intervening (intron) sequences are shown. (Adapted from L. Stryer, *Biochemistry*, 2nd ed. (San Francisco: W. H. Freeman, 1981), p. 634.)

SUMMARY

The genetic information stored in the DNA molecule is read in a triplet, degenerate, nonoverlapping, and commaless code. The genetic information is carried from the DNA molecule to the site of protein synthesis by messenger RNA (mRNA). The mRNA molecule is actually a sequence of codons (each codon is a set of three bases). These codons (61) code for each of the 20 amino acids. Three "stop" codons are also known to exist. The mRNA code is read sequentially, one codon at a time.

Transfer RNA (tRNA) molecules transfer amino acids to the site of polypeptide chain assembly. There are 20 different types of tRNA molecules, each specific for one of the 20 amino acids. The structure of phenylalanine-tRNA is shown in Figures 18–1 and 18–2. The tRNA molecule contains 3 major stems with hairpin loops. The acceptor stem (or arm) binds the specific amino acid. The anticodon loop contains a specific 3-base sequence (anticodon) that is complementary to a 3-base sequence (codon) in the mRNA molecule.

Translation is the process in which the sequence of codons within mRNA is read and the resulting polypeptide chain is assembled into an intact, functioning protein molecule. Protein synthesis requires: (1) an mRNA molecule; (2) ribosomes; (3) specific aminoacyl-tRNAs; and (4) the enzymes and protein factors responsible for the initiation, elongation, and termination of protein synthesis. A protein is assembled by the sequential addition of amino acids to a growing polypeptide chain bound to the mRNA/ribosome complex. The mRNA is read from the $5' \rightarrow 3'$ direction, and the protein is synthesized from the amino terminal to the carboxyl terminal.

The major steps of protein synthesis include: (1) formation of activated aminoacyl-tRNA molecules; (2) initiation of polypeptide chain synthesis (formation of the mRNA/ribosome/fMet-tRNA initiation complex); (3) polypeptide chain synthesis and elongation (including binding of the aminoacyl-tRNAs to the ribosomes, codon recognition, peptide bond formation, and tRNA translocation); (4) termination of protein chain synthesis (recognition of the stop codon and binding of the release factors that dissociate the protein from the ribosome/mRNA complex); and (5) posttranslational modification of the protein (removal of the fMet and, possibly, disulfide bond formation, attachment of various groups to amino-acid R groups, and proteolytic cleavage).

The *lac* operon is a section of the bacterial DNA molecule that consists of a regulatory gene (*i*), an operator gene (*o*), and structural genes (*z*, *y*, and *a*). The regulatory gene codes for a repressor protein that binds to the operator gene and prevents the transcription of the structural genes (noninduced state). In the induced state, an inducer (lactose) binds to the repressor protein, preventing the repressor from binding to the operator gene. This allows the structural genes to be transcribed, resulting in the production of the enzymes responsible for lactose transport and metabolism.

The structure of eukaryotic DNA is different from that of prokaryotic DNA. In eukaryotes, the DNA double helix is wound around histones to form a nucleosome. These nucleosomes, in turn, are tightly coiled into a complex 3-dimensional structure called a chromosome. Eukaryotic DNA contains large regions of repeated base sequences (satellite DNA) and multiple copies of genes that code for rRNA and for histones. In addition, it contains single-copy genes that are composed of exons coding for protein, interrupted by repetitive sequences called introns.

REVIEW QUESTIONS

1. We have presented a number of important concepts and terms in this chapter. Define (or briefly explain) each of the following terms:
   - **a.** Genetic code
   - **b.** Messenger RNA (mRNA)
   - **c.** Codon
   - **d.** Anticodon
   - **e.** Stop codon
   - **f.** Degeneracy
   - **g.** Transfer RNA (tRNA)
   - **h.** Translation
   - **i.** Ribosome
   - **j.** Amino acyl-tRNA
   - **k.** Initiation complex
   - **l.** P and A sites
   - **m.** Translocation step
   - **n.** Lac operon
   - **o.** Induction
   - **p.** Histones
   - **q.** Exon and intron regions

2. We have discussed a number of important enzymes in this chapter. For each enzyme listed, describe what the enzyme does and note why it is important. If possible, diagram the enzymatic reaction catalyzed by the enzyme.
   a. Aminoacyl-tRNA synthetase
   b. Peptidyl transferase

3. Consider the following section of DNA:

5′—GAT—AGA—CCT—TCT—GGT—3′ Strand 1

3′—CTA—TCT—GGA—AGA—CCA—5′ Strand 2

   Assume that strand 2 is transcribed into mRNA and that the mRNA, in turn, in translated into a polypeptide chain.
   a. What is the base sequence of the mRNA produced?
   b. What is the amino acid sequence of the polypeptide chain produced?

*4. **a.** Suggest a possible base sequence for the segment of mRNA that codes for the hormone oxytocin. Include the initiation codon (AUG) and a termination codon in the message.

S—S (linking the two Cys residues)

N-Gly-Leu-Pro-Cys-Asn-Gln-Ileu-Tyr-Cys-C

Oxytocin

   b. What is the base sequence of the DNA that codes for this segment of mRNA?

5. Protein X has a molecular weight of 30 000 daltons. Assuming that the average molecular weight of an amino acid is 120 daltons, answer the following questions:
   a. How many amino acids are present in protein X?
   b. How many codons are found in the mRNA that codes for this protein? (Include fMet and stop codons).
   c. How many nucleotide bases does this represent?
   ***d.** Assume that protein X (MW = 30,000d) is used to construct a tetramer (made from 4 *identical* molecules of protein X). How many bases are required to code for the tetrameric protein? Explain your answer.

SUGGESTED READING

Bhagavan, N. V. *Biochemistry*, 2nd ed., Ch. 5. Philadelphia: J. B. Lippincott, 1978.

Lehninger, A. L. *Biochemistry*, 2nd ed., Chs. 33–35. New York: Worth, 1975.

Stryer, L. *Biochemistry*, 2nd ed., Chs. 26–31. San Francisco: W. H. Freeman, 1981.

Watson, J. D. *Molecular Biology of the Gene*, 3rd ed. Menlo Park, CA: W. A. Benjamin, 1976.

NOTES

1. Armstrong, F. B., and Bennett, T. P., *Biochemistry* (New York: Oxford University Press, 1979), p. 424.
2. Stryer, L., *Biochemistry*, 2nd ed. (San Francisco: W. H. Freeman, 1981), p. 697.

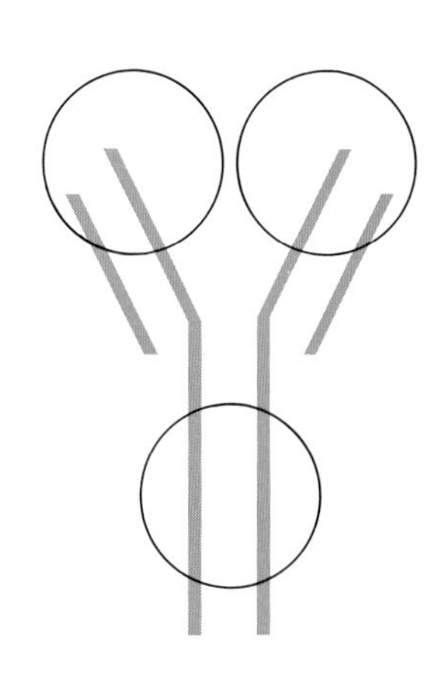

Chapter 19

Hormones and Hormone Action

19.1 INTRODUCTION

hormones

endocrine glands

A study of biochemistry would not be complete without discussing the structure and biological function of the **hormones**. Hormones are a class of chemical substances that are synthesized and secreted by the **endocrine glands**. The endocrine glands (ductless glands) are specialized organs of the body that secrete hormones directly into the blood stream (*endocrine* means "internally secreting"). The hormones are secreted by the endocrine glands in *very minute* quantities ($\sim 10^{-6}$g/100 ml blood, for the steroid hormones). Hormones evoke a particular biochemical or physiological response in what is known as the *target tissue*. In essence, endocrine hormones are *extracellular chemical messengers* (the term *hormone* means "to excite, to stimulate"). Some hormones are intracellular chemical messengers and evoke a response from inside the cell.

Hormones stimulate a particular biochemical process by affecting (1) the *rate of synthesis* of certain enzymes or proteins, (2) the *rate of enzymatic catalysis*, or (3) the *permeability of cell membranes* towards other substances. Thus, hormones *act by regulating preexisting cellular processes.*[1]

In this chapter, we will discuss the different classes of endocrine glands and hormones. The anatomy and physiology of the pituitary gland will be presented, along with a discussion of how this gland controls the other endocrine glands. The mechanisms of both the polypeptide hormones and the steroid hormones will be presented. Finally, the regulation of the female reproductive system (menstrual cycle) will be discussed to illustrate the hormonal regulation of multiple endocrine glands.

19.2 THE CLASSES OF ENDOCRINE GLANDS AND HORMONES

exocrine gland

The endocrine glands are a class of ductless organs that synthesize hormones and secrete them into the bloodstream. A number of endocrine glands exist within the body, including the pituitary gland, the thyroid gland, the parathyroid glands, the adrenal glands, the testes, the ovaries, and the pancreas, to name a few. The pancreas is often considered an **exocrine gland**, because it secretes pancreatic juice into the intestinal tract via a duct. However, the pancreas also has an endocrine function—specialized aggregations of cells (the Islets of Langerhans) synthesize insulin and glucagon and secrete them directly into the bloodstream. The pituitary gland is considered the *master* endocrine gland, because it secretes a number of hormones that *control other endocrine glands* within the body. As we shall see in Section 19.3, the pituitary gland is controlled by the brain.

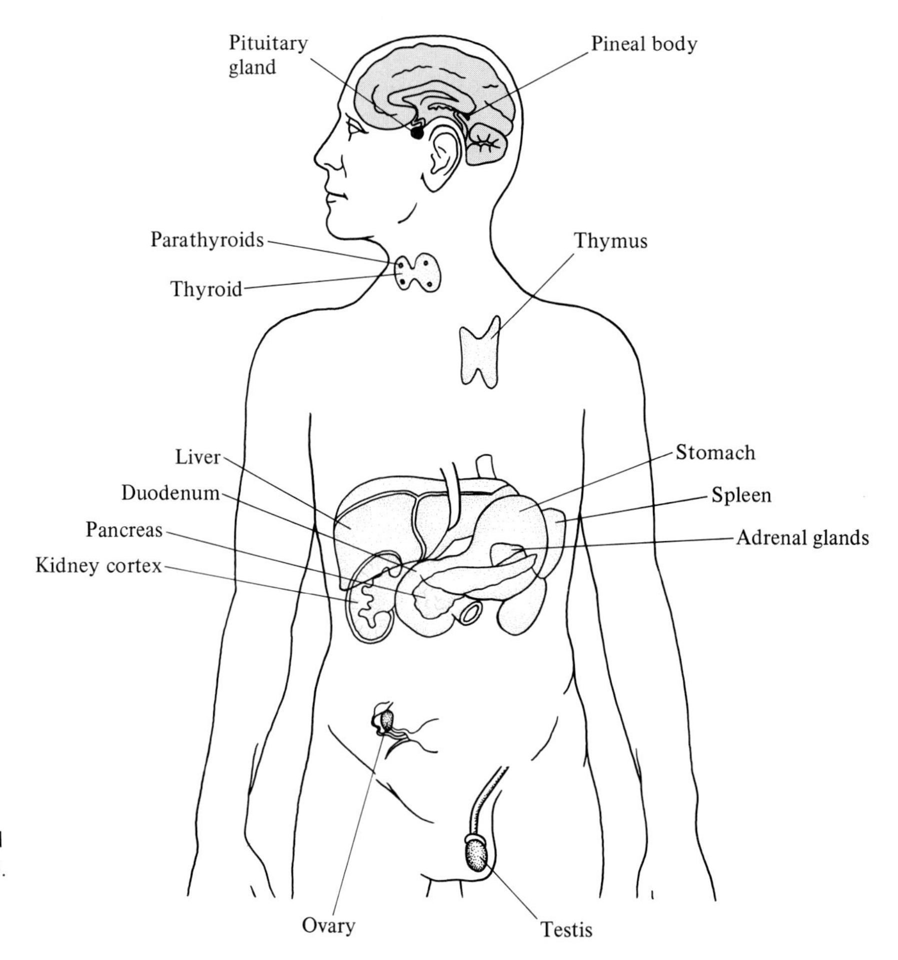

FIGURE 19–1
The major endocrine glands of the body. (Adapted from N. Tietz (ed.), *Fundamentals of Clinical Chemistry*. 2nd ed. (Philadelphia: W. B. Saunders, 1976), p. 700.)

Figure 19–1 depicts the major endocrine glands and their location in the body. In addition, Table 19–1 lists the endocrine glands, the hormones secreted by each gland, the site of action (target tissue) of each hormone, and the primary biochemical or physiological responses of the body to each hormone. The list is by no means complete—a number of other substances (enkephalins, prostaglandins, and vitamin D, to name a few) exhibit hormonelike activity.

Hormones can be classified according to their chemical structure. Three major classes are known: (1) derivatives of amino acids, (2) peptide or protein hormones, and (3) steroid hormones. The chemical structures of a representative hormone from each class is shown:

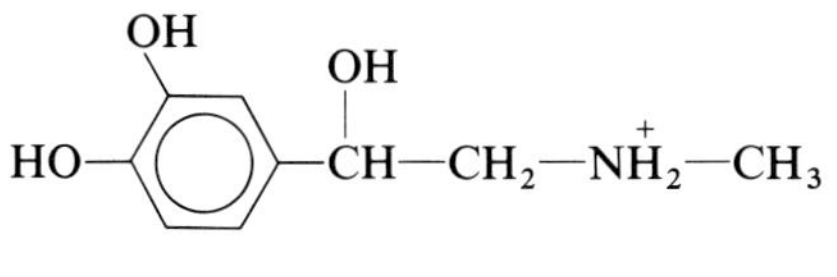

OH, OH, HO—(ring)—CH—CH_2—$\overset{+}{N}H_2$—CH_3

Epinephrine (adrenaline), a derivative of the amino acid tyrosine

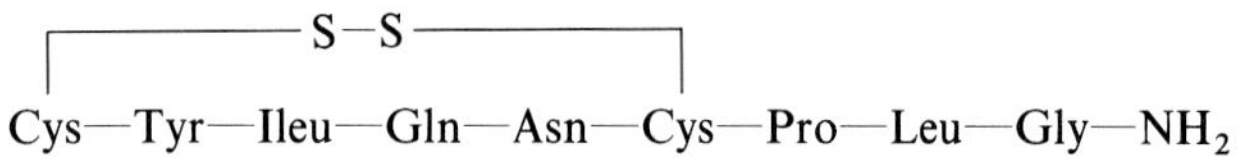

S—S (disulfide bridge between the two Cys residues)

Cys—Tyr—Ileu—Gln—Asn—Cys—Pro—Leu—Gly—NH_2

Oxytocin, a peptide hormone

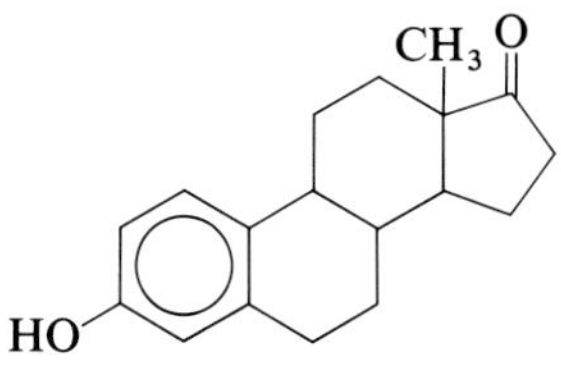

CH_3, O, HO

Estrone, a steroid hormone

The major hormones that are derivatives of amino acids include *epinephrine* (adrenaline), *norepinephrine* (noradrenaline), and *thyroxine* (T_4). There are many peptide hormones (see Table 19–1), ranging from 2–3 amino acids to 20–30 amino acids in length. Some peptide hormones are so big, they are actually classified as proteins. A number of the protein hormones are glycoproteins because they contain covalently bound sugar residues. The steroid hormones are derivatives of cholesterol and, as their name implies, contain the 4-ring steroid nucleus. Literally dozens of important steroid hormones are made in our bodies. It should also be noted that substances called *prostaglandins*, derived from a 20-carbon fatty acid, exhibit hormonelike properties.

PROHORMONES

Many endocrine glands first synthesize an *inactive prohormone* that is converted into the biologically active hormone. Prohormone conversion is carried out by two mechanisms in the body.

TABLE 19–1 The Major Hormones of the Body

Endocrine Gland and Hormone	Nature of Hormone	Site of Action	Principal Actions
Hypothalamus			
Various releasing factors	Polypeptides	Anterior pituitary	Release of trophic hormones
Anterior pituitary			
Somatotrophin, growth hormone (STH, GH)	Protein	Body as a whole	Growth of bone and muscle
Adrenocorticotrophin (ACTH)	Polypeptide	Adrenal cortex	Stimulates formation and secretion of adrenocortical steroids
Melanophore-stimulting hormone (MSH)	Polypeptide	Skin	Dispersion of pigment granules; darkening of skin
Thyrotrophin (TSH)	Glycoprotein	Thyroid	Stimulates formation and secretion of thyroid hormone
Follicle-stimulating hormone (FSH)	Glycoprotein	Ovary	Growth of follicles with LH, secretion of estrogens, and ovulation
		Testis	Development of seminiferous tubules; spermatogenesis
Luteinizing or interstitial cell–stimulating hormone (LH or ICSH)	Glycoprotein	Ovary	Formation of corpora lutea; secretion of progesterone
		Testis	Stimulation of interstitial tissue; secretion of androgens
Prolactin (lactogenic hormone, luteotrophin)	Protein	Mammary gland	Proliferation of mammary gland and initiation of milk secretion
Posterior pituitary			
Vasopressin (ADH, antidiuretic hormone)	Nonapeptide	Arterioles	Elevates blood pressure Antidiuretic activity
		Renal tubules	Water reabsorption
Oxytocin	Nonapeptide	Smooth muscle (uterus, mammary gland)	Contraction, action in parturition and in sperm transport; ejection of milk
Thyroid			
Thyroxine and triiodothyronine	Iodoamino acids	General body tissue	Stimulates oxygen consumption and metabolic rate of tissues
Calcitonin (thyrocalcitonin)	Polypeptide	Skeleton	Inhibits calcium resorption; lowers plasma calcium and phosphate
Parathyroid			
Parathyroid hormone (PTH)	Polypeptide	Skeleton, kidney, gastrointestinal tract	Regulates calcium and phosphorus metabolism

Source: Adapted from Tietz, N. (ed.), *Fundamentals of Clinical Chemistry*, 2nd ed. (Philadelphia: W. B. Saunders, 1976), pp. 700–01.

TABLE 19–1 *(continued)*

Endocrine Gland and Hormone	Nature of Hormone	Site of Action	Principal Actions
Adrenal cortex			
Adrenal cortical steroids—cortisol, aldosterone	Steroids	General body tissue	Carbohydrate, protein, and fat metabolism; salt and water balance; inflammation, resistance to infection; hypersensitivity
Adrenal medulla			
Norepinephrine and epinephrine	Aromatic amines	Sympathetic receptor	Mimics sympathetic nervous system
		Liver and muscle	Glycogenolysis
		Adipose tissue	Release of lipid
Ovary			
Estrogens	Phenolic steroids	Female accessory sex organs	Development of secondary sex characteristics
Progesterone	Steroids	Female accessory reproductive structures	Preparation for ovum implantation; maintenance of pregnancy
Relaxin	Polypeptide	Symphysis pubis, uterus	Relaxation; aids in parturition
Testis			
Testosterone	Steroid	Male accessory sex organs	Development of secondary sex characteristics, maturation and normal function
Pancreas			
Insulin	Polypeptide	Most cells	Regulation of carbohydrate metabolism; lipogenesis
Glucagon	Polypeptide	Liver	Glycogenolysis
Placenta			
Estrogens, progesterone, gonadotrophins (HCG)	Same as above	Same as above	Same as above
Placental lactogen, relaxin			
Gastrointestinal tract			
Secretin	Protein	Pancreas	Secretion of alkaline fluid and digestive enzymes
Cholecystokinin-pancreozymin	Protein	Gallbladder	Contraction and emptying
Enterogastrone	Protein	Stomach	Inhibition of motility and secretion
Gastrin	Protein	Stomach	Secretion of acid

In the first conversion process, an endocrine gland synthesizes a polypeptide prohormone (with little or no hormone activity). This prohormone is processed *within* the endocrine cells by the proteolytic clipping of the polypeptide chain and, possibly, other subsequent chemical modifications such as disulfide bond formation. These modifications allow the hormone molecule to assume a different 3-dimensional shape, one that is necessary for hormone activity. The synthesis of *insulin* is a nice example of this process, and is summarized in Figure 19–2.

A polypeptide chain called *preproinsulin* is first made by pancreatic β-cells (on the membranes of the endoplasmic reticulum). The N-terminal region (consisting of 16 amino acids) is then cleaved from the preproinsulin.

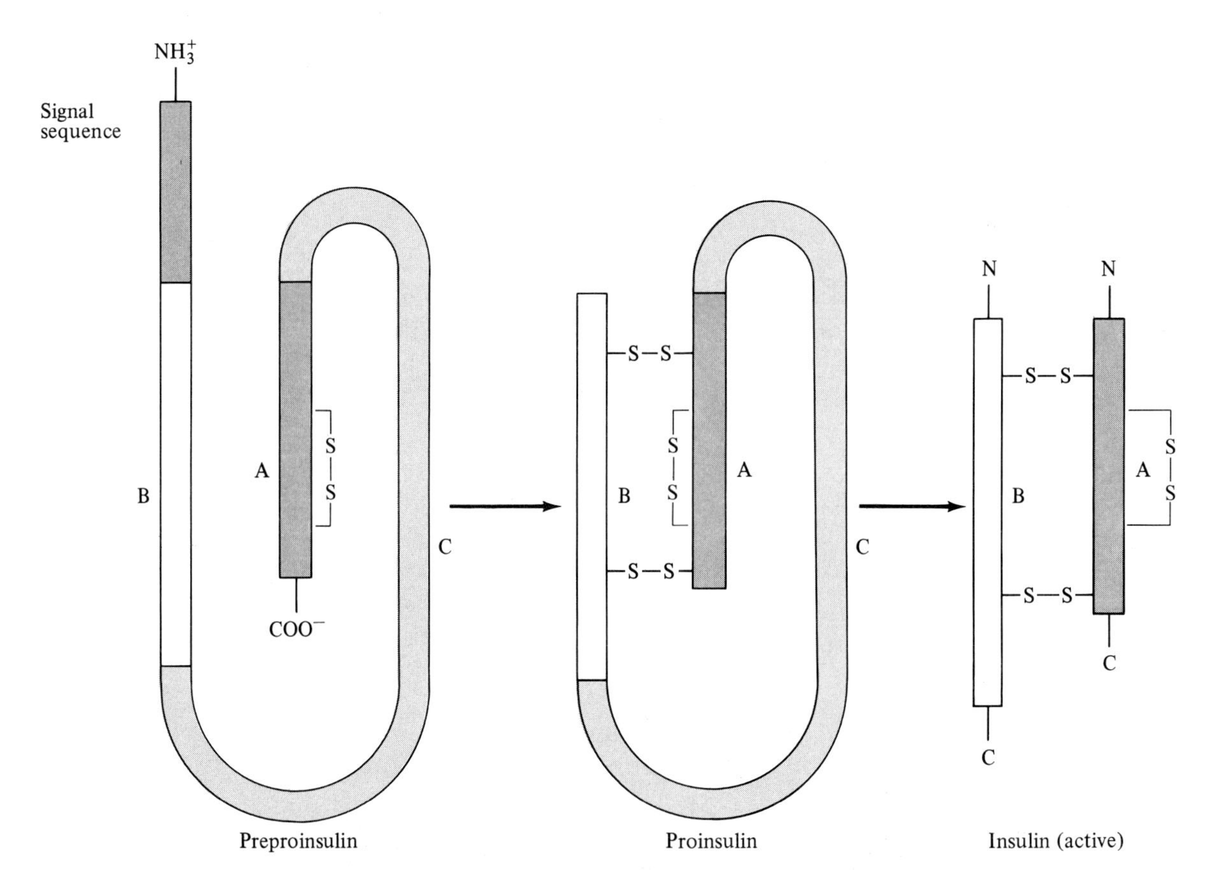

FIGURE 19–2 The conversion of preproinsulin into active insulin. A signal sequence is enzymatically removed from preproinsulin to generate proinsulin. A connecting peptide is then removed from proinsulin to yield active insulin. The A and B refer to the A and B chains of insulin; the C refers to the connecting peptide. (Adapted from L. Stryer, *Biochemistry,* 2nd ed. (San Francisco: W. H. Freeman, 1981), p. 848.)

This generates *proinsulin*, a single polypeptide chain of 83 amino acids, which is stored in storage granules in the pancreatic β-cells. There, disulfide bonds are formed, and a 33 amino acid connecting sequence (C peptide) is cleaved from the molecule, producing an active insulin molecule. You will notice that the insulin molecule now contains two polypeptide chains (noted as the A and B chains) connected by 2 sets of disulfide bonds.

The second prohormone conversion process takes place *outside* the secreting endocrine gland, in another part of the body. For example, the thyroid gland secretes thyroxine (T_4), which contains 4 iodine atoms covalently bound to the two aromatic rings (see structure). The T_4 molecule is not very biologically active. The T_4 molecule, however, can be transformed into the biologically active form 3,5,3′-*triiodothyronine* (T_3). This conversion is catalyzed in the liver by a *deiodinase*, an enzyme that removes one iodine atom from the 5′ position of T_4.

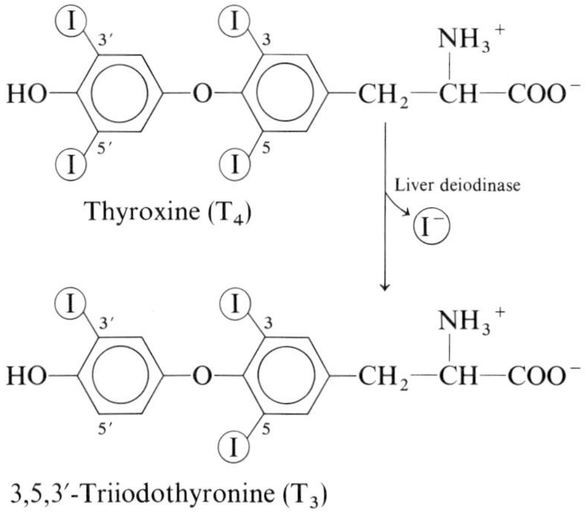

Thyroxine (T_4)

3,5,3′-Triiodothyronine (T_3)

It is believed that T_3 travels to the target tissue cells, is transported into the cell nucleus, and there stimulates the production of certain types of mRNA molecules. This stimulates the total metabolism of the cell and raises the basal metabolic rate.

19.3 THE ANATOMY AND PHYSIOLOGY OF THE PITUITARY GLAND

The pituitary gland is a very small gland (0.5 g) located at the base of the brain near the *hypothalamus*. It is connected to the hypothalamus by the *hypophyseal stalk*. The pituitary gland is divided into three physiologically

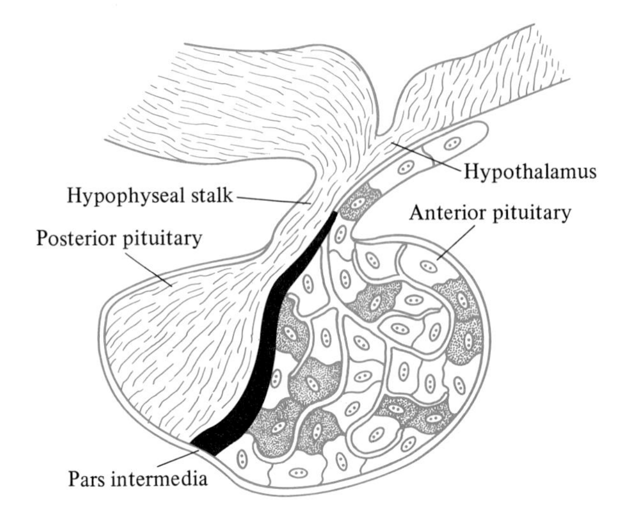

FIGURE 19–3
The pituitary gland. (Adapted from A. C. Guyton, *Textbook of Medical Physiology*, 5th ed. (Philadelphia: W. B. Saunders, 1976), p. 992.)

distinct regions: the *anterior pituitary* (*adenohypophysis*), the *posterior pituitary* (*neurohypophysis*) and the *pars intermedia*. A diagram of the pituitary gland is shown in Figure 19–3.

A variety of polypeptide hormones are synthesized and secreted by the pituitary gland (see Table 19–1, p. 366, and Figure 19–4). These pituitary hormones travel via the bloodstream to other endocrine glands and elicit the secretion of still other hormones from those glands. Thus, the pituitary gland is often referred to as the master endocrine gland of the body. Actually, the hypothalamus (which is controlled by the higher centers of the brain) should be called the master endocrine gland, because it controls the pituitary gland.

releasing factors

The hypothalamus is responsible for the synthesis and secretion of a number of small peptides called **releasing factors**. These substances are synthesized in certain groups of nerve cells within the hypothalamus, and travel to the ends of the axons for secretion. Once secreted from the nerve endings, they travel to the pituitary gland via a special hypothalamic-pituitary circulatory system and stimulate the release of pituitary hormones (hence the name "releasing factors"). Each of the pituitary hormones (with the exception of *oxytocin* and *vasopressin*) is released from the pituitary gland in response to a specific releasing factor. For example, *thyrotropin-releasing factor* (*TRF*) stimulates the release of *thyrotropic hormone*, whereas *prolactin-releasing factor* (*PRF*) stimulates the release of *prolactin*. *Luteinizing hormone* and *follicle-stimulating hormone* are both released from the anterior pituitary gland in response to a releasing factor called *luteinizing hormone and follicle-stimulating hormone releasing factor* (*LHRF/FSHRF*). The structure of the thyrotropin-releasing factor is shown in the following diagram:

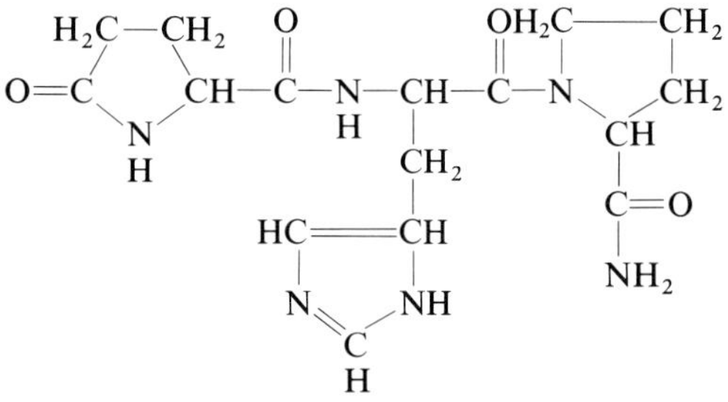

```
   H2C—CH2    O              OH2C———CH2
    /     \   ||             ||  |    |
O=C        CH—C—N—CH—C—N      CH2
    \    /       H   |           \  /
     N               CH2          CH
     H               |            |
                HC===CH          C=O
                |     |           |
                N     NH         NH2
                 \\  /
                   C
                   H
```

Thyrotropic-releasing factor

To understand how the pituitary gland controls the function of other endocrine glands, let us examine the sequence of biochemical processes that leads to the synthesis of the thyroid hormone thyroxine (T_4). First, as stated previously, the hypothalamus secretes thyrotropic-releasing factor (TRF), which travels to the anterior pituitary gland via the hypothalamic-pituitary portal system. This stimulates the release of thyroid-stimulating hormone (also known as TSH or thyrotropin). Thyroid-stimulating hormone is secreted from the pituitary gland and, entering the bloodstream, travels to the thyroid gland. There, it stimulates the synthesis and secretion of thyroxine by the thyroid follicular cells. Thyroxine next travels to the liver and is converted into triiodothyronine (T_3). Both T_4 and T_3 travel to a variety of *target tissues* and are taken into their cells, eliciting a particular biochemical response. Notice the flow of chemical information via these chemical messengers:

Brain (Hypothalamus) ⟶ Releasing factor ⟶ Pituitary hormone ⟶ Secondary hormone ⟶ Response in target tissue

Adrenocorticotropic hormone (ACTH), follicle-stimulating hormone (FSH), luteinizing hormone (LH), prolactin (PRL), and growth hormone (GH) are also secreted by the anterior pituitary gland and travel to a secondary endocrine gland, as shown in Figure 19–4. *Melanocyte-stimulating hormone*, MSH (not shown in Figure 19–4) is released by the pars intermedia in response to melanotropin-releasing hormone. Oxytocin and vasopressin are actually made by the hypothalamus and travel to the posterior portion of the pituitary gland for storage and release.

The *release* of the hypothalamic-pituitary hormones is subject to *feedback inhibition*. As an example, let us consider the regulation of thyroid hormone synthesis. As blood T_4 and T_3 levels rise in response to the sequence of events previously described, the thyroid hormones travel not only to the target tissues but also back to the hypothalamus and pituitary gland. The cells in these glands contain receptors specific for T_4 and T_3. The binding of

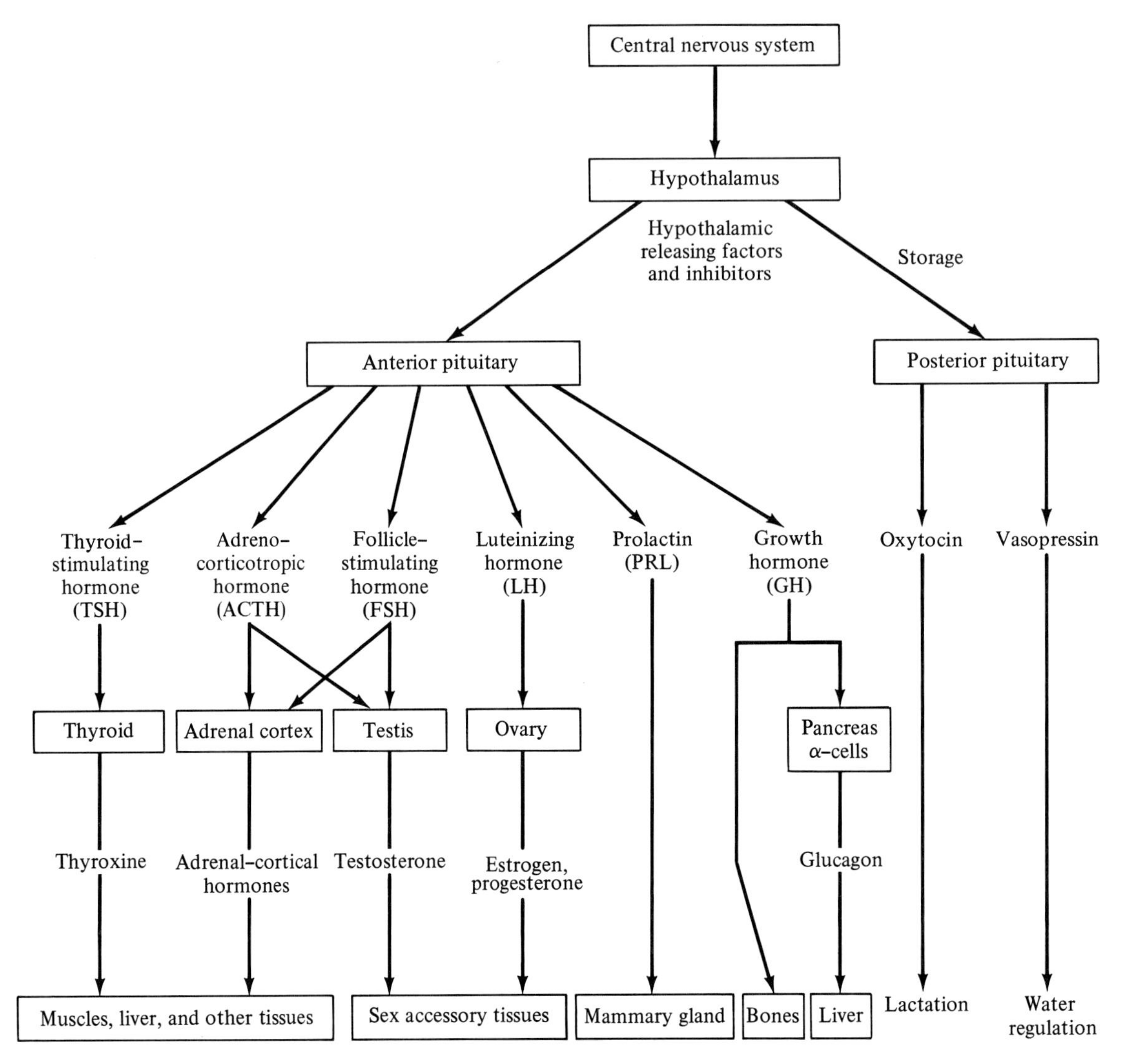

FIGURE 19–4 The organization and sequence of hormone secretion controlled by the hypothalamic-pituitary system.

thyroid hormones to these receptor sites inhibits the release of thyroid-stimulating hormone (and, possibly, thyrotropic-releasing factor). Consequently, thyrotropic hormone is not released from the pituitary gland and T_4 production by the thyroid gland decreases. If the levels of T_4 and T_3 drop below normal, the inhibition of the hypothalamic-pituitary system is removed and thyroid-stimulating hormone is secreted, triggering the thyroid gland into action. Thus, thyroxine is again synthesized in response to the need. The mechanism of feedback inhibition is summarized in Figure 19–5.

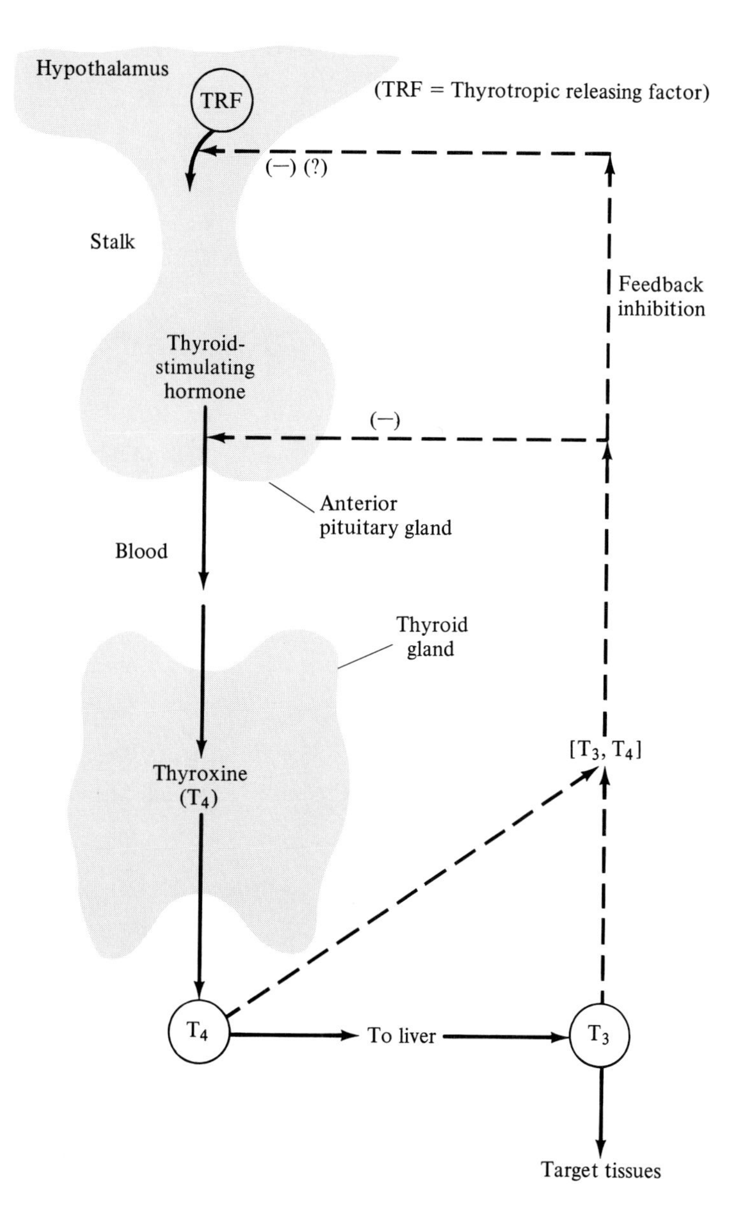

FIGURE 19–5
Regulation of thyroid hormone synthesis and secretion. Circulating T_3 and T_4 inhibit (by negative feedback) the release of thyrotropic hormone from the anterior pituitary gland.

19.4 THE MECHANISMS OF HORMONE ACTION

HORMONES THAT USE CYCLIC AMP AS CHEMICAL MESSENGER

We have already discussed the importance of cyclic AMP in carbohydrate metabolism (see Chapter 10). To review: Certain hormones travel from the secreting endocrine gland to target tissues, where they bind to membrane-

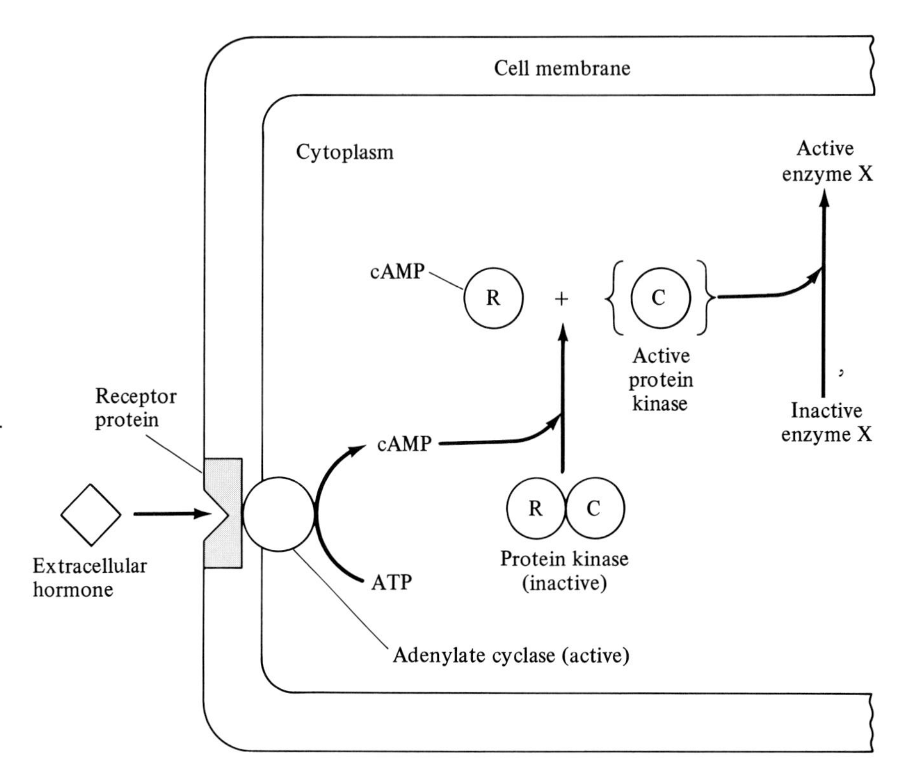

FIGURE 19–6
The activation of adenylate cyclase and the formation of cyclic AMP (cAMP). The extracellular hormone binds to the receptor protein, activating adenylate cyclase and causing cAMP to form. Cyclic AMP activates protein kinase, which in turn can activate other enzymes. *Key*: R = Regulatory subunit; C = Catalytic subunit.

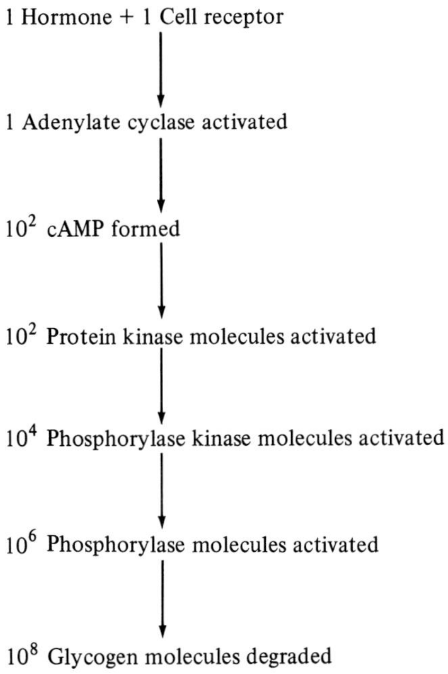

1 Hormone + 1 Cell receptor

↓

1 Adenylate cyclase activated

↓

10^2 cAMP formed

↓

10^2 Protein kinase molecules activated

↓

10^4 Phosphorylase kinase molecules activated

↓

10^6 Phosphorylase molecules activated

↓

10^8 Glycogen molecules degraded

FIGURE 19–7
The amplification of a chemical signal mediated by cyclic AMP. The numbers are hypothetical. (Adapted from R. Montgomery et al., *Biochemistry: A Case-Oriented Approach,* 3rd ed. (St. Louis: C. V. Mosby, 1980), p. 580.)

receptor proteins
adenylate cyclase

bound **receptor proteins** specific for each hormone. When an extracellular hormone binds to a receptor protein, membrane-bound **adenylate cyclase** associated with the receptor protein is activated. This adenylate cyclase catalyzes the formation of cyclic AMP (cAMP) from ATP in the cytoplasm, as shown in Figure 19–6. Cyclic AMP is known as the **second hormone messenger**, because it acts *within* the cell to regulate other cellular processes. Specifically, cAMP binds to the regulatory subunit of cytoplasmic *protein kinase* and causes the dissociation and subsequent *activation* of the protein kinase. The activated protein kinase can then activate other enzymes to elicit a particular biochemical response, such as glycogen degradation—(see Chapter 10). This sequence of events is called a cascade activation sequence. Note that hormones that interact via cAMP do so by activating enzymes.

second hormone messenger

One should note that the chemical signal carried by an extracellular hormone is tremendously amplified when a cascade activation sequence is triggered. This is nicely demonstrated in Figure 19–7, in which hypothetical numbers of molecules are given for each step of a cascade sequence beginning with the binding of 1 hormone molecule to a cell receptor and ending with the degradation of 10^8 molecules of glycogen.

Table 19–2 lists a few hormones that elicit a response via the adenylate cyclase–cAMP–protein kinase activation mechanism.

TABLE 19–2 Some Hormones that Regulate Cellular Processes via Cyclic AMP.

Hormone	Type of Compound
Epinephrine	Derivative of tyrosine
Norepinephrine	Derivative of tyrosine
Follicle-stimulating hormone	Polypeptide
Glucagon	Polypeptide
Luteinizing hormone	Polypeptide
Thyroid-stimulating hormone	Polypeptide
Vasopressin	Oligopeptide

Note: This list is not complete. Many other hormones also mediate biochemical or physiological functions via cAMP.

STEROID HORMONE ACTION: STIMULATION OF GENE EXPRESSION

steroid hormones

The **steroid hormones** (*testosterone*, *estrogen*, and *progesterone*, to name a few) cause a biochemical response in their target tissues by stimulating gene expression. As a result of this gene expression, a new protein (or enzyme) is manufactured within the cells of the target tissue and produces the appropriate biochemical response.

A steroid hormone is first secreted by a particular endocrine gland. For example, estrogen and progesterone are made by the ovarian follicle (see Section 19.5, The Menstrual Cycle). The hormone travels to the target tissue (for estrogen and progesterone, the uterus) via the blood stream and is transported from the outside of the cell to the cytoplasm. Here, the steroid hormone binds to a cytoplasmic receptor protein specific for that particular hormone, forming a steroid hormone/receptor protein complex. This complex is modified in the cytoplasm and *travels from the cytoplasm into the cell nucleus*, where it binds to proteins associated with DNA. The binding of the hormone/receptor complex to the protein/DNA complex initiates the transcription of a portion of the DNA molecule (see Chapter 17) to yield messenger RNA (mRNA). The mRNA is processed in the nucleus and then migrates out of the nucleus to the cytoplasm. Ribosomes attach to the processed mRNA and synthesize a *new* gene product (a protein or enzyme). This new gene product alters other biochemical processes within the target cell, thus eliciting the particular hormonal response. These events are summarized in Figure 19–8.

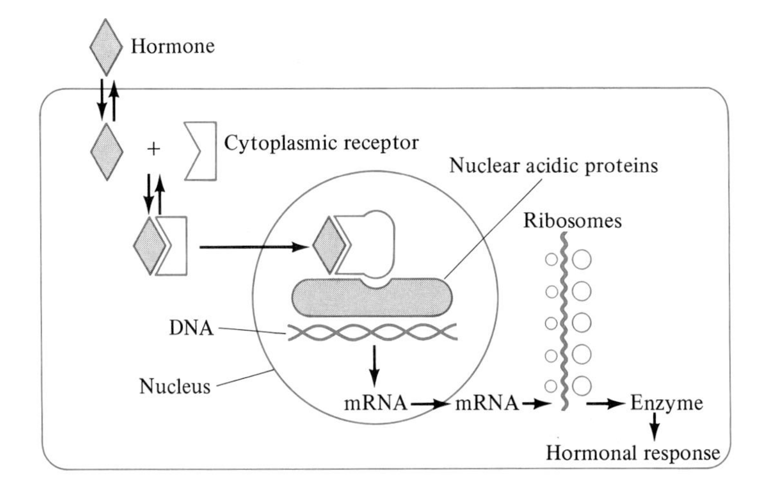

FIGURE 19–8
A proposed mechanism for steroid hormone action. (Adapted from R. Montgomery et al., *Biochemistry: A Case-Oriented Approach*, 3rd ed. (St. Louis: C. V. Mosby, 1980), p. 583.)

PROSTAGLANDINS: SUBSTANCES THAT MODULATE HORMONE ACTIVITY

prostaglandins

Prostaglandins were first isolated from sheep *prostate* glands. These substances are 20-carbon fatty acids derived from *prostanoic acid*. They resemble a hairpin in shape and contain a 5-membered ring. A number of prostaglandins have been isolated from a variety of tissues and classified into four major subgroups: PGA (prostaglandin A), PGB, PGE, and PGF. They evoke a number of different biochemical and physiological responses in various organ systems (see Section 13.10 and especially Table 13–5, p. 245). The structures of two representative prostaglandins are shown:

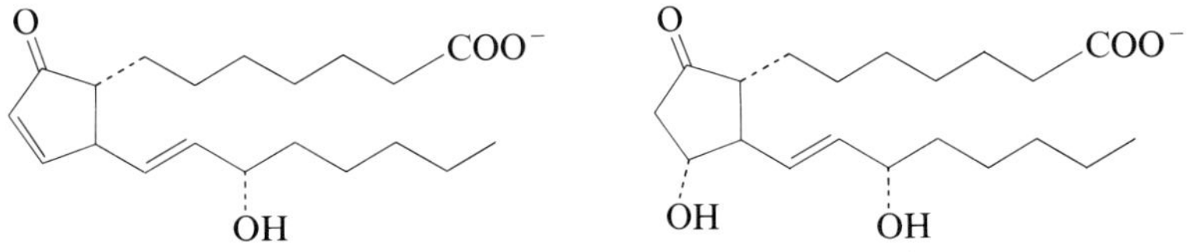

PGA$_1$ PGE$_1$

A great deal of research has been conducted on these very interesting substances. We summarize here a few important findings: (1) prostaglandins can evoke a cellular response at *very low* concentrations (often as low as 10^{-7}–10^{-8} M); (2) they seem to *modulate* the action of hormones and do not act as hormones themselves; (3) they often *modulate cellular processes in the cells in which they are made*; and (4) they often exhibit *competing* biochemical effects in different tissues. Some prostaglandins modulate the intracellular levels of cAMP and, thus, may regulate the membrane receptor protein/adenylate cyclase complex of certain cell types.

19.5 THE MENSTRUAL CYCLE

The female reproductive system is designed to secrete a fertilizable egg cell on a regular basis. The uterus must also be made ready to accept the fertilized egg cell (implantation) and must then be able to nourish the rapidly dividing cell mass (blastula) during the early stages of pregnancy. The physiology of this delicate system is controlled by four pituitary and gonadal hormones. In fact, the pituitary gland and the reproductive organs of the female exert a unique *cyclic* and *reciprocal* control on each other mediated via these hormones. The production of these hormones during the menstrual cycle is shown in Figure 19–9.

The length of the menstrual cycle is variable among women, but usually averages 28 days. The first day of the cycle corresponds to the first day on which vaginal bleeding is observed. The cycle is divided into two distinct phases, (1) the **follicular phase** in which the ovarian follicle matures in preparation for ovulation; and (2) the **luteal phase** in which the ruptured follicle matures into the *corpus luteum*. *Ovulation* usually occurs midway through the cycle (at day 15, in this example) and is defined as the time at which the mature follicle ruptures and releases the ovum for transport through the fallopian tubes to the uterus.

follicular phase

luteal phase

Four different lines are shown in Figure 19–9, each corresponding to the level of a different hormone in the blood. Starting at day 1 of the cycle

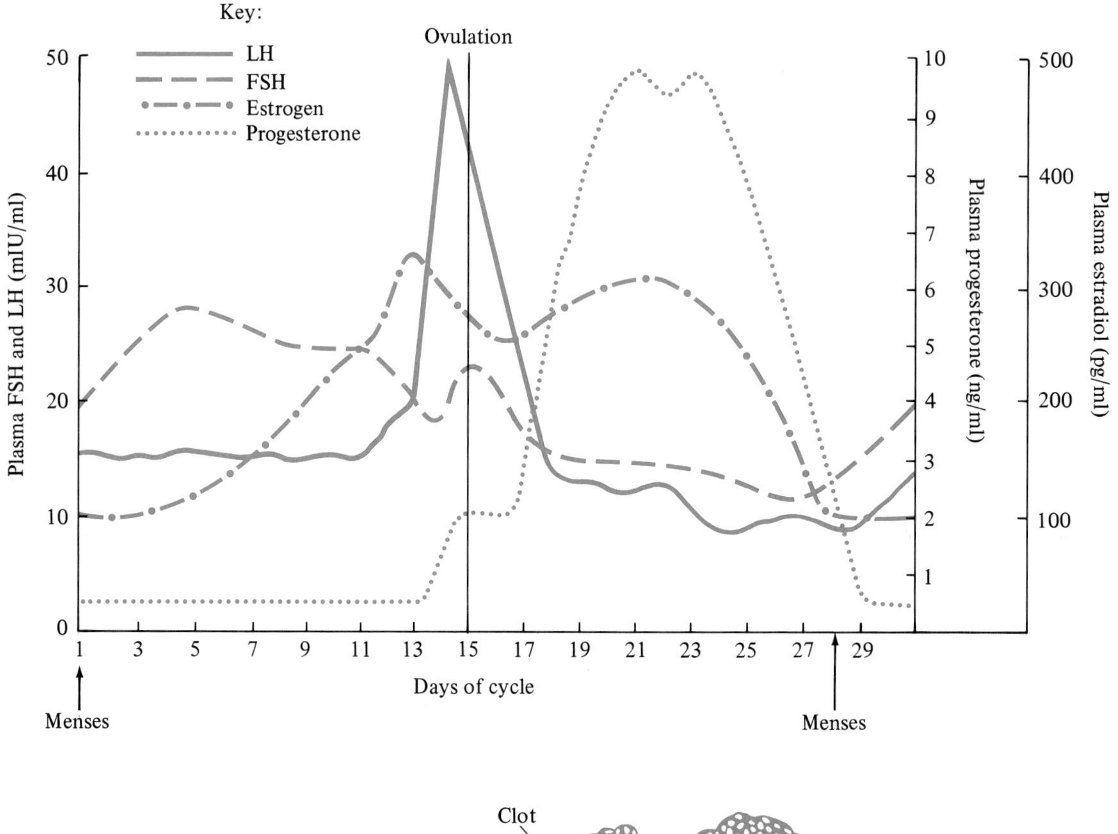

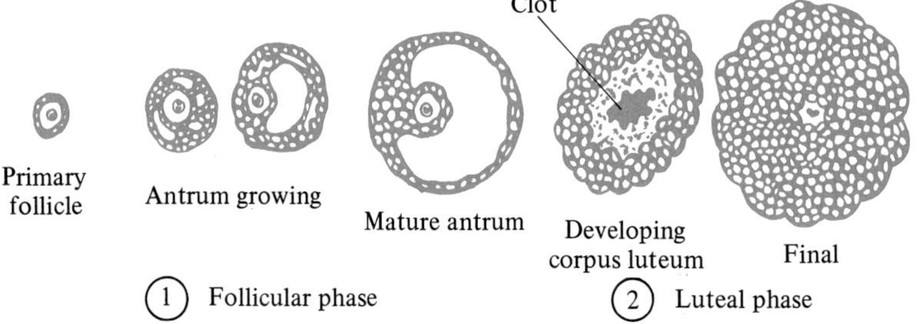

FIGURE 19–9 Variations in blood hormone levels during the menstrual cycle. Abbreviations: FSH = follicle-stimulating hormone; LH = luteinizing hormone; mIU/ml = milli-international units per milliliter; ng = nanograms; pg = picograms.
(Adapted from N. V. Bhagavan, *Biochemistry*, 2nd ed. (Philadelphia: J.B. Lippincott, 1978), p. 1243.)

(menses), you will note that the blood levels of the steroid hormones (progesterone and estrogen) are low. The level of luteinizing hormone (LH) is also low and constant at 15 mIU/ml (15 milli-international units per milliliter). The level of follicle-stimulating hormone (FSH) is 20 mIU/ml.

Between days 1 and 5, the level of FSH increases from 20 to about 28 mIU/ml. This is because the hypothalamus is now secreting FSH-releasing factor, which stimulates the release of FSH from the anterior pituitary gland. As the name implies, follicle-stimulating hormone (FSH) is responsible for the maturation of the ovarian follicle prior to ovulation. As the follicle matures, it begins to secrete estrogen into the blood. The blood estrogen level slowly, but steadily, rises from days 5 to 13. The increasing estrogen level signals the pituitary gland to secrete luteinizing hormone (LH). Luteinizing hormone travels from the pituitary gland to the ovaries and the developing follicle and stimulates the production of the steroid hormone progesterone. Notice the rapid rise in the LH level at days 13 and 14 (just before ovulation); as well as the rise in blood progesterone at days 13–15. The increase in the progesterone level (originally caused by the rapid increase in the LH level) induces ovulation at day 15. Ovulation (the rupture of the follicular wall) follows the sequence of events depicted in Figure 19–10.

Ovulation and the high level of luteinizing hormone (LH) stimulate the formation of the corpus luteum (the modified follicle after ovulation). The corpus luteum is responsible for the synthesis and secretion of progesterone and estrogen (note the increased levels of these steroid hormones between days 16 and 24). These hormones act on two different organs. First, they *inhibit* the production of the pituitary hormones FSH and LH (note the low levels of these hormones during this time). Thus, the *corpus luteum modulates pituitary function*. Second, *progesterone and estrogen act on the uterus* and cause significant biochemical and physiological changes in this organ. These changes prepare the uterus for the implantation of the fertilized

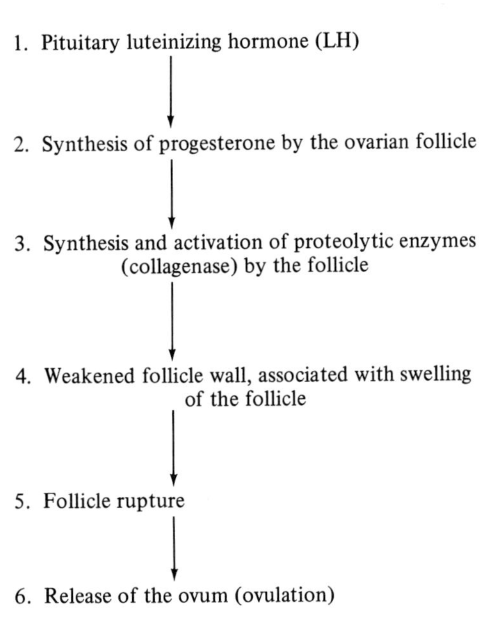

1. Pituitary luteinizing hormone (LH)
2. Synthesis of progesterone by the ovarian follicle
3. Synthesis and activation of proteolytic enzymes (collagenase) by the follicle
4. Weakened follicle wall, associated with swelling of the follicle
5. Follicle rupture
6. Release of the ovum (ovulation)

FIGURE 19–10
Suggested sequence of events prior to and during ovulation.

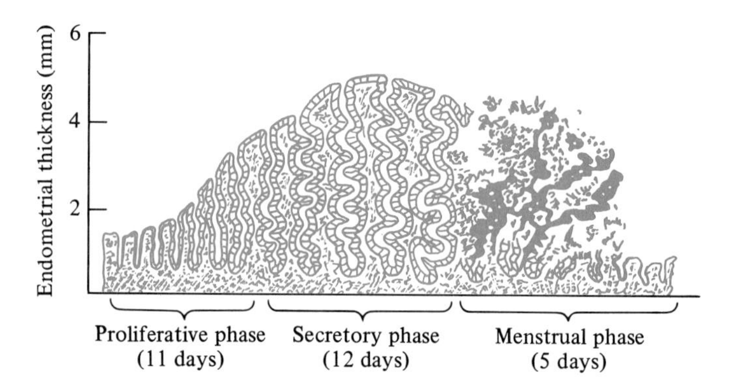

FIGURE 19–11
Changes in uterine endometrial thickness during the menstrual cycle.

egg cell, and are summarized in Figure 19–11. Because the pituitary gland is not secreting as much FSH and LH, the corpus luteum begins to degenerate between days 21 and 28. Thus, we observe a dramatic decrease in the levels of both progesterone and estrogen between days 23 and 28. Menstruation occurs because of the decrease in steroid hormone levels. Because these steroid hormones inhibit pituitary function, LH and FSH levels have dropped. Once the levels of progesterone and estrogen decrease, they no longer inhibit the pituitary and FSH and LH levels begin to rise, signaling the beginning of a second cycle.

chorionic gonadotropin

It should be noted that if fertilization and pregnancy take place, the *placenta* forms and synthesizes **chorionic gonadotropin** (*CG*). This hormone maintains the corpus luteum in a normal state, in the process keeping the levels of progesterone and estrogen high. These hormones thus continue to inhibit FSH and LH secretion by the pituitary. Thus, the menstrual cycle is broken during pregnancy.

endometrium

Figure 19–11 depicts the thickness of the **endometrium** (the lining of the uterus) during the various phases of the cycle. After menstruation has occurred, the endometrium is rather thin (1–2 mm). During the follicular phase of the menstrual cycle, estrogen and progesterone are produced by the ovarian follicle. These hormones stimulate increased endometrial cellular growth and division, increasing the thickness of the endometrium from 2 mm to 4–5 mm. The increased progesterone level also increases the production of cellular secretory products and promotes the vascularization of the endometrium. These conditions are necessary for the proper implantation of the fertilized egg cell and for its nutrition. If pregnancy does not occur, the *decreasing* levels of progesterone and estrogen cause the highly vascularized endometrial lining to slough (*menstruation*). Menstrual cramps may be due to the increased production of prostaglandins (see Section 19.4). These substances are powerful smooth muscle contractors and can cause muscle cramps. Drugs that inhibit prostaglandin synthesis are currently being tested, and some are already on the market.

SUMMARY

Hormones are chemical substances that are secreted by endocrine glands into the bloodstream. They elicit a specific biochemical or physiological response in the target tissue. Hormones can affect (1) the rate of synthesis of a protein or enzyme, (2) the rate of enzymatic catalysis, or (3) the permeability of cell membranes.

Endocrine glands are ductless organs that secrete hormones into the bloodstream. The pituitary gland is often called the master endocrine gland of the body, because it controls the function of many other endocrine glands. The endocrine glands are summarized in Figure 19–1, p. 364. Three major classes of hormones are known: (1) derivatives of amino acids, (2) polypeptide hormones, and (3) steroid hormones. The major hormones were listed in Table 19–1, pp. 366–367.

A number of endocrine glands first synthesize an inactive prohormone that is then converted into a biologically active hormone. This conversion process can include (1) chemical modification within the endocrine gland (such as proteolytic clipping of a polypeptide chain), or (2) chemical modification outside the endocrine gland (such as T_4 to T_3 conversion in the liver).

The pituitary gland is divided into three distinct regions: the anterior pituitary, the posterior pituitary, and the pars intermedia. Releasing factors are secreted by the hypothalamus, travel to the pituitary gland, and stimulate the release of pituitary hormones. These hormones travel to other endocrine glands and stimulate the secretion of secondary hormones. The secondary hormones then travel to the target tissue and evoke the necessary biochemical response. The sequence of these events is depicted in Figure 19–4, p. 372. The release of hypothalamic-pituitary hormones is controlled by feedback inhibition.

Certain hormones evoke a response in their target tissues by binding to a membrane-bound receptor protein that is specific for that particular hormone. The binding event activates adenylate cyclase, which catalyzes the formation of cyclic AMP (cAMP). Cyclic AMP activates protein kinase, which in turn can activate other enzyme systems within the cell. This sequence, called a cascade mechanism, produces very large amplifications in the original chemical signal.

The steroid hormones are transported into the cytoplasm of the target cell and bind to a cytoplasmic receptor protein that is specific for that hormone. This hormone/receptor protein complex migrates into the nucleus, binds to the DNA, and induces the production of mRNA. The mRNA, in turn, is the template for a new gene product (a protein or enzyme). This new protein can then alter other cellular processes.

The menstrual cycle is depicted in Figure 19–9, p. 378. Briefly, FSH-releasing factor stimulates the release of follicle-stimulating hormone (FSH) from the pituitary. This hormone stimulates the maturation of the ovarian follicle. The follicle synthesizes estrogen and progesterone. The rising estrogen level stimulates the release of luteinizing hormone (LH) from the pituitary. Luteinizing hormone stimulates the production of progesterone by the follicle. Ovulation takes place, followed by the development of the corpus luteum. The corpus luteum continues to secrete the steroid hormones estrogen and progesterone, and these hormones inhibit the release of FSH and LH by the pituitary gland. As the corpus luteum degenerates, the steroid hormone level falls, triggering menstruation. Since the steroid hormone level has dropped, the pituitary gland is no longer inhibited and begins again to secrete FSH, starting a new cycle. Estrogen and progesterone are also responsible for maintaining the uterine lining before and during the implantation of the fertilized egg cell.

REVIEW QUESTIONS

1. We have presented a number of important concepts and terms in this chapter. Define (or briefly explain) each of the following terms:

- **a.** Hormone
- **b.** Endocrine gland
- **c.** Polypeptide hormone
- **d.** Steroid hormone
- **e.** Prohormone
- **f.** Prostaglandin
- **g.** Hypothalamus-pituitary system
- **h.** Releasing factors
- **i.** Target tissue
- **j.** Adenylate cyclase
- **k.** Cyclic AMP (second messenger hypothesis)

- **l.** Hormone-induced amplification
- **m.** Steroid-specific receptor protein
- **n.** Follicle-stimulating hormone (FSH)
- **o.** Ovulation
- **p.** Phases of menstrual cycle

*2. Referring to the structure of thyroxine (T_4), suggest how this hormone might be synthesized in the thyroid gland.

3. What would happen if:
   - **a.** the anterior pituitary gland did *not* respond to feedback inhibition mediated by high levels of T_4 and T_3?
   - **b.** the release of thyrotropic hormone (TSH) inhibited the release of thyrotropic-releasing factor (TRF)?
   - **c.** the thyroid gland did *not* respond to thyrotropic hormone?
   - **d.** both T_4 and T_3 were immediately degraded in the blood after their synthesis and secretion?

*4. Compare and contrast the two mechanisms of hormone action described in Section 19.4. What features of each mechanism are the same, and what features are different?

*5. Figure 19–10, p. 379, depicts the sequence of biochemical events that occur before ovulation. From what you know about the mechanisms of hormonal action, suggest how follicular collagenase is (1) made or (2) activated.

6. From what you know about ovarian and pituitary function, suggest how birth control pills act. (These pills contain steroid hormones.)

SUGGESTED READING

Bhagavan, N. V. *Biochemistry*, 2nd ed., Ch. 13. Philadelphia: J. B. Lippincott, 1978.

Guyton, A. C. *Textbook of Medical Physiology*, 5th ed., Chs. 75–82. Philadelphia: W. B. Saunders, 1976.

McGilvery, R. W., and Goldstein, G. *Biochemistry: A Functional Approach*, Ch. 38. Philadelphia: W. B. Saunders, 1979.

Tietz, N., ed. *Fundamentals of Clinical Chemistry*, 2nd ed., Chs. 13 and 14. Philadelphia: W. B. Saunders, 1976.

Williams, R. H., ed. *Textbook of Endocrinology*, 5th ed. Philadelphia: W. B. Saunders, 1974.

NOTE

1. Stryer, L., *Biochemistry*, 2nd ed. (San Francisco: W. H. Freeman, 1981), p. 839.

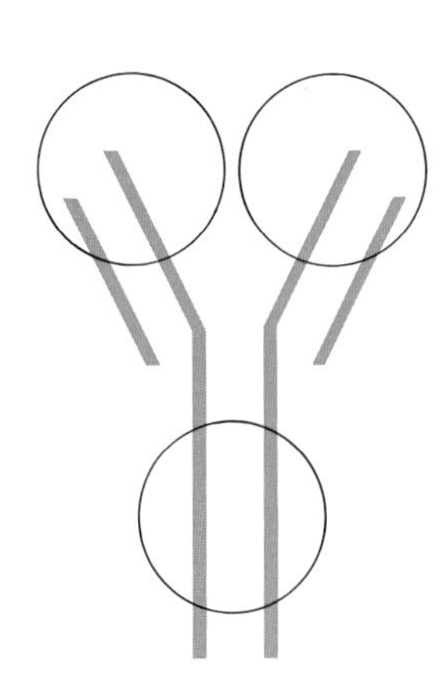

Chapter 20
Nutrition

20.1 INTRODUCTION

nutrients

Nutrition is the study of the substances **(nutrients)** that we obtain from the foods we eat and that we require for normal growth and maintenance. More specifically, nutrition deals with a number of important topics such as (1) the nutrients that are required; (2) the foods where these nutrients are found; (3) the amount of each nutrient necessary for optimal growth and health; and (4) the metabolism of each nutrient and its biochemical or physiological function. The study of human nutrition is rather complex. First, a variety of biochemical substances (proteins, carbohydrates, and lipids are metabolized (degraded and synthesized) via rather complex metabolic pathways. A complete understanding of biochemistry is, therefore, essential. Second, a large number of different nutrients are required for normal health, including proteins, carbohydrates, lipids, inorganic elements (minerals), and vitamins. For example, thirteen vitamins and some fifteen minerals are required by the body. Third, the required *amount* of each nutrient *varies*, and can differ for men and women, for children and adults—even from one person to another. Fourth, the variety of cooking procedures complicates matters further, because a number of nutrients (vitamins) can be destroyed or modified by these procedures. Finally, a bewildering array of (often contradictory) diets, health foods, and exercise plans add confusion to an already complex subject.

In this chapter, we will define the nutritional calorie and the recommended daily allowance (RDA), as well as the human nutritional requirements for the major cellular fuels and dietary fiber. A recommended dietary plan (based upon U.S. guidelines) will be presented. In addition, we will discuss the structures, biochemical functions, and dietary sources of vitamins. The nutritional role of important minerals and trace elements will also be discussed.

20.2 DEFINITIONS OF IMPORTANT TERMS

NUTRITIONAL CALORIES

Carbohydrates, fats, and proteins are nutrients that can be catabolized to yield energy (ATP). This energy can be used to perform work, such as biochemical synthesis, chemical transport, and gross mechanical work such as lifting a book. The energy in these cellular fuels is expressed in terms of **calories**. The energy content of a particular food (the number of calories per unit weight) can be determined with a bomb calorimeter. A known quantity of a particular food is burned in a chamber (containing sufficient oxygen) immersed in a larger tank containing water. During this process, the food is oxidized to CO_2 and water and a significant amount of energy is released. This energy heats the surrounding water, and the increase in water temperature can be accurately measured.

calories

A **calorie** (cal) is a unit commonly used in chemistry. It is defined as the amount of heat energy needed to raise 1 gram of water 1 °C. (This is also known as the *small calorie*.) The **kilocalorie** (kcal) is defined as 1000 small calories. The **nutritional calorie** (large calorie) is also defined as 1000 small calories. It is abbreviated as the Cal. The nutritional calorie is equal to the kcal.

calorie
kilocalorie
nutritional calorie

Different foods yield different amounts of heat energy when completely metabolized by the body. For example, carbohydrates yield 4 kcal/g, proteins 4 kcal/g, and fats 9 kcal/g.

MINIMUM DAILY REQUIREMENT AND RECOMMENDED DAILY ALLOWANCE

The United States Food and Nutrition Board publishes dietary guidelines that represent the best available biochemical information regarding the amounts of vitamins (and other nutrients) that should be ingested each day to maintain optimal health.

The **minimum daily requirement (MDR)** is defined as the *smallest amount* of a nutrient that, if ingested, will prevent a nutritional deficiency.[1] The **Recommended Daily Allowance (RDA)**, on the other hand, is defined as the daily amount needed to maintain *good nutrition and health* for the average healthy person. The RDA is greater than the MDR for each nutrient. The amounts found in nutritional tables and on food labels are often ex-

minimum daily requirement (MDR)
Recommended Daily Allowance (RDA)

pressed in specific units. For example, grams (g), milligrams (mg), and micrograms (μg) are common units. International Units (IU) are often used for nutrients that exist in different forms in different foods. The International Unit is a term that defines the *biological potency* of a particular nutrient. For example, the amount of vitamin D in milk is expressed as IU, not as mg or μg of vitamin D. Biological potencies are often expressed as a percentage (%) of the RDA. The label on a box of Whufflewheats, for instance, might inform the consumer that each 1 cup serving contained 10% of the RDA for thiamine, riboflavin, and iron.

20.3 NUTRIENT ABSORPTION

In Figure 20–1 is given a schematic diagram of the gastrointestinal tract and the sites where a variety of nutrients are absorbed.

The small intestine is composed of the duodenum, jejunum, and ileum. It should be noted that the jejunum and ileum are very important—carbohydrates, fats, amino acids, and a number of vitamins are absorbed in these regions.

Jejunileal bypass surgery has sometimes been performed on extremely obese individuals. This controversial procedure involves bypassing most of the small intestine. The jejunum is connected directly to the lower portion of the ileum, just before the ileum joins the colon. The absorption of carbohydrates and fats is restricted and thus, the individual can lose significant amounts of weight—often as much as 60–100 pounds in one year. Unfortunately, the absorption of important nutrients is also severely restricted. Cirrhosis of the liver, hypocalcemia, hypomagnesemia, and protein deficiency are serious complications that can occur as a result of this procedure.

WATER

Although water cannot be considered a nutrient, an adequate water intake is essential for the maintenance of the body. The body of an average 70 kg adult male contains about 40 liters of water (55 percent water, 19 percent protein, 19 percent fat, and 7 percent inorganic substances). A significant amount of body water (2.4 liters) is lost each day—1400 ml in urine, 700 ml by evaporation from the skin and lungs, 200 ml in feces, and 100 ml in sweat. Therefore, about 2400 ml of water should be ingested each day by drinking liquids and eating foods containing water. When water loss exceeds water intake, the body becomes dehydrated. Vomiting and severe diarrhea often cause serious dehydration. Water is also lost during hot weather and during strenuous exercise, as the body attempts to cool itself via sweating and evaporation. Prolonged heavy exercise can result in a total water loss of 5000 to 7000 ml (5–7 liters). These losses must be recouped via adequate water intake.

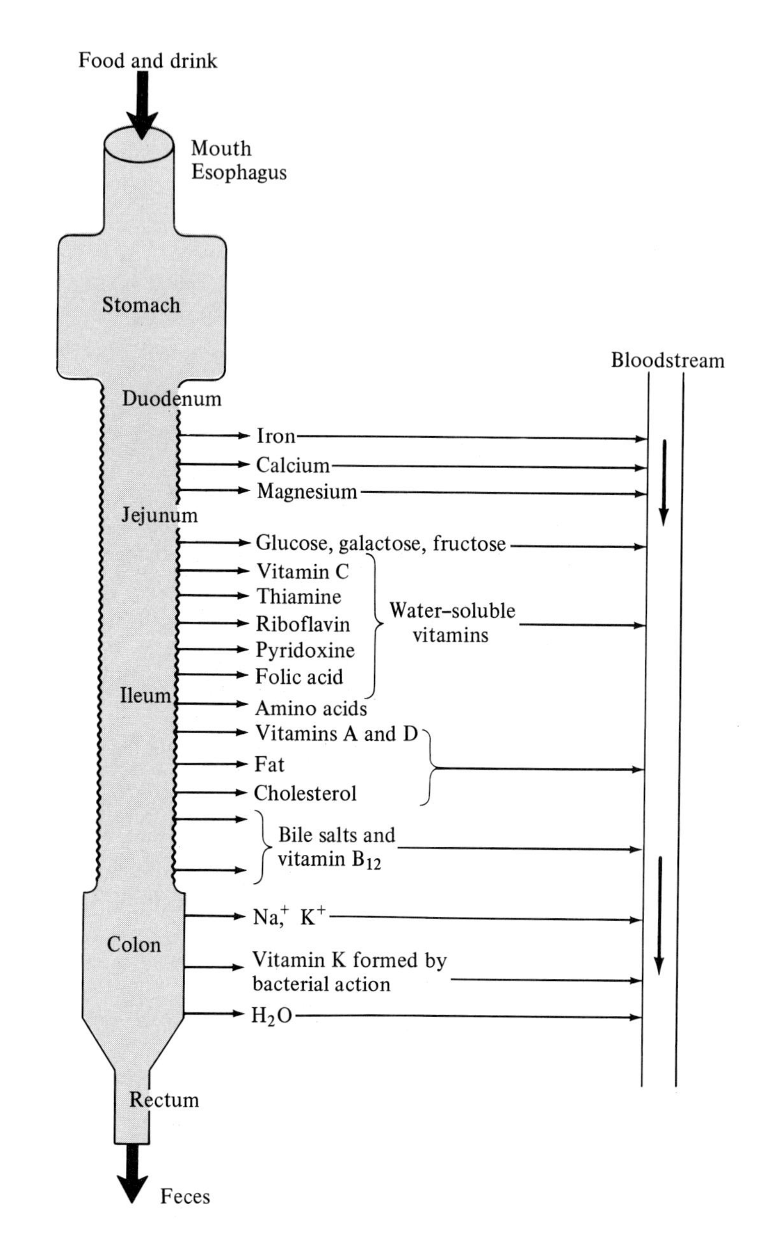

FIGURE 20–1
The absorption of important nutrients by the gastrointestinal tract. (Adapted from M. V. Krause and L. K. Mahan, *Food, Nutrition and Diet Therapy*, 6th ed. (Philadelphia: W. B. Saunders, 1979), p. 104.)

20.4 MAJOR CELLULAR FUELS AND BUILDING BLOCKS

CARBOHYDRATES

Carbohydrates are the major source of energy for most of the world population. Cereals, sugars, starchy fruits (dates and figs), and roots and tubers provide most of the carbohydrates consumed; cereals alone account for nearly 40 percent of the total human intake of food.[2] Complex carbohydrates (polysaccharides, such as starch) are found in many food staples. These polysaccharides must be hydrolyzed into simple monosaccharides before they can be absorbed and utilized by the body.

Carbohydrates supply 40–60 percent of the total number of calories in our diets. *Starch* (amylose and amylopectin) and *sucrose* (cane sugar) constitute approximately 90 percent of the digestible dietary carbohydrates; glucose, lactose, and other simple sugars constitute the remaining 10 percent. *Glucose, fructose, and galactose* are the principal monosaccharides utilized by the body. The absorption, digestion, and metabolism of both simple and complex carbohydrates was discussed in detail in Chapters 8–11; therefore, it will not be covered in this chapter. One should note, however, the *central role* of glucose in the production of energy by the cell.

DIETARY FIBER

dietary fiber

Dietary fiber (crude fiber) can be loosely defined as the *undigestible* carbohydrates in cereals and vegetables. It consists of substances that constitute the wall of the plant cell, including cellulose, hemicelluloses, lignin, and pectins. These substances enter the small intestine and colon and are partially digested by intestinal and colonic bacteria.[3] We do not absorb these colonic degradation products and, therefore, cannot utilize them for energy. Cellulose is partially digested (40–80 percent) in the large intestine by bacteria normally present there. Hemicelluloses are almost completely digested in the small intestine; the lignins are not digested.

It has been suggested that these substances affect gastrointestinal function through important physical and chemical mechanisms. For instance, these substances hold water, forming a gel; they may also bind bile salts and potentially carcinogenic products of intestinal bile-salt metabolism. The major effects of dietary fiber can be summarized as follows: *increased* stool bulk/weight (400–500 g on a fiber diet versus 100 g on a nonfiber diet); *decreased* transit time of intestinal contents through the gastrointestinal tract; and binding of reaction products normally produced in the intestinal lumen. Because the dietary fiber binds potentially harmful substances and decreases transit time, there is less chance for these substances to react with the cells of the gastrointestinal tract. Experimental evidence (from animals) and studies of certain human populations suggest a correlation between the amount of dietary fiber ingested in the diet and the incidence of colonic cancer. A *decreased* incidence of colonic cancer was observed in rats fed a diet high in dietary fiber.

Foods that are recommended for a high-fiber diet include bran-type cereals, whole wheat (and whole grain) breads and crackers, nuts, fresh fruits, and raw or steamed vegetables such as cauliflower, carrots, celery, lettuce, spinach, and tomatoes.[3] The *excess* intake of crude fiber can bind important trace minerals and, thus, induce a mineral deficiency. The absorption of zinc and iron by the GI tract, for instance, is *decreased* by high-fiber diets. How much fiber intake is enough, and how much is too much? Unfortunately, this is not an easy question to answer. However, 12–35 g of total fiber per day has been considered an optimal recommended amount.[3]

FATS

Fats (triglycerides) supply 40–45 percent of the total calories in the average American diet. Triglycerides are hydrolyzed to fatty acids, which are then degraded into acetyl-CoA molecules via β-oxidation. Large numbers of ATP molecules can be generated from each acetyl-CoA molecule. Triglycerides containing medium- and long-chain fatty acids are, thus, excellent sources of biochemical energy, because the carbon atoms of the fatty acid chains are highly reduced. Fats can release approximately 9 kcal of energy per gram when metabolized in the body, as compared to only 4 kcal/gram for carbohydrates and proteins. The absorption, catabolism, and anabolism of fats was discussed in detail in Chapter 14.

A number of medical studies have implicated the *excess* consumption of fats (and also the consumption of the *wrong* kinds of fats) with specific and very serious diseases, such as atherosclerosis and cancer. The levels of the various lipoproteins (VLDL, LDL, and HDL) are extremely important and can significantly influence the development of potentially lethal diseases of the circulatory system (see Section 14.4). A high level of serum high-density lipoproteins (HDL) is probably beneficial; they are associated with a low risk of heart disease. The HDL particle entraps cholesterol, thus preventing it from being deposited in plaques on the arterial wall. A high level of serum low-density lipoprotein (LDL) is harmful, because this type of lipoprotein contains large concentrations of cholesterol and may bind to arterial wall cells, inducing atherosclerosis.

A reduction in the amount of fats (especially saturated animal fats) is highly recommended by most medical authorities. Table 20–1, emphasizing this point, recommends that the total number of calories contributed by fats be reduced from 42 percent to 30 percent. A reduction in the consumption of foods high in saturated fats (and in cholesterol) is necessary to accomplish this goal. This can be done by employing a few simple alterations in one's eating habits and cooking procedures.[4] For instance, substitute *fish and poultry* for beef and pork. If you must eat beef, make it a *lean* cut. Broil or bake meats, don't fry them. Avoid high fat meats such as bacon, sausage, and fatty luncheons meats. Cut down the number of eggs consumed per week (remember that eggs are often used in baking) and be aware of hidden sources of cholesterol. Use nonfat milk and vegetable oils in cooking.

The total elimination of fats from the diet is not recommended (and is probably impossible). Dietary fats provide an important source of calories; furthermore, certain fatty acids are essential and can only be obtained from the food we eat. An excellent review of dietary fats is found in *Contemporary Developments in Nutrition*, by Bonnie S. Worthington-Roberts (see the Suggested Reading section at the end of this chapter).

essential fatty acid

Linoleic acid (an unsaturated fatty acid with 2 double bonds) cannot be synthesized by the body; therefore, it must be obtained from the diet. Linoleic acid is, thus, an **essential fatty acid** (in fact, the only essential fatty acid). The structure of linoleic acid is as follows:

$$H_3C{-}(CH_2)_4{-}CH{=}CH{-}CH_2{-}CH{=}CH{-}(CH_2)_7{-}COO^-$$ (18:2 fatty acid)

Linoleic acid, an essential fatty acid

An average diet should supply 2–3 grams of linoleic acid per day. Vegetable oils are an excellent source of this fatty acid. Linolenic acid (18:3) and arachidonic acid (20:4) can be synthesized in the body from linoleic acid. Linoleic acid and linolenic acid are both used in constructing cell membrane phospholipids. Arachidonic acid is the precursor to the prostaglandins, 20-carbon fatty acids that control the action of many hormones (see Section 19.4).

PROTEINS (AMINO ACIDS)

When proteins are ingested, they are first hydrolyzed by gastric and intestinal proteolytic enzymes into their constituent amino acids. These amino acids are transported from the intestine (see Figure 20–1) into the bloodstream for transport to other regions of the body. Amino acids can be used to make new proteins and to supply nitrogen for the synthesis of molecules containing nitrogen, such as heme and nucleic acids. Amino acids can be degraded into TCA cycle intermediates for energy production. In addition, during periods of starvation, amino acids can be converted either into glucose (glucogenic amino acids) or into ketone bodies (ketogenic amino acids) for energy production (see Chapter 15).

A number of health experts claim that our intake of beef and pork is too high and advocate that fish and poultry be substituted. It must be understood that there is nothing wrong with the protein in beef and pork; rather, the problem is that these meats contain a high percentage of saturated

TABLE 20–1 The Amino Acid Composition of Certain Foods.

Food(s)	Essential Amino Acids	
Meat (beef)	Complete	
Milk (whole)	Complete	
Eggs	Complete	
Cheese	Complete	
	*Insufficient	Increased Amounts:
Cereal grains	Lysine, threonine	Methionine
Corn	Lysine, threonine, and tryptophan	
Legumes (beans and peas)	Methionine, tryptophan	Threonine
Nuts and seed oils	Lysine	Methionine, Tryptophan
Green leafy vegetables	Methionine	

* Notice that most of the plant-derived foods are deficient in either lysine, methionine, threonine, or trytophan.

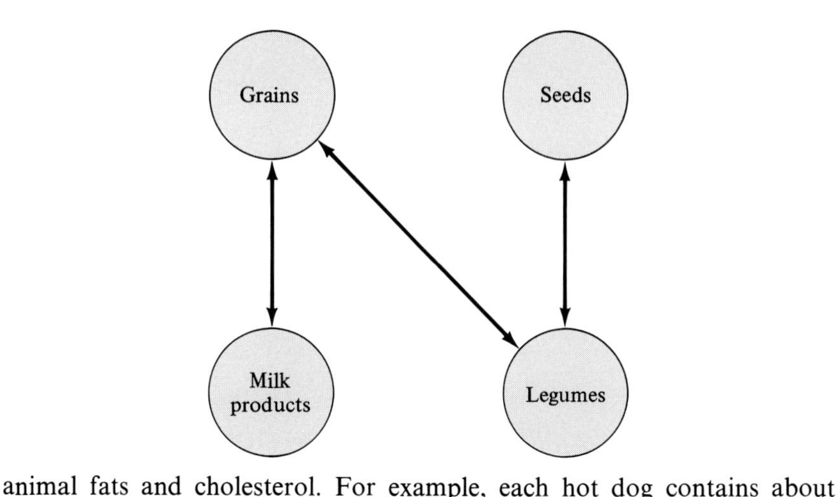

FIGURE 20–2
Protein complementarity. The double arrows indicate combinations of foods from specific groups that can be eaten to obtain all of the essential amino acids.

animal fats and cholesterol. For example, each hot dog contains about 7 grams of protein, but 15 grams of animal fat; each 3-ounce hamburger contains about 20 grams of protein, but 17 grams of animal fat. The excess consumption of *animal fats and cholesterol* is deemed unhealthy, not the consumption of beef and pork per se. Plant-derived foods can be significantly more healthy: The Food and Nutrition Board of the National Research Council has established that 0.6 g protein/Kg (ideal) body weight should be ingested every day to provide enough high quality protein to insure optimal health. This amounts to around 42 g/day of high quality protein for a 70 kg adult male, or about that contained in six hot dogs or two 3 ounce hamburgers. Most affluent Americans consume not only more protein than this, but the protein they consume is high quality animal protein. This high quality protein contains all of the **essential amino acids** (isoleucine, leucine, lysine, methionine, phenylalanine, threonine, tryptophan, and valine). The essential amino acids cannot be synthesized in our bodies and must be obtained from our diets. An amino acid deficiency is, therefore, unlikely if one consumes an adequate amount of high quality protein.

essential amino acids

Adequate amounts of protein and significantly larger amounts of carbohydrates can be obtained from plant-derived foods. Unfortunately, most plant proteins are of low quality—they are deficient in one or more of the essential amino acids (Table 20–1). A wise choice of plant-derived foods is therefore important to avoid a deficiency in essential amino acids. This can be accomplished by complementing one type of protein with another. If a given food contains proteins deficient in a particular amino acid, other proteins (that contain the deficient amino acid) are included in the meal as well. The concept of **protein complementarity** is diagrammed in Figure 20–2.

protein complementarity

SUMMARY OF DIETARY GOALS

Besides noting the major cellular fuels and building blocks, it is important to define how much of these nutrients are needed for a well-balanced diet and for optimal health. It is becoming apparent that the eating habits of many Americans are just plain unhealthy, contributing significantly to the risk of coronary heart disease, cancer, and other serious ailments. It is important to ask how *your* diet can be effectively changed (if a change is warranted) to minimize serious health risks. Most medical experts advise that fad diets and other short-term diets (1) do not work, and (2) can often lead to serious biochemical and physiological alterations—in short, that they are unhealthy. However, a scientifically designed, long-term alteration in eating habits and lifestyle can significantly enhance one's health, resistance to disease, and general well-being.

Figure 20–3 summarizes the average American diet with respect to carbohydrates, fats, and proteins. Clearly, this represents an average; your

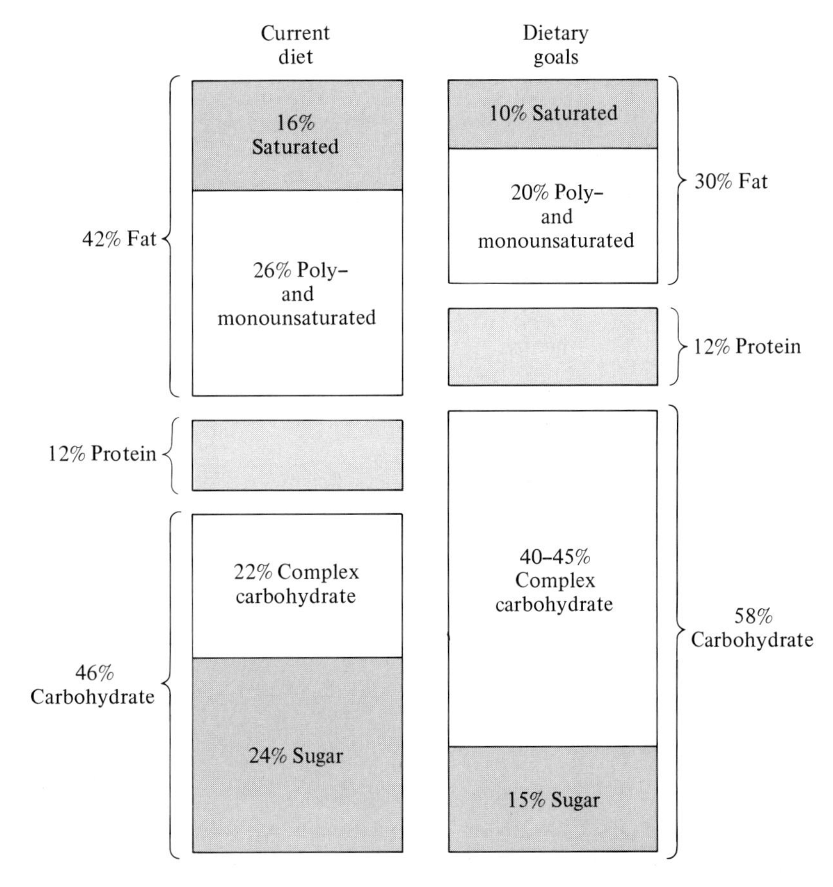

The Goals Suggest the Following Changes in Food Selection and Preparation

1. Increase consumption of fruits, vegetables, and whole grains.
2. Decrease consumption of meat and increase consumption of poultry and fish.
3. Decrease consumption of foods high in fat and partially substitute polyunsaturated fat for saturated fat.
4. Substitute nonfat milk for whole milk.
5. Decrease consumption of butterfat, eggs, and other foods high in cholesterol.
6. Decrease consumption of sugar and foods high in sugar content.
7. Decrease consumption of salt and foods high in salt content.

FIGURE 20–3
The current "average" American diet and suggested dietary goals.

particular eating habits may not be reflected in this figure. The figure also summarizes the dietary goals suggested by the U.S. Senate Select Committee, along with a list of suggested changes in food selection and preparation. The principal recommended alterations are (1) decreasing fat intake (especial-

TABLE 20–2 Simplified Self-Scoring Test of Heart Attack and Stroke Risk

Risk Habit or Factor	Increasing Risk				
Smoking Cigarettes	None	Up to 9 per day	10 to 24 per day	25 to 34 per day	35 or more per day
Score	0	1	2	3	4
Body Weight	Ideal Weight	Up to 9 lbs. excess	10 to 19 lbs. excess	20 to 29 lbs. excess	30 lbs. or more excess
Score	0	1	2	3	4
Salt Intake	$\frac{1}{5}$ average	$\frac{1}{3}$ average	U.S. average	Above average	Far above average
or	hard to achieve; no added salt, no convenience foods	no use of salt at table, spare use of high-salt foods	salt in cooking, some salt at table	frequent salt at table	frequent use of salty foods
Blood Pressure Upper Reading (if known)	Less than 110	110 to 129	130 to 139	140 to 149	150 or over
Score	0	1	2	3	4
Saturated Fat and Cholesterol Intake	$\frac{1}{5}$ average almost total vegetarian; rare egg yolk, butterfat & lean meat	$\frac{1}{3}$ average 2 meatless days/week, no whole milk products, lean meat only	$\frac{1}{2}$ average meat (mostly lean), eggs, cheese 12 times/week, nonfat milk only	U.S. average meat, cheese, eggs, whole milk 24 times/week	Above average meat, cheese, eggs, whole milk over 24 times/week
or Blood Cholesterol Level (if known)	Less than 150	150 to 169	170 to 199	200 to 219	220 or over.
Score	0	1	2	3	4
Self-Rating of Physical Activity	Vigorous exercise 4 or more times/week 20 min each	Vigorous exercise 3 times/week 20 min each	Vigorous exercise 1 to 2 times/ week	U.S. average occasional exercise	Below average exercises rarely
or Walking Rating	Brisk walking 5 times/week 45 min each	Brisk walking 3 times/week 30 min each	Brisk walking 2 times/week 30 min each or Normal walking $4\frac{1}{2}$ to 6 miles daily	Normal walking $2\frac{1}{2}$ to $4\frac{1}{2}$ miles daily	Normal walking less than $2\frac{1}{2}$ miles daily
Score	0	1	2	3	4

Adapted from Farquhar, J. W., *The American Way of Life Need Not be Hazardous to Your Health* (New York: W. W. Norton, 1978).

ly saturated animal fats); (2) decreasing sugar intake (glucose and sucrose); (3) increasing complex carbohydrate intake; and (4) decreasing salt intake.

Table 20–2 is a simplified self-scoring test for determining the risk of heart attack and stroke. A high score indicates that *significant alterations* in one's eating habits and lifestyle *must* be made if one is to reduce the risk of

TABLE 20–2 *(continued)*

VI. Self-rating of stress and tension	Rarely tense or anxious	Calmer than average	U.S. average	Quite tense	Extremely tense
			Feel tense or anxious 2 to 3 times/day	Usually rushed	
		Feel tense about 3 times/week	Frequent anger or hurried feelings	Occasionally take tranquilizer	Take tranquilizer 5 times/week or more
Score	0	1	2	3	4

Enter your total score here __________.

Notes:

1. Subtract 1 point if dietary fiber intake is high (almost all cereals whole grain, almost no sugar, and considerable fruit and vegetable intake).
2. Add 1 point if all exercise is competitive.
3. If you are a female taking estrogen or birth control pills, add 1 point if score is 12 or below, 2 points if risk score is 13 or above (especially if you smoke, are overweight, or have high blood pressure or high blood cholesterol).

Interpretation

Maximum points = 24

Zone	Score	
F	21–24	The probability of having a premature heart attack or stroke is about four to five times the U.S. average. Action is urgent. Try to drop four points within a month and three more points within six months.
E	17–20	Incidence of heart attack or stroke is about twice the U.S. average. Action is urgent. Try to drop four points within six months and continue reduction.
D	13–16	The U.S. average is 14. This is an uncomfortable and readily avoidable zone. Careful planning can result in a five- to-six-point reduction within a year.
C	9–12	The likelihood of having a heart attack or stroke is about one-half the U.S. average. This is a zone rather easily achieved by most people within a year if they are now in Zone D or E. Careful planning can result in a four- to-six-point reduction within a year.
B	5–8	Incidence of heart attack or stroke about one quarter of the U.S. average. This goal is achievable by many but often takes one or two years to reach.
A	0–4	Incidence of heart attack or stroke rates very low, averaging less than one-tenth the rate in the U.S. 35–65 age group. This goal requires diligent effort, considerable family support, and often takes three to four years to reach. Individuals in this range should be proud and gratified (and will often find themselves acting as models and teachers for the many who have not achieved this very low risk zone).

coronary heart disease. It should be noted that many young people are overweight, do not exercise, and consume too much fat, cholesterol, and salt. These individuals have initiated an unhealthy lifestyle at a very early age, and will reap the consequences later in life.

20.5 THE VITAMINS

We have already discussed the chemistry of a number of vitamin-derived enzyme cofactors. Here, we will briefly review vitamin chemistry by listing (1) the vitamins, (2) the chemical structure of each vitamin, (3) the enzymatic reactions each vitamin participates in; and (4) the pertinent nutritional aspects of each vitamin.

vitamin

The term **vitamin** was derived from *vita* (meaning "having life or vitality") and *amine* (since the first vitamin—*thiamine*, or *vitamin* B_1)—was identified as an amine. Not all vitamins are amines; therefore, the term "vitamin" is rather inaccurate. The history of the vitamins is beyond the scope of this book. However, a few salient observations can be made. It was recognized centuries ago that the symptoms of certain diseases could be eliminated or alleviated by administering various foods. For example, British sailors who did not eat fruits and vegetables developed the disease *scurvy*. The symptoms of the disease were eliminated by eating citrus fruits. A compound (later named vitamin C) was found to exist in a very high concentration in citrus fruits. Vitamin deficiencies can be classified into two main categories: (1) *nutritional vitamin deficiency* and (2) *secondary vitamin deficiency* (not due to dietary insufficiency). A nutritional vitamin deficiency results from inadequate intake of one or more of the vitamins, and can be corrected by simply ingesting foods rich in that particular vitamin. A secondary vitamin deficiency becomes manifest if other biochemical or physiological defects are present. For example, if vitamin malabsorption occurs, ingested vitamins are simply passed out of the body in the feces. Certain drugs can also affect vitamin absorption and metabolism in the body.

The vitamins are classified into two major categories: (1) the *water-soluble* vitamins (the B-complex vitamins and vitamin C) and (2) the *fat-soluble* vitamins (vitamins A, D, E, and K). The water-soluble vitamins are freely soluble in water and, therefore, can be easily lost during extended cooking. They are not stored in the body, but are easily excreted—hence, constant intake is necessary. The fat-soluble vitamins are very hydrophobic and, hence, are insoluble in water. Fat-soluble vitamins are absorbed in the intestine (along with fats) and are stored in fatty tissues. Because the fat-soluble vitamins are stored in the body, they can accumulate in excess amounts, causing a condition called **hypervitaminosis**. Hypervitaminosis

hypervitaminosis

can cause serious clinical problems, depending on the vitamin involved. The water-soluble and fat-soluble vitamins are presented in the next few pages along with their chemical structures, properties, RDA values, toxicities, and deficiency symptoms).[5,6] Portions of each vitamin molecule are circled, or noted with an arrow. These regions indicate the sites of biological action.

THE WATER-SOLUBLE VITAMINS

Biotin. The vitamin *biotin* is found in beef liver, milk, eggs, and yeast. Its chemical formula—and that of a common derivative—are as follows:

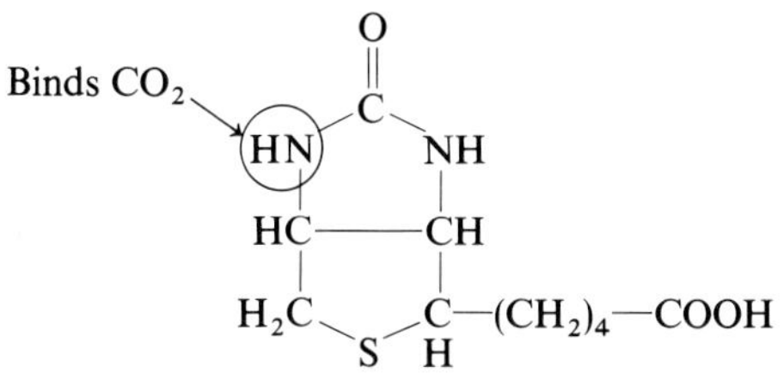

Binds CO_2 → HN; ring: O=C, NH, HC—CH, H_2C, S, C(H)—$(CH_2)_4$—COOH

Biotin

Bound CO_2 → ^-O—C(=O)—N; ring: O=C, NH, HC—CH, H_2C, S, C(H)—$(CH_2)_4$—C(=O)—N(H)—$(CH_2)_4$—C(H)(NH_3^+)—COO^-

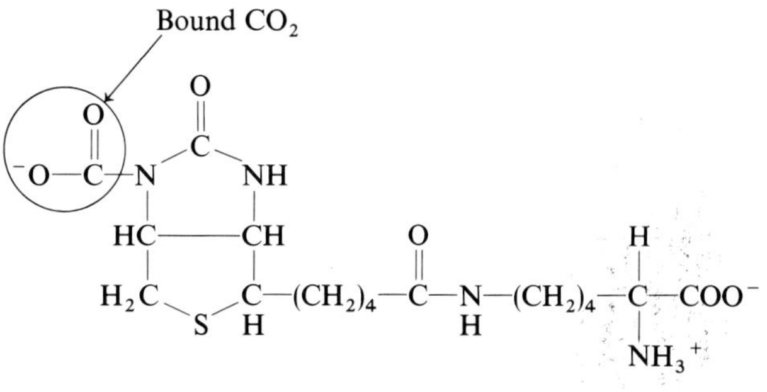

N-Carboxybiotinyllysine
(*N*-Carboxybiocytin)

The coenzyme form of biotin is *ε-N-biotinyllysyl* (*biocytin*). It participates in carboxylation reactions (CO_2 group transfers) such as the anaplerotic TCA cycle reaction catalyzed by pyruvate carboxylase.

The recommended daily allowance of biotin is not known, but nutritionists have estimated that the body requires about 150 μg/day. Excess biotin is not toxic to the body. A biotin deficiency can produce the following symptoms: dermatitis, anorexia (loss of appetite), insomnia, muscle pains, nausea, and depression. A deficiency has been experimentally induced in animals by feeding the animals large quantities of raw egg whites, which contain the protein *avidin* that binds biotin.

Biotin is chemically stable at normal cooking temperatures.

Vitamin B_{12} (Cyanocobalamin). *Vitamin B_{12}*, or *cyanocobalamin*, is synthesized only by microorganisms, either free or in symbiotic relation to other organisms. Plants and animals themselves cannot make vitamin B_{12}. It is found in fermented cheeses, organ meats, (liver) and such seafoods as clams, oysters, lobsters, scallops, and haddock. Its chemical formula is as follows:

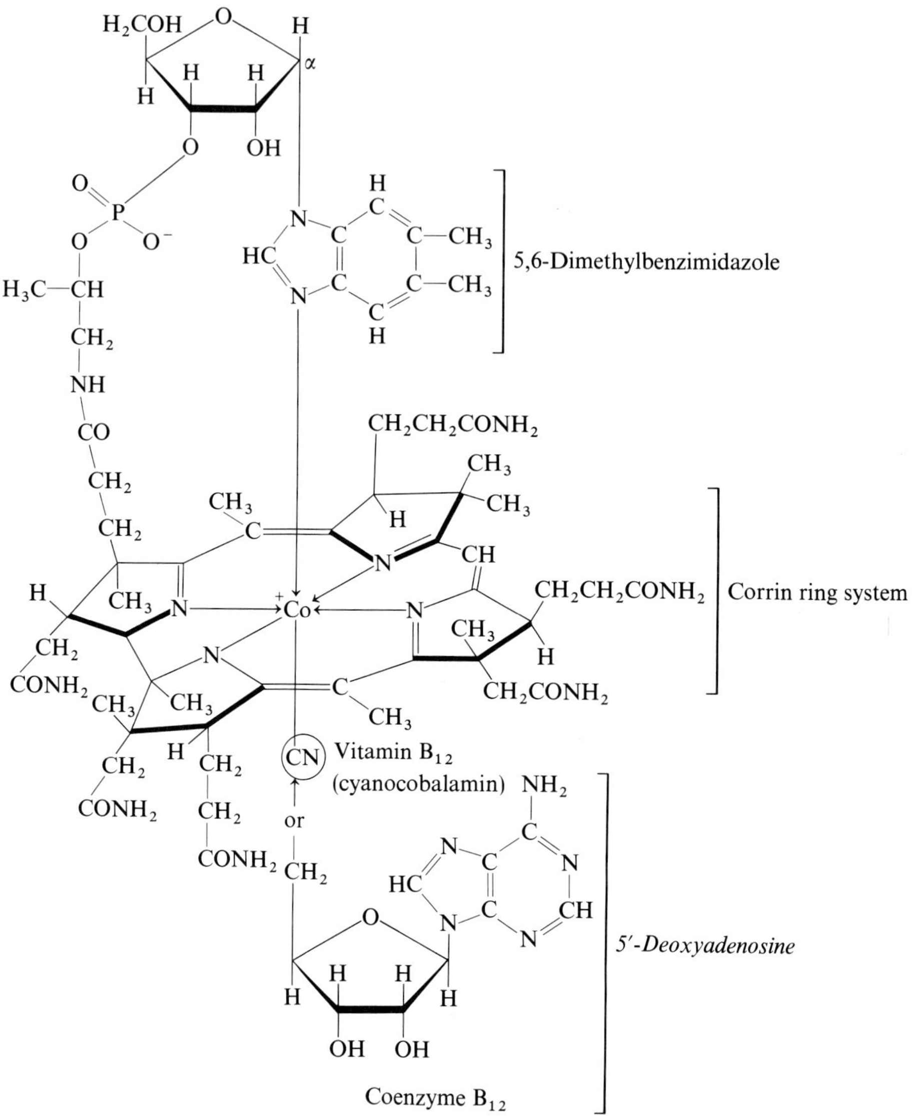

5,6-Dimethylbenzimidazole

Corrin ring system

Vitamin B_{12} (cyanocobalamin)

5′-Deoxyadenosine

Coenzyme B_{12}

An enzyme containing vitamin B_{12} removes a methyl group from methyltetrahydrofolate, which is used in synthesizing thymidine, one of the nucleoside bases of DNA. The enzyme *methylmalonyl-CoA mutase*, which also requires vitamin B_{12}, catalyzes an intramolecular rearrangement (methylmalonyl-CoA to succinyl-CoA).

The recommended daily allowance of vitamin B_{12} is 3 μg. Excess vitamin B_{12} is not toxic to the body. A vitamin B_{12} deficiency can produce the following symptoms: impaired cell division abnormal bone cells (megaloblasts), demyelination of neurons, and abnormal carbohydrate and fat metabolism. Some patients ingest an adequate amount of vitamin B_{12}, but nonetheless suffer from vitamin B_{12} deficiency because of a deficiency in *intrinsic factor*, a protein that binds vitamin B_{12} and promotes its absorption from the intestine. Lack of intrinsic factor leads to **pernicious anemia**, a condition characterized by a reduction in the number of red blood cells and an increase in their size.

pernicious anemia

Vitamin B_{12} is relatively stable during most cooking procedures; however, 30 percent of the vitamin B_{12} present in raw milk is lost when the milk is boiled.

Folic Acid. The vitamin *folic acid* is found in green vegetables, whole grain cereals, nuts, legumes (peas and beans), and liver. Its chemical formula is as follows:

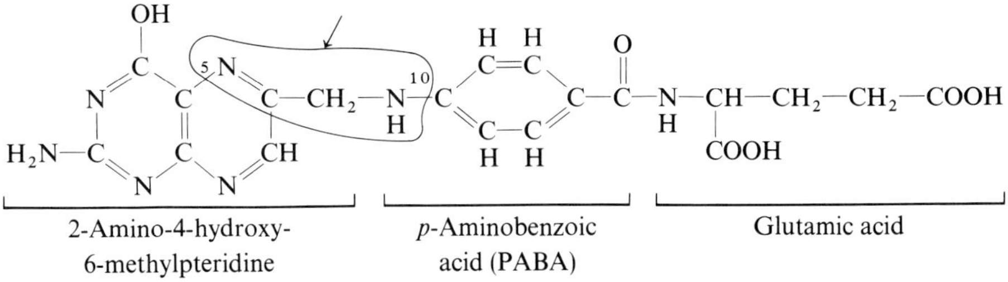

2-Amino-4-hydroxy-6-methylpteridine | *p*-Aminobenzoic acid (PABA) | Glutamic acid

Folic acid (pteroylglutamic acid)

The coenzyme forms of folic acid are the reduced intermediates *dihydrofolic acid* (FH_2) and *tetrahydrofolic acid* (FH_4). Folic acid is involved in the metabolism of 1-carbon groups (amino acid metabolism).

The recommended daily allowance of folic acid is 0.4 mg. Excess folic acid is not toxic to the body, but it can block folic acid absorption. A folic acid deficiency can produce the following symptoms: gastrointestinal disturbances, various anemias, and possible mental deterioration.

Folic acid is destroyed rapidly at normal cooking temperatures. Up to 50 percent of the total amount present in a food can be lost in cooking.

Niacin (Nicotinic Acid). *Niacin*, or *nicotinic acid*, is found in red meat, liver, fish, eggs, green vegetables, and cereals grains. Its chemical formula—and those of some common derivatives—are as shown on p. 398. The coenzyme form of niacin is *nicotinamide adenine dinucleotide* (NAD^+); a phosphate group is sometimes added to produce $NADP^+$. These coenzymes participate in oxidation-reduction reactions catalyzed by dehydrogenases, such as lactic acid dehydrogenase and the TCA cycle dehydrogenases.

The recommended daily allowance of niacin is 20 mg. Excess niacin is not toxic to the body. A niacin deficiency can produce the following symptoms: gastrointestinal tract disturbances, anorexia (loss of appetite), diarrhea, and red, swollen tongue. In regions where corn is a staple item

Nicotinic acid
(Niacin)

Location of the additional phosphate group of $NADP^+$

Nicotinamide
(Niacinamide)

of the diet, the niacin-deficiency disease *pellagra* is found. The symptoms of pellagra include changes in the skin, severe nerve dysfunction, and diarrhea.

Niacin is moderately stable at normal cooking temperatures; however, some is lost by leaching into cooking water.

Pantothenic Acid. *Pantothenic acid* is found in liver, eggs, beef, legumes, and wheat bran. Its chemical formula—and that of Coenzyme A, of which it is a part—are as shown on p. 399. Pantothenic acid is a structural component of Coenzyme A, which is a carrier of acyl groups. Coenzyme A is a component of acetyl-CoA (an important TCA cycle intermediate) and fatty acyl-CoA.

The recommended daily allowance of pantothenic acid is not known, but 5–10 mg/day is adequate. No toxic effects of excess pantothenic acid have been observed. A pantothenic acid deficiency can produce gastrointestinal tract disturbances and depression.

Pantothenic acid is stable during moderate cooking, but it is destroyed at high temperatures and after prolonged cooking at moderate temperatures.

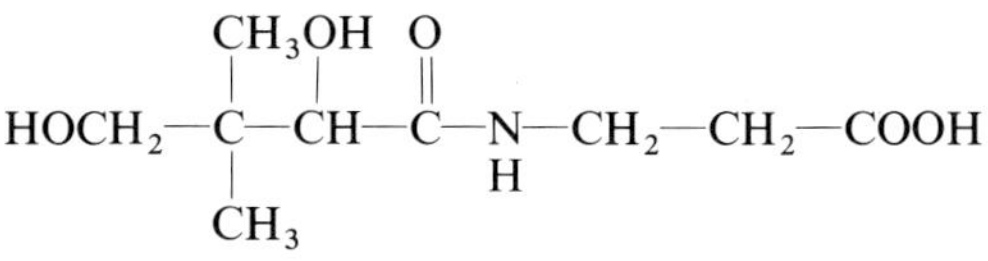

$$HOCH_2-C(CH_3OH)(CH_3)-CH-C(=O)-NH-CH_2-CH_2-COOH$$

Pantothenic acid

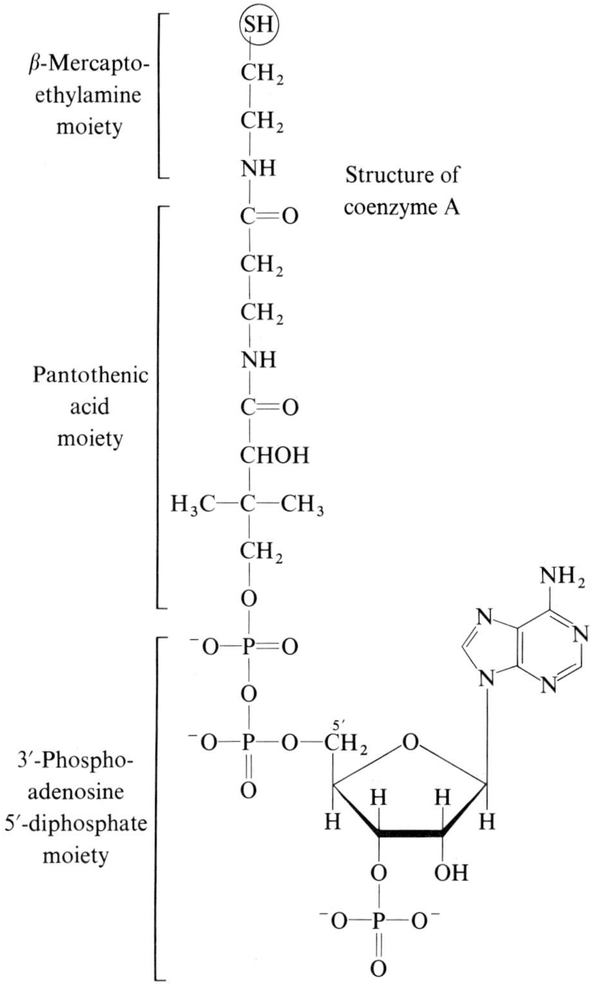

$$HS-CH_2-CH_2-NH-C(=O)-CH_2-CH_2-NH-C(=O)-CHOH-C(CH_3)_2-CH_2-O-P(=O)(O^-)-O-P(=O)(O^-)-O-{}^{5'}CH_2\text{-ribose(3'-}O-PO_3^{2-})\text{-adenine}$$

β-Mercapto-ethylamine moiety (SH, CH_2, CH_2, NH)

Pantothenic acid moiety (C=O, CH_2, CH_2, NH, C=O, CHOH, $H_3C-C-CH_3$, CH_2, O)

3′-Phospho-adenosine 5′-diphosphate moiety

Structure of coenzyme A

Vitamin B_6 (Pyridoxal). *Vitamin B_6*, or *pyridoxal*, is found in meat, eggs, liver, legumes, wheat, and corn. Its chemical formula—and that of its coenzyme form—are as follows:

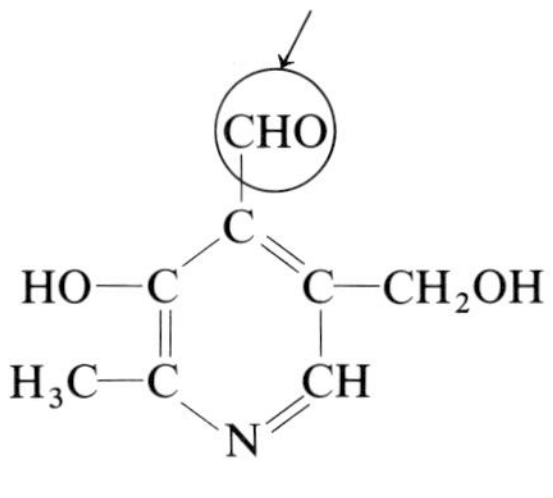

Pyridoxal: pyridine ring with CHO (circled, arrow), HO—C, H_3C—C, C—CH_2OH, CH, N

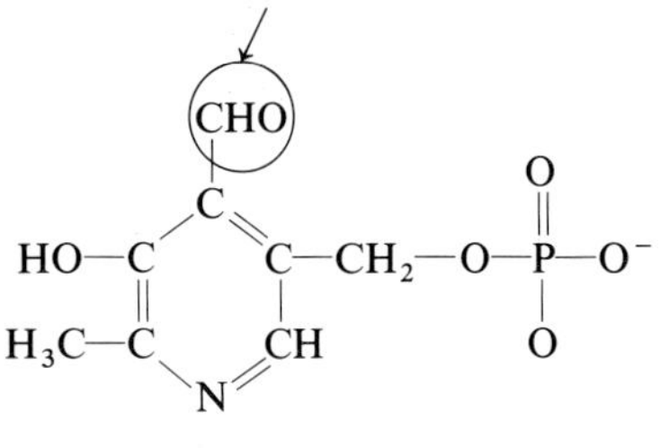

Pyridoxal

Pyridoxal phosphate: pyridine ring with CHO (circled, arrow), HO—C, H_3C—C, C—CH_2—O—P(=O)(O)—O^-, CH, N

Pyridoxal phosphate

The coenzyme form of vitamin B_6 is *pyridoxal phosphate*. Pyridoxal phosphate participates in transamination and decarboxylation reactions (for example, with amino acid transaminases).

The recommended daily allowance of vitamin B_6 is 2 mg. Excess vitamin B_6 produces no known toxic effects. A vitamin B_6 deficiency can produce the following symptoms: dermatitis, increased susceptibility to infection, insomnia, muscular weakness, nervousness, and anemia.

Vitamin B_6 is stable at normal cooking temperatures, but it is sensitive to light. Much is lost during industrial food-processing procedures.

Riboflavin (Vitamin B_2). *Riboflavin*, or *vitamin B_2*, is found in egg whites, yeast, milk, red meat, liver, green vegetables, and fish. Its chemical formula—and that of its two coenzyme forms—are as shown on page 401. The coenzyme forms of riboflavin are *flavin mononucleotide* (*FMN*) and *flavin adenine dinucleotide* (*FAD*). These take part in oxidation-reduction reactions catalyzed by flavin-linked dehydrogenases, such as succinate dehydrogenase.

The recommended daily allowance for riboflavin is 1.1–1.6 mg. Excess riboflavin produces no known toxic effects. A riboflavin deficiency can produce the following symptoms: dermatitis, inflammation of the tongue, anemia, photophobia (intolerance to light), and loss of visual acuity.

Riboflavin is stable at ordinary cooking temperatures. However, it is very sensitive to light. Milk exposed to direct sunlight in clear glass containers loses 50 percent of its riboflavin content in 2 hours.

Thiamine (Vitamin B_1). *Thiamine*, or *vitamin B_1*, is found in cereal grains, legumes, nuts, milk, beef, and pork. Its chemical formula—and that of its coenzyme form—are as follows:

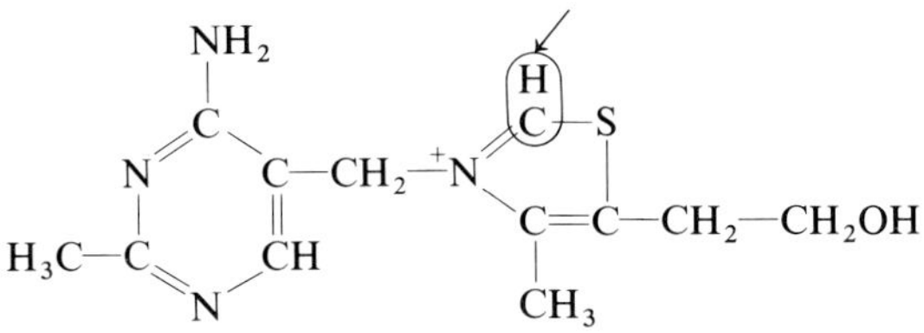

Thiamine

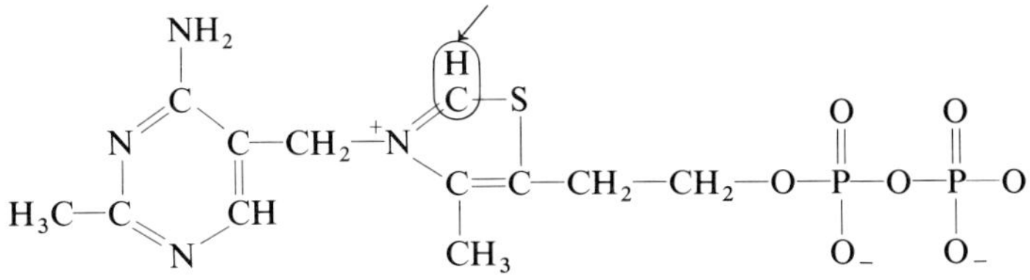

Thiamine pyrophosphate

Riboflavin

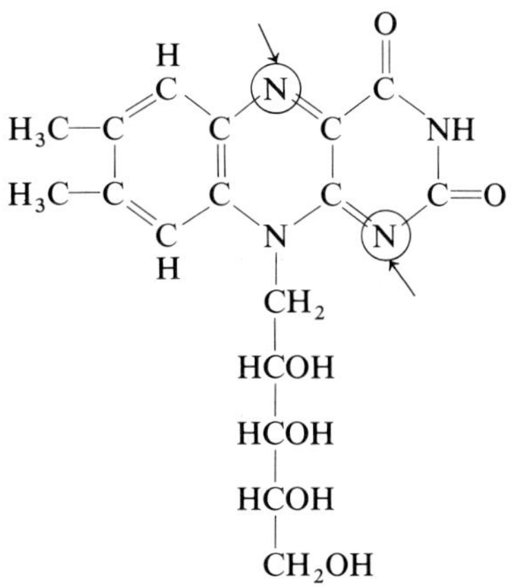

Flavin adenine dinucleotide (FAD)

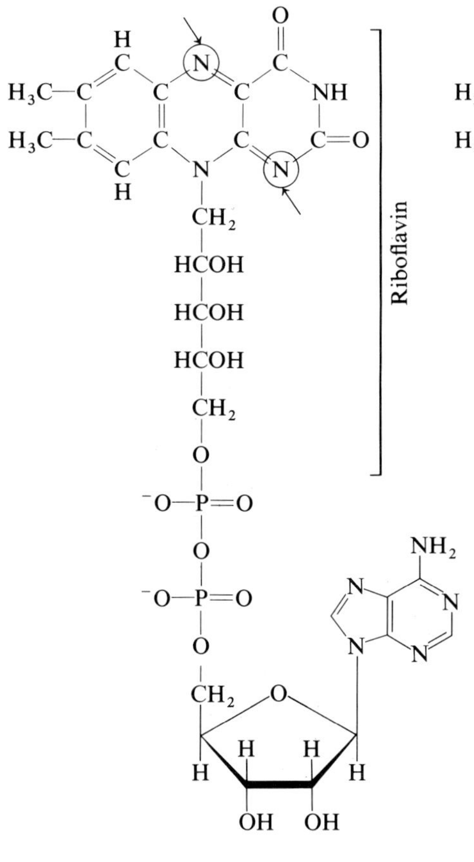

Riboflavin

Flavin mononucleotide (FMN)

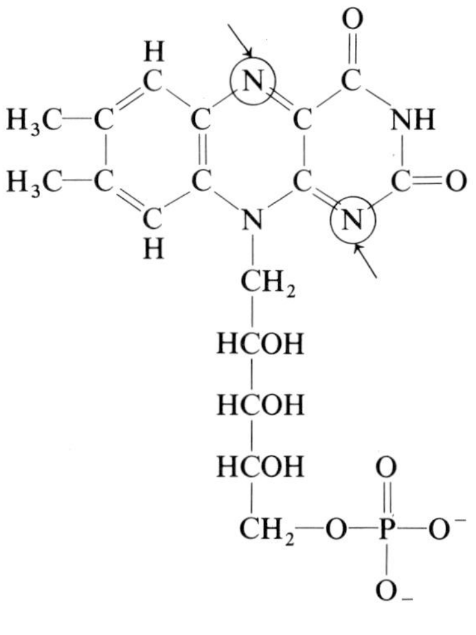

The coenzyme form of thiamine is *thiamine pyrophosphate* (*TTP*), which participates in oxidative decarboxylation reactions (for instance, with the pyruvate dehydrogenase complex).

The recommended daily allowance of thiamine is 1.5 mg. Excess thiamine produces no known toxic effects. A thiamine deficiency can produce the following symptoms: depression, irritability, failure to concentrate, muscular weakness, and altered reflexes. In Southeast Asia, some diets contain antithiamine factors, producing a thiamine-deficiency disease known as *beriberi*. The symptoms of beriberi include pain and paralysis in the extremities, muscular weakness, and swelling.

Thiamine is destroyed in foods that are baked, toasted, fried, or otherwise, for prolonged periods above 100 °C.

Vitamin C (Ascorbic Acid). *Vitamin C*, or *ascorbic acid*, is found in citrus fruits, Brussels sprouts, broccoli, cauliflower, cantaloupe, and and potatoes. Its chemical formula is as follows:

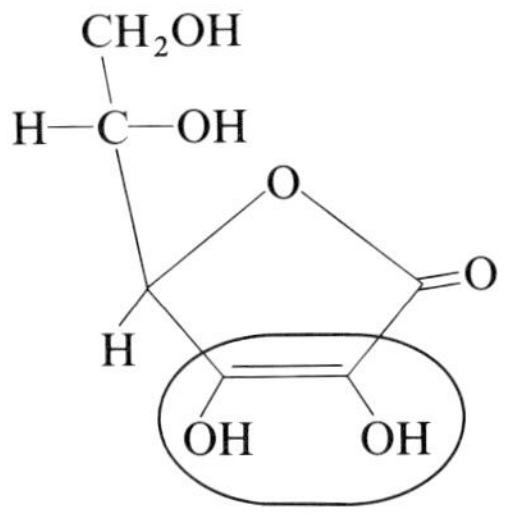

CH_2OH

H—C—OH

O

=O

H

OH OH

Vitamin C is a strong reducing agent. It acts as a cofactor in hydroxylation reactions (for example, the proline to hydroxyproline conversion required for collagen synthesis). It is also needed to form neurotransmitters and red blood cells.

The recommended daily allowance of vitamin C is 60 mg. Certain researchers advocate ingesting *megadoses* of vitamin C—very large amounts, 1000 mg or more per day—to prevent colds and lessen the risk of cancer. Research into the megadose question has produced inconclusive results. Excess vitamin C can produce the following toxic effects: oxalic acid stones, increased mobilization of bone calcium, and gastrointestinal tract abnormalities (nausea, cramps, diarrhea). A deficiency of vitamin C induces the disease *scurvy*, which has the following symptoms: swollen gums, coiled hairs and abnormalities of the hair follicles, and hyperkeratosis (inflammation and thickening of the skin). Ingestion of megadoses of vitamin C by pregnant mothers sometimes produces postnatal scurvy in the newborn.

Vitamin C is unstable under heat and prolonged cooking. It is also destroyed by light.

THE FAT-SOLUBLE VITAMINS

Vitamin A (Retinol). Vitamin A, or retinol, is found in milk, butter, fish (and fish liver), beef liver, and eggs. Its provitamin, *β*-carotene, is found in carrots, sweet potatoes, and green beans. *β*-Carotene is transformed into vitamin A in the wall of the intestine; vitamin A that reaches the bloodstream is stored in the liver. The chemical formulas of vitamin A and its provitamin are as follows:

Cleavage here produces two vitamin A molecules

β-Carotene

Two enzymatic steps

Two Vitamin A (retinol) molecules

visual cycle

The primary function of vitamin A is as a participant in the **visual cycle**. Vitamin A (*cis*-retinol) is bound to the protein *opsin*, forming rhodopsin. When the rhodopsin molecule absorbs light energy, the *cis* double bond of the bound vitamin A molecule is converted to *trans*, in the process dissociating the vitamin A molecule from the opsin molecule. This dissociation triggers a nerve impulse to the visual system of the brain. Other possible functions of vitamin A include the regulation of Ca^{2+} transport, the growth of osteoblasts (bone-forming cells), and matters related to reproduction.

The recommended daily allowance of vitamin A is 5000 IU (6 μg/kg body weight). Amounts in excess of 250–2000 IU/kg per day can induce hypervitaminosis A, which has the following symptoms: liver and spleen enlargement, abnormal fetal development, fragile bones, skin problems, and loss of appetite. A vitamin A deficiency can produce the following symptoms: night blindness, inflammation of the eyes, growth retardation, and sterility.

Because vitamin A is fat soluble, little or none is lost in normal cooking procedures.

Vitamin D (Cholecalciferol). *Vitamin D*, or *cholecalciferol*, is found in butter, fish liver oils, and vitamin-D fortified milk. When human skin is irradiated by sunlight, the body uses the energy input to turn 7-dehydrocholesterol into vitamin D_3. The chemical formula of vitamin D—and of its precursor and active form—are as follows:

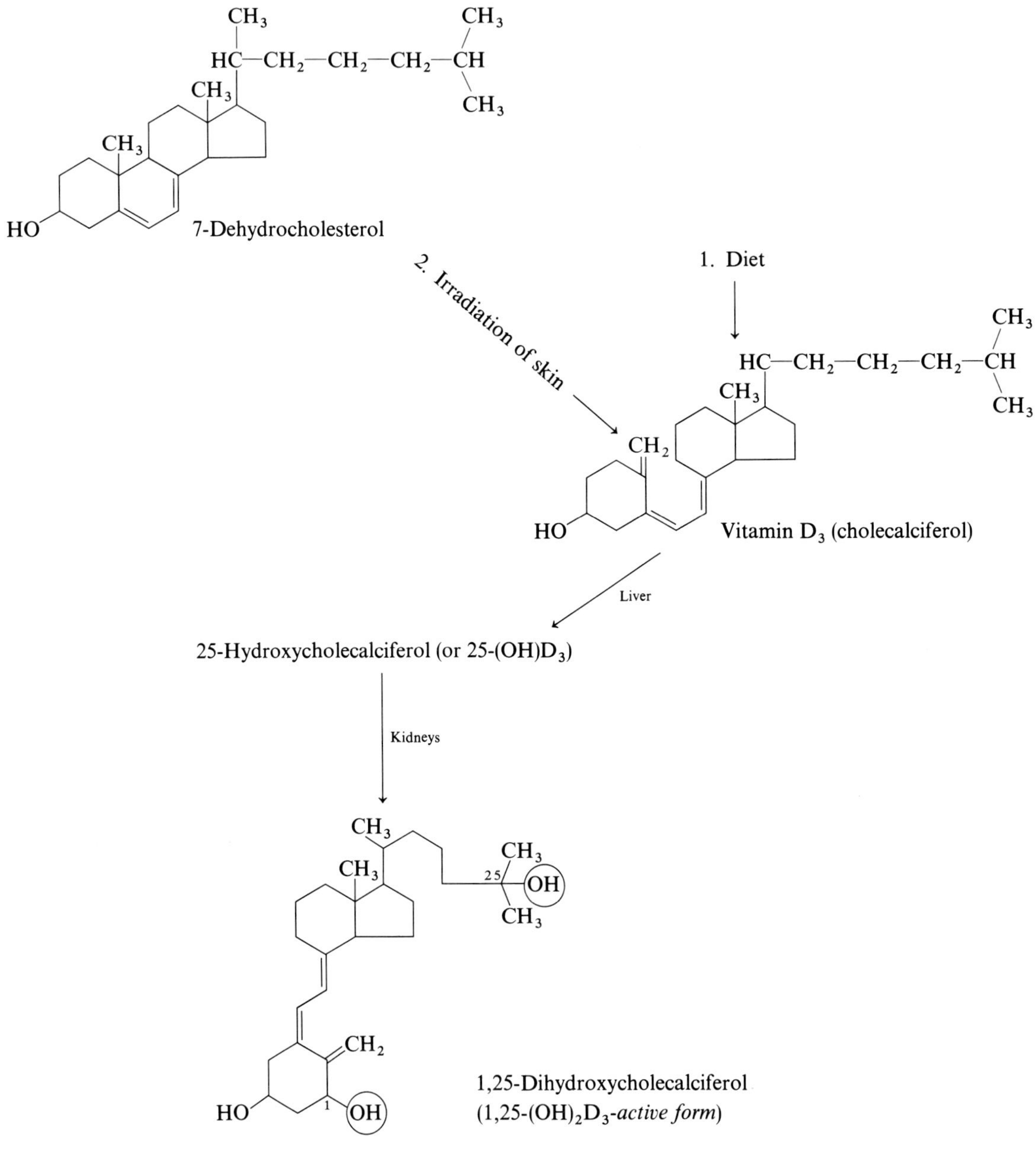

7-Dehydrocholesterol

2. *Irradiation of skin*

1. Diet

Vitamin D_3 (cholecalciferol)

Liver

25-Hydroxycholecalciferol (or 25-(OH)D_3)

Kidneys

1,25-Dihydroxycholecalciferol
(1,25-$(OH)_2D_3$-*active form*)

The biologically active form of vitamin D is *1,25-dihydroxycholecalciferol*, which is formed in the kidneys. Vitamin D apparently acts as a true hormone in the body—it promotes the synthesis of calcium-binding protein in the intestinal mucosal cell and thereby regulates the uptake of calcium from the intestine.

The recommended daily allowance of vitamin D is 400 IU (10 μ g). Doses of 6–10 times the RDA are toxic to small children, with the following symptoms: hypercalcemia (excess calcium in the blood), calcification of soft tissues, and demineralization of bones. A vitamin D deficiency can produce the following symptoms: in children, rickets (soft and pliable bones, leading to bending and distortion); celiac disease (a nutritional disease leading to defective digestion, fatty stools, and abdominal distention); and idiopathic steatorrhea (fatty diarrhea). In adults, vitamin D deficiency can lead to idiopathic steatorrhea and osteomalacia (softening of bones).

Vitamin D is stable under normal cooking procedures; however, some may be lost during the preparation and storage of *dried* milk.

Vitamin E (Tocopherol). *Vitamin E*, or *tocopherol*, is found in vegetable oils, eggs, milk, and green leafy vegetables. Its chemical formula is as follows:

$$HO-C,\ H_3C-C,\ CH_3,\ CH_3,\ H_2C,\ CH_2,\ O,\ C(CH_3)-$$

$$CH_2-CH_2-CH_2-\underset{}{\overset{CH_3}{\overset{|}{CH}}}-CH_2-CH_2-CH_2-\overset{CH_3}{\overset{|}{CH}}-CH_2-\underbrace{CH_2-CH_2-\overset{CH_3}{\overset{|}{CH}}-CH_3}_{\text{Isoprenoid unit}}$$

A number of tocopherols have been isolated and identified, of which α-tocopherol is the most abundant and biologically active. Its precise function is not known, but it may serve as an antioxidant, preventing the oxidation of cell membrane phospholipids.

The recommended daily allowance of vitamin E is 12–15 IU. The vitamin probably has a low toxicity, but because it is soluble in fat it can become concentrated in the body, producing hypervitaminosis. In animals, the following symptoms of vitamin E deficiency have been observed: retarded growth, heart damage, and infertility. Symptoms in humans have not been fully identified and classified.

Vitamin E is destroyed when it is frozen and when fats containing it become rancid or are used for deep-fat frying.

Vitamin K (Koagulation Factor). *Vitamin K*, or *coagulation factor*, is synthesized in the colon by intestinal bacteria. It is also found in green leafy vegetables, such as spinach. Its chemical formula is as follows:

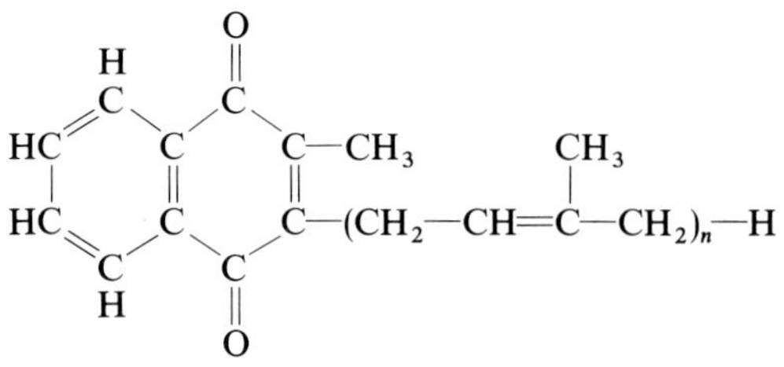

$$
\begin{array}{l}
\text{Naphthoquinone ring: } HC{=}CH \text{ ring fused, } C{=}O \text{ at 1,4; } C{-}CH_3 \\
C{-}(CH_2{-}CH{=}\overset{\displaystyle CH_3}{\overset{|}{C}}{-}CH_2)_n{-}H
\end{array}
$$

Vitamin K_2 with n being six to ten *isoprenoid* units

Vitamin K participates in the carboxylation of specific blood clotting factors.

The normal daily requirement is made by intestinal bacteria. Excess vitamin K produces no known toxic effects. A vitamin K deficiency can produce increased clotting time and bleeding under skin and in muscles.

Very little vitamin K is lost during normal food preparation.

20.6 ESSENTIAL MINERALS

essential minerals

bulk minerals

trace elements

A number of **essential minerals** are required in our diets. An essential mineral is defined as an inorganic substance that is necessary (in small quantities) for normal growth, development, and health. Two major categories of essential minerals have been established: the **bulk minerals** (macrominerals) and the **trace elements** (microminerals). Larger amounts of bulk minerals are found in the body than of trace elements. The bulk minerals include calcium, phosphorus, magnesium, sulfur, sodium, potassium, and chloride (see Table 20–3). The trace elements are required in only very minute amounts (see Table 20–3).

SUMMARY

Nutrition is the study of the nutrients required by the body, the foods where these nutrients are found, the amount of each nutrient necessary for optimal health, and the biochemical functions of each nutrient.

The small calorie (cal) is the amount of heat energy needed to raise the temperature of 1 g of water 1 °C. The large, or nutritional, calorie (kcal or Cal) is 1000 small calories.

The recommended daily allowance (RDA) of a nutrient is defined as the amount that should be ingested daily to maintain good nutrition and optimal health for the average healthy person.

Nearly all essential nutrients are absorbed by the small intestine and colon (see Figure 20–1). The jejunum and ileum absorb carbohydrates, fats, amino acids, and a number of vitamins.

Approximately 2.4 liters of water is lost each day. To prevent dehydration, this loss must be restored through an adequate intake of water and foods containing water.

Carbohydrates, fats, and proteins constitute the major cellular fuels and provide building blocks for other biomolecules. Simple sugars (glucose, fructose, and galactose), sucrose, and complex carbohydrates (starches) provide 40–60 percent of the total calories ingested in our diets. Dietary fiber (undigestible carbohydrates found in whole wheat, bran, fruits, and raw vegetables) are important because they increase stool weight, decrease intestinal transit time, and bind harmful substances formed in the gastrointestinal tract. Fats (triglycerides) supply 40–45 percent of the total calories in our diets. Decreased intake of saturated fats (and cholesterol) is highly recommended because excessive consumption of these substances has been linked to coronary heart disease and cancer. Linoleic acid is the only essential fatty acid. It is found in membrane phospholipids and is also required in synthesizing linolenic acid and arachidonic acid. Approximately 0.6 grams of high quality protein per kilogram of body weight (ideal) should be ingested daily for optimal health. High quality proteins contain all of the essential amino acids. Many plant proteins are deficient in one or more of the essential amino acids. This deficiency can be restored by consuming foods that contain more of the needed amino acid. The average American diet is summarized in Figure 20–3, p. 391. Specific dietary goals defined in this figure include increased consumption of complex carbohydrates and decreased consumption of total fats, saturated fats, cholesterol, and salt.

Vitamins are essential factors obtained in our diets that are required for specific biochemical functions. Most of the vitamins are used to construct important coenzymes. Vitamins are classified according to their solubility properties (water soluble or fat soluble). The water-soluble vitamins are biotin, vitamin B_{12}, folic acid, niacin, pantothenic acid, vitamin B_6, riboflavin, thiamine, and vitamin C. The fat-soluble vitamins are vitamins A, D, E, and K. On pages 395–406 are summarized the dietary sources, structures, functions, RDA values, and symptoms of overdose and deficiency for the known vitamins.

A number of essential minerals are required for growth and optimal health. The bulk minerals are required in larger amounts than the trace minerals and include calcium, phosphorus, magnesium, sodium, potassium, and chloride (see Table 20–3, pp. 408–411). Only very small amounts of the trace minerals are required by the body. Specific trace minerals include iron, iodine, zinc, copper, manganese, cobalt, molybdenum, selenium, and chromium.

REVIEW QUESTIONS

1. We have introduced a number of important concepts and terms in this chapter. Define (or briefly explain) each of the following terms:
   a. Nutrient
   b. Nutrient absorption
   c. Nutritional calories
   d. Recommended daily allowance (RDA)
   e. Dietary fiber
   f. Essential fatty acid
   g. Essential amino acid
   h. High quality protein
   i. Protein complementarity
   j. Vitamin
   k. Water-soluble vitamins (define and list)
   l. Fat-soluble vitamins (define and list)
   m. Secondary vitamin deficiency
   n. Trace element

2. What nutritional deficiencies would occur if:
   a. the duodenum was surgically removed?
   b. the ileum was surgically removed?
   c. the colon was surgically removed?

3. What percentage of the total water contained in a 70 kg adult male is lost and regained each day?

*4. A medium-sized cookie contains 150 calories (nutritional calories, Cal).
   a. How many small calories (cal) does this represent?
   b. How many grams of water can be heated 10 °C (from 20°C to 30°C) by this amount of energy, if the heat capacity of water is 1 cal/g·°C?
   c. From the data supplied in Section 20.3, how many kilocalories of energy does a 70-kg man contain in reserve fat? ($\Delta G^{0\prime}_{\text{fat}} \approx 9$ kcal/g.)
   d. How much would this same person weigh if the same number of kilocalories was stored as carbohydrates? ($\Delta G^{0\prime}_{\text{carbo}} \approx 4$ kcal/g.)

5. **a.** What are the biochemical functions of: (i) linoleic acid, (ii) linolenic acid, and (iii) arachidonic acid?
   *** b.** Suggest how linolenic acid and arachidonic acid can be synthesized from linoleic acid.

TABLE 20–3 A Summary of Essential Minerals.

Element and Total Amount in Human Body	Best Food Sources	RDA (1974)	Absorption and Metabolism	Principal Metabolic Functions	Clinical Manifestations of Deficiency
			1. The Bulk Minerals		
Sodium (Na^+) 1.8 g/kg	Table salt, salty foods, animal foods, milk, baking soda, baking powder, some vegetables	About 4–6 g[a]	Readily absorbed, extracellular, excreted in urine and sweat; aldosterone increases reabsorption in renal tubules	Buffer constituent, acid-base balance, water balance, osmotic pressure, CO_2 transport, cell membrane permeability, muscle irritability	Dehydration; acidosis; tissue atrophy; excess leads to edema, hypertension
Potassium (K^+) 2.6 g/kg	Vegetables, fruits, whole grains, meat, milk, legumes	About 1.5–4.5 g[a]	Readily absorbed, intracellular; secreted by kidney	Buffer constituent, acid-base balance, water balance, CO_2 transport, neuromuscular irritability	Acidosis; renal damage; cardiac arrest
Calcium (Ca^{2+}) 22 g/kg	Milk, milk products, fish bones (cooked)	800 mg	Poorly absorbed (20–40%) according to body need; absorption aided by vitamin D, lactose, acidity; hindered by excess fat, phytate, oxalate; excreted in feces; parathormone mobilizes bone Ca^{2+}	Formation of apatite in bones, teeth; blood clotting; cell membrane permeability; neuromuscular irritability	Rickets (child), poor growth; osteoporosis (adult), hyperexcitability

Phosphorus (PO_4^{3-}) 12 g/kg	Milk, milk products, egg yolk, meat, whole grains, legumes, nuts	800 mg	Readily absorbed; excreted by kidney	Constituent of bones, teeth; constituent of buffers; constituent of ATP, NAD, FAD, etc.; constituent of metabolic intermediates, nucleoproteins, phospholipids, phosphoproteins	Osteomalacia (rare), renal rickets; cardiac arrhythmia
Chloride (Cl^-) 50 meq/kg.	Animal foods, table salt	Intake 5–10 g as NaCl[a]	Rapid absorption; excreted in urine; high renal threshold; not stored	Electrolyte, osmotic balance; gastric HCl; acid-base	Hypochloremic alkalosis (pernicious vomiting)
Sulfur (SO_4^{2-})	Plant and animal proteins, as Cys and Met	2–3 gm.[a]	Derived from metabolism of Cys and Met; excreted in urine	Constituent of proteins, mucopolysaccharides, heparin, thiamine, biotin, lipoic acid; detoxication	Cystinuria; cystine renal calculi
Magnesium (Mg^{2+}) 0.5 g/kg	Chlorophyll, nuts, legumes, whole grains	350 mg	Absorbed readily; competes with Ca^{2+} for transport	Cofactor for PO_4-transferring enzymes; constituent of bones, teeth; decreases neuromuscular irritability	Magnesium-conditioned deficiency, muscular tremor, choreiform movements, confusion; vasodilatation, hyperirritability

Source: Adapted from Orten, J. M., and Neuhaus, O. W., *Human Biochemistry*, 9th ed. (St. Louis: C. V. Mosby, 1975), pp. 554–56.
Note: The inorganic elements included are those for which evidence exists that they are *essential* for man. The amounts of the element present in the entire human body are averages from the literature (Dairy Council Digest 39(1968):26. They are expressed as grams or milligrams per kilogram of body weight (*fat-free basis*) or as milligrams in entire body.

The RDA is the recommended dietary allowance per day, established by the Food and Nutrition Board, National Research Council, 1974. The values given are for a normal adult male, 19 to 22 years of age.

[a] An estimated value is given if no RDA value has been established. The estimated value is the average daily dietary intake of a normal adult.

TABLE 20–3 *(continued)*

Element and Total Amount in Human Body	Best Food Sources	RDA (1974)	Absorption and Metabolism	Principal Metabolic Functions	Clinical Manifestations of Deficiency
		2. The Trace Minerals			
Iron (Fe^{2+} or Fe^{3+}) 75 mg/kg	Liver, meats, egg yolk, green leafy vegetables, whole grains, enriched bread and cereals	10 mg male; 18 mg female	Absorbed according to body need; aided by HCl, ascorbic acid; regulated by apoferritin; stored by liver	Constituent of hemoglobin, myoglobin, catalase, ferredoxin, cytochromes; electron transport, enzyme cofactor	Anemia, hypochromic; pregnancy demands; excess produces hemochromatosis
Iodine (I^-) 20–50 mg	Seafoods, iodized salt	140 μg male; 100 μg female	Concentrates in thyroid; transported as PBI	Constituent of thyroxin, triiodothyronine; regulator of cellular oxidations	Endemic (simple) goiter (hypothyroidism); cretinism
Fluoride (F^-)	Seafoods, some drinking water	0.7–3.4 mg.[a] (1 p.p.m. in drinking water)	Easily absorbed; excreted in urine; deposited in bones and teeth	Constituent of fluproapatite—tooth enamel; strengthens bones and and teeth	Dental caries; osteoporosis; excess (5–8 p.p.m. in water) produces mottled enamel
Zinc (Zn^{2+}) 28 mg/kg	Liver, pancreas, shellfish, most animal tissues, wheat germ, legumes	15 mg	1–2 mg absorbed; phytate decreases absorption	Constituent of insulin, carbonic anhydrase, carboxy peptidase, lactic dehydrogenase, alcohol dehydrogenase, alkaline phosphatase	Anemia; stunted growth; hypogonadism in male; hypogeusia
Copper (Cu^{2+}) 2 mg/kg	Liver, kidney, egg yolk, whole grains	2–5 mg.[a]	Limited absorption; transport by ceruloplasmin; stored	Formation of hemoglobin (increases iron utilization);	Hypochromic anemia; aneurysms; CNS lesions;

			in liver; excretion via bile	constituent of 11 oxidase enzymes (tyrosinase, cytochrome oxidase, ascorbic acid oxidase, ferroxidase, etc.)	achromotrichia; excessive hepatic storage in Wilson's disease
Cobalt (Co^{2+}) 3 mg	Liver, pancreas; vitamin B_{12} in animal proteins	0.15–1 mg[a]	Limited absorption; stored in liver; excreted via bile	Constituent of vitamin B_{12}	Anemia; deficiency as vitamin B_{12} produces pernicious anemia; excess produces polycythemia
Manganese (Mn^{2+}) 20 mg	Liver, kidney, wheat germ, legumes, nuts, tea	2–3 mg[a]	Stored in liver mitochondria and bone; excreted via bile	Cofactor for number of enzymes—arginase, carboxylase, kinases, mucopolysaccharides	In animals, produces sterility, weakness, perosis, congenital ataxia
Molybdenum (Mo) 5 mg	Liver, kidney, whole grains, legumes, leafy vegetables	0.1–0.4 mg[a]	Readily absorbed; excreted in urine and bile	Constituent of xanthine oxidase, aldehyde oxidase	Unknown
Chromium (Cr^{3+})	Liver, animal, and plant tissue	0.05–0.12 mg[a]	Poorly absorbed; traces excreted in urine	Necessary for glucose utilization	Unknown; deficiency in diabetes claimed; decreased glucose tolerance in rats; possible relation to cardiovascular disease
Selenium (Se)	Liver, kidney, heart; grains, vegetables (varies with Se in soil)	60–70 μg[a]	Trace excreted in urine	Constituent of factor 3; acts with vitamin E to prevent liver necrosis and muscular dystrophy in animals; inhibits lipid peroxidation	Unknown; excess produces alkali disease in cattle, sheep

6. List important reasons for:
   a. Decreasing dietary fats and cholesterol.
   b. Increasing complex carbohydrate intake (including dietary fiber).
   c. Maintaining proper vitamin and mineral intake.
7. a. Referring to Table 20–2, pp. 392–393, calculate your average score and assess *your* risk. (Be honest!)
   b. Calculate average scores for family members.
   c. What do you plan to do about it? (If your answer is "nothing," I recommend that you read a book on heart ailments for an idea of what you are asking for.)
8. Prepare a simple table that lists each vitamin and the kinds of foods that contain a large amount of each vitamin. Use the example below as a guide in preparing your table; mark foods that contain the vitamin with an X.

Vitamin	Citrus	Cereals	. . .
C	X	—	
⋮			

* 9. Referring to Figure 20–2, p. 390, Figure 20–3, p. 391, Table 20–1, p. 389, and Table 20–2, pp. 392–393, plan a complete 3-meal menu, including breakfast, lunch, and dinner, that will do the following:
   a. supply sufficient calories
   b. follow the nutrition guidelines recommended in Figure 20–3 and Table 20–2
   c. provide adequate amounts of essential amino acids (preferably from plant protein)
   d. provide vitamin and mineral requirements

   Does this menu contain a variety of foods that a person would want to eat and would enjoy?

SUGGESTED READING

Bhagavan, N. V. *Biochemistry*, 2nd ed., Chs. 9 and 10. Philadelphia: J. B. Lippincott, 1978.

Guthrie, H. A. *Introductory Nutrition*, 3rd ed. St. Louis: C. V. Mosby, 1975.

Lehninger, A. L. *Biochemistry*, 2nd ed., Chs. 13 and 30. New York: Worth, 1975.

McGilvery, R. W., and Goldstein, G. *Biochemistry: A Functional Approach*, 2nd ed., Chs. 40–43. Philadelphia: W. B. Saunders, 1979.

Orten, J. M., and Neuhaus, O. W. *Human Biochemistry*, 9th ed., Chs. 13–15. St. Louis: C. V. Mosby, 1975.

Worthington-Roberts, B. S. *Contemporary Developments in Nutrition*. St. Louis: C. V. Mosby, 1981.

NOTES

1. Bhagavan, N. V., *Biochemistry*, 2nd ed. (Philadelphia: J. B. Lippincott, 1978), pp. 1029–30.
2. Worthington-Roberts, B. S., *Contemporary Developments in Nutrition*, (St. Louis: C. V. Mosby, 1981), p. 1.
3. Worthington-Roberts, *Contemporary Developments in Nutrition*, pp. 44–76.
4. Worthington-Roberts, *Contemporary Developments in Nutrition*, pp. 99–100.
5. Worthington-Roberts, *Contemporary Developments in Nutrition*, pp. 135–239.
6. Sanders, H. J., "Nutrition and Health," *Chemical and Engineering News* 57 (1979): 27.

Appendix A

Photosynthesis

A.1 INTRODUCTION

A discussion of carbohydrate metabolism would not be complete without describing *photosynthesis*, the process by which plants trap the energy of sunlight and use it to synthesize carbohydrates. The newly synthesized carbohydrates can, in turn, be used by the plant for its own energy needs, or stored. Animals (and humans) can use these carbohydrates for *their* energy needs by consuming various plant foods. In fact, the energy requirements of most organisms are ultimately met by utilizing solar energy that has been trapped and stored in the form of carbohydrates.

The important steps in the *transformation* of solar energy are as follows: (1) Sunlight is absorbed by the plant, and *some of the energy of sunlight is conserved as ATP*. (2) The ATP made is used (along with CO_2 and the reducing agent *NADPH*) to make simple carbohydrates, such as fructose and glucose. (3) These carbohydrates are used to make more complex plant polysaccharides, such as starch and cellulose, that are incorporated in plant tissues. (4) Plant tissues are eaten by a ruminant organism, such as a steer. (5) Microorganisms in the rumen of the steer digest the plant carbohydrates, some of which are metabolized by the microorganism and the remainder of which are absorbed by the steer. (6) The steer metabolizes some of the carbohydrates to take care of its energy needs, transforms the rest into lipids and amino acids, and incorporates them in its muscle tissue. (The carbon skeletons of these cellular fuels have become highly reduced—that is, they contain extra hydrogen atoms.) (7) Beef muscle tissue is eaten by a meat eating organism, such as a human. (8) The human absorbs and

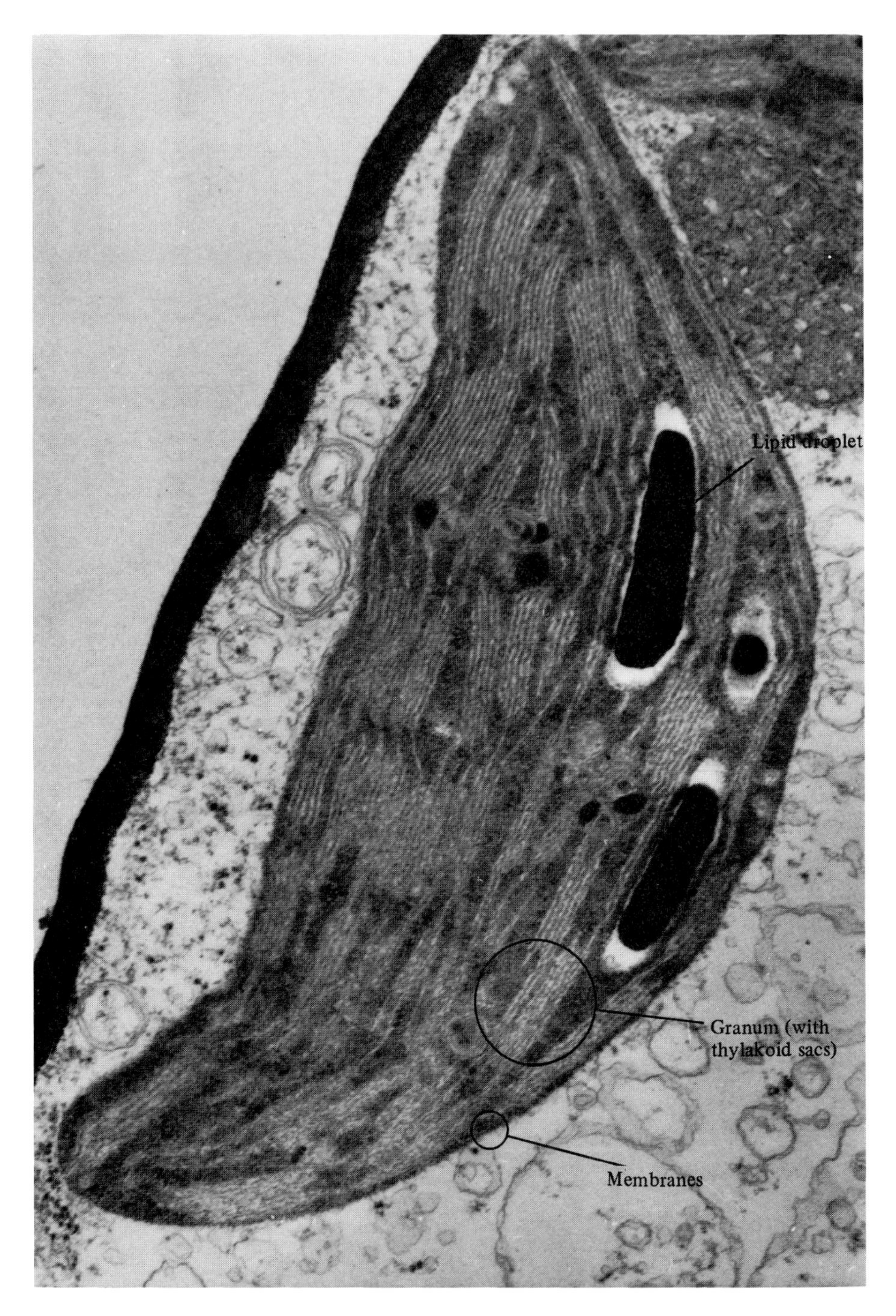

FIGURE A–1 Representation of an electron micrograph of a whole chloroplast from a spinach leaf. (Courtesy of Dr. Kenneth Miller.) (Adapted from L. Stryer, *Biochemistry*, 2nd ed. (San Francisco: W. H. Freeman, 1981), p. 430.)

digests the proteins and lipids in the beef muscle tissue, along with carbohydrates from various plant foods (for instance, starches from potatoes, corn, beans, and so on). (9) From these cellular fuels, the human generates the ATP required for muscular contraction, biosynthetic work, and other important cellular functions, as well as the raw materials for polypeptide, lipid, and nucleic acid synthesis.

Thus, we see that energy is transformed and that the energy used by living systems ultimately comes from the sun. Light energy is trapped by plants, the energy is used to generate ATP, the ATP is used to synthesize carbohydrates, the carbohydrates are either used directly for energy or transformed into other cellular fuels, and these fuels are metabolized to yield ATP that can be used to perform useful work in our bodies.

A.2 A DESCRIPTION OF THE CHLOROPLAST

When one first looks at a living plant cell through a microscope, one is often impressed by the complexity and beauty of the sight. The nucleus and mitochondria float in the cytoplasm, while tiny green chloroplasts flow in a circular manner around the perimeter.

chloroplasts

Chloroplasts are specialized cellular organelles found in plant cells. They contain the components necessary for photosynthesis, especially *chlorophyll*. Figure A–1 is an electron micrograph of a typical chloroplast. An artist's rendition of the chloroplast is shown in Figure A–2.

The chloroplast has an outer membrane surrounding an inner membrane. The space within the inner membrane, called the *stroma*, contains a solution of proteins and enzymes. Within the stroma are flat, disk-shaped sacs called *thylakoids*. A stack of these thylakoids is called a *granum*. Individual grana are interconnected by membranous structures. The membranes of the individual thylakoid sacs contain the photosynthetic components. It is here that photosynthesis actually takes place.

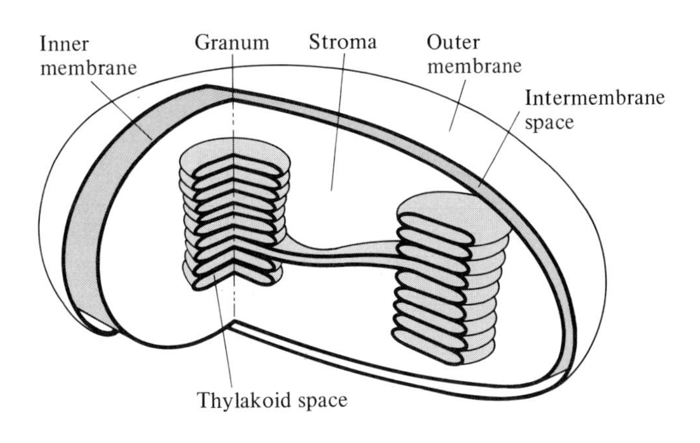

FIGURE A–2
Diagram of a chloroplast. (Adapted from L. Stryer, *Biochemistry*, 2nd ed. (San Francisco: W. H. Freeman, 1981), p. 434.)

A.3 THE PHOTOSYNTHETIC PROCESS

chlorophyll The thylakoid membranes contain the green pigment **chlorophyll.** Chlorophyll is a porphyrinlike compound with a structure similar to the heme group of hemoglobin (see Chapter 4). Chlorophyll is a substituted tetrapyrrole ring system with a magnesium ion at the center of the molecule. The porphyrin ring system is planar and has a long, hydrophobic isoprenoid tail (phytol) linked to one of the pyrrole rings. The structure of chlorophyll is shown in Figure A–3. Two different chlorophylls commonly exist in plants. *Chlorophyll a* has a methyl group substituted on one of the pyrrole rings, whereas *chlorophyll b* has a formyl group instead.

Because chlorophyll has many alternating double and single bonds (that is, it is an *extended conjugated system*), its bonding structure can efficiently absorb light energy. Chlorophyll readily absorbs blue and red wavelengths and transmits green wavelengths, as shown in Figure A–4. This is why chlorophyll is green. In effect, the chlorophyll molecule is constructed to absorb light energy (photons), thus serving as a tiny photoreceptor (light antenna) for incoming light. When a chlorophyll molecule absorbs light

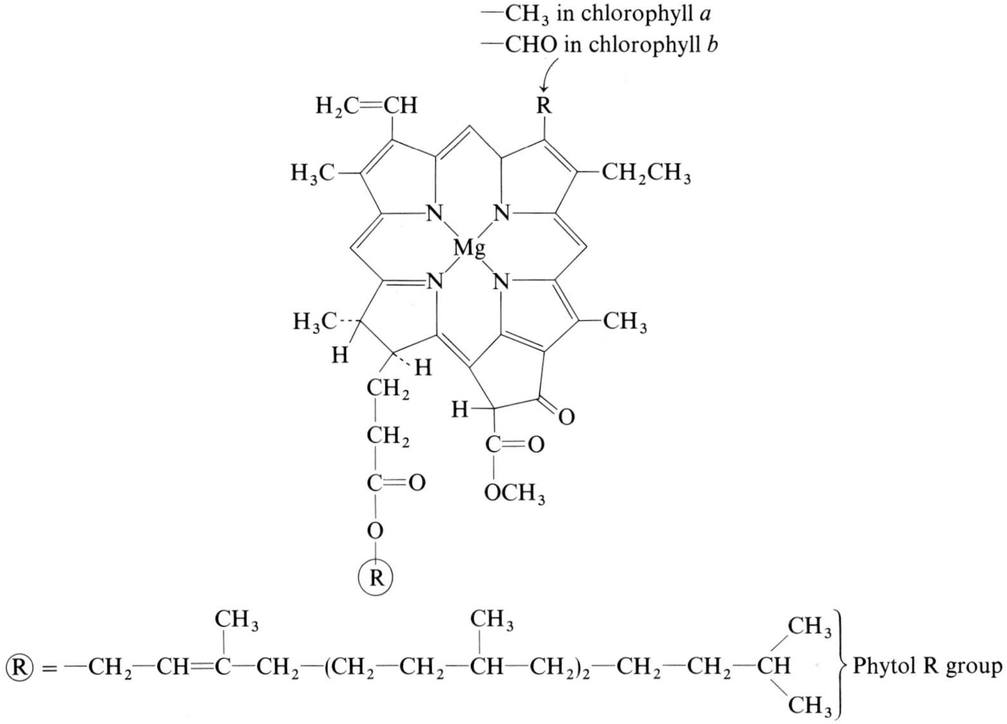

FIGURE A–3 The structure of chlorophyll. Chlorophyll *a* differs from chlorophyll *b* only in the R group bound to the upper right pyrrole ring.

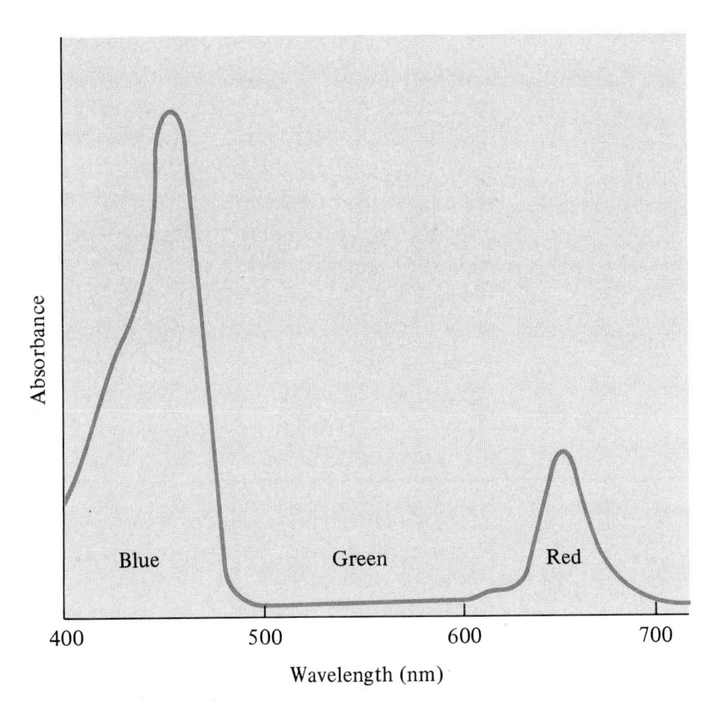

FIGURE A–4
Absorption spectrum of chlorophyll *b*. Chlorophyll *b* absorbs light energy in the blue and red regions of the visible spectrum. It does not absorb light in the green region; hence, chlorophyll appears green.

energy, the energy is transferred to a site (called a *reaction center*) within the thylakoid membrane. Here, important reactions take place in which the energy of sunlight is used to drive the photosynthetic process.

light phase

dark phase

We now have a fairly good understanding of the photosynthetic process. Photosynthesis can be divided into two main phases: (1) a **light phase** (or *light reaction*), in which light energy is converted to chemical energy in the form of ATP and NADPH; and (2) a **dark phase** (or dark reaction), in which simple sugars are made from CO_2 using the ATP and NADPH made during the light phase.

1. THE LIGHT PHASE

During the light phase of photosynthesis, *plants use light energy to split up a water molecule* according to the following reaction:

$$2\,H_2O + A \xrightarrow[(e^-\ \text{flow})\rightarrow]{\text{Light}} 2\,AH_2 + O_2$$

Hydrogen/electron acceptor (A) — Reduced hydrogen acceptor (AH_2)

This reaction is also known as the *Hill reaction*, after the biochemist who first discovered it. In essence, light energy is used to move excited electrons *away* from water to an electron acceptor, evolving molecular oxygen in the process.

Two photosystems (*photosystem I and photosystem II*) have been isolated from the thylakoid membranes and extensively studied in the laboratory. Photosystem I absorbs light at wavelengths shorter than 700 nm, generating the reducing agent NADPH. Photosystem II absorbs light at wavelengths shorter than 680 nm, forming molecular oxygen. Both photosystems contain a number of chlorophyll molecules, as well as a number of other light-absorbing components, enzymes, and electron carriers. Photosystems I and II are linked to each other by a series of electron carriers. When photosystems I and II absorb light energy, a number of events take place. These events are diagrammed in Figure A–5.

Photosystem II absorbs light energy and produces a strong oxidizing agent that extracts electrons from water (the Hill reaction). Excited electrons are next transferred to substance Z, also known as *plastoquinone*. Plastoquinone is similar in structure to coenzyme Q of the mitochondrial electron-transport system. The electrons are transferred from plastoquinone to photosystem I via a series of cytochromes and mobile electron carriers. As electrons flow from photosystem II to photosystem I, a *proton gradient* is

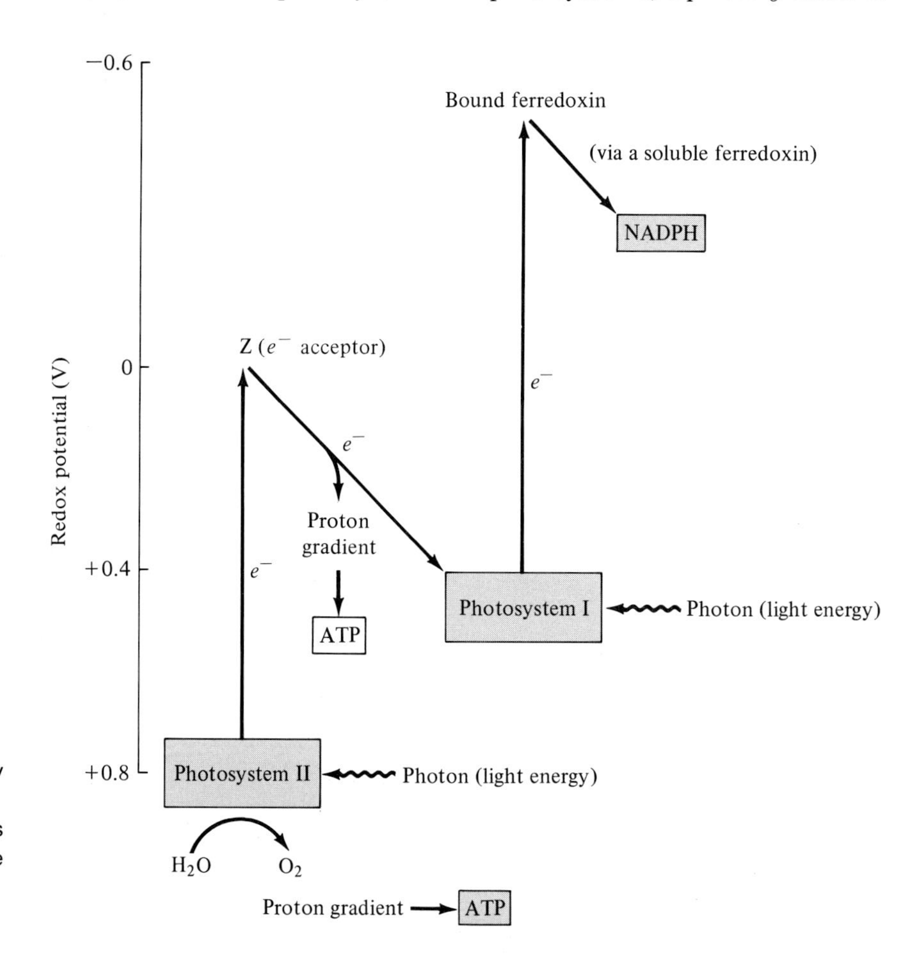

FIGURE A–5
The light reaction of photosynthesis. Light energy is absorbed by photosystems I and II, which excite electrons (e^-) and thus promote the synthesis of ATP and NADPH.

formed across the membranes of the thylakoid sacs. This gradient drives the synthesis of ATP from ADP + P_i—a situation analogous to mitochondrial electron transport and oxidative phosphorylation. The synthesis of ATP within the chloroplast is called **photophosphorylation**. As the name implies, photophosphorylation requires the input of light energy to drive electron transport and, thus, ATP synthesis.

photophosphorylation

Photosystem I, in turn, absorbs light energy and produces an excited electron. This excited electron is transferred to an iron-sulfur (FeS) protein called *ferredoxin*. This protein is *bound* to the thylakoid membranes. The electron is next transferred to a *soluble* form of ferredoxin that reduces $NADP^+$ to NADPH. Two electrons from 2 ferredoxin molecules are required to reduce $NADP^+$. This process is catalyzed by the enzyme *ferredoxin-NADP reductase*.

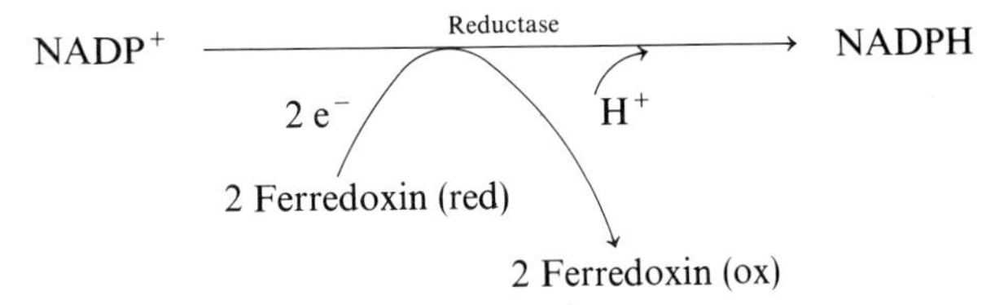

One should note that as photosystem I absorbs light energy, it produces an excited electron that is transferred to membrane-bound ferredoxin. Thus, photosystem I becomes *oxidized* (deficient in an electron). Photosystem I is *reduced* when an electron is transferred from photosystem II. This electron originally came from a water molecule.

2. THE DARK PHASE

We now must explain exactly how the chloroplast synthesizes carbohydrates. As we have explained, light energy is used in the light phase to generate ATP and NADPH. A water molecule is split up, and molecular oxygen is evolved as a waste product. The chloroplast can now use the ATP and NADPH to fix carbon dioxide and make simple carbohydrates, a process known as the dark phase of photosynthesis. The dark phase is catalyzed by a set of enzymes that constitute the Calvin cycle, named after the biochemist who first elucidated the pathway. The Calvin cycle is shown (in somewhat abbreviated form) in Figure A–6.

The essential features of the pathway are summarized as follows: A molecule of carbon dioxide (CO_2) first condenses with an intermediate called *ribulose 1,5-diphosphate*. This reaction is catalyzed by the enzyme *ribulose 1,5-diphosphate carboxylase*, which is located on the inner surface of the thylakoid membranes in contact with the stroma (the soluble portion of the chloroplast). A transient intermediate is formed that is rapidly hydrolyzed into two molecules of *3-phosphoglycerate*. Next, two molecules of ATP and two molecules of NADPH are used to produce two molecules of *glyceraldehyde 3-phosphate* (a 3 C sugar). The enzyme *aldolase* is then used to

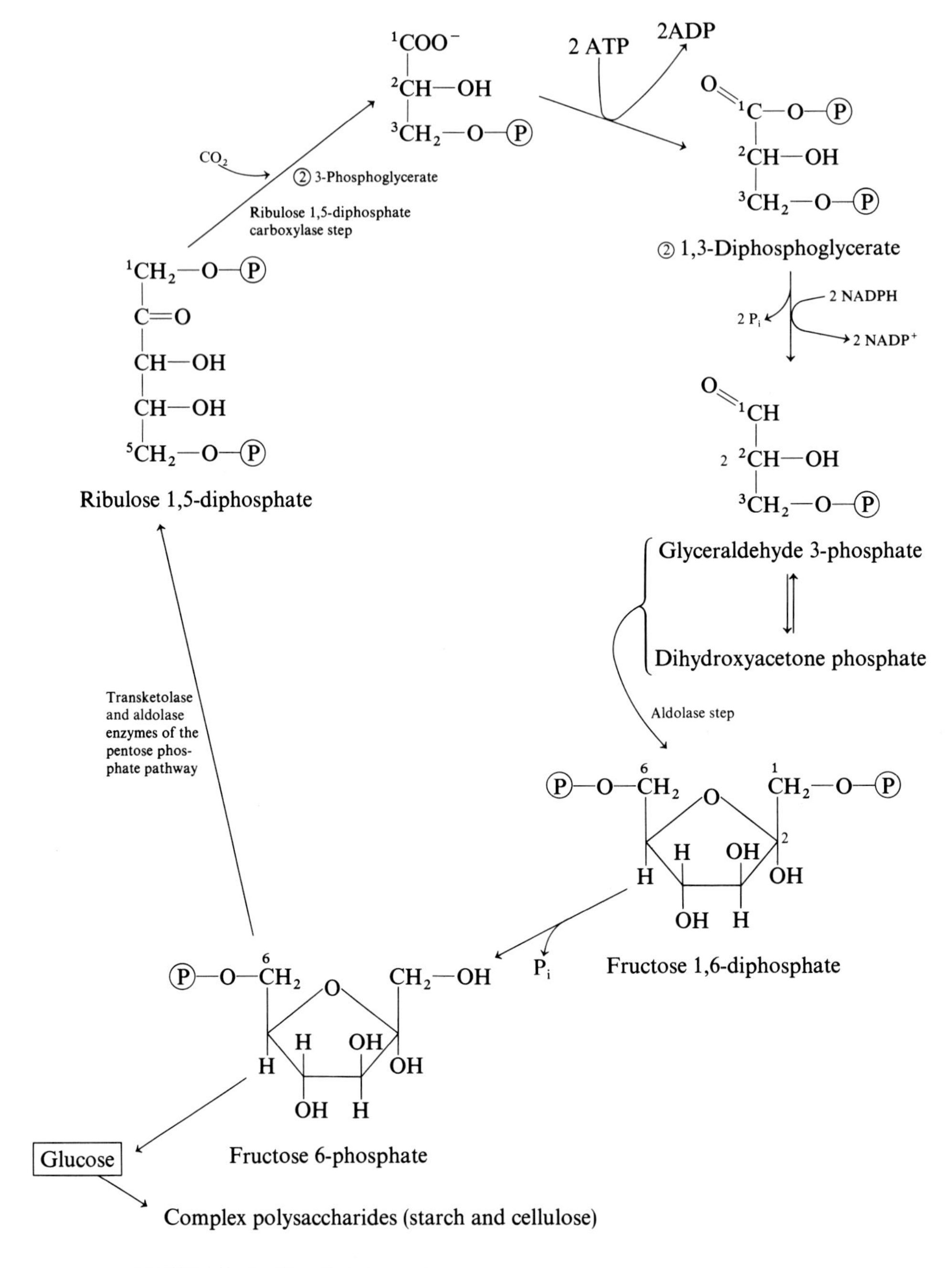

FIGURE A–6 The Calvin cycle.

combine glyceraldehyde 3-phosphate (3 C) and dihydroxyacetone phosphate (3 C) to produce *fructose 1,6-diphosphate* (6 C). The fructose 1,6-diphosphate generated by this process can be used to construct other monosaccharides (glucose) and, finally, complex polysaccharides. Some of the fructose 1,6-diphosphate is used to regenerate the ribulose 1,5-diphosphate to close the cycle. This is done by enzymes of the pentose phosphate pathway (see Chapter 10).

In summary, photosynthesis is a process by which light energy is efficiently absorbed and used to make ATP and the reducing agent NADPH. Using these energy-rich molecules, carbon dioxide is fixed into ribulose 1,5-diphosphate, from which, in turn, fructose 1,6-diphosphate is made. This 6-carbon sugar is then interconverted into other monosaccharides, such as glucose, which can be used to construct more complex polysaccharides, such as starch and cellulose.

Appendix B

Mathematical Review

Proficiency in manipulating numbers is extremely advantageous (and often necessary) in understanding chemical principles. This short review section will hopefully refresh your memory, cement important concepts in your mind, and eliminate an unjustified fear of math and numbers.

B.1 ROUNDING OFF NUMBERS

Numbers can be rounded off by following a few simple rules.

RULE 1

When the first digit after those being retained is less than 5, the digits retained remain the same.

15.263 Round off to 15.26

Retain these 4 digits

RULE 2

When the first digit after those being retained is larger than 5, the last digit retained is increased by one.

3.566 Round off to 3.57

RULE 3

When the first digit after those being retained is 5 and all others beyond it are zeros, the last digit retained remains the same if it is an *even* number and is increased by one if it is an *odd* number.

3.565 Round off to 3.56
↑
Even number

3.575 Round off to 3.58
↑
Odd number

B.2 A FEW SIMPLE ALGEBRAIC MANIPULATIONS

1. $\frac{a}{b} = x$

a. Solve for x: Divide a by b.

b. Solve for a: Multiply both sides by b.

$$\frac{a}{b} = x$$

$$\frac{a}{\cancel{b}} \cdot \cancel{b} = x \cdot b$$

$$\therefore a = x \cdot b$$

c. Solve for b: Multiply both sides by b.

$$\frac{a}{b} = x$$

$$\frac{a}{\cancel{b}} \cdot \cancel{b} = x \cdot b$$

$$\therefore a = x \cdot b$$

Now, divide both sides by x.

$$\frac{a}{x} = \frac{\cancel{x} \cdot b}{\cancel{x}}$$

$$\therefore \frac{a}{x} = b$$

2. $y = mx + b$ (equation for a straight line)
Solve for x (when $y = 2$, $m = 0.5$, and $b = 1.5$)

Step 1: Manipulate equation and solve for x.

$$y = mx + b$$

Subtract b from both sides.

$$y - b = mx + b - b$$

$$\therefore y - b = mx$$

Divide both sides by m.

$$\therefore \frac{y - b}{m} = x$$

Step 2: Substitute values given into the equation and solve for x.

$$\frac{y - b}{m} = x$$

$$\therefore \frac{2 - 1.5}{0.5} = \frac{0.5}{0.5} = 1.0 = x$$

Check:

$$y = mx + b$$

$$2 = (0.5)(1) + 1.5$$

$$2 = 0.5 + 1.5$$

$$2 = 2$$

B.3 A REVIEW OF SCIENTIFIC NOTATION

Very large numbers (1 000 000 000) or very small numbers (0.000 000 001) are often encountered in science. These numbers are difficult to write or to use in routine scientific calculations. Such numbers are more conveniently expressed in a form called *scientific* (*or exponential*) *notation*. In this form, a number is expressed as

$$n \times 10^{x} \longleftarrow \text{Exponent}$$

(Coefficient: n; Base: 10)

where n (*coefficient*) is a number between one and ten (1.000 and 9.999) and x is the *exponent* (a positive or negative whole number such as +2, +5, −2, or −5). The exponent tells us how many times 10 (the *base*) is to be multiplied by itself—that is, it gives the location of the decimal point. A few examples will illustrate this concept.

$$100. = 1.0 \times 10^2 \qquad (10 \times 10 = 100)$$

Move the decimal point 2 digits to the left to obtain the coefficient (n). Note that when the decimal point is moved to the left, the exponent is positive and indicates a large number, greater than 10.

$$1000. = 1.0 \times 10^3$$

Move decimal point 3 digits to the left.

$$0.0010 = 1.0 \times 10^{-3}$$

Move decimal point 3 digits to the right. The negative sign indicates a very small number.

Let us examine a few more examples:
Express 5256.0 in scientific notation.

$$5256.0 = 5.2560 \times 10^3$$

Express 0.001234 in scientific notation.

$$0.001234 = 1.234 \times 10^{-3}$$

By using scientific notation, very large and very small numbers can easily be multiplied and divided.

Example 1: Multiply 400×200

1. Convert each number to scientific notation.
2. Multiply coefficients and *add* exponents.

$$(4 \times 10^2) \times (2 \times 10^2) = 8 \times 10^4$$

Example 2: Multiply 300×0.02

1. Convert both numbers to scientific notation.
2. Multiply coefficients and *add* exponents.

$$\begin{aligned}(3 \times 10^2) \times (2 \times 10^{-2}) &= 6 \times 10^{[2+(-2)]} \\ &= 6 \times 10^0 \\ &= 6\end{aligned}$$

Example 3: Divide 400 by 0.002

1. Convert to scientific notation.
2. Divide the cofficients and *subtract* exponents.

$$\begin{aligned}\frac{400}{0.002} = \frac{4 \times 10^2}{2 \times 10^{-3}} &= 2 \times 10^{[2-(-3)]} \\ &= 2 \times 10^5\end{aligned}$$

Example 4: Very complex calculations can be performed. Multiply 2237×0.835

$$(2.237 \times 10^3) \times (8.35 \times 10^{-1}) = 18.679 \times 10^2$$
$$= 1.8679 \times 10^3$$
$$= 1.87 \times 10^3 \text{ (round off to 3 significant digits)}$$

Note: 18.679×10^2 is *not* in correct scientific notation—the number n is greater than 10. Therefore, it must be converted to correct scientific notation before rounding off.

B.4 A REVIEW OF LOGARITHMS

The common logarithm (log) of a number (n) is the exponent (x) to which the base (10) must be raised to equal n.

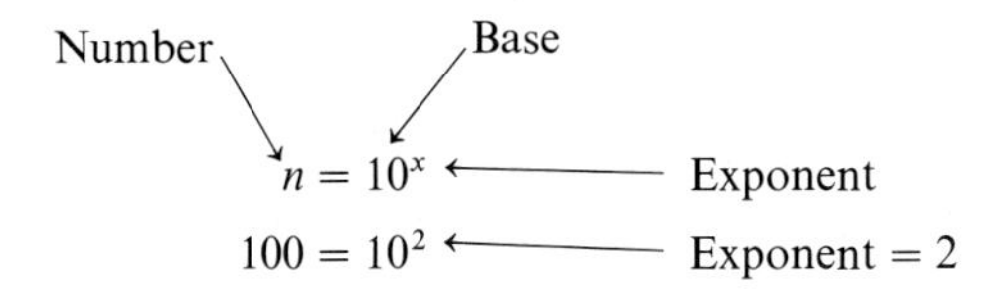

Therefore:

$$\log n = x$$
$$\log 100 = 2$$

and

$$\log (0.01) = \log (10^{-2}) = -2$$

Since logs are exponents, *logs of products are added and logs of quotients are subtracted.*

$$\log (a \times b) = \log a + \log b$$
$$\log \left(\frac{a}{b}\right) = \log a - \log b$$

Many numbers are more complicated than those given in the examples cited. The logs of these numbers can be found by looking them up in a table of logarithms (see Table B–1) or by using a hand-held calculator. For example, let us find the logarithms of the numbers 20, 200, 201, and 2010. First the number must be rewritten in scientific notation.

$$20 = 2 \times 10^1$$

Base (the 10); Exponent (characteristic) (the superscript 1); Log of Coefficient (mantissa) (the 2)

The mantissa is found by consulting a log table, as shown here:

Number n			
This means 1.0 → 10	0000	0043	. . .
11	0414	0453	. . .
⋮	⋮	⋮	
This means 2.0 → 20	3010	3032 ← This means 2.01	. . .
21	3222	3243	. . .

The log of 2 = *0.3010*. The log of $10^1 = 1$. Therefore, the log of 20 is

$$\begin{aligned}\log(2 \times 10^1) &= \log 2 + \log(10^1)\\ &= 0.3010 + 1\\ &= 1.3010\end{aligned}$$

Characteristic (the 1) Mantissa (the .3010)

Let us calculate the log of 200:

$$\begin{aligned}200 &= 2 \times 10^2\\ \log(200) &= \log(2 \times 10^2)\\ &= \log 2 + \log(10^2)\\ &= 0.3010 + 2\\ &= 2.3010\end{aligned}$$

Now, however, let us calculate the log of 201 (a number with more significant digits):

$$\begin{aligned}201 &= 2.01 \times 10^2\\ \log(201) &= \log(2.01 \times 10^2)\\ &= \log(2.01) + \log(10^2)\\ &= 0.3032 + 2\\ &= 2.3032\end{aligned}$$

Likewise, the log of $2010 = \log(2.01) + \log(10^3) = 3.3032$.

Notice how increasing the exponent increases the characteristic, leaving the mantissa constant, whereas adding significant digits to the right of the decimal point changes the mantissa and leaves the characteristic the same:

Number (n)	$\log n$
20	1.3010
200	2.3010
201	2.3032
2010	3.3032

What would the log of 20 000 be? Of 201 000?

TABLE B–1 Four-Place Logarithms.

n	0	1	2	3	4	5	6	7	8	9
10	0000	0043	0086	0128	0170	0212	0253	0294	0334	0374
11	0414	0453	0492	0531	0569	0607	0645	0682	0719	0755
12	0792	0828	0864	0899	0934	0969	1004	1038	1072	1106
13	1139	1173	1206	1239	1271	1303	1335	1367	1399	1430
14	1461	1492	1523	1553	1584	1614	1644	1673	1703	1732
15	1761	1790	1818	1847	1875	1903	1931	1959	1987	2014
16	2041	2068	2095	2122	2148	2175	2201	2227	2253	2279
17	2304	2330	2355	2380	2405	2430	2455	2480	2504	2529
18	2553	2577	2601	2625	2648	2672	2695	2718	2742	2765
19	2788	2810	2833	2856	2878	2900	2923	2945	2967	2989
20	3010	3032	3054	3075	3096	3118	3139	3160	3181	3201
21	3222	3243	3263	3284	3304	3324	3345	3365	3385	3404
22	3424	3444	3464	3483	3502	3522	3541	3560	3579	3598
23	3617	3636	3655	3674	3692	3711	3729	3747	3766	3784
24	3802	3820	3838	3856	3874	3892	3909	3927	3945	3962
25	3979	3997	4014	4031	4048	4065	4082	4099	4116	4133
26	4150	4166	4183	4200	4216	4232	4249	4265	4281	4298
27	4314	4330	4346	4362	4378	4393	4409	4425	4440	4456
28	4472	4487	4502	4518	4533	4548	4564	4579	4594	4609
29	4624	4639	4654	4669	4683	4698	4713	4728	4742	4757

TABLE B–1 *(continued)*

n	0	1	2	3	4	5	6	7	8	9
30	4771	4786	4800	4814	4829	4843	4857	4871	4886	4900
31	4914	4928	4942	4955	4969	4983	4997	5011	5024	5038
32	5051	5065	5079	5092	5105	5119	5132	5145	5159	5172
33	5185	5198	5211	5224	5237	5250	5263	5276	5289	5302
34	5315	5328	5340	5353	5366	5378	5391	5403	5416	5428
35	5441	5453	5465	5478	5490	5502	5514	5527	5539	5551
36	5563	5575	5587	5599	5611	5623	5635	5647	5658	5670
37	5682	5694	5705	5717	5729	5740	5752	5763	5775	5786
38	5798	5809	5821	5832	5843	5855	5866	5877	5888	5899
39	5911	5922	5933	5944	5955	5966	5977	5988	5999	6010
40	6021	6031	6042	6053	6064	6075	6085	6096	6107	6117
41	6128	6138	6149	6160	6170	6180	6191	6201	6212	6222
42	6232	6243	6253	6263	6274	6284	6294	6304	6314	6325
43	6335	6345	6355	6365	6375	6385	6395	6405	6415	6425
44	6435	6444	6454	6464	6474	6484	6493	6503	6513	6522
45	6532	6542	6551	6561	6571	6580	6590	6599	6609	6618
46	6628	6637	6646	6656	6665	6675	6684	6693	6702	6712
47	6721	6730	6739	6749	6758	6767	6776	6785	6794	6803
48	6812	6821	6830	6839	6848	6857	6866	6875	6884	6893
49	6902	6911	6920	6928	6937	6946	6955	6964	6972	6981
50	6990	6998	7007	7016	7024	7033	7042	7050	7059	7067
51	7076	7084	7093	7101	7110	7118	7126	7135	7143	7152
52	7160	7168	7177	7185	7193	7202	7210	7218	7226	7235
53	7243	7251	7259	7267	7275	7284	7292	7300	7308	7316
54	7324	7332	7340	7348	7356	7364	7372	7380	7388	7396
55	7404	7412	7419	7427	7435	7443	7451	7459	7466	7474
56	7482	7490	7497	7505	7513	7520	7528	7536	7543	7551
57	7559	7566	7574	7582	7589	7597	7604	7612	7619	7627
58	7634	7642	7649	7657	7664	7672	7679	7686	7694	7701
59	7709	7716	7723	7731	7738	7745	7752	7760	7767	7774
60	7782	7789	7796	7803	7810	7818	7825	7832	7839	7846
61	7853	7860	7868	7875	7882	7889	7896	7903	7910	7917
62	7924	7931	7938	7945	7952	7959	7966	7973	7980	7987
63	7993	8000	8007	8014	8021	8028	8035	8041	8048	8055
64	8062	8069	8075	8082	8089	8096	8102	8109	8116	8122

TABLE B–1 *(continued)*

n	0	1	2	3	4	5	6	7	8	9
65	8129	8136	8142	8149	8156	8162	8169	8176	8182	8189
66	8195	8202	8209	8215	8222	8228	8235	8241	8248	8254
67	8261	8267	8274	8280	8287	8293	8299	8306	8312	8319
68	8325	8331	8338	8344	8351	8357	8363	8370	8376	8382
69	8388	8395	8401	8407	8414	8420	8426	8432	8439	8445
70	8451	8457	8463	8470	8476	8482	8488	8494	8500	8506
71	8513	8519	8525	8531	8537	8543	8549	8555	8561	8567
72	8573	8579	8585	8591	8597	8603	8609	8615	8621	8627
73	8633	8639	8645	8651	8657	8663	8669	8675	8681	8686
74	8692	8698	8704	8710	8716	8722	8727	8733	8739	8745
75	8751	8756	8762	8768	8774	8779	8785	8791	8797	8802
76	8808	8814	8820	8825	8831	8837	8842	8848	8854	8859
77	8865	8871	8876	8882	8887	8893	8899	8904	8910	8915
78	8921	8927	8932	8938	8943	8949	8954	8960	8965	8971
79	8976	8982	8987	8993	8998	9004	9009	9015	9020	9025
80	9031	9036	9042	9047	9053	9058	9063	9069	9074	9079
81	9085	9090	9096	9101	9106	9112	9117	9122	9128	9133
82	9138	9143	9149	9154	9159	9165	9170	9175	9180	9186
83	9191	9196	9201	9206	9212	9217	9222	9227	9232	9238
84	9243	9248	9253	9258	9263	9269	9274	9279	9284	9289
85	9294	9299	9304	9309	9315	9320	9325	9330	9335	9340
86	9345	9350	9355	9360	9365	9370	9375	9380	9385	9390
87	9395	9400	9405	9410	9415	9420	9425	9430	9435	9440
88	9445	9450	9455	9460	9465	9469	9474	9479	9484	9489
89	9494	9499	9504	9509	9513	9518	9523	9528	9533	9538
90	9542	9547	9552	9557	9562	9566	9571	9576	9581	9586
91	9590	9595	9600	9605	9609	9614	9619	9624	9628	9633
92	9638	9643	9647	9652	9657	9661	9666	9671	9675	9680
93	9685	9689	9694	9699	9703	9708	9713	9717	9722	9727
94	9731	9736	9741	9745	9750	9754	9759	9763	9768	9773
95	9777	9782	9786	9791	9795	9800	9805	9809	9814	9818
96	9823	9827	9832	9836	9841	9845	9850	9854	9859	9863
97	9868	9872	9877	9881	9886	9890	9894	9899	9903	9908
98	9912	9917	9921	9926	9930	9934	9939	9943	9948	9952
99	9956	9961	9965	9969	9974	9978	9983	9987	9991	9996

TABLE B–2 The Metric System.

1. Length

1 meter (m) = 100 *centi*meters (cm)

1 centimeter (cm) = 10 *milli*meters (mm)

1 millimeter (mm) = 1000 *micro*meters (μm)

1 angstrom (Å) = 10^{-1} *nano*meters (nm)

= 10^{-8} cm

= 10^{-10} m

Conversions: 1 inch (in.) ≈ 2.54 cm

2. Volume

1 liter (1) = 1000 *milli*liters (ml)

≈ 1000 *cubic centi*meters (cm^3)

1 milliliter (ml) ≈ 1 cubic centimeter (cm^3 or cc)

= 1000 *micro*liters (μl)

Conversions: 1 quart = 0.946 l

1 fluid ounce (oz) = 29.6 ml

3. Mass

1 *kilo*gram (kg) = 1000 grams (g)

1 gram (g) = 1000 *milli*grams (mg)

1 milligram (mg) = 1000 *micro*grams (μg)

Conversions: 1 pound (lb) = 0.454 kg = 454 g

or 1 kg = 2.2 lbs

1 ounce (oz) = 28.4 g

Appendix C

Clinical Chemistry Values for Body Fluids

Table C–1 lists the concentration ranges of various biological constituents in body fluids. The concentration of a particular substance is considered normal if the value is within the normal range. However, these concentration ranges should not be considered *absolute*; normal values often vary depending upon the assay method used. Differences related to age and gender are noted where appropriate. Often, the concentration of a particular biological substance will change depending upon age, diet, and other related factors.

TABLE C–1 Clinical Chemistry Values for Body Fluids.

Substance	Body Fluid	Concentration
Acetoacetic acid + acetone (ketone bodies)	Serum	0.5–3.0 mg/100 ml
Albumin	Serum	3.8–5.0 g/100 ml
Ammonia nitrogen	Serum Urine (24 h)	40–80 μg/100 ml 500–1200 mg/day
Amylase	Serum	15–90 U/L
Aspartate amino*trans*ferase (AST). Old name: *G*lutamate *o*xaloacetate *t*ransaminase (GOT).	Serum	12–29 U/L
Barbiturate	Whole blood	Therapeutic concentration: 0.1–0.2 mg/100 ml Coma induction concentration: 1–5 mg/100 ml
Bicarbonate (HCO_3^-)	Arterial blood Venous blood	21–28 mmol/L 22–29 mmol/L
Bilirubin	Serum	Total: 0.2–1.0 mg/100 ml Conjugated: 0.0–0.2 mg/100 ml Unconjugated: 0.2–0.8 mg/100 ml Infants <1 week of age (total): 1–12 mg/100 ml
Bile salts	Feces (24 h)	0.8 g/day
Blood pH	Blood (arterial)	7.35–7.45
Blood volume	Whole blood (adults)	60–70 ml/kg
Calcium	Serum	8.5–10.5 mg/100 ml
Carbon dioxide (total CO_2)	Arterial blood Venous blood	19–25 mmol/L 22–27 mmol/L
Carbon dioxide pressure (pCO_2)	Arterial blood Venous blood	35–45 mm Hg 38–50 mm Hg
Chloride	Serum	98–106 mmol/L

Source: Adapted from Tietz, N. (ed.), *Fundamentals of Clinical Chemistry*, 2nd ed. (Philadelphia: W. B. Saunders, 1976), pp. 1198–1227.

Abbreviations: ng = *nano*grams = 10^{-9} grams; μg = *micro*grams = 10^{-6} grams; mg = *milli*grams = 10^{-3} grams; g = gram = 10^0 grams; kg = kilogram = 10^3 grams; 100 ml = often abbreviated as dl in other texts; L = liter; mmol/L = millimoles/liter; U/L = units/liter (1 U = 1 μmole product formed/min = 1×10^{-6} mol/min); IU = International Units; Hb = Hemoglobin; mm Hg = millimeters of mercury (also known as torr); h = hour; d = day; yr = year.

TABLE C–1 *(continued)*

Substance	Body Fluid	Concentration
Cholesterol (total)	Serum	140–250 mg/100 ml
Cholesterol esters	Serum	65–75% of total cholesterol value
Creatine kinase (CK or CPK)	Serum	Males: 12–65 U/L Females: 10–50 U/L
Creatinine	Serum	Males: 0.6–1.2 mg/100 ml Females: 0.5–1.0 mg/100 ml
Cyanide	Whole blood	Normal: Negative (no concentration) Lethal: >0.1 mg/100 ml
Epinephrine	Plasma	18–26 ng/100 ml
Estrogens (total)	Urine (24 h)	Men: 5–18 μg/day Females (a) Nonpregnant Preovulatory phase: 4–25 μg/day Ovulation peak: 28–99 μg/day Luteal peak: 22–105 μg/day (b) Pregnant: Up to 45 000 μg/day (c) Postmenopausal: 1.4–19.6 μg/day
Fat	Feces (24 h)	Adults: <6 g/day
Fat (triglycerides)	Serum	Males: 40–160 mg/100 ml Females: 35–135 mg/100 ml
Fatty acids (free)	Serum	0.30–0.90 mmol/L
Fibrinogen	Plasma	0.20–0.40 g/100 ml
Follicle stimulating hormone (FSH)	Plasma	Females Premenopausal: 4–30 mIU/L
Gastric juice	Stomach fluids	Total acids: 10–50 mmol/L pH: 1.5–3.5 Volume: 20–100 ml
Glucose	Serum	Normal: 70–105 mg/100 ml Diabetics: >140 mg/100 ml
Hematocrit	Whole blood	Males: 42–50% Females: 37–47%

TABLE C–1 *(continued)*

Substance	Body Fluid	Concentration
Hemoglobin	Whole blood	Males: 14–17 g/100 ml Females: 12–15 g/100 ml Newborns: 16–20 g/100 ml
Insulin	Plasma (fasting)	4–24 μIU/L
Iron (total)	Serum	Males: 60–150 μg/100 ml Females: 50–130 μg/100 ml
17-Ketogenic steroids (total)	Urine (24 h)	Males: 5–23 mg/day Females: 3–15 mg/day
Ketone bodies (see acetoacetic acid + acetone)		
Lactate	Whole blood	Arterial: 0.36–0.75 mmol/L Venous: 0.5–1.3 mmol/L
Lactate Dehydrogenase (LDH) (pyruvate → lactate conversion)	Serum	85–190 U/L
Lecithin/spingomyelin ratio (L/S ratio)	Amniotic fluid	2.0–5.0 (no units) (probably indicates lung maturity)
Lipase	Serum	28–280 U/L
Oxygen capacity	Whole blood (arterial)	1.34 ml/gm Hb
Oxygen pressure (pO_2)	Whole blood (arterial)	12–40 yr: 83–108 mm Hg 40–80 yr: 72–104 mm Hg
Phosphate (inorganic)	Serum	Adults: 3.0–4.5 mg/100 ml
Phosphatase (acid)	Serum	Males: 2.5–11.7 U/L Prostatic fraction: 0.2–5.0 U/L Females: 0.3–9.2 U/L
Phosphatase (alkaline)	Serum	Adults: 25–90 U/L Children: 20–150 U/L
Potassium	Serum	3.5–5.3 mmol/L
Progesterone	Plasma	Females (a) Follicular phase: 40–60 ng/100 ml (b) Luteal phase: 1000–2000 ng/100 ml (c) During pregnancy: increases steadily, 9th–32nd week

TABLE C–1 *(continued)*

Substance	Body Fluid	Concentration
Proteins	Serum	Total: 6.0–8.0 g/100 ml Fibrinogen: 0.2–0.4 g/100 ml Albumin: 3.8–5.0 g/100 ml Globulins: 2.3–3.5 g/100 ml *Protein concentration by electrophoresis* Albumin: 52–67% of total α_1-Globulin: 2.8–4.6% of total α_2-Globulin: 6.6–13.6% of total β-Globulin: 9.1–14.7% of total γ-Globulin: 9.0–20.6% of total
Protein	Urine (24 h)	50–100 mg/day
Sodium	Serum	135–148 mmol/L
Testosterone	Plasma	Males: 500–860 ng/100 ml Females: 26–54 ng/100 ml
Thyroxine (T_4)	Serum	4.4–10.7 μg/100 ml
Triiodothyronine (T_3)	Serum	160–270 ng/100 ml
Transaminase (see aspartate transaminase, AST)		
Triglycerides (see fats, triglycerides)		
Urea nitrogen (*b*lood *u*rea *n*itrogen BUN)	Serum	15–39 mg urea/100 ml (or 7–18 mg urea-nitrogen/100 ml)
Urea	Urine (24 h)	12–20 g/day
Uric acid	Serum	Males: 3.5–7.2 mg/100 ml Females: 2.6–6.0 mg/100 ml
	Urine (avg. diet)	250–750 mg/day
Volume	Blood Urine (varies)	60–70 ml/kg Males: 800–1800 ml/day Females: 600–1600 ml/day

Appendix D

Answers to Selected Problems

Chapter 1, page 11

*4. Carbon cannot easily lose its electrons because these electrons are attracted to the positive nucleus. A great deal of energy is required to remove an electron from a carbon atom.

*5. The cell is an open system because it can freely exchange matter and energy with the environment.

*6. Animal cells normally exist in a regulated environment.

Chapter 2, page 28

3. **a.** 2 **b.** 0.3010 **c.** 0.7451 **d.** 4.6021
4. **a.** 3 **b.** 7 **c.** 6.96 **d.** 3.52
5. **a.** 4.74 **b.** 9.24

*6. **a.** 3.16×10^{-7} M **b.** 5.0×10^{-3} M **c.** 1.58×10^{-7} M

7. The pH 3 solution has a hydrogen-ion concentration 10^5 times that of the pH 8 solution.

*8. $K_a = 1.8 \times 10^{-16}$ M

9. Assume $[A^-] = [HA]$, or $(0.5) = (0.5)$. The Henderson-Hasselbach equation then reduces to $pH = pK_a$.
10. pH = 7.51 (metabolic alkalosis)

Chapter 3, page 51

2. **a.** $R{-}CH(NH_3^+){-}COOH$ **b.** $^+H_3N{-}(CH_2)_4{-}CH(NH_3^+){-}COOH$

c. $^-OOC{-}(CH_2)_2{-}CH(NH_2){-}COO^-$

3. Net charge of isoleucine: **a.** +1 **b.** 0 **c.** −1
Net charge of lysine: **a.** +2 **b.** +1 **c.** −1
Net charge of glutamic acid: **a.** +1 **b.** −1 **c.** −2

4. $^{+}H_3N{-}(CH_2)_4{-}\underset{\displaystyle |}{\overset{\displaystyle NH_3^{+}}{CH}}{-}COOH \longrightarrow {}^{+}H_3N{-}(CH_2)_4{-}\overset{NH_3^{+}}{CH}{-}COO^{-} \longrightarrow$

$^{+}H_3N{-}(CH_2)_4{-}\overset{NH_2^{0}}{CH}{-}COO^{-} \longrightarrow {}^{0}H_2N{-}(CH_2)_4{-}\overset{NH_2^{0}}{CH}{-}COO^{-}$

5. **a.** See text for structure of amino acids and peptides.
b. Net charge at pH 1 = +2; at pH 7 = 0; at pH 12 = −3.
6. Glycine does not have four different groups bound to the α-carbon atom.
7. 490 amino acids.
8. Bradykinin contains a large number of proline residues.
9. **a.** Tyr—Glu **d.** Asp—Ser
10. **b.** Lys—Asp **d.** Asp—Arg

Chapter 4, page 75

2. 2.5×10^{13} RBC
*3. **a.** 2.54×10^{8} molecules of Hb in RBC
b. 1.02×10^{9} O_2 molecules bound
*4. **a.** Large surface area **b.** Efficient gas exchange
5. Hydrophobic amino acids and histidine
6. No
7. Hyperbolic curve
*8. The oxygen saturation curve would be shifted to the left of the Earth hemoglobin.
9. Answer (b), respiratory alkalosis
10. Chlorophyll would absorb the blue and red wavelengths and transmit the green.
11. Kinked secondary structure, caused by proline.
12. Provides physical strength.

Chapter 5, page 89

3. See Figure 5–1 (page 79).
4. No
*5. Products C and D have a higher final energy level; therefore, the reverse reaction is favored.
6. Random binding sequence—either substrate A or substrate B could bind to the active site first.
7. **a.** Urease; glucose 6-phosphatase; adenosine triphosphatase
b. Peptidase; glycosidase
c. Hydrolase; oxidoreductase; transferase

Chapter 6, page 114

*2. **a.** 6.67×10^{2} IU **b.** 2.22×10^{2} IU/mg
*3. 1.72×10^{-1} IU
4. **a.** Optimum temperature = 37 °C **b.** 2 °C **c.** 90 °C
5. See Figure 6–3 (page 96) for typical optimum pH curve.

6. a. See Figure 6–5 (page 98). b. $V_{max} = 3.0\ \mu$mol/min and $K_m = 5$ mM
   c. TN = 6.0×10^4 mol prod/min · mol enz
7. a. Dimer; AA, AB, and BB
   b. Trimer; AAA, AAB, ABB, and BBB

Chapter 7, page 134

2. a. aldotetrose b. aldopentose c. aldohexose d. ketopentose
   e. ketohexose
3. a. Carbon #1 and carbon #3 do not have four different groups.
   b. Dihydroxyacetone does not contain an asymmetric carbon.
10. a. No color observed
    b. No. Iodine only produces the blue color when it is complexed with the helical amylose.
11. See Chapter 5 (linkage specificity).
*12. Polar amino-acid residues in the active site and hydrogen bonding of the amylose molecule.

Chapter 8, page 154

4. a. Nonspontaneous b. +5.45 kcal/mol
*5. a. $K_{eq} = 850$ b. Spontaneous reaction
*6. a. $\Delta G' = -0.8$ kcal/mol b. Spontaneous in cell
7. Phosphocreatine + ADP ⟶ creatine + ATP
   Glucose + ATP ⟶ Glucose 6-phosphate + ADP
8. The Mg^{2+} will bind to the last two phosphate groups of the adenosine triphosphate tail.
9. a. See Chapter 7 and Figure 8–6, (page 148).
10. a. The blood glucose level is high.
    b. The blood glucose level is low and constant (hypoglycemic).
    c. The blood glucose level would decrease (enter the hypoglycemic state).

Chapter 9, page 177

*3. a. The label would appear on the #3 carbon of L-lactate.
    b. The label would appear on the #2 carbon atom of ethanol (the methyl carbon).
*4. a. 3 ATP used b. 8 ATP made c. 0 net NADH made
*5. (1) Glycerol + ATP ⟶ Glycerol phosphate + ADP
    (2) Phosphoenolpyruvate + ADP ⟶ pyruvate + ATP
    (3) Pyruvate + NADH ⟶ Lactate + NAD^+
    Therefore, determine the decrease in the NADH concentration by determining the decrease in the absorbance at 340 nm.
6. Anaerobic glycolysis is favored; therefore, LDH_4 and LDH_5 will be present in high concentrations.

Chapter 10, page 192

*3. a. The label would not appear in the glucose molecule.
    b. The label would appear on the #1 and #6 carbon atoms of the glucose molecule.
4. Six high-energy bonds are required.
5. See Section 9.5.
6. Answer (e)
*7. Glucose ⟶ Glucose 6-phosphate ⟶ Glucose 1-phosphate
    Glucose 1-phosphate + UTP ⟶ UDP-glucose + PP_i

Chapter 11, page 214

4. See section 11.3. The mechanism is similar to that of the pyruvate dehydrogenase complex (oxidative decarboxylation, transfer of succinyl group to CoA via swinging arm mechanism).
*5. See Figure 11–7 (page 207).
6. Malonate inhibits succinate dehydrogenase (competitive inhibition). Therefore, the conversion of succinate to fumarate is blocked.
   a. The TCA cycle would cease to operate.
   b. The succinate concentration would increase.
   c. Yes; the added fumarate would be converted into malate.
   d. Inhibition can be relieved by adding excess succinate.
7. Avidin binds biotin; therefore, pyruvate carboxylase would be inhibited, and the concentration of TCA cycle intermediates would decrease.

Chapter 12, page 226

*3. Sequence: DAN, Delta Xi, Q, protein Y, protein X, O_2
4. CN^- and CO bind to the cytochrome oxidase heme group; therefore, electron transport and mitochondrial respiration would stop.
*5. Snake venom phospholipases destroy cell membranes; therefore, the mitochondrial membranes will be destroyed. Along with them is destroyed the pH gradient, which is essential for oxidative phosphorylation.

Chapter 13, page 249

2. **a** and **d** are false (Section 13.3).
6. **a.** Main chemical features of cholesterol molecule: hydrophobic 4-ring system, hydrophobic R chain, and OH group on the #3 position of the A ring.
   b. Chemical differences: bile salts contain more OH groups and have a modified R group on the D ring.
   c. See Table 13–4 and compare the structures.
*7. The prostaglandin molecule has a hairpin structure, polar groups on the 5-membered ring, and hydrophobic R groups extending from the ring. Therefore, the binding site would have (a) an elongated hairpin structure and (b) polar groups to hydrogen bond with the polar groups on the prostaglandin molecule.

Chapter 14, page 276

3. Defective triglyceride degradation in the intestine; defective pancreatic lipase.
4. Stearic acid can generate 146 ATP.
*5. Similarities: Both employ sulfhydryl-containing compounds (CoA-SH and ACP-SH); acetyl-CoA is used or made; 2 carbons are processed at a time; similar intermediates are formed; similar cofactors are used.
6. Biotin, pantothenic acid, riboflavin, and niacin.
7. See Section 14.8.
*8. The label would appear on the #2 and #4 carbons of acetoacetyl-CoA.

Chapter 15, page 299

3. **a.** decreased rate: inadequate pepsin activation,
   b. decreased rate: inactivation of pancreatic enzymes by acidic pH;
   c. decreased rate: trypsin not activated; therefore, other pancreatic zymogens not activated.

*4. Features: large binding site, positively charged site, negatively charged site, hydrophobic regions to associate with the hydrophobic R group of the bound amino acid.

5. See other biochemistry texts.

*6. Pyridoxine ⟶ (1) oxidation ⟶ (2) phosphorylation ⟶ pyridoxal phosphate

7. See Table 15–2 (page 284) and Chapter 9 (a high level indicates liver disease—cirrhosis, hepatitis, or cancer).

*8. **a.** 30 ATP made **b.** 4 ATP used **c.** Net ATP = 26

9. Ornithine is similar to lysine.

10. Depletion of NADH; decreased ATP production

11. **a.** Dehydration leads to concentration of blood constituents—therefore, to increased BUN levels.
    b. Pregnancy leads to increased demand for maternal amino acids—therefore, to decreased maternal BUN levels.

*12. See Figure 15–7 (page 293). Label would appear on one of the carboxyl groups of fumarate.

Chapter 16, page 320

3. See Section 16.2.

4. Hydrolysis drives a chemical reaction to completion.

5. **a.** Glycine, aspartic acid, and glutamine.
   b. Aspartic acid is derived from oxaloacetate; glutamine is derived from glutamic acid (which is derived from α-ketoglutarate).

*6. Inosinate, AMP, and GMP act as negative allosteric effectors (feedback inhibition).

7. Common elements of purine biosynthesis and pyrimidine biosynthesis: phosphorylated ribose is used; ATP is required; amino acids are used; the nucleotide is synthesized first, then modified to yield other derivatives. Differences: the purine ring is assembled while attached to PRPP; the pyrimidine ring is assembled first, then attached to ribose.

8. See Section 16.2 and Figure 16–5 (page 312).

9. **a.** Sequence of complementary strand; 3′—T—T—A—C—G—T—A—G—5′
   b. Sequence of complementary strand; 3′—C—C—G—A—A—T—G—G—C—T—A—T—C—G—5′

10. Sequence of mRNA strand: 3′—U—C—C—G—A—U—C—C—G—G—G—U—U—C—C—G—A—U—5′

Chapter 17, page 340

3. See Figure 17–1 (page 322).

4. **a** and **b.** See Sections 17.2 and 17.4.

*5. Common elements of DNA replication and DNA transcription: DNA is unwound; one strand is used as a template for the other; enzymes are required; pyrophosphate cleavage occurs; the strands are synthesized in the 5′ → 3′ direction. Differences: both DNA strands are replicated, but only one strand is transcribed; deoxyribonucleotides are used in replicating DNA, but ribonucleotides are used in transcribing DNA; an RNA primer is required in replication, but no primer is required in transcription.

6. Palindrome sequence: 5′—A—A—G—C—T—T—3′
   3′—T—T—C—G—A—A—5′

Chapter 18, page 361

3. **a.** mRNA transcript sequence: 5′—GAU—AGA—CCU—UCU—GGU—3′
   b. Amino acid sequence (Table 18–1): N—Asp—Arg—Pro—Ser—Gly—C

*4. **a.** One possible sequence: 5′—GGU—CUU—CCA—UGU—AAU—CAA—AUU—UAU—UGU—3′
   b. Sequence of DNA: 3′—CCA—GAA—GGT—ACA—TTA—GTT—TAA—ATA—ACA—5′

5. **a.** 250 amino acids = 250 codons
   b. 252 codons (including start and stop codons)
   c. 252 codons = 756 bases
   d. 1 gene = 252 codons (tetramer is constructed from 4 identical chains)

Chapter 19, page 381

*2. (1) iodination of tyrosyl residues, forming diiodinated tyrosyl residues; (2) coupling of diiodinated tyrosyl residues, forming thyroxine (T_4)

3. **a.** Continued production of T_4 and T_3
   b. No TSH synthesis or secretion and, therefore, no T_4 production
   c. No synthesis of T_4
   d. Low blood T_4 (and T_3) levels imply that no feedback inhibition occurs, therefore synthesis of TSH and stimulation of the thyroid gland continue.

*4.

	cAMP/adenylate cyclase		Steroid Hormones
1.	Hormone binds to receptor on outside surface of cell.	1.	Hormone binds to receptor protein inside cell.
2.	Second messenger (cAMP) is formed.	2.	No second messenger is formed.
3.	Events occur in cytoplasm.	3.	Events occur in both cytoplasm and nucleus.
4.	Primary events do not involve DNA or RNA.	4.	Events involve DNA and RNA.
5.	Preexisting enzymes are activated.	5.	New proteins (or enzymes) are synthesized.

*5. Progesterone (a steroid hormone) probably induces the synthesis of collagenase via the mechanism depicted in Figure 19–8.

6. See Figure 19–9 (page 378). Birth control pills contain estrogen progesterone, or derivatives of these hormones. High levels of these hormones in the blood inhibit the secretion of FSH and LH by the pituitary, thus inhibiting follicle maturation and ovulation.

Chapter 20, page 407

2. See Figure 20–1 (page 386).
   a. Fe^{2+}, Ca^{2+}, and Mg^{2+} deficiencies
   b. water-soluble vitamin deficiencies and fat-soluble vitamin deficiencies
   c. Na^{+}, K^{+}, and vitamin K deficiencies

3. 6% total body water is lost and regained each day.

*4. **a.** 150 Cal = 150 000 cal (small calories) **b.** 15 000 ml, or 15 liters
   c. $\sim 1.2 \times 10^5$ kcal energy
   d. 1.2×10^5 kcal energy can be stored in 13.3 kg fat. To store the same amount of energy, 30 kg of carbohydrates would be required, representing 16.7 kg of extra weight.

5. **a.** Linoleic acid is a precursor to linolenic acid and arachidonic acid. Linoleic acid and linolenic acids are structural components of cell membrane lipids; arachidonic acid is the precursor of the prostaglandins.
   b. Oxidation reactions to form additional double bonds, together with the addition of 2 extra carbon atoms to turn the 18-carbon linoleic acid into a 20-carbon fatty acid.
6. See Sections 20.4, 20.5, and 20.6; Figure 20–3 (page 391), and Table 20–2 (page 392).
7. See Table 20–2 (page 392).
8. See Section 20.5 and summarize the data about each vitamin.
*9. See Figures 20–2 (page 390) and 20–3 (page 391); Tables 20–1 (page 389) and 20–2 (page 392).

Index